D1674207

**Phase Transformations
in Multicomponent Melts**

*Edited by
Dieter M. Herlach*

Further Reading

Hirsch, J., Gottstein, G., Skrotzki, B. (eds.)
Aluminium Alloys
Their Physical and Mechanical Properties

2 Volumes
2008
ISBN: 978-3-527-32367-8

Krupp, U.
Fatigue Crack Propagation in Metals and Alloys
Microstructural Aspects and Modelling Concepts

2007
ISBN: 978-3-527-31537-6

Pfeiler, W. (ed.)
Alloy Physics
A Comprehensive Reference

2007
ISBN: 978-3-527-31321-1

Lipowsky, H., Arpaci, E.
Copper in the Automotive Industry

2007
ISBN: 978-3-527-31769-1

Kainer, K. U. (ed.)
Metal Matrix Composites
Custom-made Materials for Automotive and Aerospace Engineering

2006
ISBN: 978-3-527-31360-0

Phase Transformations in Multicomponent Melts

Edited by
Dieter M. Herlach

WILEY-VCH Verlag GmbH & Co. KGaA

The Editor

Prof. Dr. Dieter M. Herlach
Deutsches Zentrum für Luft- und Raumfahrt (DLR)
Institut für Materialphysik im Weltraum
Linder Höhe
D-51147 Köln
Germany

and

Ruhr-Universität Bochum
Fakultät für Physik und Astronomie
Universitätsstraße 150
D-44780 Bochum
Germany

www.dieter-herlach.de

Cover Description

The figure shows a dendrite in a slightly off-eutectic AlAgCu alloy cast into a cold aluminium mould. The dendrite consists of two intermetallic phases (Al_2Cu and Ag_2Cu) and the aluminium solid solution phase (Photo courtesy by Lorenz Ratke).

■ All books published by Wiley-VCH are carefully produced. Nevertheless, authors, editors, and publisher do not warrant the information contained in these books, including this book, to be free of errors. Readers are advised to keep in mind that statements, data, illustrations, procedural details or other items may inadvertently be inaccurate.

Library of Congress Card No.:
applied for

British Library Cataloguing-in-Publication Data
A catalogue record for this book is available from the British Library.

Bibliographic information published by the Deutsche Nationalbibliothek
The Deutsche Nationalbibliothek lists this publication in the Deutsche Nationalbibliografie; detailed bibliographic data are available on the Internet at <http://dnb.d-nb.de>.

© 2008 WILEY-VCH Verlag GmbH & Co. KGaA, Weinheim

All rights reserved (including those of translation into other languages). No part of this book may be reproduced in any form—by photoprinting, microfilm, or any other means—nor transmitted or translated into a machine language without written permission from the publishers. Registered names, trademarks, etc. used in this book, even when not specifically marked as such, are not to be considered unprotected by law.

Typesetting Laser Words, Madras, Madras
Printing Strauss GmbH, Mörlenbach
Binding Litges & Dopf GmbH, Heppenheim
Cover Design Adam Design, Weinheim

Printed in the Federal Republic of Germany
Printed on acid-free paper

ISBN: 978-3-527-31994-7

Foreword

The traditional research field of phase transformations in liquid metals is of great importance in materials production. It attracted new interest in research since the development of novel methods of modeling such as the phase field approach, and the application of new experimental techniques such as levitation processing to directly observe phase transformations on freely suspended liquid samples. The exciting progress in this field stimulated innovative work in describing microstructure evolution as a function of processing parameters even in multicomponent alloys, which is a mandatory boundary condition for a predictive capability in materials design.

The recent efforts have led to a multidisciplinary cooperation of mathematicians, theoretical and experimental physicists, chemists, and materials scientists of mechanical engineering. The Deutsche Forschungsgemeinschaft has essentially enforced the interdisciplinary cooperation by approving a new priority programme SPP1120 on *Phase Transformations in Multicomponent Melts*. The priority programme was funded over six years and was found to be very successful at its end.

More than 20 research groups from German universities and research centers were combined within the priority programme. Phase field models were employed and extended for the description of solidification modes in Aachen, Jülich, Karlsruhe, Köln, Magdeburg, and Regensburg. These mesoscopic modeling were complemented by microscopic modeling as molecular dynamics in Göttingen and Mainz, and modeling by a hard sphere system in Dresden. However, reliable modeling of phase transformations require careful measurements of thermophysical parameters of selected sample systems that came from groups in Berlin, Clausthal-Zellerfeld, Chemnitz, Köln, and München. The priority programme made it possible to directly verify the predictions of modeling with results of experiments of growth dynamics and microstructure evolution, which were performed in Aachen, Dresden, Göttingen, and Köln. Even dedicated systems were investigated as nanoscale systems in Karlsuhe, Münster, and Saarbrücken, which play an essential role in the young research field of nanomaterials. Last but not least, a complete new approach was started in the priority programme SPP1120 to test colloidal systems as model systems for metals and alloys in Köln and Mainz.

Phase Transformations in Multicomponent Melts. Edited by D. M. Herlach
Copyright © 2008 WILEY-VCH Verlag GmbH & Co. KGaA, Weinheim
ISBN: 978-3-527-31994-7

The latter activity has led to a new priority programme SPP1296 that is supported by the Deutsche Forschungsgemeinschaft since 2007.

During the course of the running priority programme, it was very active in reporting the results to the scientific community during national and international conferences in Hamburg, Dresden, Regensburg, Lausanne, München, and Prag. Even presentations were realized to the public during exhibitions in Dresden. It is also evident in many publications of referred scientific journals and many invited talks of its members at international congresses. Not to forget its merit to educate young scientists to establish a new generation of scientists in research and industry, which is very important to hold and even increase the high standard of research in the field of phase transformations from which our industrial society may benefit in the future. During the six years of its existence new full chairs in materials physics and science were occupied by young members of the priority programme in Aachen, Bochum, Dresden, Karlsruhe, Köln, Leoben, and Münster.

The priority programme worked so well also thanks to the engagement of its coordinator, Professor Dr Dieter Herlach from Köln, who did an excellent job. We congratulate him and all the members of the priority programme for the enormous outcome of this network.

Göttingen, 30 June 2008   *Professor Dr. Reiner Kirchheim*

Contents

Foreword *V*
Preface *XVII*
List of Contributors *XXI*

Part One Thermodynamics 1

1 **Phase Formation in Multicomponent Monotectic Al-based Alloys** *3*
Joachim Gröbner, Djordje Mirković, and Rainer Schmid-Fetzer
1.1 Introduction *3*
1.2 Experimental Methods *4*
1.3 Systematic Classification of Ternary Monotectic Phase Diagrams *4*
1.4 Selected Ternary Monotectic Alloy Systems *6*
1.4.1 Al–Bi–Zn: Type 2a Monotectic System *6*
1.4.2 Al–Bi–Sn: Type 1b Monotectic System *6*
1.4.3 Al–Sn–Cu: Type 3a Monotectic System *8*
1.4.4 Al–Bi–Cu: Type 2b Monotectic System *9*
1.4.5 Bi–Cu–Sn: Type 3a Monotectic System *11*
1.5 Quaternary Monotectic Al–Bi–Cu–Sn System *11*
1.6 Conclusion *15*

2 **Liquid-liquid Interfacial Tension in Multicomponent Immiscible Al-based Alloys** *19*
Walter Hoyer and Ivan G. Kaban
2.1 Introduction *19*
2.2 Measurement Technique *21*
2.3 Experimental Details *24*
2.4 Results *25*
2.5 Discussion *30*
2.5.1 Composition Dependences of the l–l Interfacial Tension *30*
2.5.2 Adsorption at the l–l Interface *32*
2.5.3 Temperature Dependence of the l–l Interfacial Tension *34*

Phase Transformations in Multicomponent Melts. Edited by D. M. Herlach
Copyright © 2008 WILEY-VCH Verlag GmbH & Co. KGaA, Weinheim
ISBN: 978-3-527-31994-7

2.5.4	Wetting Phenomena	34
2.6	Summary	36

3 Monotectic Growth Morphologies and Their Transitions 39
Lorenz Ratke, Anja Müller, Martin Seifert, and Galina Kasperovich

3.1	Introduction	39
3.2	Experimental Procedures	40
3.2.1	Alloys	40
3.2.2	ARTEMIS Facility	41
3.2.3	Evaluation Procedures	42
3.3	Experimental Results	42
3.3.1	Microstructures	42
3.3.2	Jackson–Hunt Plot	45
3.3.3	Stability Diagrams	46
3.4	Discussion	47
3.4.1	Fibrous Monotectic Growth	47
3.4.2	Transition from Fibers to String of Pearls	48
3.4.3	Origin of Irregular Drops	50
3.5	Outlook	53

4 Thermal Expansion and Surface Tension of Multicomponent Alloys 55
Jürgen Brillo and Ivan Egry

4.1	Introduction	55
4.1.1	General	55
4.1.2	Density and Thermal Expansion	56
4.1.3	Surface Tension	57
4.2	Experimental	58
4.2.1	Levitation	58
4.2.2	Density and Thermal Expansion	59
4.2.3	Surface Tension	59
4.3	Results	60
4.3.1	Density	60
4.3.2	Surface Tension	64
4.4	Conclusion and Summary	69

5 Measurement of the Solid-Liquid Interface Energy in Ternary Alloy Systems 73
Annemarie Bulla, Emir Subasic, Ralf Berger, Andreas Bührig-Polaczek, and Andreas Ludwig

5.1	Introduction	73
5.2	Experimental Procedure	74
5.2.1	The Radial Heat Flow Apparatus	74
5.2.2	Equilibration of the Sample	75
5.2.3	Quenching	75
5.3	Evaluation of the Local Curvature of the Grain Boundary Grooves	75

5.3.1	Preparation of the Sample 75
5.3.2	Geometrical Correction of the Groove Coordinates 76
5.3.3	Determination of the Local Undercooling 77
5.3.4	Determining the Interface Energy 78
5.4	Results and Discussion 79
5.4.1	Al–Cu System with Eutectic Composition 80
5.4.2	Al–Cu–Ag System with Invariant Eutectic Composition 81
5.4.3	Concentration Dependence of σ_{SL} 83
5.5	Summary and Conclusions 84

6 Phase Equilibria of Nanoscale Metals and Alloys 87
Gerhard Wilde, Peter Bunzel, Harald Rösner, and Jörg Weissmüller

6.1	Introduction 87
6.2	Phase Stability and Phase Transformations in Nanoscale Systems 88
6.2.1	Single-Phase Material: External Interfaces 88
6.2.2	Binary Nanoalloys: Internal Heterophase Interfaces 97
6.3	Summary 105

Part Two Microscopic and Macroscopic Dynamics 109

7 Melt Structure and Atomic Diffusion in Multicomponent Metallic Melts 111
Dirk Holland-Moritz, Oliver Heinen, Suresh Mavila Chathoth, Anja Ines Pommrich, Sebastian Stüber, Thomas Voigtmann, and Andreas Meyer

7.1	Introduction 111
7.2	Experimental Details 113
7.2.1	Quasi elastic Neutron Scattering 113
7.2.2	Elastic Neutron Scattering 115
7.3	Results and Discussion 115
7.3.1	Atomic Dynamics in Liquid Ni 115
7.3.2	Atomic Dynamics in Ni–P-based Glass-forming Alloy Melts 118
7.3.3	Atomic Dynamics in Zr–Ti–Ni–Cu–Be and $Zr_{64}Ni_{36}$ Alloy Melts 119
7.3.4	The Short-Range Order of Liquid $Zr_{64}Ni_{36}$ 120
7.3.5	Analysis Within Mode Coupling Theory 124
7.4	Conclusions 125

8 Diffusion in Multicomponent Metallic Melts Near the Melting Temperature 131
Axel Griesche, Michael-Peter Macht, and Günter Frohberg

8.1	Introduction 131
8.2	Experimental Diffusion Techniques 132
8.2.1	Long-Capillary Method 132
8.2.2	Long-Capillary Method Combined with X-ray Radiography 134
8.3	Influence of Thermodynamic Forces on Diffusion 135

| 8.3.1 | Systems with Mixing Tendency: Al–Ni *136* |
| 8.3.2 | Systems with Demixing Tendency: Pd–Cu–Ni–P *137* |

9   Phase Behavior and Microscopic Transport Processes in Binary Metallic Alloys: Computer Simulation Studies *141*
Subir K. Das, Ali Kerrache, Jürgen Horbach, and Kurt Binder

9.1	Introduction *141*
9.2	Transport Coefficients *143*
9.3	A Symmetric LJ Mixture with a Liquid–Liquid Demixing Transition *144*
9.4	Structure, Transport, and Crystallization in Al–Ni Alloys *148*
9.5	Summary *154*

10   Molecular Dynamics Modeling of Atomic Processes During Crystal Growth in Metallic Alloy Melts *157*
Helmar Teichler and Mohamed Guerdane

10.1	Introduction *157*
10.2	Entropy and Free Enthalpy of Zr-rich Ni_xZr_{1-x} Melts from MD Simulations and Their Application to the Thermodynamics of Crystallization *158*
10.2.1	Survey *158*
10.2.2	Results and Their Meaning *158*
10.2.2.1	The Method that Works *158*
10.2.2.2	Free Enthalpy Results for Zr-rich Ni_xZr_{1-x} Melts *159*
10.2.2.3	Zr-rich Part of the (x, T) Phase Diagram *161*
10.3	Bridging the Gap between Phase Field Modeling and Molecular Dynamics Simulations. Dynamics of the Planar $[Ni_xZr_{1-x}]_{liquid} - Zr_{crystal}$ Crystallization Front *162*
10.3.1	Survey *162*
10.3.2	Results and Their Meaning *162*
10.3.2.1	MD-Generated Input Parameter for PF Modeling *162*
10.3.2.2	Comparison of MD and PF Results for the Concentration Profiles and Propagation of the Crystallization Front *163*
10.4	Entropy and Free Enthalpy in Ternary $Al_yNi_{0.4-y}Zr_{0.6}$ Alloy Melts *165*
10.4.1	Survey *165*
10.4.2	Results and Their Meaning *166*
10.4.2.1	The Method: Test of Its Numerical Reliability *166*
10.4.2.2	Results for the Entropy Change in the $Al_yNi_{0.4-y}Zr_{0.6}$ Melt Series at 1700 K *167*
10.5	Concluding Remarks *169*

11   Computational Optimization of Multicomponent Bernal's Liquids *171*
Helmut Hermann, Antje Elsner, and Valentin Kokotin

| 11.1 | Introduction *171* |
| 11.2 | Methods *172* |

11.2.1	Force Biased Algorithm *172*
11.2.2	The Nelder–Mead Algorithm *173*
11.2.3	Voronoi Tessellation *174*
11.3	Results and Discussion *175*
11.3.1	Monoatomic Liquids *175*
11.3.2	Multicomponent Liquids *177*
11.4	Conclusion *180*

12	**Solidification Experiments in Single-Component and Binary Colloidal Melts** *185*
	Thomas Palberg, Nina Lorenz, Hans Joachim Schöpe, Patrick Wette, Ina Klassen, Dirk Holland-Moritz, and Dieter M. Herlach
12.1	Introduction *185*
12.2	Experimental Procedure *186*
12.2.1	Tunable Interactions in Charged Colloidal Suspensions *186*
12.2.2	Instrumentation for Time-Resolved Static Light Scattering *190*
12.3	Results *193*
12.3.1	The Full Phase Diagram of Charge Variable Systems *193*
12.3.2	Shapes of Phase Diagrams of Charged Sphere Mixtures *196*
12.3.3	Growth of Binary Colloidal Crystals *199*
12.3.4	Quantitative Determination of Nucleation Kinetics and Extraction of Key Parameters *203*
12.3.5	Investigations on the Structure of Undercooled Melts *206*
12.4	Conclusions *208*

Part Three   Nd–Fe based Alloys   213

13	**Phase-Field Simulations of Nd–Fe–B: Nucleation and Growth Kinetics During Peritectic Growth** *215*
	Ricardo Siquieri and Heike Emmerich
13.1	Introduction *215*
13.2	Phase-Field Model with Hydrodynamic Convection *217*
13.3	Investigating Heterogeneous Nucleation in Peritectic Materials via the Phase-Field Method *220*
13.4	Conclusion *223*

14	**Investigations of Phase Selection in Undercooled Melts of Nd–Fe–B Alloys Using Synchrotron Radiation** *227*
	Thomas Volkmann, Jörn Strohmenger, and Dieter M. Herlach
14.1	Introduction *227*
14.2	Description of the Investigations *228*
14.3	Experimental Results and Discussion *231*
14.4	Analysis Within Models of Nucleation and Dendrite Growth *236*
14.5	Summary and Conclusion *240*

15	**Effect of Varying Melt Convection on Microstructure Evolution of Nd–Fe–B and Ti–Al Peritectic Alloys** *245*
	Regina Hermann, Gunter Gerbeth, Kaushik Biswas, Octavian Filip, Victor Shatrov, and Janis Priede
15.1	Introduction *245*
15.2	Methods Developed *246*
15.2.1	Forced Rotation Technique *246*
15.2.1.1	Experimental *246*
15.2.1.2	Numerical Simulation *247*
15.2.2	Floating Zone Facility with Additional Magnetic Field *248*
15.2.2.1	Experimental *248*
15.2.2.2	Numerical Simulation *250*
15.3	Sample Preparation *251*
15.4	Results and Discussion *252*
15.4.1	Nd–Fe–B Alloys *252*
15.4.2	Ti–Al Alloys *256*
15.5	Conclusion *258*

16	**Nanosized Magnetization Density Profiles in Hard Magnetic Nd–Fe–Co–Al Glasses** *263*
	Olivier Perroud, Albrecht Wiedenmann, Mihai Stoica, and Jürgen Eckert
16.1	Introduction *263*
16.2	SANS with polarized neutrons in unsaturated magnetic systems *265*
16.3	Experimental Procedure *267*
16.3.1	Sample Preparation *267*
16.3.2	SANSPOL Measurements *267*
16.4	Results and Discussion *268*
16.4.1	Microstructure *268*
16.4.2	Magnetic Behavior *268*
16.5	Conclusion *275*

17	**Microstructure and Magnetic Properties of Rapidly Quenched $(Nd_{100-x}Ga_x)_{80}Fe_{20}$ ($x = 0, 5, 10,$ and 15 at%) alloys** *277*
	Mihai Stoica, Golden Kumar, Mahesh Emmi, Olivier Perroud, Albrecht Wiedenmann, Annett Gebert, Shanker Ram, Ludwig Schultz, and Jürgen Eckert
17.1	Introduction *277*
17.2	Sample Preparation and Experimental Investigations *279*
17.3	Binary $Nd_{80}Fe_{20}$ Rapidly Quenched Alloys *280*
17.3.1	Structure and Cooling Rate *280*
17.3.2	The Metastable A1 "Phase" *283*
17.3.3	Magnetic Properties *287*
17.4	Ternary $(Nd_{100-x}Ga_x)_{80}Fe_{20}$ ($x = 5, 10,$ and 15) Rapidly Quenched Alloys *288*
17.4.1	XRD Studies *288*

| 17.4.2 | Tuning the Metastable Hard Magnetic A1 Zones *290* |
| 17.5 | Conclusions *292* |

Part Four Solidification und Simulation 297

18 **Solidification of Binary Alloys with Compositional Stresses— A Phase-Field Approach** *299*
Bo Liu and Klaus Kassner
18.1	Introduction *299*
18.2	Equations of Motion *301*
18.3	Neutral Curves *304*
18.4	Phase-Field Model *305*
18.5	Simulation Results *306*
18.6	Conclusions *308*

19 **Elastic Effects on Phase Transitions in Multi-component Alloys** *311*
Efim A. Brener, Clemens Gugenberger, Heiner Müller-Krumbhaar, Denis Pilipenko, Robert Spatschek, and Dmitrii E. Temkin
19.1	Melting of Alloys in Eutectic and Peritectic Systems *312*
19.1.1	Isothermal Melting in Eutectic System *313*
19.1.2	Isothermal Melting in Peritectic Systems *315*
19.2	Combined Motion of Melting and Solidification Fronts *316*
19.3	Continuum Theory of Fast Crack Propagation *319*
19.4	Summary *323*

20 **Modeling of Nonisothermal Multi-component, Multi-phase Systems with Convection** *325*
Harald Garcke and Robert Haas
20.1	Introduction *325*
20.2	Phase-field Models for Multicomponent, Multiphase Systems *326*
20.3	Multiphase Ginzburg–Landau Energies *327*
20.3.1	Some Examples of Ginzburg–Landau Energies *329*
20.4	Convective Phase-Field Models *330*
20.4.1	Conservation Laws and Entropy Inequality *330*
20.4.2	Exploitation of the Entropy Principle *332*
20.4.2.1	Example *336*
20.5	Mathematical Analysis *337*

21 **Phase-field Modeling of Solidification in Multi-component and Multi-phase Alloys** *339*
Denis Danilov and Britta Nestler
21.1	Introduction *339*
21.2	Phase-field Model for Multicomponent and Multiphase Systems *339*
21.3	Modeling of Dendritic Growth *341*

21.4	Solute Trapping During Rapid Solidification 345
21.5	Comparison of Molecular Dynamics and Phase-field Simulations 345
21.6	Modeling of Eutectic Growth 346

22 Dendrite Growth and Grain Refinement in Undercooled Melts 353
Peter K. Galenko and Dieter M. Herlach

22.1	Introduction 353
22.2	Solidification of Pure (One-Component) System 354
22.2.1	Diffuse Interface Model 354
22.2.2	Sharp-Interface Model 356
22.2.3	Results of Nominally Pure Ni 357
22.2.4	Results on Congruently Melting Intermetallic Alloy $Ni_{50}Al_{50}$ 359
22.3	Solidification of Binary Alloys with Constitutional Effects 362
22.3.1	Diffuse Interface Model 362
22.3.2	Sharp-Interface Model and Growth Velocities of Binary Ni–Zr Alloys 362
22.3.3	Grain Refinement Through Undercooling 364
22.4	Solidification of a Ternary Alloy 366
22.4.1	Diffuse Interface Model 366
22.4.2	Sharp-Interface Model and Dendrite Growth Velocities of Ni–Zr–Al 367
22.5	Summary and Conclusions 369

23 Dendritic Solidification in the Diffuse Regime and Under the Influence of Buoyancy-Driven Melt Convection 373
Markus Apel and Ingo Steinbach

23.1	Introduction 373
23.2	The Multiphase-field Model 373
23.2.1	Extension of the Karma corrections to Multicomponent Alloys 374
23.2.2	Fluid Flow Coupling for Multicomponent Alloys 375
23.3	Rapid Solidification in $Ni_{98}Al_1Zr_1$ 376
23.4	Directional Solidification with Buoyancy-driven Interdendritic Flow 378
23.4.1	Spacing Selection in Binary Alloy 379
23.4.2	Buoyancy-driven Fluid Flow in a Ternary Alloy 381
23.5	Summary and Conclusion 383

24 Stationary and Instationary Morphology Formation During Directional Solidification of Ternary Eutectic Alloys 387
Bernd Böttger, Victor T. Witusiewicz, Markus Apel, Anne Drevermann, Ulrike Hecht, and Stephan Rex

24.1	Introduction 387
24.1.1	About the Project 387
24.1.2	General Aspects of Ternary Eutectic Systems 388
24.2	Investigations on the Eutectic System Ag–Cu–Zn 390

24.2.1	Thermodynamic Properties and Thermodynamic Assessment	390
24.2.2	Bridgman Experiments	391
24.3	Investigation on the Ternary Alloy System In–Bi–Sn	395
24.3.1	Measurement of Thermodynamic Properties and Thermodynamic Assessment	395
24.3.2	Micro-Bridgman Assembly	397
24.3.3	Stationary Coupled Growth	398
24.4	Transient Growth	399
24.4.1	Solidification Path During Transient Solidification	401
24.4.2	Quantitative Comparison to Simulation: Calibration of Diffusion Data	403
24.4.3	Stationary Univariant Growth: Calibration of Interfacial Energies	403
24.5	Summary and Conclusion	404

25 Dendritic Microstructure, Decomposition, and Metastable Phase Formation in the Bulk Metallic Glass Matrix Composite $Zr_{56}Ti_{14}Nb_5Cu_7Ni_6Be_{12}$ 407

Susanne Schneider, Alberto Bracchi, Yue-Lin Huang, Michael Seibt, and Pappannan Thiyagarajan

25.1	Introduction	407
25.2	Experimental Procedures	409
25.3	Results and Discussion	409
25.4	Summary	418

Index 421

Preface

More than 90% of all metallic materials in daily human life are prepared from the liquid state as their parent phase. Nowadays, efforts are focused to find a comprehensive understanding of physical properties of liquid metallic alloys and to develop physically relevant models of solidification experimentally verified to describe all processes of relevance such as nucleation and crystal growth. All of these investigations may form a solid basis to evolve computer assisted modelling of solidification as needed in casting and foundry industry to reduce time and energy consuming post solidification treatment and to shorten the production route from the as cast material to the final product with its desired properties. The present book may contribute to a milestone to approach this vision of computer assisted materials design.

Most of the metallic materials consist of several components and belong to binary, ternary, quaternary and even higher multicomponent alloys. However, by far most of the models to describe properties of liquids and solidification are limited in their application to pure metals and binary alloys. More recently, phase-field models were developed capable to be used also for multicomponent alloys. These models attract growing interest to analyze phase transformations in general. In addition, new experimental techniques make it possible to measure physical properties of liquids and the growth kinetics during crystallization of metallic melts with high accuracy and reliability. During the "Second Decennial Workshop on Solidification Microstructures", held in Zermatt in 1998, 26 invited experts from all over the world discussed the progress in experiment and modelling of solidification and future challenges in this traditional research field.

Stimulated by this workshop a programme committee was set up in Germany consisting of Prof. Dr. H. Müller-Krumbhaar, Prof. Dr. W. Petry, Prof. Dr. K. Samwer, later on Prof. Dr. A. Meyer, and the editor of the present book to formulate an application for a new priority programme on "Phase Transformations in Multicomponent Melts" to the German Research Foundation DFG. This proposal was approved by DFG in 2000. The priority programme SPP1120 started with a Kick off Meeting held in Bonn on 30/31 October 2000. According to the capabilities of the 25 participating groups the entire priority programme as reflected in the present book are structured into four focal points, which were leaded by Prof. Dr. Heike Emmerich, Prof. Dr. Andreas Meyer, Prof. Dr. Britta Nestler, and Prof. Dr. Rainer Schmid-Fetzer as:

1. *Thermodynamics (Schmid-Fetzer)*. Here, the essential target is the coupling of materials property data and thermodynamic functions. Specifically, these are surface/interface energy (σ), viscosity (η), and diffusion coefficient (D) on the one hand and thermodynamic functions such as Gibbs energy (G), or chemical potentials (μ_i) on the other hand. Compositional dependency of (σ, η, D) in multicomponent liquids is calculated based on the knowledge of Gibbs energy, which, in turn, is also the basis for the phase equilibria. The diffusion coefficient and mobility in e.g. Al-Ce-Ni melts are independently coupled through calculation of the thermodynamic factor.
2. *Microscopic and Macroscopic Dynamics (Meyer)*. Transport coefficients of metallic liquids play an important role in modelling nucleation, crystal growth, and vitrification. Diffusion coefficients are measured by quasielastic neutron scattering. To attain a proper understanding of transport processes in alloys and to gain insight into the microscopic transport mechanisms a combination of different experimental findings are correlated to results of molecular dynamics and Monte Carlo computer simulation. Partial static and dynamic structure factors as well as different transport coefficients such as diffusion constants and the shear viscosity are determined. Furthermore, microscopic processes in the interface region between a liquid and a solid is studied and vibrational and configurational contributions to the entropy of a melt are disentangled. The results from such investigations provide a good testing ground for microscopic theories.
3. *Nd-based Alloys (Emmerich)*. Nd-Fe-B alloys as the most important hard magnetic material are investigated. The phase selection behaviour is directly observed by applying energy dispersive X-ray scattering using synchrotron radiation during solidification of containerlessly processed melts. Phase selection diagrams are constructed revealing the importance of undercooling on the primary solidification of the hard magnetic phase. The influence of convection on solidification is investigated and hard sphere models are used for modelling microscopic processes. The as solidified

microstructure controls the magnetic properties of the final material as demonstrated for quaternary Nd-Fe-Co-Al alloys. The experiments are accompanied by phase field modelling to simulate both nucleation and growth in multicomponent Nd-based alloys.

4. *Solidification and Simulations (Nestler)*. Molecular dynamics and phase-field methods are applied to analyze effects and component treatments on the atomic and the mesoscopic scale. Liquid-solid, liquid-liquid or solid-solid phase transitions are studied. The classes of materials investigated are ranging from pure metals via binary alloys to multicomponent alloys. Heat and mass transport, atomic diffusion, the dynamics of the melt(s), elasticity, the surface tensions and the interfacial kinetics as well as anisotropy in interface energy and kinetics of the solidification front are identified as important physical quantities in solidification modelling. The theoretical work is directly related to experimental investigations of growth dynamics and phase evolution for a comparison between simulated microstructures and experimental observations.

The priority programme presented its results to the scientific community by several symposia at national conferences as annual spring conferences of the Section Condensed Matter of the German Physical Society in Dresden (2003), in Berlin (2008), and at the Materials Week of the German Materials Society in Munich (2004). Internationally, the priority programme communicated its results at EUROMAT conferences in Lausanne (2003) and in Prag (2005). In between, three internal symposia for intense discussions of all members of the SPP1120 and the members of the peer review group were held in the Physik Zentrum in Bad Honnef (2004, 2005, 2007). In addition, many informal meetings within focal groups stimulated the discussions and made the priority programme to a vivid research action.

Sincere thanks are expressed to the German Research Foundation DFG for the substantial financial support. In particular, I am very grateful to the constructive cooperation with the Programme directors at DFG, Dr. J. Tobolski during the application period, Dr. F. Fischer during the first three years and Dr. B. Jahnen for the remaining time of the running programme. We benefited a lot from the careful work of the peer review group consisting of Prof. Dr. J. Bilgram, Prof. Dr. F. Faupel, Prof. Dr. F. Haider, Prof. Dr. W. Kurz, Prof. Dr. H. Neuhäuser, Prof. Dr. H. Ruppersberg, Dr. H. Schober and chaired by Prof. Dr. R. Kirchheim. I am thankful to all members of this group for their balanced evaluation and many stimulating discussions. Also, I thank Dr. N. Oberbeckmann-Winter and Dr. M. Ottmar from Wiley-VCH for always friendly and effective cooperation from the submission of the manuscripts till the publication of the present book. My special thanks are directed to the leaders of the focal groups who have essentially contributed to make the entire programme a success. Last but not least I express my deep appreciation to all members of the SPP1120 for the always-constructive cooperation during the entire course of the priority programme.

Köln, 3 June 2008 *Professor Dr. Dieter Herlach*

List of Contributors

Markus Apel
ACCESS e.V.
Intzestrasse 5
52072 Aachen
Germany

Ralf Berger
RWTH Aachen
Foundry Institute
Intzestr. 5
52072 Aachen
Germany

Kurt Binder
Johannes Gutenberg-Universität Mainz
Institut für Physik
Staudinger Weg 7
55099 Mainz
Germany

Kaushik Biswas
Central Glass and Ceramic Research Institute (CGCRI)
196 Raja SC Mullick Road
Kolkata - 700032
India
and
Leibniz Institute for Solid State and Materials Research (IFW) Dresden
P.O. Box 270116
01171 Dresden
Germany

Bernd Böttger
ACCESS e.V.
Intzestrasse 5
52072 Aachen
Germany

Alberto Bracchi
Georg-August-Universität Göttingen
Physikalisches Institut
Friedrich-Hund-Platz 1
37077 Göttingen
Germany

Efim A. Brener
Forschungszentrum Jülich
Institut für Festkörperforschung
52425 Jülich
Germany

Jürgen Brillo
Deutsches Zentrum für Luft- und Raumfahrt (DLR)
Institut für Materialphysik im Weltraum
Linder Höhe
51147 Köln
Germany

Andreas Bührig-Polaczek
RWTH Aachen
Foundry Institute
Intzestr. 5
52072 Aachen
Germany

Annemarie Bulla
RWTH Aachen
Foundry Institute
Intzestr. 5
52072 Aachen
Germany

Peter Bunzel
Research Center Karlsruhe
Institute of Nanotechnology
PO Box 3640
76021 Karlsruhe
Germany
and
Universität des Saarlandes
FR 7.3, Technische Physik
PO Box 151150
66041 Saarbrücken
Germany

Denis Danilov
Karlsruhe University of Applied Sciences
Institute of Computational Engineering
Moltkestrasse 30
76133 Karlsruhe
Germany

Subir K. Das
Johannes Gutenberg-Universität Mainz
Institut für Physik
Staudinger Weg 7
55099 Mainz
Germany

Anne Drevermann
ACCESS e.V.
Intzestrasse 5
52072 Aachen
Germany

Jürgen Eckert
Dresden University of Technology
Institute of Materials Science
Helmholtzstr. 7
01062 Dresden
Germany

and
IFW Dresden
Institute for Complex Materials
Helmholtzstr. 20
01069 Dresden
Germany

Ivan Egry
Deutsches Zentrum für Luft- und Raumfahrt (DLR)
Institut für Materialphysik im Weltraum
Linder Höhe
51147 Köln
Germany

Antje Elsner
IFW Dresden
Institute for Complex Materials
Helmholtzstr. 20
01069 Dresden
Germany

Heike Emmerich
RWTH Aachen
Computational Materials Engineering Center
of Computational Engineering Science
Institute of Minerals Engineering
Mauerstrasse 5
52064 Aachen
Germany

Mahesh Emmi
IFW Dresden
Institute for Complex Materials
Helmholtzstr. 20
01069 Dresden
Germany
and
Indian Institute of Technology (IIT) Kharagpur
Materials Science Centre, Qtr. no. C1-123
West Bengal 721302
India

Octavian Filip
University of Erlangen-Nürnberg
Department of Material Science 6
Martensstrasse 7
91058 Erlangen
Germany

Günter Frohberg
Technische Universität Berlin
Inst. für Werkstoffwissenschaften
u.–technologien
Hardenbergstr. 36/Sekr. PN 2-3
10623 Berlin
Germany

Peter K. Galenko
Deutsches Zentrum für Luft- und Raumfahrt (DLR)
Institut für Materialphysik im Weltraum
Linder Höhe
51147 Köln
Germany

Harald Garcke
Universität Regensburg
NWF I – Mathematik
Universitätsstrasse 31
93040 Regensburg
Germany

Annett Gebert
IFW Dresden
Institute for Metallic Materials
Helmholtzstr. 20
01069 Dresden
Germany

Gunter Gerbeth
Forschungszentrum Dresden-Rossendorf
P.O. Box 510119
01314 Dresden
Germany

Axel Griesche
Hahn-Meitner-Institut Berlin
Abteilung Werkstoffe
Glienicker Str. 100
14109 Berlin
Germany
and
Deutsches Zentrum für Luft- und Raumfahrt (DLR)
Institut für Materialphysik im Weltraum
Linder Höhe
51147 Köln
Germany

Joachim Gröbner
Clausthal University of Technology
Institute of Metallurgy
Robert-Koch-Str. 42
38678 Clausthal-Zellerfeld
Germany

Mohamed Guerdane
Universität Göttingen
Institut für Materialphysik
Friedrich-Hund-Platz 1
37077 Göttingen
Germany

Clemens Gugenberger
Forschungszentrum Jülich GmbH
Institut für Festkörperforschung
52425 Jülich
Germany

Robert Haas
Universität Regensburg
NWF I – Mathematik
Universitätsstrasse 31
93040 Regensburg
Germany

Ulrike Hecht
ACCESS e.V.
Intzestrasse 5
52072 Aachen
Germany

Oliver Heinen
Deutsches Zentrum für Luft- und Raumfahrt
(DLR)
Institut für Materialphysik im Weltraum
Linder Höhe
51147 Köln
Germany

Dieter M. Herlach
Deutsches Zentrum für Luft- und Raumfahrt
(DLR)
Institut für Materialphysik im Weltraum
Linder Höhe
51147 Köln
Germany
and
Ruhr-Universität Bochum
Fakultät für Physik und Astronomie
Universitätsstraße 150
D-44780 Bochum
Germany

Helmut Hermann
IFW Dresden
Institute for Complex Materials
Helmholtzstr. 20
01069 Dresden
Germany

Regina Hermann
Leibniz Institute for Solid State and
Materials Research (IFW) Dresden
P.O. Box 270116
01171 Dresden
Germany

Dirk Holland-Moritz
Deutsches Zentrum für Luft- und Raumfahrt
(DLR)
Institut für Materialphysik im Weltraum
Linder Höhe
51147 Köln
Germany

Jürgen Horbach
Deutsches Zentrum für Luft- und Raumfahrt
(DLR)
Institut für Materialphysik im Weltraum
Linder Höhe
51147 Köln
Germany
and
Johannes Gutenberg-Universität Mainz
Institut für Physik
Staudinger Weg 7
55099 Mainz
Germany

Walter Hoyer
Chemnitz University of Technology
Institute of Physics
09107 Chemnitz
Germany

Yue-Lin Huang
Georg-August-Universität Göttingen
Physikalisches Institut
Friedrich-Hund-Platz 1
37077 Göttingen
Germany

Ivan G. Kaban
Chemnitz University of Technology
Institute of Physics
09107 Chemnitz
Germany

Galina Kasperovich
Deutsches Zentrum für Luft- und Raumfahrt (DLR)
Institut für Materialphysik im Weltraum
Linder Höhe
51147 Köln
Germany

Klaus Kassner
Otto-von-Guericke-Universität
Institut für Theoretische Physik
Postfach 4120
39016 Magdeburg
Germany

Ali Kerrache
Johannes Gutenberg-Universität Mainz
Institut für Physik
Staudinger Weg 7
55099 Mainz
Germany

Ina Klassen
Deutsches Zentrum für Luft- und Raumfahrt (DLR)
Institut für Materialphysik im Weltraum
Linder Höhe
51147 Köln
Germany

Valentin Kokotin
IFW Dresden
Institute for Complex Materials
Helmholtzstr. 20
01069 Dresden
Germany

Golden Kumar
Yale University
Faculty of Engineering
PO Box 208284
New Haven, CT
06570-8284
USA

Bo Liu
MPI für Marine Mikrobiologie
AG Mathematische Modellierung
Celsiusstrasse 1
28359 Bremen
Germany

Nina Lorenz
Johannes Gutenberg Universität Mainz
Institut für Physik
Staudinger Weg 7
55128 Mainz
Germany

Andreas Ludwig
University of Leoben
Department of Metallurgy
Buchmüllerplatz 2/ Jahnstraße 4
8700 Leoben
Austria

Michael-Peter Macht
Hahn-Meitner-Institut Berlin
Abteilung Werkstoffe
Glienicker Str. 100
14109 Berlin
Germany

Suresh Mavila Chathoth
Technische Universität München
Physik Department E13
85747 Garching
Germany

Andreas Meyer
Deutsches Zentrum für Luft- und Raumfahrt (DLR)
Institut für Materialphysik im Weltraum
Linder Höhe
51147 Köln
Germany
and
Technische Universität München
Physik Department E13
85747 Garching
Germany

Djordje Mirković
Clausthal University of Technology
Institute of Metallurgy
Robert-Koch-Str. 42
38678 Clausthal-Zellerfeld
Germany

Anja Müller
Deutsches Zentrum für Luft- und Raumfahrt (DLR)
Institut für Materialphysik im Weltraum
Linder Höhe
51147 Köln
Germany

Heiner Müller-Krumbhaar
Forschungstentrum Jülich GmbH
Institut für Festkörperforschung
52425 Jülich
Germany

Britta Nestler
Karlsruhe University of Applied Sciences
Institute of Computational Engineering
Moltkestrasse 30
76133 Karlsruhe
Germany

Thomas Palberg
Johannes Gutenberg Universität Mainz
Institut für Physik
Staudinger Weg 7
55128 Mainz
Germany

Olivier Perroud
Hahn Meitner Institut Berlin GmbH
Glienicker Strasse 100
14109 Berlin
Germany
and
Université Paul Cézanne
Faculté des Sciences et Techniques
Avenue Escadrille Normandic Niemen
13397 Marseille Cedex 20
France

Denis Pilipenko
Forschungstentrum Jülich GmbH
Institut für Festkörperforschung
52425 Jülich
Germany

Anja Ines Pommrich
Deutsches Zentrum für Luft- und Raumfahrt (DLR)
Institut für Materialphysik im Weltraum
Linder Höhe
51147 Köln
Germany

Janis Priede
Coventry University
Applied Mathematics Research Centre
Coventry, CV1 5FB
UK

Shanker Ram
Indian Institute of Technology (IIT) Kharagpur
Materials Science Centre, Qtr. no. C1-123
West Bengal 721302
India

Lorenz Ratke
Deutsches Zentrum für Luft- und Raumfahrt (DLR)
Institut für Materialphysik im Weltraum
Linder Höhe
51147 Köln
Germany

Stephan Rex
ACCESS e.V.
Intzestrasse 5
52072 Aachen
Germany

Harald Rösner
University of Münster
Institute of Materials Physics
Wilhelm-Klemm-Str. 10
48149 Münster
Germany

Rainer Schmid-Fetzer
Clausthal University of Technology
Institute of Metallurgy
Robert-Koch-Str. 42
38678 Clausthal-Zellerfeld
Germany

Susanne Schneider
Georg-August-Universität Göttingen
Physikalisches Institut
Friedrich-Hund-Platz 1
37077 Göttingen
Germany

Hans Joachim Schöpe
Johannes Gutenberg Universität Mainz
Institut für Physik
Staudinger Weg 7
55128 Mainz
Germany

Ludwig Schultz
Dresden University of Technology
Mathematic and Natural Science Faculty
Physik-Bau
Zellescher Weg 16
01069 Dresden
Germany
and
IFW Dresden
Institute for Metallic Materials
Helmholtzstr. 20
01069 Dresden
Germany

Michael Seibt
Georg-August-Universität Göttingen
Physikalisches Institut
Friedrich-Hund-Platz 1
37077 Göttingen
Germany

Martin Seifert
Deutsches Zentrum für Luft- und Raumfahrt (DLR)
Institut für Materialphysik im Weltraum
Linder Höhe
51147 Köln
Germany

Victor Shatrov
Forschungszentrum Dresden-Rossendorf
P.O. Box 510119
01314 Dresden
Germany

Ricardo Siquieri
RWTH Aachen
Computational Materials Engineering Center
of Computational Engineering Science
Institute of Minerals Engineering
Mauerstrasse 5
52064 Aachen
Germany

Robert Spatschek
Forschungszentrum Jülich GmbH
Institut für Festkörperforschung
52425 Jülich
Germany

Ingo Steinbach
Ruhr-University Bochum
ICAMS
Stiepelerstrasse 129
44780 Bochum
Germany

Mihai Stoica
IFW Dresden
Institute for Complex Materials
Helmholtzstr. 20
01069 Dresden
Germany

Jörn Strohmenger
Deutsches Zentrum für Luft- und Raumfahrt (DLR)
Institut für Materialphysik im Weltraum
Linder Höhe
51147 Köln
Germany

Sebastian Stüber
Technische Universität München
Physik Department E13
85747 Garching
Germany

Emir Subasic
RWTH Aachen
Foundry Institute
Intzestr. 5
52072 Aachen
Germany

Helmar Teichler
Universität Göttingen
Institut für Materialphysik
Friedrich-Hund-Platz 1
37077 Göttingen
Germany

Dmitrii E. Temkin
Forschungszentrum Jülich GmbH
Institut für Festkörperforschung
52425 Jülich
Germany

Pappannan Thiyagarajan
Intense Pulsed Neutron Source
Argonne National Laboratory
Argonne
IL 60439
USA

Thomas Voigtmann
Deutsches Zentrum für Luft- und Raumfahrt (DLR)
Institut für Materialphysik im Weltraum
Linder Höhe
51147 Köln
Germany

Thomas Volkmann
Deutsches Zentrum für Luft- und Raumfahrt (DLR)
Institut für Materialphysik im Weltraum
Linder Höhe
51147 Köln
Germany

Jörg Weissmüller
Research Center Karlsruhe
Institute of Nanotechnology
PO Box 3640
76021 Karlsruhe
Germany
and
Universität des Saarlandes
FR 7.3, Technische Physik
PO Box 151150
66041 Saarbrücken
Germany

Patrick Wette
Deutsches Zentrum für Luft- und Raumfahrt (DLR)
Institut für Materialphysik im Weltraum
Linder Höhe
51147 Köln
Germany

Albrecht Wiedenmann
Institut Laue-Langevin (ILL) Grenoble
BP. 152
38042 Grenoble Cedex 09
France

Gerhard Wilde
University of Münster
Institute of Materials Physics
Wilhelm-Klemm-Str. 10
48149 Münster
Germany

Victor T. Witusiewicz
ACCESS e.V.
Intzestrasse 5
52072 Aachen
Germany

Part One
Thermodynamics

1
Phase Formation in Multicomponent Monotectic Al-based Alloys
Joachim Gröbner, Djordje Mirković, and Rainer Schmid-Fetzer

1.1
Introduction

The motivation of this work arises since very little is known about phase transformations in *multicomponent* liquid alloys exhibiting demixing in two liquid phases, L′ and L″. This is in distinct contrast to binary monotectic alloys that are more or less well established. Especially, the possibility to form a ternary monotectic invariant reaction involving two liquid and *two* solid phases is a unique feature. It will be shown that the simultaneous occurrence of liquid demixing and abundant intermetallic phase formation is another distinct feature of *ternary* monotectic systems. In that context the understanding of phase relations, solidification paths, and transformation reactions in monotectic Al-based alloys was the focus of our study. State-of-the-art thermodynamic modeling using the Calphad method is vital for a quantitative treatment of both the multicomponent phase diagrams and the thermodynamic properties in a consistent manner. It will also be shown that this enables the calculation of complex "solidification" paths that may start with liquid demixing, followed by "primary" crystallization or vice versa, depending on the alloy composition. This aspect has been more specifically elaborated in a separate paper [1], where the solidification of three ternary monotectic alloy systems, Al–Bi–Zn, Al–Sn–Cu, and Al–Bi–Cu have been studied. Such examination of solidification paths and microstructure formation may be relevant for advanced solidification processing of multicomponent monotectic alloys.

A potential application of such fundamental knowledge to the development of advanced bearing alloys may be noted. Such alloys may benefit from the Al base (light weight) and the absence of lead, a toxic component being still present in a variety of bearing alloys. Knowing the intricacies of phase formation in such alloys is vital for focused development of suitable microstructures and also for various aspects of processing such alloys.

Basic implications of multiple demixing were also studied in the course of the current project by calculation of so-called rose diagrams in highly symmetrical systems. The apparent violation of the phase rule in such systems were resolved and a quantitative determination of degrees of freedom under such conditions was developed by Chen et al. [2]. The focus of the present report is on the thermodynamics of demixing and phase formation in the quaternary alloy system Al–Bi–Cu–Sn, including the ternary subsystems. It is noted that ternary liquid demixing had been detected in a number of alloy systems using approximate thermodynamic calculations as detailed in an earlier study [3]. Moreover, unexpected liquid demixing, occurring in ternary alloys only but not in any of the binary subsystems, was also studied in detail in the Al–Mg–Sn system [4] and more examples are given in the following text for the Al–Sn–Cu and Cu–Bi–Sn alloy systems.

Since the current chapter also serves as a final report on our DFG-sponsored project within the six years of Priority Programme SPP 1120, what follows is added for the sake of completeness. In the initial period of this project, the thermodynamics and phase formation pertinent to grain refinement of Al alloys was investigated. The binary Al–B system was studied experimentally and thermodynamic inconsistency in a published assessment was identified and removed [5]. Thermodynamic aspects of constitution, grain refinement, and solidification enthalpies of alloys in the Al–Ce–Si system were given in Reference [6]. Finally a consistent thermodynamic description of quaternary Al–Si–Ti–B alloys was generated and applied to the calculation of Al–Si–Ti–B phase diagram sections for practically relevant temperatures and compositions of Al–Si alloys, covering the range from Al-rich to typical Al–Si foundry alloys [7].

1.2
Experimental Methods

Samples for validation of the calculated monotectic phase diagrams were prepared from pure elements and sealed in a Ta capsule, which was specially developed in our lab to be inserted into the differential scanning calorimetry (DSC) equipment [8]. Metallographic preparation, scanning electron microscopy (SEM) with energy dispersive X-ray (EDX) microanalysis, and DSC were used for characterization. Details of the experimental investigations are given in particular publications [9–11]. An enhancement of the encapsulation technique using BN (boron nitride) inlays to avoid chemical reaction, observed here especially between Cu and Al in the melt and the Ta capsule, is described in [11].

1.3
Systematic Classification of Ternary Monotectic Phase Diagrams

A systematic classification of ternary monotectic alloys was presented focusing on Al-based systems including a larger number of real alloy examples [3]. Following this classification, the current ternary monotectic Al systems may be categorized

Table 1.1 Systematic overview of ternary monotectic Al–X–Z systems.

Type		Binary liquid miscibility gap			Ternary monotectic four-phase reactions	System studied in this work
		Al–X	Al–Z	X–Z		Al–X–Z
Type 1: The ternary miscibility gap (L′ + L″), terminating at one or two binary systems, intersects only *one* primary crystallization field	1a	Stable	Stable	None	No	—
	1b	Stable	Metastable	None	No	Al–Bi–Sn
Type 2: A similar ternary miscibility gap (L′ + L″), but intersecting *more than one* primary crystallization fields, thus forming invariant monotectic four-phase reaction(s)	2a	Stable	None	Stable	Yes	Al–Bi–Zn
	2b	Stable	None	Metastable	Yes	Al–Bi–Cu
	2c	Stable	None	None	Yes	—
Type 3: A miscibility gap (L′ + L″) occurring in the ternary system only	3a	Metastable	None	Metastable	Yes	Al–Sn–Cu
	3b	None	None	None	Yes	—

into three different types as given in Table 1.1. The occurrence of ternary monotectic four-phase reactions (L′ + L″ + solid$_1$ + solid$_2$), not present in type 1, is the most distinctive feature. Unexpected liquid demixing, occurring in ternary alloys only but not in any of the binary subsystems, is the key feature of type 3. The Bi–Cu–Sn system, belonging to type 3, is not listed since it does not contain Al.

1.4
Selected Ternary Monotectic Alloy Systems

1.4.1
Al–Bi–Zn: Type 2a Monotectic System

The first example system, Al–Bi–Zn, is result of a systematic search for monotectic four-phase reactions in ternary Al alloys, which is not complicated by the occurrence of intermetallic phases. Extensive thermodynamic phase diagram calculations were used for this search, excluding the alloying elements Cd and Hg. The finally identified Al–Bi–Zn system is a rare occasion where such a reaction, $L' = L'' + (Al)' + (Zn)$, actually occurs. Experimental work could be focused on five key samples in that system and involved DSC, for thermal analysis, and calorimetry, and also metallographic analysis using SEM/EDX. The existence of the monotectic reaction $L' = L'' + (Al)' + (Zn)$, predicted through thermodynamic calculations, was experimentally confirmed. This truly ternary invariant reaction was used for a quantitative thermodynamic modeling of the Al–Bi–Zn system. Only one small ternary parameter for the Gibbs energy of the liquid phase was sufficient to quantitatively reproduce all the ternary experimental data. Solidification paths and microstructures of Al–Bi–Zn alloys are shown to be rather complex. Using thermodynamic calculations, these rich details involving up to three invariant reactions and unusual monovariant reaction types can be clearly revealed and understood. Details are given in a separate paper [9].

The calculated liquidus surface and a calculated phase diagram section at constant 18 at% Al including DSC signals from experimental samples are given in Figures 1.1 and 1.2, respectively. Figure 1.3 shows a micrograph using back-scattered electron (BSE) contrast of the sample $Al_{18}Bi_{26}Zn_{56}$ solidified at $1 \, K \, min^{-1}$. The demixing of the two liquids, Bi poor (L', dark area) and Bi- rich (L'', light area), is obvious even in the solidified sample. In the dark area, the dendritic growth of primary (Al)' crystals during the monovariant reaction $L' = (Al)' + L''$ is visible. The residual liquid L' subsequently decomposes in the interdendritic space into $(Zn) + (Al)' + L''$ in the truly ternary monotectic reaction at 376 °C.

1.4.2
Al–Bi–Sn: Type 1b Monotectic System

In this system no ternary *invariant* monotectic reaction is observed, since the liquid miscibility gap, originating at the Al–Bi edge, intersects with the primary crystallization field of (Al) only. The monotectic is just a monovariant *three-phase* equilibrium, $L' + L'' + (Al)$, which ends in a critical point at about $Al_{66}-Bi_{02}-Sn_{32}$ (at%) and 611 °C. This closing of the ternary miscibility gap is related to the fact that the liquid gap in the binary Al–Sn edge is metastable with a critical point of the $L' + L''$ equilibrium at $Al_{78}-Sn_{22}$ (at%) and 536 °C. The calculated liquidus surface of the ternary Al–Bi–Sn system is given [3].

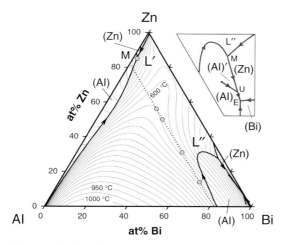

Fig. 1.1 Calculated liquidus surface. The ternary liquid miscibility gap dominates the central region, joining the binary Al–Bi and Bi–Zn gaps. Circles denote the compositions of five key samples. Dotted line indicates the vertical section of Figure 1.2. The monovariant line L″ + (Al)′ + (Al) occurring between 316 and 278 °C is due to an additional solid state demixing and is depicted in a schematic enlargement of the Bi-rich corner. It almost coincides with the line L″ + (Al)′ + (Zn).

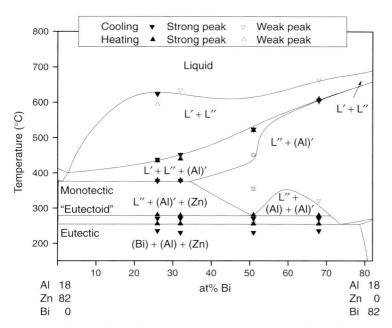

Fig. 1.2 Calculated phase diagram section at constant 18 at% Al including DSC signals.

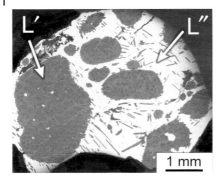

Fig. 1.3 Electron micrograph (BSE) of the sample $Al_{18}\ Bi_{26}\ Zn_{56}$ solidified at $1\,K\,min^{-1}$.

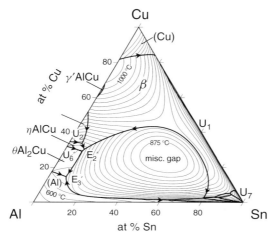

Fig. 1.4 The calculated liquidus surface of the Al–Sn–Cu system exhibits a large liquid miscibility gap occurring in ternary alloys only; the dome of that gap summits just below 900 °C as indicated by the isotherms.

1.4.3
Al–Sn–Cu: Type 3a Monotectic System

The complex features of the Al–Sn–Cu phase diagram are revealed by a combination of thermodynamic modeling and experimental studies [10]. The liquidus surface is dominated by ternary liquid demixing as shown in Figure 1.4. Nine ternary alloys were selected to cover all essential features involving the liquidus surface and the invariant solidification reactions. These were analyzed by differential thermal analysis (DTA) as well as metallography and local chemical analysis of solidified microstructures. The ternary liquid miscibility gap intersects the primary solidification fields of (Al) and the intermetallic phases ΘAl_2Cu, $\eta AlCu$, and β. This produces three different monotectic invariant reactions as follows:

$E_2 : \beta = L' + L'' + \eta AlCu$ at 581 °C
$U_6 : L' + \eta AlCu = L'' + \Theta Al_2Cu$ at 569 °C
$E_3 : L' = L'' + (Al) + \Theta Al_2Cu$ at 524 °C

Small changes in alloy composition may produce distinctly different microstructures with primary crystallization and secondary demixing, or vice versa. The microstructure of sample 8 (Al_{30} Cu_{10} Sn_{60}, wt%) in Figure 1.5 is a perfect example of primary liquid demixing. The higher density Sn-rich liquid L'' accumulates at the bottom of this vertically cross-sectioned sample while the lower density L' floats on top. The dark gray area of L' essentially consists of the fine monotectic microstructure, (Al) + ΘAl_2Cu with tiny white spots of L'', which is formed in E_3.

1.4.4
Al–Bi–Cu: Type 2b Monotectic System

The ternary Al–Bi–Cu system is even more complex than the systems described so far. Seven monotectic reactions are observed as follows:

$E_1: L' = \beta + (Cu) + L''$ at 1016.5 °C
$U_1: L'' + \beta = L' + \gamma'AlCu$ at 1012.9 °C
$U_2: L'' + \gamma'AlCu = \beta' + L'$ at 953.2 °C
$U_6: L' + \varepsilon AlCu = \eta AlCu + L''$ at 624.9 °C
$U_7: L' + \eta AlCu = \Theta Al_2Cu + L''$ at 595.9 °C
$E_2: L' = \Theta Al_2Cu + (Al) + L''$ at 547.6 °C

The liquidus surface is dominated by the immiscibility of the liquid phase (Figure 1.6). The stunning feature is the drastic widening of the comparably narrow binary Al–Bi liquid gap; when Cu is added, it covers essentially the entire ternary system. This is also reflected in the huge dome of the ternary liquid miscibility gap rising up to more than 1600 °C. The extraordinary compositional extent

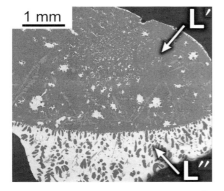

Fig. 1.5 The electron micrograph (BSE) of sample 8 (Al_{30} Cu_{10} Sn_{60}, wt%) solidified at 1 K min^{-1} shows the primary liquid demixing, with the heavier Sn-rich liquid L'' at the bottom.

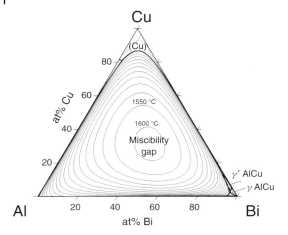

Fig. 1.6 Calculated liquidus projection of the Al–Bi–Cu system.

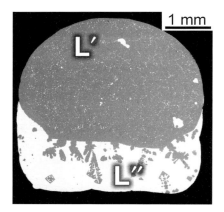

Fig. 1.7 Electron micrograph (BSE) of the sample Al$_{14}$–Bi$_{15}$–Cu$_{71}$ solidified at 2 K min^{-1}. The demixing of the two liquids, Bi-poor (L', dark area) and Bi-rich (L'', light area), is obvious even in the solidified sample.

of this gap is validated by experiments and is even larger than that predicted by the thermodynamic extrapolation from the binary datasets. The microstructures of all ten studied samples clearly show this demixing as given, for example, in Figure 1.7 for the sample (Al$_{14}$ Bi$_{15}$ Cu$_{71}$, at%) solidified at 2 K min^{-1}. The demixing of the two liquids, Bi-poor L' (dark area) and Bi-rich L'' (light area) is obvious. The various Al–Cu intermetallic phases with slightly different compositions in the dark area are hardly distinguishable. During equilibrium solidification the phases β, γ AlCu, γ' AlCu, δAl$_2$Cu$_3$, εAlCu, ηAlCu, ΘAl$_2$Cu, and ζAlCu are formed. The phases β, γ' AlCu, and εAlCu transform or decompose at a lower temperature and are therefore not found in the microstructure at room temperature. Details of the experimental work and the calculation will be published soon [11].

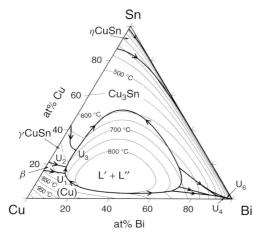

Fig. 1.8 Calculated liquidus surface of the Bi–Cu–Sn system.

1.4.5
Bi–Cu–Sn: Type 3a Monotectic System

As a result of thermodynamic extrapolation of binary data sets into the Bi–Cu–Sn system, an unexpected miscibility gap in the liquid phase of this Al-free ternary was observed. To check whether this is real or a mathematical artifact of the extrapolation, seven sample compositions were investigated experimentally. The results confirm the existence of a miscibility gap in the ternary liquid phase with an even larger extension as given by the extrapolation. The liquid miscibility gap becomes metastable in the binary edge systems Cu–Bi and Cu–Sn with critical temperatures of 668 and 248 °C, respectively.

Three monotectic invariant reactions are calculated for this system and confirmed by DSC signals as follows:

$U_1: L' + (Cu) = \beta + L''$ at 728 °C
$U_2: L'' + \beta = L' + \gamma CuSn$ at 708 °C
$U_3: L'' + \gamma CuSn = L' + Cu_3Sn$ at 672 °C.

All three are transition-type reactions. The calculated liquidus surface of the Bi–Cu–Sn system is shown in Figure 1.8 The calculated vertical section at 30 wt% Bi in Figure 1.9 is compared with measured thermal signals of four samples.

1.5
Quaternary Monotectic Al–Bi–Cu–Sn System

The basis for a calculation of the quaternary Al–Bi–Cu–Sn system has been worked out by the generation of validated thermodynamic descriptions of all four ternary subsystems. This enables a reasonable extrapolation of the Gibbs energy of

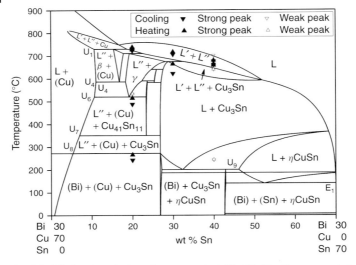

Fig. 1.9 Calculated vertical section at constant 30 wt% Bi with measured thermal signals.

all solution phases, especially for the liquid phase, into the quaternary composition space using the Redlich–Kister/Muggianu-type extrapolation; the equations are given in [12]. The calculation of the quaternary liquidus projection reveals the existence of 18 invariant five-phase reactions. However, most of them can be regarded as degeneration of a ternary invariant reaction, see Table 1.2, generally with a close invariant temperature and a liquid composition very close to a ternary edge system. Only the reaction $L + \gamma \text{AlCu} = \beta + \delta \text{Al}_2\text{Cu}_3 + (\text{Bi})$ at 192 °C appears as a real quaternary reaction, even though the Al content of that liquid phase is very small. This invariant five-phase reaction is connected to five monovariant four-phase reactions, but none of them leads to a close-by invariant reaction of a ternary edge system.

Only one of the 18 invariant reactions comprises two liquids, and thus presents a quaternary monotectic reaction, calculated at 674 °C. This reaction can be derived from the reaction $U_1: \beta + \gamma \text{CuSn} = L + \text{Cu}_3\text{Sn}$ of the ternary Al–Cu–Sn with small amounts of the second melt L' and also a low Al content. The significance of this finding is that one cannot expect to produce a microstructure with appreciable phase amounts to qualify this as a truly quaternary monotectic involving *two liquid* and *three solid* phases, which are in equilibrium according to $L' + L'' + \gamma \text{CuSn} + \text{Cu}_3\text{Sn} + \beta$.

Two vertical quaternary phase diagram sections at constant 40 at% Cu and 5 at% Bi (Figure 1.10) and at constant 50 at% Sn and 5 at% Bi (Figure 1.11) were calculated. The measured DTA signals of six samples with different compositions are also included. The dominating miscibility gap in the liquid phase in the quaternary is observed in both sections. The DTA signals confirm some of the monovariant lines in the sections. Some signals may be caused by reactions not located exactly in the section due to nonequilibrium solidification.

1.5 Quaternary Monotectic Al–Bi–Cu–Sn System

Table 1.2 Calculated invariant reactions in the Al–Bi–Cu–Sn system and their relations to ternary reactions.

Quaternary invariant reaction	Calculated temperature in °C	Relation to ternary reaction
$L'' + \gamma\text{CuSn} = L' + \text{Cu}_3\text{Sn} + \beta$	674	U_3 of Bi–Cu–Sn at 672 °C
$L'' + \beta + \gamma\text{CuSn} + \text{Cu}_3\text{Sn} = \text{Cu}_{10}\text{Sn}_3$	640	D_1 of Bi–Cu–Sn at 640 °C
$L'' + \beta + \text{Cu}_{10}\text{Sn}_3 + \gamma\text{CuSn} = \text{Cu}_{41}\text{Sn}_{11}$	590	D_2 of Bi–Cu–Sn at 590 °C
$L'' + \beta + \text{Cu}_{10}\text{Sn}_3 = \text{Cu}_3\text{Sn} + \text{Cu}_{41}\text{Sn}_{11}$	582	U_5 of Bi–Cu–Sn at 582 °C
$L'' + \beta + \gamma\text{CuSn} = \text{Cu}_{41}\text{Sn}_{11} + (\text{Cu})$	524	U_6 of Bi–Cu–Sn at 519 °C
$L'' + \beta + \text{Cu}_{41}\text{Sn}_{11} = (\text{Cu}) + \text{Cu}_3\text{Sn}$	461	U_7 of Bi–Cu–Sn at 350 °C
$L'' + (\text{Cu}) + \text{Cu}_3\text{Sn} = \beta + (\text{Bi})$	271	U_8 of Bi–Cu–Sn at 272 °C
$L'' + (\text{Cu}) = \beta + (\text{Bi}) + \gamma\text{AlCu}$	270	U_{11} of Al–Bi–Cu at 271.8 °C
$L'' + \beta + \text{Cu}_3\text{Sn} = (\text{Bi}) + \eta'\text{CuSn}$	201	U_9 of Bi–Cu–Sn at 201 °C
$L'' + \gamma\text{AlCu} = \beta + \delta\text{Al}_2\text{Cu}_3 + (\text{Bi})$	192	not assigned
$L'' + \beta + \eta'\text{CuSn} + (\text{Sn}) = \eta\text{CuSn}$	187.5	D_3 of Bi–Cu–Sn at 187 °C
$L'' + \beta + \eta'\text{CuSn} = (\text{Bi}) + \eta\text{CuSn}$	187.2	U_{10} of Bi–Cu–Sn at 187 °C
$\beta + (\text{Sn}) = L'' + \eta\text{CuSn} + \delta\text{Al}_2\text{Cu}_3$	140.647	
$L'' + \beta = \eta\text{CuSn} + \delta\text{Al}_2\text{Cu}_3 + (\text{Bi})$	140.643	
$L'' + \delta\text{Al}_2\text{Cu}_3 = (\text{Sn}) + (\text{Bi}) + \zeta\text{AlCu}$	140.638	
$L'' + \zeta\text{AlCu} = \eta\text{AlCu} + (\text{Sn}) + (\text{Bi})$	140.6	E_1 of Bi–Cu–Sn at 140 °C
$L'' + \eta\text{AlCu} = (\text{Sn}) + \Theta\text{Al}_2\text{Cu} + (\text{Bi})$	140.5	
$L'' + \Theta\text{Al}_2\text{Cu} = (\text{Al}) + (\text{Sn}) + (\text{Bi})$	140.3	

E_1 relates to all reactions between 140.647 °C and 140.3 °C.

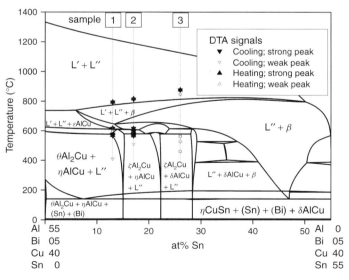

Fig. 1.10 Calculated vertical section of the Al–Bi–Cu–Sn system at constant 40 at% Cu and 5 at% Bi including the measured DTA signals of samples 1–3.

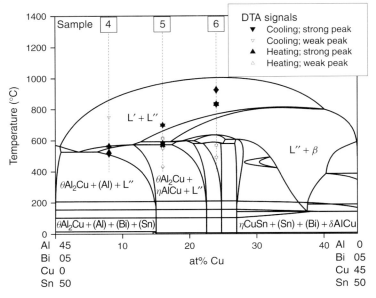

Fig. 1.11 Calculated vertical section of the Al–Bi–Cu–Sn system at constant 50 at% Sn and 5 at% Bi including the measured DTA signals of samples 4–6.

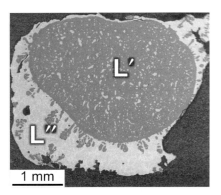

Fig. 1.12 Electron micrograph (BSE) of sample 3 (Al$_{29}$ Bi$_5$ Cu$_{40}$ Sn$_{26}$, at%) solidified at 2 K min^{-1}.

The microstructure of sample 3 (Al$_{29}$ Bi$_5$ Cu$_{40}$ Sn$_{26}$, at%) in Figure 1.12 clearly shows the significant liquid demixing into a Bi-rich (light, L″) and a Bi-poor (dark gray, L′) region, as predicted by the thermodynamic calculation in Figure 1.10. The essentially contiguous dark gray region, denoted as L′, solidifies first to the Al–Cu-rich phase β. Dendrites of β grow subsequently into the higher density Bi-rich liquid L″, beneath the L′-region in the bottom part of the sample, as predicted by the phase field L″ + β in the calculated phase diagram. The β phase transforms during cooling into the Al–Cu-rich phases ηAlCu and ζAlCu. Therefore, the solid

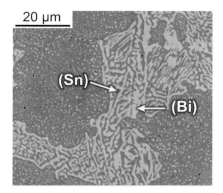

Fig. 1.13 Blow-up of Figure 1.12, showing the eutectic microstructure in the bright region of L" in magnification.

phases in the darker region in Figure 1.12 consist of tiny intergrowth of these two phases. Because of their very similar compositions, they are not discernible in the BSE micrograph because of the minute mass contrast.

Subsequent solidification of the Bi-rich liquid L" terminates in a eutectic-like microstructure, as shown in Figure 1.13. A fine structure of (Sn) and (Bi) is present, formed in one of the reactions around 140 °C in Table 1.2. These reactions are very close to the ternary eutectic E_1 of Bi–Cu–Sn at 140 °C that, in turn, is very close to the binary Bi–Sn edge at a liquid composition with less than 0.02 at% Cu. This is the reason why neither Al- nor Cu-containing phases are visible in the last liquid to solidify.

1.6 Conclusion

Using a combination of computational thermodynamics, involving the Calphad method, and selected key experiments reveals intricate details and stunning features of phase transformations in ternary and quaternary alloys with demixing in the liquid phase. Only in the most simple case does the well-known binary monotectic reaction $L' = L'' + \text{solid}$ extends into the ternary without producing any invariant monotectic four-phase reaction and the Al–Bi–Sn system is given as an example. All other alloy systems in the present study exemplify the distinct feature of ternary monotectic reaction(s), involving $L' + L'' + \text{solid}_1 + \text{solid}_2$. A simple example is Al–Bi–Zn that is unique since it does not involve intermetallic phases and comprises only one ternary monotectic invariant. The Al–Sn–Cu system comprises three and Al–Bi–Cu even seven such monotectic invariant reactions. Also, the microstructures become more complex in these alloys, because of the appearance of the numerous intermetallic Al–Cu phases. All of the four ternary subsystems of the quaternary Al–Bi–Cu–Sn system were studied experimentally and through thermodynamic calculations. Based on that, the phase relations of the quaternary Al–Bi–Cu–Sn system could be quantitatively obtained without any additional fitting parameter. These phase relations were then verified by an experimental study of key samples.

Each of these monotectic systems shows some peculiarity. The Al–Bi–Cu system is unique because of the stunning intensification of liquid demixing of Al–Bi upon addition of Cu. The occurrence of unexpected liquid demixing, not present in any of the binary systems, is detected in the Al–Sn–Cu system. This particular feature of an isolated ternary miscibility gap is, surprisingly, also found in the Al-free Bi–Cu–Sn system, where it produces three invariant monotectic reactions.

Only one of the 18 invariant five-phase reactions in the quaternary Al–Bi–Cu–Sn system involves *two liquid* and *three solid* phases, which are in equilibrium according to $L' + L'' + \gamma CuSn + Cu_3Sn + \beta$. However, this invariant monotectic reaction, calculated at 674 °C, is degenerate to the Al–Cu–Sn system with a very small fraction of the melt L'. The significance of this finding is that one cannot expect to produce a microstructure with appreciable phase amounts to qualify this as a truly quaternary monotectic.

Acknowledgment

This study is supported by the German Research Foundation (DFG) in the Priority Programme "DFG-SPP 1120: Phase transformations in multicomponent melts" under grant no. Schm 588/24.

References

1 Mirković, D., Gröbner, J. and Schmid-Fetzer, R. (**2008**) "Solidification paths of multicomponent monotectic aluminum alloys". *Acta Materialia*, accepted.

2 Chen, S.-L., Zhang, J.-Y., Lu, X.-G., Chou, K.-C., Oates, W.A., Schmid-Fetzer, R. and Austin Chang, Y. (**2007**) "Calculation of rose diagrams". *Acta Materialia*, **55**, 243–50.

3 Gröbner, J. and Schmid-Fetzer, R. (**2005**) "Phase transformations in ternary monotectic aluminum alloys". *Journal of Metals*, 9, 19–23.

4 Doernberg, E., Kozlov, A. and Schmid-Fetzer, R. (**2007**) "Experimental Investigation and thermodynamic calculation of Mg-Al-Sn phase equilibria and solidification microstructures". *Journal of Phase Equilibria and Diffusion*, **28**, 523–35.

5 Mirković, D., Gröbner, J., Schmid-Fetzer, R., Fabrichnaya, O. and Lukas, H.L. (**2004**) "Experimental study and thermodynamic re-assessment of the Al-B system". *Journal of Alloys and Compounds*, **384**, 168–74.

6 Gröbner, J., Mirković, D. and Schmid-Fetzer, R. (**2004**) "Thermodynamic aspects of constitution, grain refining and solidification enthalpies of Al-Ce-Si alloys". *Metallurgical and Materials Transactions A*, **35A**, 3349–62.

7 Gröbner, J., Mirković, D. and Schmid-Fetzer, R. (**2005**) "Thermodynamic aspects of grain refinement of Al-Si alloys using Ti and B". *Materials Science and Engineering A*, **395**, 10–21.

8 Mirković, D. and Schmid-Fetzer, R. (**2007**) "Solidification curves for commercial Mg alloys determined from differential scanning

calorimetry with improved heat transfer modeling". *Metallurgical and Materials Transactions A*, **38A**, 2575–92.
9 Gröbner, J., Mirković, D. and Schmid-Fetzer, R. (**2005**) "Monotectic four-phase reaction in Al-Bi-Zn alloys". *Acta Materialia*, **53**, 3271–80.
10 Mirković, D., Gröbner, J. and Schmid-Fetzer, R. (**2008**) "Liquid demixing and microstructure formation in ternary Al-Sn-Cu alloys". *Materials Science and Engineering A*, **487**, 456–67.
11 Mirković, D., Gröbner, J., Schmid-Fetzer, R., Kaban, I. and Hoyer, W. *Phase Relations, Solidification, Density and Thermodynamic Calculation in the Al-Bi-Cu System*, to be published.
12 Schmid-Fetzer, R. and Gröbner, J. (**2001**) "Focused development of magnesium alloys using the calphad approach". *Advanced Engineering Materials*, **3**, 947–61.

2
Liquid-liquid Interfacial Tension in Multicomponent Immiscible Al-based Alloys

Walter Hoyer and Ivan G. Kaban

2.1
Introduction

Al-based alloys containing soft metals such as Bi, In, or Pb are considered as very promising candidates to be used as self-lubricating bearings [1–3]. In this connection, ternary and quaternary systems are more interesting than binaries because there might exist more phases in equilibrium, and more microstructures could be produced [4–6]. The main obstacle in the way of practical applications of these alloys (called *monotectic alloys*) is their demixing. In the liquid state above the critical temperature T_C, all constituents of the monotectic alloys are completely miscible. However, the components start to demix at T_C, and the initially homogeneous liquid separates into two liquid phases that coexist in equilibrium down to the monotectic temperature T_M. They solidify into two layers separated by a distinct interface at normal gravity conditions. As an example, the $(Al_{0.345}Bi_{0.655})_{90}Sn_{10}$ alloy solidified in a graphite crucible is shown in Figure 2.1. (Here and in the following, numerical indices indicate mass percent).

Initially, it was supposed that the suppression of gravity-dependent phenomena such as sedimentation or buoyancy-driven convection will result in a homogeneous solid structure. However, investigations under reduced gravity revealed that gravity force alone does not govern the phase separation in monotectic alloys. The microstructure observed is determined by a complex interplay of nucleation, growth, as well as Stokes and Marangoni motions [1–3].

The nucleation of the minor phase during cooling within the miscibility gap is essential to the description of the liquid–liquid (l–l) decomposition kinetics [2, 3]. In the classical nucleation theory, the stationary nucleation rate is given by the expression [7]:

$$I^{hom} = N_0 O \Gamma \exp\left(-\frac{\Delta G_c}{k_B T}\right) \tag{2.1}$$

Phase Transformations in Multicomponent Melts. Edited by D. M. Herlach
Copyright © 2008 WILEY-VCH Verlag GmbH & Co. KGaA, Weinheim
ISBN: 978-3-527-31994-7

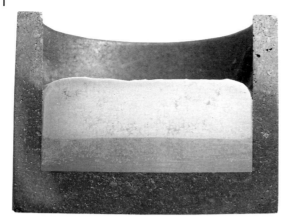

Fig. 2.1 Cross section of the $(Al_{0.345}Bi_{0.655})_{90}Sn_{10}$ (wt %) sample solidified in a graphite crucible.

with

$$N_0 = (x_A \Omega_A + x_B \Omega_B)^{-1}$$

$$O = 4n_c^{2/3}$$

$$\Gamma = 6D/\lambda$$

$$Z = \left(\frac{\Delta G_c}{3\pi k_B T n_c^2}\right)^{1/2}$$

$$\Delta G_c = \frac{16}{3}\pi \frac{\sigma_{\alpha\beta}^3}{\Delta G_v^2}$$

Here N_0 is the number density, Ω_A and Ω_B are the atomic volumes of components A and B, x_A and x_B are the mole fractions of components, n_c is the number of atoms in a droplet of critical radius $R^* = 2\sigma_{\alpha\beta}/\Delta G_v$, $\sigma_{\alpha\beta}$ is the interfacial tension between the two liquids, ΔG_v is the gain of free energy per volume on nucleation, D is the diffusion coefficient, λ is the average jump distance of a solute atom due to diffusion, Z is the Zeldovich factor, k_B is the Boltzmann constant, T is the absolute temperature, and ΔG_c is the energy barrier for nucleation.

Another process contributing to the phase separation in monotectic alloys and directly related to the interfacial tension (more exactly, to its temperature dependence) is the Marangoni (thermocapillary) motion [2, 3]. Temperature gradients that occur during cooling of the melt induce variation of the interfacial tension along the liquid–liquid interface. The droplets nucleated move upon the tangential stress with the velocity u_M, which can be determined from the equation [3, 8]:

$$u_M = -\frac{2R}{[(2\lambda_m + \lambda_d)/\lambda_m](2\eta_m + 3\eta_d)} \frac{\partial \sigma_{\alpha\beta}}{\partial T} \nabla T \qquad (2.2)$$

where λ_m and λ_d are the thermal conductivities and η_m and η_d are the viscosities of the matrix liquid and the droplet, respectively, and R is the radius of the droplet.

Thus, the interfacial tension between coexisting liquid phases and its temperature dependence are indispensable for the description of microstructure evolution during the solidification of monotectic alloys.

In this chapter we summarize the results of interfacial tension measurements in Al–Bi binary [9]; Al–Bi–Cu, Al–Bi–Si, Al–Bi–Sn ternary [10–13]; and Al–Bi–Cu–Sn quaternary [14] alloys.

2.2 Measurement Technique

There exist a number of experimental techniques for the measurement of the surface tension of liquids (a review of methods can be found, e.g., in Ref. 15) and theoretically it should not be a problem to apply them for the determination of the interfacial tension between two liquids. However, this is mainly true for transparent liquids and at normal temperatures. Investigation of interfaces in metallic liquids is substantially hampered because of their nontransparency. This practically prevents the use of methods based on the direct analysis of the interface shape (profile).

Tensiometry, such as Du Noüy ring or Wilhelmy plate methods, is one of the simplest techniques often applied to liquid surfaces and interfaces. In this technique, surface or interfacial tension is determined from the maximum force exerted on a body (plate, ring, or rod) that is withdrawn from the liquid surface or interface. There is, however, a great problem since the l–l interfacial tension at the monotectic temperature is significantly less than the surface tension of the upper phase and it decreases continuously with increasing temperature. To reduce the surface contribution to the measured force, Merkwitz *et al.* [16, 17] suggested the application of a cylindrical stamp as shown schematically in Figure 2.2. The technique is based on the relation between the force exerted on an alumina stamp withdrawn from the liquid–liquid interface (measured experimentally) and the shape of meniscus (modeled by numerical solution of the Young–Laplace equation of capillarity). A whole set of the experimental data (measured force as a function of the stamp's height), and not at just one point (maximum force), is analyzed here. This method was used for the measurements in the binary monotectic alloys Ga–Hg, Ga–Pb [16, 17], Al–In, and Al–Pb [16, 18], as also in the Al–Bi-based multicomponent systems [9–14].

Principles of the experimental setup and the measurement process can be understood from Figures 2.2 and 2.3. The weight of the meniscus formed at the interface is measured as a function of the stamp's height. The experimental curve for the liquid $(Al_{34.5}Bi_{65.5})_{90}Sn_{10}$ alloy at 660°C is shown as an example. After correcting for the experimental data for the buoyancy force, geometry (crucible, stamp), and hysteresis (if necessary), the volume of meniscus V_{Men}^{exp} is determined as a function of the height of the contact line x_0 from the relation:

$$F_{Men}^{exp}(x_0) = V_{Men}^{exp}(x_0) g \Delta \rho_{\alpha\beta} \qquad (2.3)$$

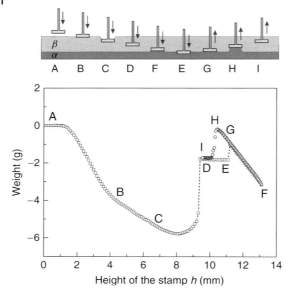

Fig. 2.2 Main stages of the measurement process and the experimental curve obtained from liquid $(Al_{0.345}Bi_{0.655})_{90}Sn_{10}$ at 660°C.

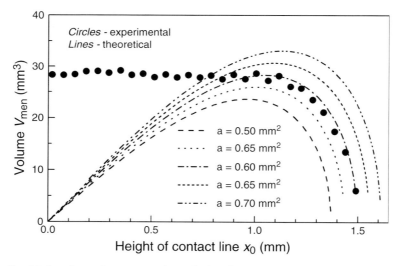

Fig. 2.3 Experimental meniscus volume for liquid $(Al_{0.345}Bi_{0.655})_{90}Sn_{10}$ at 660°C obtained from the measured curve plotted in Figure 2.2 and the theoretical volume curves calculated for different capillary constants.

where $F_{Men}^{exp}(x_0)$ is the force caused by the meniscus in the vertical direction (which is equal to the weight of the liquid displaced by the meniscus [19]), $\Delta\rho_{\alpha\beta}$ is the difference in macroscopic densities of two coexisting liquid phases, and g is the acceleration due to gravity.

On the other hand, the meniscus is modeled by a numerical solution of the Young–Laplace equation, which describes the pressure difference Δp across the curved interface:

$$\sigma_{\alpha\beta}\left(\frac{1}{R_1}+\frac{1}{R_2}\right)=\Delta p=p_1-p_2 \qquad (2.4)$$

where $\sigma_{\alpha\beta}$ is the interfacial tension between liquids α and β; R_1 and R_2 are the principal radii of curvature of the interface; and p_1 and p_2 are the pressures on the concave and convex sides of the interface, respectively. Owing to the cylindrical form of the stamp, solution of Equation 2.4 is reduced to a one-dimensional problem, so that the Young–Laplace equation contains the radii of curvature and a linear gravitational component:

$$\sigma_{\alpha\beta}\left(\frac{1}{R_1}+\frac{1}{R_2}\right)=\Delta\rho_{\alpha\beta}\,g\,x \qquad (2.5)$$

where x is a coordinate in the direction of the gravitational force. Substituting the principal radii of curvature by the differential geometry expressions, one obtains the following equation [20, 21]:

$$\frac{x'}{r(1+x'^2)^{1/2}}+\frac{x''}{(1+x'^2)^{3/2}}=\frac{\Delta\rho_{\alpha\beta}\,g\,x}{\sigma_{\alpha\beta}} \qquad (2.6)$$

where r is the radial distance in a horizontal plane from the axis of symmetry ($r^2=y^2+z^2$), $x'=dx/dr$, and $x''=d^2x/dr^2$.

Numerical solution of Equation 2.6 allows modeling of the radially symmetrical meniscus and calculation of its volume as a function of the height of the contact line $x_0 - V_{Men}^{model}(x_0)$—for various values of the capillary constant a, which is defined as

$$a=\frac{\sigma_{\alpha\beta}}{\Delta\rho_{\alpha\beta}\,g} \qquad (2.7)$$

Comparison of the experimental volume curves $V_{Men}^{exp}(x_0)$ with the volumes of menisci modeled for the stamp used $V_{Men}^{model}(x_0)$ results in the determination of the capillary constant for a studied alloy as is shown, for example, in Figure 2.3.

Difference in macroscopic densities of the coexisting liquid phases $\Delta\rho_{\alpha\beta} = \rho_\alpha - \rho_\beta$ in a monotectic alloy is needed for the calculation of interfacial tension with Equation 2.7. It is an advantage of the tensiometric technique described that $\Delta\rho_{\alpha\beta}$ can be found from the same experimental data that are used for the determination of the capillary constant.

2.3
Experimental Details

Monotectic alloys were prepared from high-purity Al, Bi, Cu, Si, and Sn (99.999%). The measurements were performed in the tensiometer shown in Figure 2.4. Before heating, the chamber was evacuated to better than 1×10^{-5} mbar and filled with a gas mixture of Ar–10H$_2$ (vol %) with total pressure of $\sim$1 bar. A niobium getter was used to reduce the amount of oxygen that gets into the chamber through leaks during measurements. The heating system consisted of a concentric heater outside the chamber, a power supply, and an electronic temperature control device. The graphite crucible (5 cm inner diameter, 4 cm height) was moved by an ultrahigh vacuum manipulating system. The force exerted on the stamp was measured by a balance with an accuracy of $\pm$1 mg. The error of interfacial tension is 5–8% near the monotectic temperature.

Phase transitions in the monotectic alloys were studied with a differential scanning calorimeter (DSC) NETZSCH DSC 404 C. Temperature calibration of the calorimeter was performed using different elements of high purity. The samples for calorimetric measurements ($\sim$170 mg of mass) were prepared by arc-melting under argon atmosphere after initial evacuation of the furnace down to about 10^{-3} mbar. Samples with total weight loss in preparation below 2 mg were chosen for the measurements. DSC scans were obtained during non-isothermal heating and cooling at the rate of 20 K min^{-1} under an Ar flow. The heating–cooling cycles were carried out several times with several different samples, and good reproducibility of the peak positions (within $\pm$5 K) was observed.

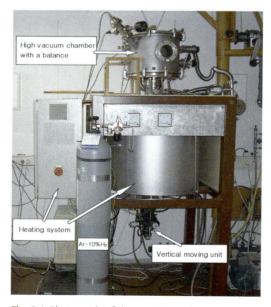

Fig. 2.4 Photograph of the experimental setup.

2.4
Results

DSC scans measured during cooling at a constant rate of 20 K min^{-1} [13, 14] are shown in Figure 2.5. The phase separation is manifested as a small step in the exothermic direction on the DSC curves for all alloys studied except the alloy containing Si. The onset temperatures of phase separation T_{sep} and monotectic reaction T_M extracted from DSC curves are collected in Table 2.1.

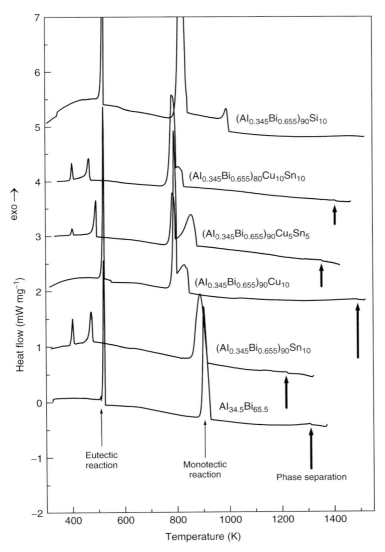

Fig. 2.5 DSC curves obtained at 20 K min^{-1} cooling rate.

Table 2.1 Phase separation temperature T_{sep} and monotectic temperature T_M extracted from differential scanning calorimetry cooling scans. Experimental uncertainty is ±5 K. Temperature dependences of the liquid–liquid interfacial tension $\sigma_{\alpha\beta}$: fits to the experimental data. T is the absolute temperature, T_C is the critical temperature, μ is the critical point exponent, and σ_0 is a constant. Uncertainties are determined by the fitting.

Alloy composition (wt %)	DSC cooling 20 K min^{-1}		Fit $\sigma_{\alpha\beta} = \sigma_0 \times (1 - T/T_C)_\mu$ (mN m^{-1})			
	T_M (K)	T_{sep} (K)	fitting range (K)	T_C (K)	σ_0 (mN m^{-1})	μ
Al$_{34.5}$Bi$_{65.5}$ 9	—	—	933–1213	1310	289	1.29[a]
Al$_{34.5}$Bi$_{65.5}$ 13	928	1312	933–1213	1310 ± 4	290.6 ± 3.9	1.3[b]
(Al$_{0.345}$Bi$_{0.655}$)$_{90}$Cu$_{10}$ 13	849	1491	863–1213	1493 ± 7	227.6 ± 2.6	1.3[b]
Al$_{23.25}$Bi$_{65.5}$Cu$_{11.25}$ 11	—	—	828–1213	1673 ± 15	230.1 ± 3.1	1.3[b]
(Al$_{0.345}$Bi$_{0.655}$)$_{95}$Si$_5$ 11	—	—	873–1238	1383 ± 8	265.5 ± 2.5	1.3[b]
(Al$_{0.345}$Bi$_{0.655}$)$_{90}$Si$_{10}$ 13	1004	—	1023–1193	1529 ± 14	226.8 ± 6.9	1.3[b]
(Al$_{0.345}$Bi$_{0.655}$)$_{95}$Sn$_5$ 11	—	—	933–1208	1273 ± 2	236.6 ± 1.7	1.3[b]
(Al$_{0.345}$Bi$_{0.655}$)$_{90}$Sn$_{10}$ 13	902	1216	933–1203	1238 ± 5	208.7 ± 9.4	1.31 ± 0.04[c]
(Al$_{0.345}$Bi$_{0.655}$)$_{90}$Cu$_5$Sn$_5$ 14	878	1357	893–1193	1367 ± 4	238.5 ± 2.2	1.3[b]
(Al$_{0.345}$Bi$_{0.655}$)$_{80}$Cu$_{10}$Sn$_{10}$ 14	829	1408	843–1193	1409 ± 4	214.8 ± 1.8	1.3[b]

a $T_C = 1310$ K was fixed.
b $\mu = 1.3$ was fixed by the fitting.
c All parameters (σ_0, T_C, and μ) were free.

Al$_{34.5}$Bi$_{65.5}$ is a critical composition in the Al–Bi binary system with the critical temperature of 1310 K [4]. DSC study of this alloy was carried out in Refs. [13, 14]. T_{sep} for Al$_{34.5}$Bi$_{65.5}$ determined in Ref. 13 is 1312 K, and T_{sep} established in Reference [14] is 1306 K. Taking into account that different samples were investigated at different times, the good agreement with the literature data observed in both cases suggests reliability of the phase separation temperatures for the ternary and quaternary samples measured in the same way.

Composition dependence of the l–l interfacial tension in the Al–Bi–Cu, Al–Bi–Si, and Al–Bi–Sn systems is shown in Figure 2.6, while that for the Al–Bi–Cu–Sn system is shown in Figure 2.7. The temperature dependences of the interfacial tension in the binary, ternary, and quaternary alloys are plotted in Figures 2.8–2.11.

It is seen that additions of Cu, Si, or Sn strongly affect the interfacial tension. The addition of either Cu or Si to the Al$_{34.5}$Bi$_{65.5}$ binary causes a remarkable increase of the interfacial tension between Al-rich and Bi-rich liquid phases. In contrast, the addition of Sn to Al$_{34.5}$Bi$_{65.5}$ decreases $\sigma_{\alpha\beta}$ over the whole temperature interval where immiscibility is observed. Simultaneous addition of Cu and Sn to the Al$_{34.5}$Bi$_{65.5}$ binary results in a slight decrease (5–6%) of the interfacial tension at low temperatures. Beginning from ~1050 K, the interfacial tension in both four-component alloys is larger than that in liquid Al$_{34.5}$Bi$_{65.5}$ (Figures 2.7 and 2.11).

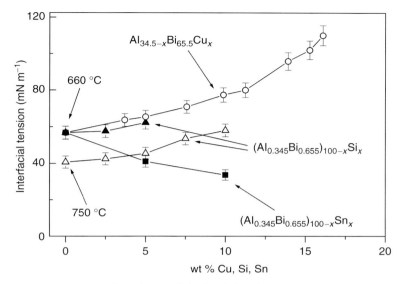

Fig. 2.6 Composition dependences of the liquid–liquid interfacial tension in Al–Bi–Cu, Al–Bi–Si, and Al–Bi–Sn ternary alloys.

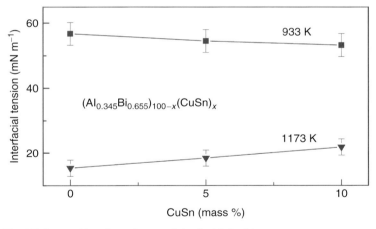

Fig. 2.7 Composition dependences of the liquid–liquid interfacial tension in the quaternary alloys $(Al_{0.345}Bi_{0.655})_{100-x}(CuSn)_x$ at 933 and 1173 K [14].

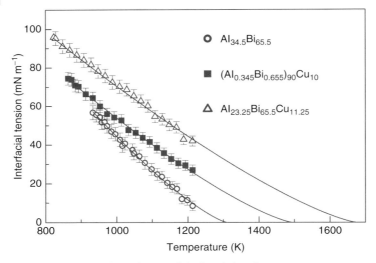

Fig. 2.8 Temperature dependences of the liquid–liquid interfacial tension in $Al_{34.5}Bi_{65.5}$ binary [9] and $Al_{23.25}Bi_{65.5}Cu_{11.25}$ and $(Al_{0.345}Bi_{0.655})_{90}Cu_{10}$ ternaries [11, 13] (wt %). The lines are fits to the experimental data: $\sigma_{\alpha\beta} = \sigma_0 \times (1 - T/T_C)^{1.3}$ mN m^{-1} (see text and Table 2.1 for details).

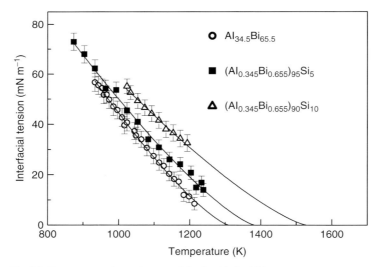

Fig. 2.9 Temperature dependences of the liquid–liquid interfacial tension in $Al_{34.5}Bi_{65.5}$ binary [9] and $(Al_{0.345}Bi_{0.655})_{95}Si_5$ and $(Al_{0.345}Bi_{0.655})_{90}Si_{10}$ ternaries [10, 11] (wt %). The lines are fits to the experimental data: $\sigma_{\alpha\beta} = \sigma_0 \times (1 - T/T_C)^{1.3}$ mN m^{-1} (see text and Table 2.1 for details).

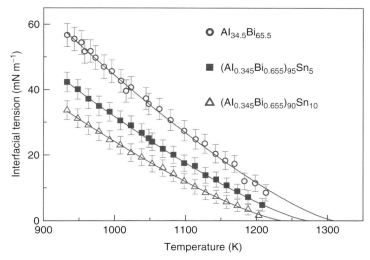

Fig. 2.10 Temperature dependences of liquid–liquid interfacial tension in $Al_{34.5}Bi_{65.5}$ binary [9] and $(Al_{0.345}Bi_{0.655})_{95}Sn_5$ and $(Al_{0.345}Bi_{0.655})_{90}Sn_{10}$ ternaries [11, 13] (wt %). The lines are fits to the experimental data: $\sigma_{\alpha\beta} = \sigma_0 \times (1 - T/T_C)^{1.3}$ mN m^{-1} (see text and Table 2.1 for details).

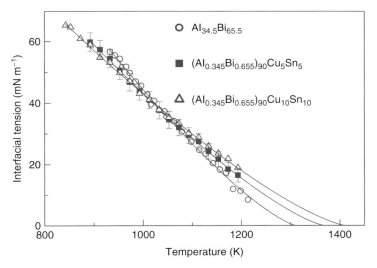

Fig. 2.11 Temperature dependences of the liquid–liquid interfacial tension in $Al_{34.5}Bi_{65.5}$ binary [9] and $(Al_{0.345}Bi_{0.655})_{90}Cu_5Sn_5$ and $(Al_{0.345}Bi_{0.655})_{80}Cu_{10}Sn_{10}$ quaternaries [14] (wt %). The lines are fits to the experimental data: $\sigma_{\alpha\beta} = \sigma_0 \times (1 - T/T_C)^{1.3}$ mN m^{-1} (see text and Table 2.1 for details).

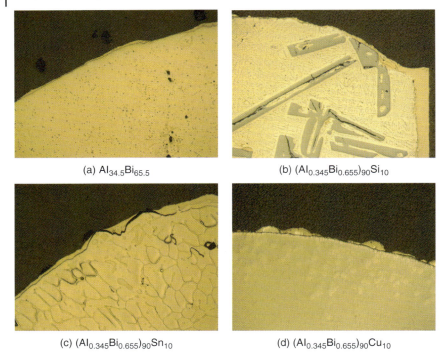

(a) $Al_{34.5}Bi_{65.5}$
(b) $(Al_{0.345}Bi_{0.655})_{90}Si_{10}$
(c) $(Al_{0.345}Bi_{0.655})_{90}Sn_{10}$
(d) $(Al_{0.345}Bi_{0.655})_{90}Cu_{10}$

Fig. 2.12 Optical micrographs of the solidified samples showing Al-rich phase/surface layer (micrograph width corresponds to 2.74 mm).

Figure 2.12 shows the light microscope images of the upper phase in the binary and ternary alloys studied.

2.5
Discussion

2.5.1
Composition Dependences of the l–l Interfacial Tension

Interfacial tension is defined as the difference, per unit area of interface, between the free energy of the whole system with the interface and the free energy of a homogeneous system, which could be formed by the same atoms with the same chemical potentials. Therefore, the interfacial tension can be calculated by the methods of statistical thermodynamics.

In a very simple model suggested by Becker [22], a demixing system is presented as a regular solution with a sharp interface between two phases, and the interactions between the atoms restricted to the nearest neighbors. Then, the interfacial tension is equal to additional energy resulting from the asymmetry of the bonds across the interface and can be calculated by the expression:

$$\sigma_{\alpha\beta} = n_{\alpha\beta} n_b w \left(x_{i\alpha} - x_{i\beta}\right)^2 \tag{2.8}$$

where $n_{\alpha\beta}$ is the atomic interfacial density (in atoms per interface); n_b is the number of bonds of an atom across the interface; w is an interaction energy; and $x_{i\alpha}$ and $x_{i\alpha\beta}$ are the concentrations of the ith component in the phase α and β, respectively.

In the model of Cahn and Hilliard [23], there is no sharp interface between coexisting phases, and the concentration profile normal to the interface changes continuously. Cahn and Hilliard applied the van der Waals model of diffusive interfaces [24] to a binary immiscible system and obtained the following expression for the interfacial energy at a flat interface:

$$\sigma_{\alpha\beta} = N_V \int_{-\infty}^{+\infty} \left[\Delta f(c_i) + \kappa (dc_i/dx)^2\right] dx \tag{2.9}$$

N_V is the number of molecules per unit volume, c_i is the mole fraction of one of the two components, κ is a gradient energy coefficient associated with the gradient of composition dc_i/dx. $\Delta f(c_i)$ is regarded as a free energy referred to a standard state of an equilibrium mixture. Hoyt [25] extended the Cahn and Hilliard theory [23] to multicomponent systems and derived the expression for the interfacial tension similar to Equation 2.9:

$$\sigma_{\alpha\beta} = N_V \int_{-\infty}^{+\infty} \left[\Delta f(c_1, c_2, \ldots, c_{n-1}) + \sum_i^{n-1} \sum_j^{n-1} \kappa_{ij} \nabla c_i \nabla c_j\right] dx \tag{2.10}$$

Here, n stands for the number of components, and $\nabla = \partial/\partial x$.

Thus, interfacial tension depends on the composition of the coexisting phases and the concentration profile at the interface. Obviously, the smaller the variation in composition across the interface, the smaller the contribution of the gradient energy term to the interfacial tension. It could be expected that the concentration gradient decreases with narrowing miscibility gap and increases with widening miscibility gap. This suggests that the interfacial tension will also increase if the miscibility gap grows and decrease if it narrows.

It is known that the addition of either Cu or Si to the Al–Bi binary significantly enhances the l–l separation. The miscibility region covers almost the entire concentration triangle in both Al–Bi–Cu [5, 6] and Al–Bi–Si [26] ternary systems, and the critical temperature remarkably increases. Thus, the increase of interfacial energy between Al-rich and Bi-rich liquid phases with the addition of either Cu or Si to the $Al_{34.5}Bi_{65.5}$ (Figures 2.6, 2.8 and 2.9) can be explained by the extension of the l–l miscibility gap. On the other hand, the miscibility gap decreases when Sn is added to the Al–Bi binary [5, 6]. This results in the decrease of the l–l interfacial tension in the ternary Al–Bi–Sn alloys in contrast to the Al–Bi binary (Figures 2.6 and 2.10).

Two different effects are observed when Cu and Sn are added to the Al–Bi binary simultaneously [6]. On one hand, the width of the miscibility gap decreases, and on the other, its height increases upon addition of Cu, Sn. This explains why the isothermal composition dependences of the l–l interfacial tension are opposed in the low- and high-temperature regions (Figures 2.7 and 2.11).

2.5.2
Adsorption at the l–l Interface

Another physical phenomenon that influences the interfacial tension is Gibbs adsorption at the interface [27]. Gibbs adsorption equation relates the changes in interfacial tension $\sigma_{\alpha\beta}$ to the corresponding changes in thermodynamic variables, such as temperature and chemical potential, and involves the excess quantities associated with the interface. Variation of the interfacial tension due to adsorption at the interface in a multicomponent system at a fixed temperature is expressed by the equation:

$$d\sigma_{\alpha\beta} = -\sum_i \Gamma_i \, d\mu_i \qquad (2.11)$$

where μ_i is the chemical potential of the ith species and Γ_i is the surface excess of the component i per unit area [27–29]. The change of interfacial tension can be calculated using Equation 2.11 if the amount of adsorbed material at the interface is known. Unfortunately, this is very difficult to determine experimentally.

Chatain et al. [30, 31] suggested a very simple criterion to estimate the tensioactivity of a component C at the interface in a three-component system A–B–C by calculating the adsorption separately from A-rich (α) and B-rich (β) liquids. Only the composition of the bulk phases and partial excess free energies at infinite dilution are to be known in their approach as seen below.

According to Chatain et al. [30, 31], if the interface is presented by two layers l_1 and l_2, then the adsorption of element C (at the infinite dilution) from the volume of liquid α to the interface layer l_1 can be written as an exchange:

$$(C)_\alpha + (A)_{l_1} \to (C)_{l_1} + (A)_\alpha \qquad (2.12)$$

Under assumption that solutions AB, BC, and CA are pseudoregular solutions, the change of energy due to adsorption can be written as

$$E_{a,1}^\infty = m(\Delta \overline{G}_{C(B)}^{XS\infty} - \Delta \overline{G}_{C(A)}^{XS\infty} - \Delta \overline{G}_{A-B}^{XS\infty}) = m\tau_\alpha \qquad (2.13)$$

where m is a structural parameter, which depends on the interfacial layer, $\Delta \overline{G}_{C(A)}^{XS\infty}$ and $\Delta \overline{G}_{C(B)}^{XS\infty}$ are the partial excess free energies at infinite dilution of C in component

A and B respectively, and

$$\Delta \overline{G}_{A-B}^{XS\infty} = \frac{\overline{G}_A^{XS\infty} + \overline{G}_B^{XS\infty}}{2} \tag{2.14}$$

Similarly, adsorption energy of component C from liquid β to the interface layer l_2 is:

$$E_{a,2}^{\infty} = m(\Delta \overline{G}_{C(A)}^{XS\infty} - \Delta \overline{G}_{C(B)}^{XS\infty} - \Delta \overline{G}_{A-B}^{XS\infty}) = m\tau_\beta \tag{2.15}$$

Then the tensioactivity of C can be expressed by the parameter $\tau_{\alpha\beta}$:

$$\tau_{\alpha\beta} = \frac{K_{\alpha\beta} \cdot \tau_\alpha + \tau_\beta}{1 + K_{\alpha\beta}} \tag{2.16}$$

where $K_{\alpha\beta}$ is a partitioning coefficient that describes the distribution of component C (with concentration X_C) between liquid α and liquid β by the infinite solution:

$$K_{\alpha\beta} = \left(\frac{X_C^\alpha}{X_C^\beta}\right)_{X_C \to 0} \tag{2.17}$$

A negative value of $\tau_{\alpha\beta}$ corresponds to the segregation of component C at the interface between α and β liquid phases. The more negative $\tau_{\alpha\beta}$ is, the more significant the adsorption and the more distinctive the drop of the interfacial tension $\sigma_{\alpha\beta}$. For positive values of $\tau_{\alpha\beta}$, the solute C desorbs from the interface.

The tensioactivity parameter $\tau_{\alpha\beta}$ was determined with Equation 2.16 for Al–Bi–Cu, Al–Bi–Si, and Al–Bi–Sn ternary systems in [13]. It is found that $\tau_{\alpha\beta}/RT \approx -2$ for Sn in the Al–Bi–Sn system, $\tau_{\alpha\beta}/RT \approx 2$ for Cu in Al–Bi–Cu, and $\tau_{\alpha\beta}/RT \approx 4$ for Si in Al–Bi–Si ($T = 933$ K). Thus, the drop of interfacial tension observed in the ternary $(Al_{0.345}Bi_{0.655})_{90}Sn_{10}$ alloy as compared to the $Al_{34.5}Bi_{65.5}$ binary is caused not only by the decrease of the miscibility gap but also by adsorption of Sn at the l–l interface. Positive values of $\tau_{\alpha\beta}$ are indicative of increasing interfacial tension upon addition of either Cu or Si to the Al–Bi binary. These findings agree with the experimental observations (Figures 2.6–2.10).

If we consider the relation between interfacial tension and width of the miscibility gap, then $\sigma_{\alpha\beta}$ in the $(Al_{0.345}Bi_{0.655})_{100-x}(CuSn)_x$ alloys, as compared to the $Al_{34.5}Bi_{65.5}$ binary, is expected to be smaller below 1273–1310 K and larger at higher temperatures. Indeed, the miscibility gap in the $(Al_{0.345}Bi_{0.655})_{100-x}(CuSn)_x$ alloys is narrower than in the Al–Bi binary below 1273–1310 K [6]. However, experiments show that the interfacial tension in quaternary alloys is larger than in the Al–Bi binary at temperatures already above $\sim$1050 K (Figures 2.7 and 2.11). If we assume that the tensioactivity of Cu and Sn at the l–l interface in Al–Bi–Cu–Sn quaternary alloys is the same as in Al–Bi–Cu and Al–Bi–Sn ternaries, then the experimental observation can be explained by a predominant influence of Cu on the interfacial tension with increasing temperature.

2.5.3
Temperature Dependence of the l–l Interfacial Tension

The interface between coexistent liquid phases thickens and becomes diffuse upon heating. At the critical point, the interface disappears and the two liquids merge into one homogeneous fluid. Also, interfacial tension decreases with increasing temperature and it vanishes at the critical point. The temperature dependence of interfacial tension can be described by the power law:

$$\sigma_{\alpha\beta} = \sigma_0 \cdot (1 - T/T_C)^{\mu} \tag{2.18}$$

where σ_0 is a constant, T is the absolute temperature, T_C is the critical temperature, and μ is the so-called critical-point exponent. According to the classical (mean-field) theory [23, 24] $\mu = 1.5$, while the renormalization group theory of critical behavior [28] gives $\mu = 1.26$.

It was shown in Reference 9 that the temperature dependence of the interfacial tension for liquid $Al_{34.5}Bi_{65.5}$ is well described by the power function (Equation 2.18) with $\sigma_0 = 289$ mN m^{-1} and $\mu \approx 1.3$. Therefore, in later studies the experimental temperature dependences of the interfacial tension in ternary and quaternary alloys studied were fitted by Equation 2.18 with free parameters σ_0 and T_C and the fixed value of exponent $\mu = 1.3$ [10–14]. Taking into account that the interfacial tension for the ternary alloy $(Al_{0.345}Bi_{0.655})_{90}Sn_{10}$ was measured practically over the whole miscibility temperature range, it was fitted with the function (Equation 2.18) when all parameters (σ_0, T_C, μ) were free [13]. The exponent $\mu = 1.31 \pm 0.04$ was found. The values obtained by fitting are collected in Table 2.1, while the fitting curves are plotted in Figures 2.8–2.11.

Thus, the experimental temperature dependences of the l–l interfacial tension in the binary, ternary, and quaternary alloys studied are well described by Equation 2.18 with the critical exponent close to that predicted by the renormalization theory.

2.5.4
Wetting Phenomena

If a liquid α is in contact with another liquid β and a vapor phase v, then the contact angle Θ, which is determined by a balance of interfacial tensions $\sigma_{\alpha\beta}$ (liquid α–liquid β), $\sigma_{\alpha v}$ (liquid α–vapor), and $\sigma_{\beta v}$ (liquid β–vapor) might be either zero or nonzero [32]. If the surface tension of the denser α-phase is lower than that of the β-phase ($\sigma_{\alpha v}$ ¡ $\sigma_{\beta v}$) then: (i) $180° > \Theta \geq 90°$ indicates that the liquid α does not wet liquid β; (ii) $90° > \Theta > 0°$ corresponds to a partial wetting; and (iii) in the case $\Theta = 0°$, liquid β is completely wetted by the liquid α. The latter means that a macroscopic layer of the α-phase intrudes between the surface of the β-phase and the vapor v. In principle, any other (noninteracting) phase can be considered instead of the vapor, such as, for example, a crucible for the liquids α and β.

Cahn showed theoretically [33] that transitions from partial to complete wetting (wetting transitions) should occur in two-phase liquid systems displaying a miscibility gap. The wetting phenomenon was first proven experimentally in organic mixtures. Moldover and Cahn [34] studied wetting in a mixture of methanol and cyclohexane and established that the denser methanol-rich phase forms a thin wetting layer which separates the cyclohexane-rich phase from the glass cuvette and from the vapor. It is interesting that they did not observe the transition from complete to partial wetting at the vapor interface upon cooling, because the mixture froze before the onset of the wetting transition. On the other hand, such a transition was found in the methanol–cyclohexane–water mixtures. Later, the wetting and wetting transitions were evidenced experimentally also in metallic immiscible systems [35–37].

Chatain et al. [38] described the l–l interface for the Al–Bi binary system with a multilayer model in the approximation of regular solutions and found that more than perfect wetting should occur in the Al–Bi alloys already at the monotectic temperature. We have studied cross-sections of the solidified $Al_{34.5}Bi_{65.5}$, $(Al_{0.345}Bi_{0.655})_{90}Cu_{10}$, $(Al_{0.345}Bi_{0.655})_{90}Si_{10}$, and $(Al_{0.345}Bi_{0.655})_{90}Sn_{10}$ samples with optical microscopy and energy dispersive X-ray (EDX) spectroscopy. The compositions of some selected areas are given in Table 2.2. (It should be noted that these compositions do not necessarily represent the composition of coexisting solid phases).

An examination of the $Al_{34.5}Bi_{65.5}$ sample (Figure 2.12a) shows that the Al-rich phase is covered by a rather regular Bi-rich layer with thickness 8–125 μm. The same picture is observed on the surface of the $(Al_{0.345}Bi_{0.655})_{90}Si_{10}$ alloy (Figure 2.12b). Here, the Bi-rich layer has a thickness between 10 and 200 μm. It can be concluded that the transition from complete to incomplete wetting in these alloys does not occur in the liquid state.

The Al-rich phase of the $(Al_{0.345}Bi_{0.655})_{90}Sn_{10}$ sample is covered by a thin layer (Figure 2.12c). The composition of this layer as well as its thickness changes continuously from the crucible wall to the surface center. The layer is rich in Bi

Table 2.2 Composition analysis of some selected areas of the solidified Al–Bi–X alloys.

Alloy composition (wt %)	Alloy area	Area composition (at%)		
		Al	Bi	X
$Al_{34.5}Bi_{65.5}$	Al-rich phase	99.57 ± 4.8	0.43 ± 0.1	—
	Bi-rich phase	1.30 ± 0.1	98.70 ± 3.6	—
	Surface layer	8.10 ± 0.1	91.90 ± 3.6	—
$(Al_{0.345}Bi_{0.655})_{90}Cu_{10}$	Al-rich phase	84.62 ± 3.5	0.14 ± 0.1	15.24 ± 0.9
	Bi-rich phase	6.37 ± 0.1	93.63 ± 3.5	0
	Surface droplet	22.48 ± 0.2	77.52 ± 3.5	0
$(Al_{0.345}Bi_{0.655})_{90}Si_{10}$	Al-rich phase	86.90 ± 4.1	0	13.10 ± 0.6
	Bi-rich phase	5.76 ± 0.1	93.85 ± 3.5	0.39 ± 0.0
	Surface layer	4.64 ± 0.1	94.77 ± 3.5	0.59 ± 0.0
$(Al_{0.345}Bi_{0.655})_{90}Sn_0$	Al-rich phase	95.84 ± 4.0	1.25 ± 0.6	2.91 ± 0.7
	Bi-rich phase	3.56 ± 0.1	42.40 ± 2.3	54.04 ± 1.5
	Surface layer (at the middle)	1.47 ± 0.2	16.56 ± 1.3	81.96 ± 2.2

close to the walls, while the middle part of the surface is significantly enriched by Sn. The Bi-rich layer is 20–100 μm thick, and the thickness of the Sn-rich phase at the middle of the sample surface reaches up to 900 μm.

A Bi-rich phase is also seen on the surface of the Al–Bi–Cu alloy (Figure 2.12d), however, mainly in the form of droplets and only partly as a layer. The thickness of the layer is 8–22 μm, while the droplets are up to 110 μm high. Obviously, wetting transition in the Al–Bi–Cu and Al–Bi–Sn alloys takes place before their solidification.

2.6
Summary

Experimental measurements have shown that the l–l interfacial tension increases when either Cu or Si is added to the $Al_{34.5}Bi_{65.5}$ alloy but decreases when Sn is added. Changes of the l–l interfacial tension in the systems studied are related to the size of the miscibility gap. If the molecular miscibility decreases, the interfacial tension increases, and vice versa. This is explained by the increasing (respectively, decreasing) composition gradient across the interface upon the addition of a third element to the Al–Bi binary. Reduction of the interfacial tension between Al-rich and Bi-rich liquids in Al–Bi–Sn alloys is also due to the adsorption of Sn at the l–l interface. Simultaneous addition of Cu and Sn to $Al_{34.5}Bi_{65.5}$ results in a decrease of the interfacial tension at low temperatures and its increase at high temperatures. This also correlates with the mutual solubility of the constituents and the sizes of the miscibility gap.

The temperature dependences of l–l interfacial tension in the binary, ternary and quaternary alloys studied are well described by the power function $\sigma_{\alpha\beta} = \sigma_0 \times (1 - T/T_C)^\mu$ with the critical point exponent $\mu = 1.3$, which is close to the value ($\mu = 1.26$) predicted by the renormalization theory. The results obtained prove that the critical behavior of metallic monotectic alloys does not depend on the type of constituents.

The existence of a rather regular Bi-rich layer on the surface of the Al-rich phase in the solid $Al_{34.5}Bi_{65.5}$ and $(Al_{0.345}Bi_{0.655})_{90}Si_{10}$ indicates that complete wetting is preserved there far below the monotectic temperature. The addition of either Cu or Sn to the $Al_{34.5}Bi_{65.5}$ binary induces a wetting transition that occurs before solidification of Al–Bi–Cu and Al–Bi–Sn alloys.

Acknowledgments

This study was supported by the German Research Foundation (DFG) in the Priority Programme "DFG-SPP 1120: Phase transformations in multicomponent melts" under grant HO 1688/8. We are indebted to Prof. Rainer Schmid–Fetzer, Dr Joachim Gröbner, and Dr. Djordje Mirkovic for the valuable information and discussions of the phase equilibria in monotectic systems. We thank Mrs

Heidemarie Teichmann, Mrs Gisela Baumann, Ms Karen Friedrich, and Mr Rene Krone for their technical assistance.

References

1. Predel, B. (**1997**) "Constitution and thermodynamics of monotectic alloys: a survey". *Journal of Phase Equilibria*, **18**, 327–37.
2. Ratke, L. and Diefenbach, S. (**1995**) "Liquid immiscible alloys". *Materials Science and Engineering R-Reports*, **15**, 263–347.
3. Zhao, J., Ratke, L., Jia, J. and Li, Q. (**2002**) "Modeling and simulation of the microstructure evolution during a cooling of immiscible alloys in the miscibility gap". *Journal of Materials Science and Technology*, **18**, 197–205.
4. Massalski, T.B., Murray, J.L., Bennet, K.H. and Baker, H. (**1986**) *Binary Alloy Phase Diagrams*, American Society for Metals, Metals Park, OH.
5. Gröbner, J. and Schmid-Fetzer, R. (**2005**) "Phase transformations in ternary monotectic aluminum alloys". *Journal of Metals*, **57**, 19–23.
6. Gröbner, J., Mirkovic, D., Schmid-Fetzer, R. (**2008**) *Phase Formation in Multicomponent Monotectic Aluminum-Based Alloys*, this book.
7. Cristian, J.W. (**1965**) *The Theory of Phase Transformations in Metals and Alloys*, Pergamon Press, Oxford.
8. Young, N.O., Goldstein, J.S. and Block, M.J. (**1959**) "The motion of bubbles in a vertical temperature gradient". *Journal of Fluid Mechanics*, **6**, 350–56.
9. Kaban, I., Hoyer, W. and Merkwitz, M. (**2003**) "Experimental study of the liquid-liquid interfacial tension in immiscible Al-Bi system". *Zeitschrift fur Metallkunde*, **94**, 831–34.
10. Kaban, I. and Hoyer, W. (**2006**) "Liquid-liquid interfacial tension in the monotectic alloy $(Al_{34.5}Bi_{65.5})_{95}i_5$ (wt%)". *International Journal of Materials Research*, **97**, 362–64.
11. Hoyer, W. and Kaban, I. (**2006**) "Experimental and calculated liquid-liquid interfacial tension in demixing metal alloys". *Rare Metals*, **25**, 452–56.
12. Kaban, I., Hoyer, W. and Kehr, M. (**2007**) "Liquid-liquid interfacial tension in ternary monotectic alloys Al-Bi-Cu and Al-Bi-Si". *International Journal of Thermophysics*, **28**, 723–31.
13. Kaban, I.G. and Hoyer, W. (**2008**) "Characteristics of liquid-liquid immiscibility in Al-Bi-Cu, Al-Bi-Si, and Al-Bi-Sn monotectic alloys: differential scanning calorimetry, interfacial tension and density difference measurements". *Physical Review B*, **77** (125426), 1–7, http://dx.doi.org/10.1103/PhysRevB.77.125426.
14. Kaban, I. and Hoyer, W. (**2008**) "Effect of Cu and Sn on liquid–liquid interfacial energy in ternary and quaternary Al–Bi-based monotectic alloys". *Materials Science and Engineering A*, http://dx.doi.org/10.1016/j.msea.2007.08.097.
15. Iida, T. and Guthrie, R.I.L. (**1993**) *The Physical Properties of Liquid Metals*, Clarendon Press, Oxford.
16. Merkwitz, M. (**1997**) *Oberflächen- und Grenzflächenspannung in binären metallischen Entmischungssystemen*, Ph. D. Thesis, Technische Universität Chemnitz, http://archiv.tu-chemnitz.de/pub/1997/0036.
17. Merkwitz, M., Weise, J., Thriemer, K. and Hoyer, W. (**1998**) "Liquid-liquid interfacial tension in the demixing metal systems Ga-Hg and Ga-Pb". *Zeitschrift fur Metallkunde*, **89**, 247–55.
18. Merkwitz, M. and Hoyer, W. (**1999**) "Liquid-liquid interfacial tension in the demixing metal systems Al-Pb and Al-In". *Zeitschrift fur Metallkunde*, **90**, 363–70.

19 Keller, J.B. (**1998**) "Surface tension force on a partly submerged body". *Physics of Fluids*, **10**, 3009–10.

20 Greiner, W. and Stock, H. (**1984**) *Theoretische Physik: Hydrodynamik*, Harri Deutsch, Frankfurt am Main.

21 de Gennes, P.-G., Brochard-Wyart, F. and Quéré, D. (**2004**) *Capillarity and Wetting Phenomena: Drops, Bubbles, Pearls, Waves*, Springer Verlag, New York.

22 Becker, R. (**1938**) "Die keimbildung bei der ausscheidung in metallischen mischkristallen". *Annales de Physique*, **5**, 128–40.

23 Cahn, J.W. and Hilliard, J.E. (**1958**) "Free energy of a nonuniform system. I. interfacial free energy". *The Journal of Chemical Physics*, **28**, 258–67.

24 van der Waals, J.D. (**1894**) "Thermodynamische theorie der kapilarität unter voraussetzung stetiger dichteänderung". *Zeitschrift fur Physikalische Chemie*, **13**, 657–725.

25 Hoyt, J.J. (**1990**) "The continuum theory of nucleation in multicomponent systems". *Acta Metallurgica*, **38**, 1405–12.

26 Sun-Keun, Yu. (**1994**) *Thermodynamische untersuchung monotektischer Systeme*, Ph. D. Thesis, University Stuttgart, Stuttgart.

27 Gibbs, J.W. (**1961**) *The Scientific Papers of J. Willard Gibbs*, Vol. **1**: Thermodynamics, Dover Publications, p. 219.

28 Rowlinson, S.S. and Widom, B. (**1982**) *Molecular Theory of Capillarity*, Clarendon Press, Oxford.

29 Widom, B. (**1979**) "Remarks on the gibbs adsorption equation and the van der Waals, cahn-hilliard theory of interfaces". *Physica A*, **95**, 1–11.

30 Chatain, D., Vahlas, C. and Eustathopoulos, N. (**1984**) "Etude des tensions interfaciales liquide-liquide et solide-liquide dans les systemes à monotectique Zn-Pb et Zn-Pb-Sn". *Acta Metallurgica*, **32**, 227–34.

31 Chatain, D. and Eustathopoulos, N. (**1984**) "Tensions interfaciales liquide-liquide. I: Epaisseur de l'interface". *Journal de Chimie Physique*, **81**, 587–97.

32 Young, T. (**1856**) *Miscellaneous Works of Dr Thomas Young*, (eds G. Peacock and J. Leitch), John Murray, London.

33 Cahn, J.W. (**1977**) "Critical point wetting". *Journal of Chemical Physics*, **66**, 3667–72.

34 Moldover, M.R. and Cahn, J.W. (**1980**) "An interface phase transition: complete to partial wetting". *Science*, **207**, 1073–75.

35 Chatain, D. and Wynblatt, P. (**1996**) "Experimental evidence for a wetting transition in liquid Ga-Pb alloys". *Surface Science*, **345**, 85–90.

36 Freyland, W., Ayyad, A.H. and Mechdiev, I. (**2003**) "Wetting, prewetting and surface freezing transitions in fluid Ga-Based alloys: a surface light scattering study". *Journal of Physics. Condensed Matter*, **15**, S151–57.

37 Huber, P., Shpyrko, O.G., Pershan, P.S., Ocko, B.M., DiMasi, E. and Deutsch, M. (**2002**) "Tetra point wetting at the free surface of liquid Ga-Bi". *Physical Review Letters*, **89**, 035502, 1–4.

38 Chatain, D., Eustathopoulos, N. and Desre, P. (**1981**) "The interfacial tension and wetting in the two-liquids region of a regular solution". *Journal of Colloid and Interface Science*, **83**, 384–92.

3
Monotectic Growth Morphologies and Their Transitions

Lorenz Ratke, Anja Müller, Martin Seifert, and Galina Kasperovich

3.1
Introduction

The solidification of monotectic alloys leads to different microstructures depending on the solidification velocity and the temperature gradient ahead of the solid–liquid interface [1–9]. If an alloy of exact monotectic composition is cooled from the single-phase L_1 to the monotectic temperature, the liquid decomposes at equilibrium conditions simultaneously into a solid phase of nearly pure A and a liquid L_2 usually having a low volume fraction (Figure 3.1). This nonvariant monotectic reaction can lead to fibers of the L_2 phase embedded in the solid matrix. The microstructure is very similar to a fibrous eutectic one as observed, for instance, in Al–Ni [10]. It also is observed that the L_2 phase is arranged in the form of well-aligned strings of pearls or as droplets irregularly distributed in the matrix. Generally speaking, it is observed that at a given temperature gradient a transition from fibers to strings of pearls and irregularly arranged drops occurs with increasing solidification velocity [11].

Theoretical explanations for fibrous monotectic growth are usually based on the classical Jackson and Hunt model of eutectics [12]. The most complete description of monotectic growth was given recently by Coriell *et al.* [9, 13]. They extended the Jackson and Hunt model to treat some peculiarities of immiscible alloys, such as the large density difference between the phases, which leads to advective fluxes towards the interface [13]. Stöcker and Ratke [14, 15] developed on the basis of the Jackson and Hunt model and the extensions made by Coriell *et al.* a new model including fluid flow aspects, especially a possible Marangoni convection at the melt/fiber interface. In some recent experiments on monotectic solidification, very strong static magnetic fields (10 T) were applied to dampen possible convections in the melt ahead of the solidification front [16, 17].

The origin of the transition sequence between the three kinds of microstructures is still under discussion. Toloui *et al.* consider Rayleigh instability of the fibers as the origin of the strings of pearls [7]. Chattopadhyay discusses a special variant of this instability first described by Mullins and Nicholson [18, 19]. Johnston and

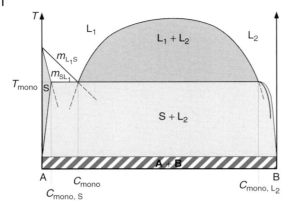

Fig. 3.1 Scheme of a typical monotectic system exhibiting a miscibility gap in the liquid state, where two liquids L_1 and L_2 are in equilibrium. The boundary curve is called bimodal curve. At T_{mono} the liquid L_1 of composition C_{mono} decomposes into a solid S and the liquid L_2. The liquidus line connecting the melting point of A with the monotectic point has the slope m_{L1S} and the solidus line the slope m_{SL1}. The partition coefficient is $k = c_{mono}/c_{mono,S}$ or m_{SL1}/m_{L1S}.

Parr think that the string of pearls is a consequence of fibers pinching off due to temperature oscillations at the solid–liquid interface [20].

3.2
Experimental Procedures

3.2.1
Alloys

With aluminum as the base metal, four different types of monotectic alloys can be studied: Al–Bi, Al–Cd, Al–In, and Al–Pb. These four alloys were studied in this work. Some relevant data are summarized in Table 3.1 showing that the monotectic concentration ranges from 1.4 to 17.3 wt%. The phase fraction of the L_2-phase changes from 0.2 to 5.6%. The partition coefficient is in all cases small compared to conventional single-phase alloys such as AlSi and AlCu, smaller by at least a factor of 10. The interface tension [21] between the two immiscible liquids varies from small values in Al–In to the highest values in Al–Pb.

The alloy samples were prepared from 99.99% pure Al (kindly provided by Hydro Aluminium Deutschland GmbH, Bonn). Bi, Cd, In, Pb were from AlfaAesar, 99.999% pure metal basis. The Al was melted and heated to 700 °C. Then the Bi, Cd, In, and Pb metal shots were introduced and stirred with a corundum rod. After homogenization at 700 °C for 30 min, samples of 8 mm diameter and 120 mm length were cast in a steel mould, yielding a finely dispersed microstructure due to a cooling rate of around 100 K s^{-1}. The Al–Cd alloy was prepared in a glove box under a protective atmosphere.

Table 3.1 Four immiscible alloys based on aluminum investigated. $c_s(B)$ denotes the maximum solid solubility of the alloying element B = Bi, Cd, In, Pb in aluminum, $c_{mono}(B)$ the monotectic composition in weight and atomic percent, $c_{L_2}(B)$ the amount of B in atomic percent of the L_2-phase at the monotectic temperature. The partition coefficient k is the ratio of c_s/c_{mono}, and σ_{exp} the interfacial tension between the two immiscible liquids at the monotectic temperature [21].

Al–B	Al–Bi	Al–Cd	Al–In	Al–Pb
$c_s(B)$ at T_m (wt %)	0.232	0.378	0.170	0.001
$c_s(B)$ at T_m (at%)	0.030	0.091	0.040	0.001
$c_{mono}(B)$ (wt %)	3.383	6.945	17.347	1.400
$c_{mono}(B)$ (at%)	0.450	1.760	4.700	0.190
$c_{L_2}(B)$ (at%)	84.000	94.900	82.500	98.500
f_{L_2} at T_m (%)	0.500	1.760	5.651	0.192
Partition coefficient k	0.069	0.054	0.010	0.001
T_m (°C)	657.000	650.000	636.000	659.000
σ_{exp} at T_m (mJ m^{-2})	57.000	31.800	25.500	125.500

3.2.2
ARTEMIS Facility

The experiments were performed in a facility called ARTEMIS (Figure 3.2, for details see [22–25]). The central part of the furnace consists of two heating elements, a top and a bottom heater, a piece of silica aerogel as a thermal insulating region, and a cooling base plate below the bottom heater. The heating elements are resistance heaters with a boron-nitride core. Using thermocouples, a computer reads the temperature of each heater, and by controlling the heating power it controls the temperature of the heating elements during the experiment.

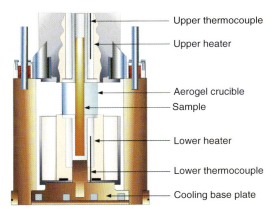

Fig. 3.2 Schematic of the aerogel furnace device ARTEMIS. Essential are the two heating elements being thermally shortcut by the sample only because the aerogel is a nearly perfect insulator.

The sample is melted and held for homogenization for a certain period at this temperature before the cooling phase takes place. The free sample length can be varied between 20 and 55 mm. Varying also the temperature of the top heater allows varying the temperature gradient from 2 to 8 K mm^{-1}. The maximum gradient is restricted by the maximum temperature the silica aerogel can withstand before sintering considerably (around 950 °C).

The progress of the solidification front is observed by an infrared line charge-coupled device (CCD) camera, which passes the retrieved intensities to a computer. The CCD line camera is calibrated such that the free sample length inside the aerogel is mapped onto the 256 pixels of the line CCD. In this way a position on the sample can be assigned to a pixel. The radiation intensity emitted from the sample surface continuously decreases with decreasing temperature. The phase change from liquid to solid is accompanied by a change in emissivity, yielding a hump in the intensity measurement by the CCD camera. The hump recorded on a pixel indicates the time the monotectic reaction front passed a certain position in the sample [22, 25]. Plotting this sample coordinate as a function of time when the hump occurred allows determination of the real solidification velocity and controlling its constancy over the processing length (it also allows verification of the actual temperature gradient).

3.2.3
Evaluation Procedures

The samples were cut with a diamond saw after processing. In the processed zone, two sections perpendicular to the samples axis and one along the axis were prepared. The samples were polished using an ultra-micromilling machine from Reichert-Jung (Leica). The polished surfaces were examined in a scanning electron microscope (SEM) (LEO VP1530) in the backscattered mode, which always yields excellent contrast. On the cross-sections we evaluated the average fiber distance and the fiber, pearl, or droplet diameter. On longitudinal sections the periodicity of the string of pearl arrangements was measured.

3.3
Experimental Results

In all alloys of monotectic compositions three microstructures were observed: fibers, strings of pearls, and irregular droplet arrangements. From the thousands of SEM pictures we first present a few illustrative examples of the microstructures and their transitions.

3.3.1
Microstructures

At low solidification velocities and high temperature gradients one can observe fibrous microstructures as in eutectic alloys. Figure 3.3(a) and (b) shows two

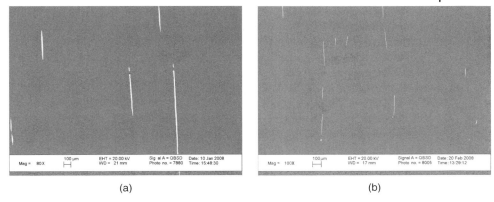

Fig. 3.3 Fibrous microstructure in Al–Bi (a) and Al–Pb (b). For Al–Bi the processing parameters were $G = 8$ K mm^{-1} and $v = 1\,\mu$m s^{-1} and for AlPb $G = 6$ K mm^{-1}, $v = 3\,\mu$m s^{-1}.

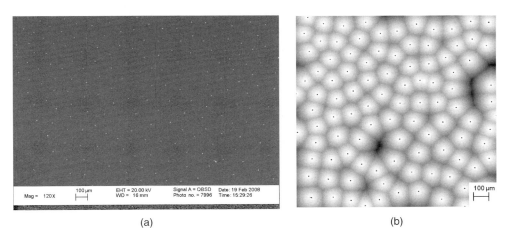

Fig. 3.4 Arrangement of the Pb fibers perpendicular to the growth direction in a monotectic Al–Pb alloy (a) and a Voronoi-like cell construction to illustrate the deformed neighborhood of each fiber deviating from the ideal hexagonal packing (b).

examples for Al–Bi and Al–Pb. In accordance with the second phase volume fraction (see Table 3.1), there are only a few fibers visible in a randomly chosen longitudinal section. The fibers have a large aspect ratio and are well aligned in the solidification direction.

Ideally the fibers would be arranged on a hexagonal lattice. This was, however, not observed over larger scales in the micrographs taken perpendicular to the growth direction. An example of such a microstructure is shown Figure 3.4a. A closer inspection reveals a structure in which fibers have locally a varying number of nearest neighbors. Their spacing is not regular and the structure shows deformations and defects. This becomes especially clear looking at Figure 3.4b, in which a Voronoi construction was made. The cells surrounding each fiber are not regular but deformed and the number of neighbors, indeed, varies between 4 and 7.

The fibrous microstructure can transform into an arrangement looking like a "string of pearls". In all alloys one can observe such a beautiful arrangement of the second phase. A few examples are shown in Figure 3.5a–d.

The pearls are well aligned in the solidification direction as in the case of the fibers. Looking at the axial position of the pearls in different strings one can see that they are all aligned, meaning they stem from a planar solidification front as expected from a nonvariant reaction. The string of pearls originate, as will be discussed later,

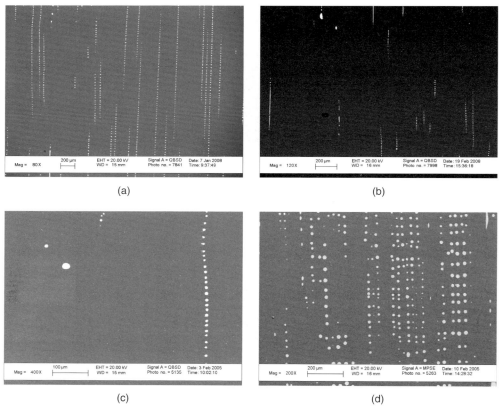

Fig. 3.5 String of pearl microstructure in Al–Bi (a), Al–Pb (b), Al–Cd (c), and AlIn (d). For Al–Bi the processing parameters were $G = 6$ K mm^{-1}, $v = 1\,\mu$m s^{-1}; for Al–Pb $G = 6$ K mm^{-1}, $v = 2\,\mu$m s^{-1}; for Al–Cd $G = 3$ K mm^{-1}, $v = 0.5\,\mu$m s^{-1}; and for Al–In $G = 3$ K mm^{-1}, $v = 3\,\mu$m s^{-1}.

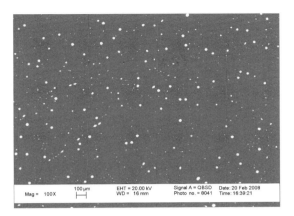

Fig. 3.6 Irregularly spaced droplets of Bi in an Al matrix processed with $G = 3.5$ K mm^{-1} and $v = 4\,\mu$m s^{-1}.

from a Rayleigh-type instability of the fibers behind the solidification front, and therefore their geometrical arrangement can be evaluated with respect to the fiber spacing as if they were fibers [8]. The thickness of the fibers and the periodicity are evaluated separately and discussed within the context of the transition from fibers to pearls.

The third microstructure is just an arrangement of irregular sized and distributed droplets of the second phase in a nearly pure Al matrix. One example is shown in Figure 3.6. The irregularity shows that there is no relation to a fibrous growth and that the occurrence of this structure type at low temperature gradients and high solidification velocities must be discussed separately from the monotectic composite growth.

3.3.2
Jackson–Hunt Plot

The average fiber distance measured on cross-sections could be determined as a function of solidification velocity for Al–Bi, Al–In, and Al–Pb. In the case of Al–Bi, the data points are collected from different temperature gradients (4.5–8 K mm^{-1}), whereas for Al–Pb the data are for one gradient, 6 K mm^{-1}. The Al–In data were taken from 3 K mm^{-1}. We show two examples of a Jackson–Hunt plot for Al–Pb and Al–Bi in Figure 3.7a and b. In both cases there is a large scatter in the fiber spacing data. Interestingly enough, the absolute spacing in Al–Pb is larger than in Al–Bi as expected from the volume fraction of second phase; the smaller it is the larger the spacing. The data can be fitted with the Jackson and Hunt relation $\lambda^2 v = C_{JH}$. Here λ denotes the fiber spacing and v the solidification velocity. For the constant C_{JH} the following values were obtained for Al–Pb, Al–In, and Al–Bi: $C_{JH}(\text{Al–Bi}) = 2.46 \times 10^{-15}$ m^3 s^{-1}, $C_{JH}(\text{Al–In}) = 4.5 \times 10^{-15}$ m^3 s^{-1}, and $C_{JH}(\text{Al–Pb}) = 3.6 \times 10^{-14}$ m^3 s^{-1}.

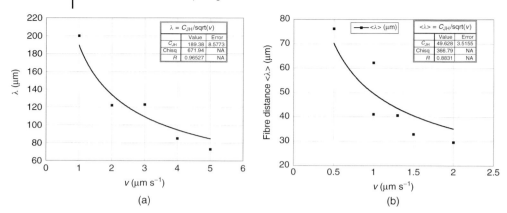

Fig. 3.7 Jackson and Hunt plot for two different alloys, Al–Pb (a) and Al–Bi (b). The drawn in line is a fit according to the Jackson and Hunt relation between fiber separation and solidification velocity.

3.3.3
Stability Diagrams

The type of microstructure observed in the metallographic sections can be plotted as a function of temperature gradient and solidification velocity in a so-called stability diagram. Figure 3.8a–c shows the results for Al–Bi, Al–In, and Al–Pb. The data points shown are a combination of our own measurements and data from the literature.

Three different microstructures are observable in these alloys and they obviously exhibit different ranges of existence. These ranges are not exclusive and distinct. There seems to be an overlap, and repeating experiments, as was done, does not always lead to exactly the same response of the microstructure. It also has to be mentioned that there are micrographs in which fibers and strings of pearls coexist. The drawn-in lines are theoretical estimates according to the model outlined below (Equation 3.6). These diagrams show that at high temperature gradients and low solidification velocities fibers are obtained; at low gradients and high velocities irregular structures; and in between the "string of pearls" structure. As drawn in, the boundary lines between the regimes seem to be curved. The square root dependence for the transition between fibers and string of pearls are calculated from Equation 3.6 (see below). A few points are interesting in these diagrams. First of all, compared to eutectic alloys the solidification velocity range in which fibers are observed is much smaller than for eutectics. Whereas in eutectics fibers can be observed at velocities up to $100\,\mu m\,s^{-1}$, depending of course on the gradient [26], in monotectic alloys very low solidification velocities in the range of $1\,\mu m\,s^{-1}$ are required. Secondly, it seems that the system with the highest interfacial tension between the two immiscible liquids, Al–Pb (see Table 3.1), exhibits fibers at lower temperature gradients than the other two alloys. There also seems to be a sequence

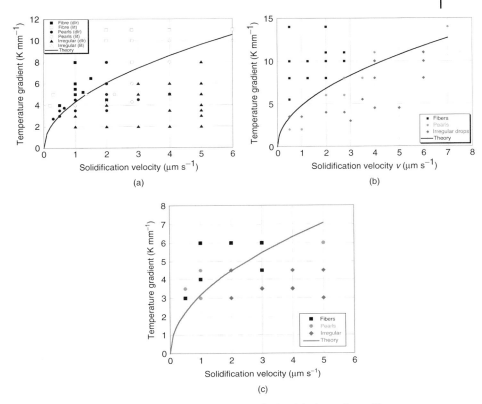

Fig. 3.8 Stability diagram of Al–Bi, (a), Al–In (b), and Al–Pb (c). The three different microstructures observable in these alloy exhibit different ranges of existence. The drawn-in lines are essentially guides to the eye to distinguish the regimes. The square root dependence for the transition between fibers and string of pearls was calculated from Equation 3.6.

with increasing interfacial tension: the higher the interfacial tension the lower the gradient needed to produce a fibrous structure. This might be fortuitous, since the volume fraction of second phase goes along with the interfacial tension, the higher the interfacial tension, the smaller the second phase volume fraction.

3.4
Discussion

3.4.1
Fibrous Monotectic Growth

Fibrous monotectic growth is usually analyzed by adopting the Jackson-Hunt model [12] of eutectic growth in a suitable manner. One of the essential differences to eutectic growth was, however, not taken into account by the work of Coriell et al. [8, 13], namely that the interface between the L_1 melt and the L_2 fibers is a liquid one. Therefore a temperature gradient at the solidification front induces a Marangoni convection in the matrix melt and the fibers. This fluid flow should modify the constitutional supercooling. Since the magnitude of the Marangoni flow is proportional to the gradient, the constitutional supercooling should vary with the gradient. Then its balance to the capillary supercooling would vary with the temperature gradient too. An analytical model was developed by Stöcker and Ratke [14], extended later by a numerical analysis of the problem [15]. Recently, the analytical model was extended to arbitrary Peclet numbers [27]. The Peclet number $Pe_D = u_M r_f / D$ is determined by the magnitude of the Marangoni flow u_M ahead of the interface, the fiber radius r_f, and the solute diffusion coefficient D. Two regimes can be distinguished. At small Peclet numbers or small Marangoni flow velocities the Jackson and Hunt relation holds and the flow gives only a second-order variation to the classical relation, namely

$$v\lambda^2[S_1 + 3S_2 G\lambda^2] = \frac{aD}{4C_0} \tag{3.1}$$

where v is the solidification velocity and λ the fiber spacing. S_1, S_2, and a are constants, G denotes the temperature gradient, and C_0 the difference between the L_2 concentration and the maximum solubility of the solid phase at the monotectic temperature [14]. This relation predicts a decreasing Jackson and Hunt constant with increasing gradient. At large Peclet numbers Pe_D a different relation holds:

$$v\lambda^2 = \frac{A}{Pe_D} \exp(B Pe_D) \tag{3.2}$$

where A and B are complicated constants. A is of the order 0.2 and B around 1.5 (for details see [27]). The Pe_D number contains the fiber spacing squared and the temperature gradient linearly. At small Peclet numbers the exponential can be approximated by 1 and Equation 3.1 can be regained. At high Peclet numbers the exponential dominates and the spacing increases with increasing gradient. To date, there is no direct measurement of the gradient dependence of the fiber spacing. The new experiments mentioned in the Introduction with large magnetic fields that damp the fluid flow [16, 17] clearly show that convections at the interface are very important in monotectic composite growth.

3.4.2
Transition from Fibers to String of Pearls

Majumdar and Chattopadhyay [18] suggested that the transition from fibers to strings of pearls is a Rayleigh-type instability behind the solidification front described by a model developed by Nichols and Mullins [19]. They discuss several mechanisms by which a cylinder can become unstable if surface perturbations appear (axial, azimuthal). They discern between different transport modes such as surface or volume diffusion. Majumdar and Chattopadhyay calculate the transition from fibers to strings of pearls using the surface diffusion model of Nichols and Mullins. We keep their basic idea, but assume that volume diffusion inside the fibers is the relevant transport mechanism, since the L_2 phase fibers change their concentration already during cooling from the monotectic to the eutectic temperature by volume diffusion of A in the L_2 fibers to an appreciably amount. Therefore it is natural to assume that volume diffusion is important during the evolution of perturbations.

Adopting the volume diffusion model of Nichols and Mullins one can calculate the time an initial perturbation of amplitude δ_0 needs to attain a value of the fiber radius R_0. Denoting this time for pinch-off as t_f^{MN}, the expression of Nichols and Mullins reads:

$$t_f^{MN} = \frac{3R_0^3}{0.118A} \quad (3.3)$$

with A being a constant

$$A = \frac{D\sigma_{SL_2}\Omega_{L_2}}{RT} \quad (3.4)$$

Here, σ_{SL_2} is the interface energy and Ω_{L_2} represents the average atomic volume of the L_2-phase. This time can be greater or lesser than the time available to cool the sample from the monotectic to the eutectic temperature. At constant gradient G and velocity v, the cooling rate $dT/dt = Gv$ is constant. The time t_f^{max} is, therefore, given by

$$t_f^{max} = \frac{T_m - T_e}{Gv} = \frac{\Delta T}{Gv} \quad (3.5)$$

with T_m the monotectic temperature and T_e the eutectic. If this time is larger than the time needed for the fibers to pinch off, a string of pearls results, or else fibers are observed in a microstructure; thus, the criterion is $t_f^{max} > t_f^{MN}$.

The fiber radius in Equation 3.3 is, however, difficult to measure, whereas the fiber spacing λ_M is readily accessible. There is a simple geometrical relation between $R_0 = \lambda_M\sqrt{f_{L_2}}$, where f_{L_2} is the volume fraction of the liquid phase. If a Jackson and Hunt relationship would be valid for monotectics as for eutectics, one could replace the fiber spacing with the solidification velocity $\lambda^2 v = C_{JH}$, ignoring thereby a possible dependence on the temperature gradient. Combining the equations we obtain a criterion for the transition from fibers to strings of pearls that can be drawn into the diagrams shown in Figure 3.8a–c:

$$G < \frac{0.0393 A \Delta T}{C_{JH}^{3/2} f_{L_2}^{3/2}} \sqrt{v} \quad (3.6)$$

The thermophysical parameters of the Al–Bi, Al–In, and Al–Pb systems are sufficiently well known that the expression of Equation 3.6 can be fitted to the experimental data. The real unknowns are the Jackson–Hunt constants C_{JH} and the diffusion coefficient of component A within the L_2 phase. Assuming a diffusion constant varying linearly with temperature [28] and being in all three L_2 liquids equal to the diffusion of Al in liquid In [28], one can determine the value of C_{JH} from the fit to the stability diagram and compare it with measurements of the fiber spacing. For Al–Bi there are several experimental results reported in the literature. Our own measurements on Al–Bi gave a value for C_{JH} of 2.46×10^{-15} m^3 s^{-1}. Derby and Favier [29] report a value of 2.5×10^{-17}, Yang et al. a value of 2.2×10^{-15} [30], Kamio et al. [31] 3.9×10^{-15}, and Grugel et al. [5] 2.5×10^{-14} m^3 s^{-1}. The fit to the stability data in Figure 3.8a yields a value of 2.2×10^{-14} m^3 s^{-1} assuming a diffusion coefficient of 10^{-9} m^2 s^{-1}.

For Al–In, a fit to the stability data yields an apparent C_{JH} value of $C_{JH} = 2.7 \times 10^{-15}$ m^3 s^{-1} compared with our own value of $C_{JH} = 4.5 \times 10^{-15}$ m^3 s^{-1} and the values of Kamio from 1991 [31] of $C_{JH} = 7.8 \times 10^{-16}$ m^3 s^{-1}, Vinet and Potard from 1983 [32] of $C_{JH} = 5.3 \times 10^{-16}$ m^3 s^{-1}, and Grugel and Hellawell [4] of $C_{JH} = 4.5 \times 10^{-16}$ m^3 s^{-1}. For Al–Pb no comparable data are reported in the literature. The direct fiber measurement gave a value of $C_{JH} = 3.6 \times 10^{-14}$ m^3 s^{-1} and the fit to the stability data would yield a value of $C_{JH} = 7.0 \times 10^{-14}$ m^3 s^{-1}. In all three cases a fit of the model to the data would require a larger Jackson and Hunt constant than measured. This discrepancy stresses at least that both the fiber spacing measurements have a large scatter, that the "fit" to the stability data is somewhat ambiguous, and, for instance, the diffusion coefficient is not really known to an accuracy better than a factor of 5.

Another important measure of stability is the periodicity of the pearls within the strings in relation to their radius. The theory outlined above would predict that disturbances of wavelengths smaller than $\lambda_p = 2\pi r_f$, with r_f the fiber radius, are stable and a maximum growth rate is obtained at a wavelength being equal to $9r_f$. The periodicity and the fiber diameter on large magnifications of the samples (cross sections) were measured. A result for Al–Bi is shown in Figure 3.9. The data show that the wavelengths that led to pearls, indeed, had a periodicity leading to positive growth rate, but the maximum growth rate could not be observed. In Al–In alloys we observed relations in the range of $\lambda_p \approx 6.9$–$8.6 r_f$. Vinet reports similar values [32], as also Majumdar and Chattopadhyay [18].

We therefore conclude that the transition from fibers to string of pearls is likely to be a Rayleigh-type instability induced by disturbances of the fibers behind the solidification front, where the fibers stay for a few hundred kelvin as liquid and change their composition drastically until the final eutectic of the lower-melting-point metal is reached. Volume diffusion of Al inside the fibers dominates the growth of pearls.

3.4.3
Origin of Irregular Drops

Let us assume that an alloy of monotectic composition is cooled down. On reaching the monotectic temperature, the equilibrium situation cannot be realized since in order to nucleate the solid phase of concentration c_s some undercooling is necessary. Therefore the melt concentration follows the metastable extension of

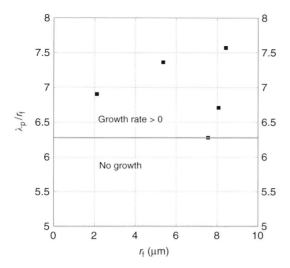

Fig. 3.9 Ratio of the periodicity and fiber radius in the string of pearls of Al–Bi as a function of the fiber radius. If the perturbations grow, the ratio must be larger than 2π.

the liquidus line. At a certain undercooling, the solid phase with concentration $c_s < c < c_s^*$ will be nucleated, and as in single-phase solidification a metastable liquid L_1 is in equilibrium with a metastable solid.

Ahead of the solid phase, a B-rich layer builds up, which could be calculated according to the well-known equation yielding an exponential decay of the solute profile ahead of the interface, if the growth would be stationary. Since we enforce a directional solidification at a fixed temperature gradient and solidification velocity v_s, the solid–liquid interface moves and the liquid enriches further in component B until ahead of the solidification front the L_2 phase can be nucleated. This nucleation can be understood as follows.

The binodal line bounds the two-phase equilibrium region of the immiscible liquids L_1 and L_2. Cooling an alloy of hypermonotectic composition into the miscibility gap does not immediately lead to the formation of two liquids, but requires an activation energy and thus a certain undercooling. It was shown experimentally by Perpezko et al. [33] and Uebber and Ratke [34] that the undercooling to induce the liquid–liquid decomposition varies with concentration in a fashion schematically drawn as the dashed line in Figure 3.10. If we extend such a line below the monotectic temperature, as already done by Perepezko, we arrive at a situation in which the metastable extension of the liquidus and this binodal-like nucleation line cross at one temperature. There we have a metastable equilibrium; the coexistence of three metastable phases, which is similar to the three-phase equilibrium at the monotecic temperature. Here, however, the phases have the compositions c_s^*, c_m^*, and $c_{L_2}^*$. This situation means that L_2 droplets nucleate ahead of the solidification front and consume rapidly the B atoms in their surrounding leading within a fraction of a second to micron-sized spherical caps. Their surrounding, however,

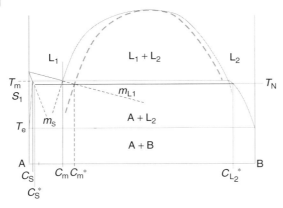

Fig. 3.10 Monotectic phase diagram with metastable extension of the liquidus, the solidus, and a nucleation curve below the binodal. This line sketches the onset of second-phase minority droplets from an undercooled liquid L_1.

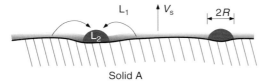

Fig. 3.11 Scheme of the nucleation and growth ahead of the advancing solidification front of a solid being nearly pure A. At the interface, a solute-rich layer of B builds up in which at a certain undercooling nuclei of the L_2-phase appear in the form of drops or spherical cups to which solute B diffuses.

still is constitutionally supercooled and therefore promotes the driving force for the lateral diffusion of solute to the droplets, which thereby grow further. This growth leads to a decrease of solute B in the whole boundary layer, which is shown schematically in Figure 3.11, and an increase of the interface temperature back to the monotectic one.

At low solidification speeds, the lateral diffusion speed is in line with the advancement of the solidification front leading to a fibrous structure. At high solidification speeds, the lateral diffusion towards the nuclei (at the given spacing essentially determined by the volume fraction second phase) is not fast enough such that the nuclei are overgrown and new nuclei appear at other places along the interface in the already supersaturated boundary layer. This competition between L_2 nucleation-growth and the advancement of the solid phase determines the irregular structure. Fluid flow will have a dramatic effect on the mass transport parallel to the interface. Marangoni convection assumed by Stöcker and Ratke [14, 15] is only one possible source. Thermal fluctuations and thermo-solutal natural convections induced by temperature or concentration gradients in radial direction modify the solute transport along the interface. Therefore, strong magnetic fields are very suitable to suppress flow and establish composite, coupled growth conditions by diffusive species transport.

3.5 Outlook

The experimental data obtained for the monotectic composite growth of Al–Bi, Al–In, and Al–Pb alloys provide an improved basis to develop and test models for fibrous growth, the transition from fibers to strings of pearls, and the understanding of irregular growth. Although the data are not sufficient at the moment to quantitatively test the new theories of flow-induced microstructures, the understanding of the transitions could be brought a step nearer. The results suggest that future experiments should utilize high magnetic fields to suppress fluid flow or microgravity conditions.

Acknowledgement

This work was funded by the German Science Foundation DFG within the priority program SPP 1120 under contract number DFG Ra 537/5.

References

1 Kamio, A., Tezuka, H., Kumai, S., Sueda, S. and Takahashi, T. (1985) *Journal of the Japan Institute of Metals*, **49**, 677–83.
2 Kamio, A. and Oya, S. (1969) *Journal of the Japan Institute of Metals*, **33**, 60–66.
3 Grugel, R. and Hellawell, A. (1981) *Metallurgical Transactions*, **12A**, 669–81.
4 Grugel, R. and Hellawell, A. (1982) in *Materials Processing in the Reduced Gravity Environment of Space*, (ed G. Rindone), Elsevier Science, Amsterdam, pp. 553–61.
5 Grugel, R., Lograsso, T. and Hellawell, A. (1984) *Metallurgical Transactions*, **15A**, 1003–12.
6 Grugel, R., Kim, S., Woodward, T. and Wang, T. (1992) *Journal of Crystal Growth*, **121**, 599–607.
7 Toloui, B., Macleod, A.J. and Double, D.D. (1982) in *Proceedings of the Conference In Situ Composites IV*, (eds F.D. Lemkey, H.E. Cline and M. Mc Lean), Elsevier Science, New York, pp. 253–66.
8 Pfefferkorn R. (1981) PhD thesis, Montanuniversität Leoben, Austria; (a) Kneissl, A., Pfefferkorn, P. and Fischmeister, H. (1983) in *4th European Symposium on Material Sciences Under Microgravity*, (ed T.D. Guyenne and J. Hunt), ESA SP-191, Noordwijk, Madrid, Netherlands, pp. 55–61.
9 Coriell, S.R., Mitchell, W.F., Murray, B.T., Andrews, J.B. and Arikawa, Y. (1997) *Journal of Crystal Growth*, **179**, 647–57.
10 Ratke, L. and Alkemper, J. (2000) *Acta Materialia*, **48**, 1939–48.

11 Ratke, L. and Diefenbach, S. (1995) *Materials Science and Engineering Reports*, **15**, 263–347.
12 Jackson, K.A. and Hunt, J.D. (1966) *Transactions of the Metallurgical Society of AIME*, **236**, 1129–42.
13 Coriell, S.R., McFadden, G.B., Mitchell, W.F., Murray, B.T., Andrews, J.B. and Arikawa, Y. (2001) *Journal of Crystal Growth*, **224**, 145–54.
14 Stöcker, C. and Ratke, L. (1999) *Journal of Crystal Growth*, **203**, 582–93.
15 Stöcker, C. and Ratke, L. (2000) *Journal of Crystal Growth*, **212**, 324–33.
16 Yasuda, H., Ohnaka, I., Fujimoto, S., Takezawa, N., Tsuchiyama, A., Nakano, T. and Uesugi, K. (2006) *Scripta Materialia*, **54**, 527–32.
17 Yasuda, H., Ohnaka, I., Fujimoto, S., Sugiyama, A., Hayashi, Y., Yamamoto, M., Tsuchiyama, A., Nakano, T., Uesugi, K. and Kishio, K. (2004) *Materials Letters*, **58**, 911–15.
18 Majumdar, B. and Chattopadhyay, K. (1996) *Metallurgical and Materials Transactions*, **27**, 2053–57.
19 Nichols, F.A. and Mullins, W. (1965) *Transactions of the Metallurgical Society of AIME*, **233**, 1840–47.
20 Parr, R.A., Johnston, M.H. (1978) *Metallurgical Transactions* **9A**, 1825–28.; (a) Schafer, C. Johnston, M.H. and Parr, R.A. (1983) *Acta Materialia*, **31**, 1221–24.
21 Hoyer, W., Kaban, I. and Merkwitz, M. (2003) *Journal of the Optoel Advanced Materials*, **5**, 1069–73.
22 Alkemper, J., Sous, S., Stöcker, C. and Ratke, L. (1998) *Journal of Crystal Growth*, **191**, 252–60.
23 Ahrweiler, S., Lacaze, J. and Ratke, L. (2003) *Advanced Engineering Materials*, **5**, 17–23.
24 Steinbach, S. and Ratke, L. (2007) *Metallurgical and Materials Transactions A*, **38A**, 1388–94.
25 Steinbach, S. and Ratke, L. (2004) *Scripta Materialia*, **50**, 1135–38.
26 Kurz, W. and Sahm, P.R. (1975) *Gerichtet Erstarrte Eutektische Werkstoffe*, Springer, Berlin.
27 Ratke, L. (2003) *Metallurgical and Materials Transactions A*, **34A**, 449–57.
28 Bräuer, P. and Müller-Voigt, G. (1998) *Journal of Crystal Growth*, **186**, 520–27.
29 Derby, B. and Favier, J.J. (1983) *Acta Metallurgica*, **31**, 1123–30.
30 Yang, S., Liu, W. and Jia, J. (2001) *Journal of Materials Science*, **36**, 5351–55.
31 Kamio, A., Kumai, S. and Tezuka, H. (1991) *Materials Science and Engineering A*, **146**, 105–21.
32 Vinet, B. and Potard, C. (1983) *Journal of Crystal Growth*, **61**, 355–61.
33 Perepezko, P., Galup, C. and Cooper, K. (1982) in *Materials Processing in the reduced gravity environment of space*, (ed G.E. Rindone), Elsevier Science, Amsterdam, pp. 491–501.
34 Uebber, N. and Ratke, L. (1991) *Scripta Materialia*, **25**, 1133–37.

4
Thermal Expansion and Surface Tension of Multicomponent Alloys

Jürgen Brillo and Ivan Egry

4.1
Introduction

4.1.1
General

Multicomponent alloys represent the vast majority of technically relevant metallic materials. They are in most cases developed empirically by variation of the components until the desired properties are obtained. In recent years, numerical simulations have brought a significant improvement of the development process. Since computers and algorithms are constantly improving, the accuracy of the input material parameters, in particular thermophysical properties of the liquid phase, has become the main limiting factor. Information on such data is sparse. While for monoatomic liquids data might exist, there is already a lack of information for binary systems. For ternary alloys, systematic data exist only in exceptional cases.

A first key step for our work is, therefore, a systematic measurement of thermophysical properties of the liquid phase, such as density, thermal expansion, and surface tension.

The subjects of our study are on one hand the ternary monotectic alloys Cu–Ni–Fe and Cu–Co–Fe, which are both interesting from technological and academic points of view. In Cu–Ni–Fe, the metastable miscibility gap closes gradually as Ni is added to the binary Cu–Fe. In Cu–Co–Fe, the gap extends from the binary Cu–Fe throughout the ternary system to the Cu–Co border.

The third system that we investigated is Cu–Al–Ag. In contrast to the monotectic alloys that form solid solutions, it is characterized by the existence of intermetallic phases as well as eutectic and peritectic points in the phase diagram. Cu–Al–Ag also serves as a model system for a ternary aluminum-based alloy.

4.1.2
Density and Thermal Expansion

The density, $\rho(T)$, of a liquid metal can be considered as a linear function of temperature, T, within a limited temperature interval that includes the melting point:

$$\rho(T) = \rho_L + \rho_T(T - T_L) \tag{4.1}$$

In this equation, ρ_L is the density at the liquidus temperature, T_L, and ρ_T is the constant temperature coefficient.

For a multicomponent liquid solution with components $A_i (i = 1, \ldots, n)$ with atomic concentrations c_i and respective molar masses M_i, the molar volume of the solution, V, is generally represented by [1]

$$V = \sum_{i=1}^{n} c_i \frac{M_i}{\rho_i} + {}^E V \tag{4.2}$$

where ρ_i is the density of the pure substance A_i at temperature T. ${}^E V$ is the excess volume. For ${}^E V = 0$, Equation 4.2 reduces to a simple linear combination of the molar volumes and is often referred to as ideal or Vegard's law [2]. Generally, ${}^E V$ depends on both temperature and concentration. A simple expression is given by the following relation [1]:

$$^E V(c_1^B, \ldots, c_3^B, T) = \sum_{i}^{2} \sum_{j>i}^{3} c_i^B c_j^B \, {}^{BE}V_{i,j} + c_1^B c_2^B c_3^B \, {}^{BE}V^T \tag{4.3}$$

In Equation 4.3, ${}^E V_{i,j}$ denotes the binary interaction parameter between components i and j, and ${}^E V^T$ is the parameter for a possible ternary interaction. It is an interesting question whether or not the ternary interaction parameter ${}^E V^T$ can be neglected [3], implying that the densities of ternary alloys could be derived from the binary phases alone. From Equation 4.2 the following expression for ρ_T can be obtained by simple derivation:

$$\rho_T = \frac{\partial \rho(T)}{\partial T} = \frac{\left[\sum_i c_i M_i\right] \times \left[\sum_i c_i \frac{M_i \rho_{T,i}}{\rho_i^2} - \frac{\partial^E V}{\partial T}\right]}{V^2} \tag{4.4}$$

In this equation, $\rho_{T,i}$ is the temperature coefficient of the density of pure component A_i. In most cases the term $\partial^E V/\partial T$ is negligible and Equation 4.4 can be used to predict the thermal expansion coefficient of an alloy from only the properties of the pure substances [4–7]. There are some systems, however, where this procedure is not possible and were a linear temperature dependence of the excess volume needs to be assumed instead, that is, $\partial^E V/\partial T = \text{const}$ [8].

4.1.3
Surface Tension

Similar to Equation 4.1, the measured surface tensions, γ, are usually also described as linear functions of temperature, T, with corresponding parameters γ_L and γ_T, the surface tension at the liquidus and the thermal coefficient:

$$\gamma(T) = \gamma_L + \gamma_T(T - T_L) \tag{4.5}$$

For a liquid ternary alloy consisting of three elements $i = 1, 2, 3$, with corresponding surface tensions, $\gamma_i(T)$, the surface tension $\gamma_{123}(T)$ of the alloy is predicted by the Butler equation [9]:

$$\gamma_{123}(T) = \gamma_1 + \frac{RT}{S_1} \ln\left(\frac{1 - c_2^S - c_3^S}{1 - c_2^B - c_3^B}\right) + \frac{1}{S_1}\left\{{}^E G_1^S(T, c_2^S, c_3^S) - {}^E G_1^B(T, c_2^B, c_3^B)\right\}$$

$$= \gamma_2 + \frac{RT}{S_2} \ln\left(\frac{c_2^S}{c_2^B}\right) + \frac{1}{S_2}\left\{{}^E G_2^S(T, c_2^S, c_3^S) - {}^E G_2^B(T, c_2^B, c_3^B)\right\} \tag{4.6}$$

$$= \gamma_3 + \frac{RT}{S_3} \ln\left(\frac{c_3^S}{c_3^B}\right) + \frac{1}{S_3}\left\{{}^E G_3^S(T, c_2^S, c_3^S) - {}^E G_3^B(T, c_2^B, c_3^B)\right\}$$

where R is the universal gas constant, T is the temperature, S_i are the surface areas in a monolayer of pure liquid i, c_i^B are the mole fractions of component i in the bulk phase, and c_i^S are the mole fractions of component i in the surface phase. ${}^E G_i^B$ denotes the partial excess Gibbs energy in the bulk and ${}^E G_i^S$ is the partial excess Gibbs energy of component i in the surface layer.

The Butler equation is based on the assumption that the surface layer can be treated as a separate thermodynamic phase. With respect to the bulk, atoms on the surface have a reduced coordination number that lowers $|{}^E G_i^S|$. It was shown by Tanaka and Iida [9], that, apart from a constant factor $\frac{3}{4}$, ${}^E G_i^S$ can be assumed to have the same functional form as the expression for the bulk phase, ${}^E G_i^B$, that is, ${}^E G_i^S \approx \frac{3}{4} {}^E G_i^B$. The surface area $S_i (i = 1, 2, 3)$ is calculated from the molar volume V_i as follows [4–8]: $S_i = 1.091\,(6.02\,10^{23})^{1/3} V_i^{2/3}$. The partial excess Gibbs energy ${}^E G_i^B$ is derived from the excess Gibbs energy ${}^E G$ by using standard thermodynamic rules.

Equation 4.6 can be solved numerically for the surface concentrations as unknown variables. For the required potentials ${}^E G(T, c^B i)$, an expression similar to Equation 4.3 is used [1, 10, 11]:

$$ {}^E G(c_1^B, \ldots, c_3^B, T) = \sum_{i}^{2} \sum_{j > i}^{3} c_i^B c_j^B {}^{BE}G_{i,j} + c_1^B c_2^B c_3^B {}^{BE}G^T(T) \tag{4.7}$$

In Equation 4.7, ${}^E G_{i,j}$ denotes the binary interaction parameters for components i and j, and ${}^E G^T$ is the parameter for the ternary interaction. The parameter ${}^E G_{i,j}$

can be written as a function of temperature and concentration according to the Redlich–Kister form [11].

4.2 Experimental

4.2.1 Levitation

The technique of electromagnetic levitation offers a convenient platform for a contactless processing of electrically conductive alloys [12]. A basic sketch of the setup used in this work is shown in Figure 4.1. The sample is located at the center of a levitation coil to which an alternating current with a frequency of approximately 250 kHz is applied. It is positioned by forces owing to interactions with the inhomogeneous magnetic field and melted by eddy currents that are induced within. In order to avoid pronounced evaporation, the sample is processed under a protecting atmosphere of He or Ar. Control of the temperature is achieved by cooling of the sample in a laminar gas flow from a nozzle from below.

The temperature, T, is measured using an infrared pyrometer aimed at the top of the sample. As the emissivity is not known for most materials, it is necessary to recalibrate the pyrometer signal with respect to the known liquidus temperature, T_L. This procedure is described in Ref. 7, and assumes that the sample emissivity at the operating wavelength of the pyrometer remains constant over the experimentally scanned range of temperature. For most metals [13] this is a good approximation.

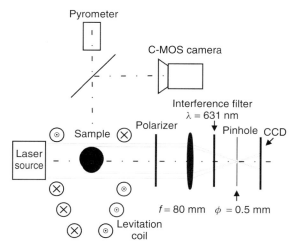

Fig. 4.1 Schematic sketch of the experimental setup used to measure density and surface tension by means of electromagnetic levitation.

4.2.2
Density and Thermal Expansion

To measure the density based on the volume, shadowgraphs are taken from the levitated sample. As Figure 4.1 shows, a polarized He–Ne laser beam, equipped with a spatial filter and a beam expander is used to illuminate the sample from behind. The shadow image is captured by means of a digital charge-coupled device (CCD) camera and analyzed by an edge detection algorithm that locates the edge curve $R(\varphi)$ where R and φ are the radius and azimuthal angle with respect to the drop center. In order to eliminate the influence of oscillations, the edge curve is averaged over 1000 frames. This average $\langle R(\varphi) \rangle$ is fitted by Legendre polynomials of order ≤ 6. An analysis of top view images [4] shows that the equilibrium shape of the sample is symmetric with respect to the vertical axis. Hence, its volume is calculated using the following integral:

$$V = \tfrac{2}{3}\pi \int_0^\pi \langle R(\varphi) \rangle^3 \sin(\varphi) d\varphi \tag{4.8}$$

When M is the mass of the sample, the density, ρ, is calculated from $\rho = M/V$. An absolute uncertainty analysis of the measurement technique is given in Ref. 7. The total error for the density was found to be $\Delta \rho / \rho \leq 1.5\%$.

In order to check whether the mass remained constant over the duration of the experiment, each sample is weighed before and after the measurement. Only those measurements are accepted in which the mass loss is less than $\approx 0.1\%$.

4.2.3
Surface Tension

Surface tension measurements are performed using the oscillating drop technique [14]. A fast recording digital video camera (400 frames per second, 1024×1000 pixels) is directed at the sample from the top, (Figure 4.1). A series of 4196 frames is recorded at each temperature and analyzed afterwards by an edge detection algorithm. The frequency spectrum of the radius R exhibits a set of five peaks ω_m, $m = -2, -1, 0, 1, 2$, which correspond to the Rayleigh surface oscillation modes. The surface tension is calculated from these frequencies following the sum rule of Cummings and Blackburn [15]:

$$\gamma = \frac{3M}{160\pi} \sum_{m=-2}^{+2} \omega_m^2 - 1.9\Omega^2 - 0.3 \left(\frac{g}{a}\right)^2 \Omega^{-2} \tag{4.9}$$

where M is the mass of the sample, g is the gravitational acceleration, and a is the radius of the sample. Ω is a correction factor that accounts for the magnetic pressure and is calculated from the three translational frequencies of the sample.

Taking into account a relative mass loss of $\Delta M/M = 0.1\%$, and an uncertainty in the determination of the frequencies of ± 0.1 Hz, the relative error of the surface tension will approximately be $\Delta \gamma / \gamma \approx 5\%$.

4.3
Results

4.3.1
Density

Results on the density are shown in Figure 4.2 for the pure elements aluminum and copper. Data are recorded for temperature range of $1100\,°C > T > 1500\,°C$ for copper and $660\,°C > T > 1300\,°C$ for pure aluminum. As is seen from Figure 4.2, there is a linear decrease of ρ with an increase of temperature. Hence Equation 4.1 can be fitted to the data. The resulting fit parameters, ρ_L and ρ_T, are shown in Table 4.1 for Cu and Al [8]. A linear decrease of the density was found in all our measurements, so that Equation 4.1 serves as a general form for representing experimental $\rho(T)$ data.

Other pure elements that we investigated are Ag, Au, Co, Fe, and Ni [4–8,16]. Their fit parameters ρ_L and ρ_T are also shown in Table 4.1. From the monoatomic systems, we moved on to the investigation of their binary combinations, which is demonstrated in the following on the example of liquid Al–Cu in Figure 4.2 [8]. The Al concentrations range from 0 to 100 at%. Owing to the good levitation stability and negligible evaporation of the Al–Cu samples, precise data could be obtained for a broad temperature range of up to $T_L > T > T_L + 1000$ K.

In order to study the concentration dependence, ρ is calculated from Equation 4.1 for a constant temperature of $T = 750\,°C$ using the corresponding parameters ρ_L and ρ_T. The result of these calculations is shown in Figure 4.3. As can be seen, the density decreases with increasing aluminum concentration. The dashed line in Figure 4.3 represents the ideal law, that is Equation 4.2 with an excess volume $^{E}V = 0$.

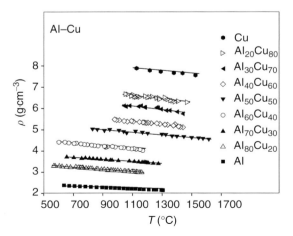

Fig. 4.2 Density plotted versus temperature for Al, Cu, and binary Al–Cu alloys with different copper concentrations.

Table 4.1 Parameters T_L, ρ_L, and ρ_T for liquid Ag, Au, Cu, Co, Cu, Fe, Ni.

System	T_L (°C)	ρ_L (g·cm^{-3})	ρ_T (10^{-4} g·cm^{-3}·K^{-1})	Reference
Ag	962	9.15	−7.4	[7]
Au	1064	17.4	−11.0	[6]
Al	660	2.35	−2.0	[8]
Co	1495	7.81	−8.85	[16]
Cu	1085	7.90	−7.65	[8]
Fe	1545	7.04	−10.8	[5]
Ni	1454	7.93	−10.1	[5]

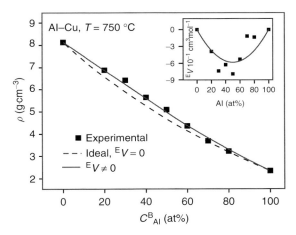

Fig. 4.3 Density of liquid Al–Cu as a function of atomic Al concentration at $T = 750°$C. The experimental data are shown in comparison to Equation 4.2 with $^EV = 0$ (dashed line) and ^EV from Equation 4.3 with $^EV_{Al,Cu}$ used as a fit parameter. The inset shows the excess volume calculated by Equation 4.2 from the experimental density data directly.

A deviation of the density from the ideal law toward larger values is clearly visible. This deviation is nearly 8% of the magnitude of the ideal density and is large compared to the experimental error. The excess volume ^EV is therefore negative in Al–Cu. The solid line in Figure 4.3 shows Equation 4.2 with a nonzero ^EV from a fit of Equation 4.3 with $^EV_{Al,Cu}$ as the fitting parameter. As can be seen, the agreement with the data is far better than for the case of the ideal solution. The experimental density values can also be used as input to Equation 4.2 to calculate the corresponding excess volumes ^EV as a function of the respective aluminum concentration. This is shown in the inset of Figure 4.3, and, in agreement with Equation 4.3, ^EV can be described as a function of concentration by a parabola.

Generally speaking, the excess volume can be zero, positive, or negative. Binary systems for which the excess volume is zero are Fe–Ni, Co–Fe, Au–Cu, Ag–Cu,

Table 4.2 Binary interaction paramerters $^EV_{i,j}$ used in Equation 4.3 for the calculation of excess volumes.

System i, j	$^EV_{i,j}$ (cm³ mole⁻¹)	Reference
Co, Fe	0	[16]
Fe, Ni	0	[5]
Ag, Cu	0	[7]
Ag, Au	0	[7]
Cu, Au	0	[6]
Cu, Fe	0.65	[5]
Cu, Co	0.45	[16]
Cu, Ni	−0.85	[5]
Al, Cu	−3.87	[8]
Al, Ag	−2.68	[8]
Al, Fe	−1.7	[18]
Al, Ni	−5.0	[18]

and Ag–Au [5–8,16]. Systems for which we found a negative excess volume are Cu–Ni, Al–Cu, Al–Ag, Al–Ni, and Al–Fe [4, 5, 8, 17]. Binary alloy systems for which a positive excess volume was found are Cu–Fe and Cu–Co [5, 16]. In all these cases Equation 4.3 served as a representation for the excess volume. The corresponding interaction parameters $^EV_{i,j}$ according to Equation 4.3 are listed for each binary system in Table 4.2.

A plot of the temperature coefficient, ρ_T, for Al–Cu [8] is shown in Figure 4.4. With increasing aluminum concentration, there is a general tendency toward larger values. A weak minimum exists at $c_{Al} \approx 30$ at%, which, however, may be negligible with respect to the uncertainty of the data.

The dashed curve in Figure 4.4 shows a calculation of ρ_T according to Equation 4.4 with the temperature derivative $\partial^E V/\partial T$ of the excess volume being set to zero. Although there are systematic deviations of the experimental data from this curve, the agreement can be regarded as good with respect to the experimental error. A better description is provided if $\partial^E V/\partial T$ is not neglected. In this case, even the small minimum at $c_{Al} \approx 20$ at% is predicted. Among the systems that we investigated, Al–Cu and Al–Ag are the only ones where $\partial^E V/\partial T$ had to be taken into account for the calculation of ρ_T [8].

The investigations on binary systems were extended to ternary alloys. For example, Figure 4.5 shows a plot of the density of $Cu_x Co_{0.5(1−x)} Fe_{0.5(1−x)}$, $0 \le x \le 1$, at $T = 1483\,°C$ [16] as a function of the atomic copper concentration. The densities are all in the range between 7.4 and 7.6 g cm⁻³ and appear significantly lower than the corresponding ideal solution. The excess volume is therefore positive. The experimental data agree well with the calculation from Equation 4.3 with the ternary interaction parameter $^EV^T = 0$ and parameters $^EV_{i,j}$ from Table 4.2. This is also true for density data measured along a second cut through the system, discussed in Ref. 16 for the same temperature. Hence, in the special case of liquid

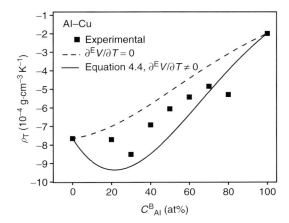

Fig. 4.4 Temperature coefficient, ρ_T, plotted for Al–Cu versus Al atomic concentration. The experimental data are shown in comparison with Equation 4.4 with $\partial^E V / \partial T = 0$ (dashed line) and $\partial^E V / \partial T$ used as a constant fitting parameter (solid line).

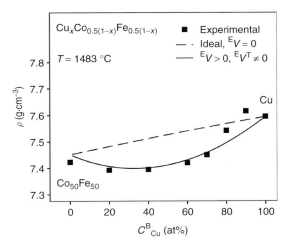

Fig. 4.5 Density of liquid $Cu_x Co_{0.5(1-x)} Fe_{0.5(1-x)}$ samples at 1483 °C versus copper concentration. The experimental data are shown together with calculations for the ideal solution (dashed line) and Equation 4.3 with $^E V_{i,i}$ taken from Table 4.2.

ternary Cu–Co–Fe, the densities can be predicted from those of the binary systems [16]. Although this is also possible for the system Al–Cu–Ag [8], $^E V^T$ is generally not zero. Our measurements on Ni–Cu–Fe [19] showed that the influence of the ternary parameter, $^E V^T$, can be quite large. In Ni–Cu–Fe, it actually dominates the density.

Table 4.3 Ternary parameters $^{E}V^{T}$ used in Equation 4.3 for the calculation of the excess volume of the liquid ternary alloys Cu–Fe–Ni, Cu–Co–Fe, and Al–Cu–Ag.

System	$^{E}V^{T}$ (cm^3 mol^{-1})	Reference
Cu–Co–Fe	0	[16]
Cu–Fe–Ni	11.5	[19]
Cu–Al–Ag	0	[21]

Results for the density measurements of the ternary systems Cu–Ni–Fe [19], Co–Cu–Fe [16], and Al–Cu–Ag [20] are summarized in Table 4.3. Together with the information shown in Tables 4.1 and 4.2, a complete set of density and thermal expansion data is established for these three liquid alloy systems.

4.3.2
Surface Tension

As with the densities, surface tensions are investigated by a stepwise extension of the work from monoatomic systems to binary and ternary alloys. Results on the surface tension obtained for liquid copper and iron are shown as a function of temperature in Figure 4.6. Obviously, $\gamma(T)$ can be described by a linear law as well, and Equation 4.5 can be fitted to the data. The corresponding parameters, γ_L and γ_T, are shown in Table 4.4 for the pure elements Cu and Fe [20]. A linear temperature dependence was found in all our measurements of the surface tension. Equation 4.5 therefore serves as a general form for representing experimental $\gamma(T)$ data. Another pure element that we investigated is Ni [20]. The surface tension parameters γ_L and γ_T are also shown in Table 4.4. Figure 4.6 also shows the surface

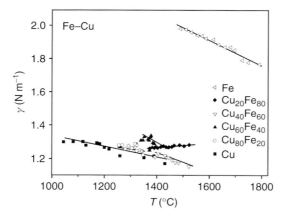

Fig. 4.6 Surface tension plotted versus temperature for Cu, Fe, and binary Cu–Fe alloys with different copper concentrations.

Table 4.4 Parameters T_L, γ_L, and γ_T for Cu, Fe, Ni, and their binary alloys [20].

System	T_L (°C)	γ_L (g·cm^{-3})	γ_T (10^{-4} g·cm^{-3}·K^{-1})
Cu	1085	1.29	−2.3
Fe	1545	1.92	−4.0
Ni	1454	1.77	−3.3
Ni$_{75}$Fe$_{25}$	1440	1.73	−2.8
Ni$_{50}$Fe$_{50}$	1440	1.91	−3.3
Ni$_{25}$Fe$_{75}$	1473	1.93	−1.7
Cu$_{10}$Ni$_{90}$	1433	1.61	−0.7
Cu$_{20}$Ni$_{80}$	1417	1.51	−0.2
Cu$_{30}$Ni$_{70}$	1387	1.43	−0.8
Cu$_{40}$Ni$_{60}$	1347	1.38	−0.5
Cu$_{50}$Ni$_{50}$	1311	1.37	−0.9
Cu$_{60}$Ni$_{40}$	1280	1.36	−1.9
Cu$_{70}$Ni$_{90}$	1235	1.32	−3.2
Cu$_{80}$Ni$_{90}$	1189	1.34	−2.2
Cu$_{90}$Ni$_{10}$	1136	1.31	−2.2
Cu$_{80}$Fe$_{20}$	1385	1.24	−3.8
Cu$_{60}$Fe$_{40}$	1424	1.22	−4.4
Cu$_{40}$Fe$_{60}$	1435	1.24	−4.9
Cu$_{20}$Fe$_{80}$	1463	1.30	+1.4

tension data of liquid binary Fe–Cu alloys with copper concentrations ranging from 20 to 80 at%, see also Table 4.4. The data fall between 1.2 and 1.4 N m^{-1}. Even for 80 at% iron, the values are close to those of pure copper. In order to study the concentration dependence, the surface tensions were calculated from Equation 4.5 using the corresponding parameters γ_L and γ_T. The result is shown in Figure 4.7 as a function of the atomic copper bulk concentration, $c^B{}_{Cu}$, at 1550 °C.

In Figure 4.7, starting with pure liquid iron the surface tension decreases from 1.88 N·m^{-1} sharply down to the value of pure copper within the first 20% increase of the copper concentration. It remains there almost constant with further $c^B{}_{Cu}$ increase. In the same figure, the calculated surface tension values for the ideal solution model, Equation 4.6 with $^E G = 0$, are shown. The experimental values are significantly lower. A far better agreement is obtained when the Butler Equation 4.6 is solved by using parameters for a nonvanishing $^E G$, taken from [10]. Figure 4.7 also shows the Butler equation solved for $T = 1750\,°C$. As can be seen, this solution predicts surface tensions that are lower than for 1550 °C as long as $c^B{}_{Fe} < 80$ at%. For larger concentrations, the Butler equation predicts $\gamma(1750\,°C) > \gamma(1550\,°C)$. This crossover of the curves is encircled in Figure 4.7 and is in agreement with the observed positive γ_T in Figure 4.6 for Cu$_{20}$Fe$_{80}$. It is related to a favoring of Cu segregation at lower temperatures.

As a by-product, the Butler equation yields the surface concentration, that is, $c^S{}_{Cu}(c^B{}_{Cu})$. It is found that copper segregates to the surface and its concentration $c^S{}_{Cu}$ is more than 80 at% for $c^B{}_{Cu} > 20$ at%.

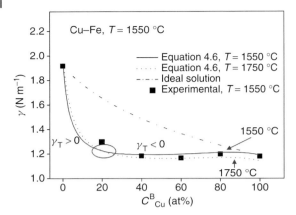

Fig. 4.7 Surface tension of liquid Cu–Fe as a function of atomic Cu concentration at $T = 1550\,°C$. The experimental data are shown in comparison to those for the ideal solution, Equation 4.6 with $^EG = 0$ (dash-dotted line), Equation 4.6 with $^EG \neq 0$ at 1550°C (solid line), and Equation 4.6 with $^EG \neq 0$ at 1750°C (dotted line). The circle marks the transition between positive and negative γ_T.

Other binary systems that we investigated are Cu–Ni, Fe–Ni, Al–Fe, and Al–Ni [16, 22], (Tables 4.4 and 4.5). The isothermal surface tension curves $\gamma(c^B_{Cu})$ for Cu–Ni and Cu–Co are basically the same as for Cu–Fe shown in Figure 4.7 [21]. The calculated segregation curves for these systems all predict a surface segregation of copper, which is demonstrated for Cu–Ni and Cu–Fe in Figure 4.8. The process of segregation can be understood as the result of a minimization of the surface free energy [1, 9]. Therefore, the element with the lower surface tension should also determine the surface tension of the alloy.

Table 4.5 Parameters T_L, γ_L, and γ_T for binary Al-based alloys [18].

System	T_L (°C)	γ_L (g·cm^{-3})	γ_T (10^{-4} g·cm^{-3}·K^{-1})
$Al_{82}Ni_{18}$	950	1.01	−5.1
Al_3Ni	1110	1.21	−8.3
$Al_{70}Ni_{30}$	1283	1.15	−8.8
Al_3Ni_2	1559	1.30	−6.3
AlNi	1638	1.44	−6.7
$Al_{37}Ni_{63}$	1540	1.55	−2.9
$AlNi_3$	1385	1.44	−2.6
$Al_{13}Ni_{87}$	1433	1.58	−5.1
$Al_{90}Fe_{10}$	1016	0.95	−4.8
$Al_{80}Fe_{20}$	1157	1.08	−5.8
Al_3Fe	1160	1.09	−3.8
Al_5Fe_2	1169	1.12	−7.1
Al_2Fe	1175	1.16	−4.4
$Al_{60}Fe_{40}$	1232	1.22	−1.5

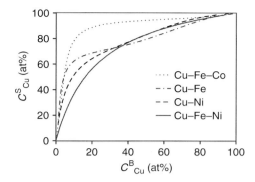

Fig. 4.8 Calculated surface concentrations c_{Cu}^S plotted versus copper bulk concentration for Cu–Ni, Cu–Fe, Cu–Fe–Ni, and Cu–Co–Fe.

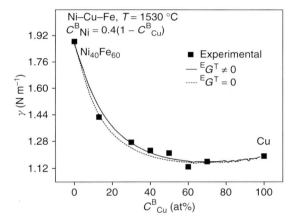

Fig. 4.9 Surface tension of liquid $Cu_x Ni_{0.4(1-x)} Fe_{0.6(1-x)}$ samples at 1530 °C versus copper concentration. The experimental data are shown with calculations from the Butler Equation 4.6 for the ideal solution (dotted line), $^E G^T \neq 0$ (solid line), and for $^E G^T = 0$ (dashed line).

The situation is slightly different for alloys with mainly attractive interactions between unlike atoms in the melt. Such alloys are Ni–Fe, Al–Fe, and Al–Ni. In these systems, we found that segregation is practically suppressed [16, 18, 20, 22, 23] and the surface tension can mainly be described by the ideal solution model.

Results on the surface tension obtained for a ternary monotectic system are shown in Figure 4.9 at $T = 1530\,°C$ for $Cu_x Ni_{0.4(1-x)} Fe_{0.6(1-x)}$, $0 \leq x \leq 1$. The parameters γ_L and γ_T are listed in Table 4.6. The general appearance is the same as for the binary systems in Figure 4.7. Again, copper is found to be the main component of the surface, see Figure 4.8. In Figure 4.9, the experimental data are plotted together with solutions of the Butler Equation 4.6 with or without the ternary term $^E G^T$ being included into the excess Gibbs free energy $^E G$. It becomes

Table 4.6 Parameters T_L, γ_L, and γ_T for ternary alloys [19, 22, 23].

System	T_L (°C)	γ_L (N m^{-1})	γ_T (10^{-4} N m^{-1} K^{-1})
$Cu_{20}Co_{40}Fe_{40}$	1423	1.30	−2.25
$Cu_{40}Co_{30}Fe_{30}$	1402	1.17	+0.75
$Cu_{60}Co_{20}Fe_{20}$	1389	1.21	−2.85
$Cu_{80}Co_{10}Fe_{10}$	1366	1.22	−2.0
$Cu_{20}Co_{20}Fe_{60}$	1438	1.30	−2.25
$Cu_{20}Co_{60}Fe_{20}$	1420	1.35	−4.8
$Ni_{33}Cu_{13}Fe_{54}$	1419	1.45	−1.88
$Ni_{28}Cu_{30}Fe_{42}$	1365	1.29	−1.09
$Ni_{25}Cu_{40}Fe_{35}$	1337	1.28	−2.95
$Ni_{20}Cu_{50}Fe_{30}$	1318	1.26	−2.36
$Ni_{16}Cu_{60}Fe_{24}$	1307	1.24	−4.93
$Ni_{17}Cu_{70}Fe_{13}$	1273	1.30	−5.3
$Ni_{15}Cu_{20}Fe_{65}$	1428	1.34	−2.9
$Ni_{32}Cu_{20}Fe_{48}$	1396	1.38	−2.1
$Ni_{45}Cu_{20}Fe_{35}$	1390	1.42	−1.8
$Ni_{60}Cu_{20}Fe_{20}$	1395	1.44	−1.9
$Cu_{90}Ag_{10}$	1037	0.89	−3.9
$Al_{20}Cu_{70}Ag_{10}$	917	1.00	−2.1
$Al_{40}Cu_{50}Ag_{10}$	607	0.84	−4.5
$Al_{50}Cu_{40}Ag_{10}$	567	1.24	−2.2

clear from this depiction that, with respect to the accuracy of the experimental data, these two cases cannot be distinguished from each other. Consequently, $^EG^T$ can be neglected for the description of the surface tension. For Ni–Cu–Fe, we also studied a second cut through the system where the copper concentration was kept constant at 20 at% and the ratio of the nickel and iron bulk concentration was varied [19]. The observed change of γ with c_{Ni}^B is so small that it is justified to consider γ as constant for this section. It is found for this cut that, with respect to the accuracy of the experimental data, $^EG^T$ can also be neglected.

Qualitatively, the same result was obtained for the surface tension of liquid ternary Cu–Co–Fe alloys, Table 4.6. As in Cu–Ni–Fe, the surface tension is determined by the concentration of copper (Figure 4.8), while it is almost insensitive for a substitution of the two transition metals (TMs) by each other.

The above results and the striking similarities with corresponding results for binary systems, Cu–Ni, Cu–Fe [20], and Cu–Co [21, 23], suggest that with respect to the surface tension Cu–Co–Fe and Cu–Ni–Fe may be described as quasi binary systems Cu–TM. Simple models, published elsewhere [23], show that this is indeed a good approach for the description of the surface tension of this type of ternary alloy.

As indicated above, the situation may be different for the surface tension of some Al-based binary and ternary alloys. Figure 4.10 shows the results measured for liquid $Al_{0.9(1-x)}Cu_{0.9x}Ag_{0.1}$ alloys as a function of atomic aluminum concentration at a constant temperature, compare Table 4.6. They agree surprisingly well with the ideal solution in this case. From the investigations on the binary systems

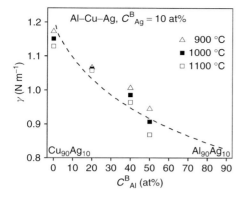

Fig. 4.10 Surface tension of liquid $Al_{0.9-x}Cu_x$–$Ag_{0.1}$ samples at 1530°C versus copper concentration. The experimental data are shown with calculations from the Butler Equation 4.6 for the ideal solution (dotted line), $^EG^T \neq 0$ (solid line), and for $^EG^T = 0$ (dashed line).

Al–Ni and Al–Fe [18, 24, 25] we find, however, that for Al-based alloys deviations occur at compositions in which the phase diagram shows features of intermetallic compound formation [24, 25] in the solid phase. In this case, other models such as [26, 27], which take compound formation into account, need to be applied.

4.4
Conclusion and Summary

Densities, thermal expansion coefficients, and surface tensions were determined by means of electromagnetic levitation for a large number of pure metals, binary and ternary copper-based alloys. A comprehensive dataset was created, which was analyzed in terms of the mixing behavior.

On the basis of our data, we found the following general trends for the density:

- Alloys consisting of elements with similar electronic structures have no excess volume. Such alloys are Ni–Fe, Co–Fe, Cu–Ag, Cu–Au, and Au–Ag.
- Copper-based alloys with a positive excess free energy containing a TM showed a positive excess volume. Alloys for which we found such a behavior are: Cu–Fe, Cu–Co, Cu–Fe–Ni, and Cu–Co–Fe. These alloys also exhibit a metastable miscibility gap. The only exception in this class may be Cu–Ni, where $^EV < 0$ was found [4].
- Al-based alloys, such as Al–Fe, Al–Ni, Al–Cu, and Al–Cu–Ag, show an excess volume which is strongly negative. These alloys are usually characterized by an exothermic mixing behavior and their phase diagrams exhibit intermetallic phases.

For all alloys, it was found that the surface tensions could be described by the Butler equation and depend strongly on surface segregation. A pronounced segregation takes place in copper-based binary and ternary alloys that contain a TM. The surface tension of the ternary alloys Cu–Ni–Fe and Cu–Co–Fe could be derived from the Butler equation using excess free energies of the constituting binary systems only. For Fe–Ni and Al-based alloys, segregation is reduced, and the surface tension is described by the ideal solution model. In particular, for Al-based alloys, models taking intermetallic compound formation into account [24–27] are required for an adequate description of the experimental data.

Acknowledgements

Financial support by the German Science Foundation (DFG) within the Priority Program 1120 "Phase transformations in multicomponent melts" under grant EG 93/4 is gratefully acknowledged. We would like to thank our collaborators R. Schmid-Fetzer, A. Griesche, M. P. Macht, G. Frohberg, and S. Rex for fruitful cooperation.

References

1 Lüdecke, C. and Lüdecke, D. (**2000**) *Thermodynamik*, Springer, Heidelberg, p. 506.
2 Vegard, L. (**1921**) *Zeitschrift für Physik*, **5**, 17.
3 Kohler, F. (**1960**) *Monatshefte für Chemie*, **91**, 738.
4 Brillo, J. and Egry, I. (**2003**) *International Journal of Thermophysics*, **24**, 1155.
5 Brillo, J. and Egry, I. (**2004**) *Zeitschrift für Metallkunde*, **95**, 691.
6 Brillo, J., Egry, I., Giffard, H.S. and Patti, A. (**2004**) *International Journal of Thermophysics*, **25**, 1881.
7 Brillo, J., Egry, I. and Ho, I. (**2006**) *International Journal of Thermophysics*, **27**, 494.
8 Brillo, J., Egry, I. and Westphal, J. (**2008**) *International Journal of Materials Research*, **99**, 162.
9 Tanaka, T. and Iida, T. (**1994**) *Steel Research*, **65**, 21.
10 Servant, C., Sundman, B. and Lyon, O. (**2001**) *Calphad*, **25**, 79.
11 Schmid-Fetzer, R.S. and Gröbner, J. (**2001**) *Advanced Engineering Materials*, **3**, 947.
12 Herlach, D.M., Cochrane, R.F., Egry, I., Fecht, H.J. and Greer, A.L. (**1993**) *International Materials Review*, **38**, 273.
13 Krishnan, S., Hansen, G.P., Hauge, R.H. and Margrave, J.L. (**1990**) *High Temperature Science*, **29**, 17.
14 Schneider, S., Egry, I. and Seyhan, I. (**2002**) *International Journal of Thermophysics*, **23**, 1241.
15 Cummings, D.L. and Blackburn, D.A. (**1991**) *Journal of Fluid Mechanics*, **224**, 395.
16 Brillo, J., Egry, I. and Matsushita, T. (**2006**) *International Journal of Materials Research*, **97**, 1526.
17 Plevachuk, Y., Egry, I., Brillo, J., Holland-Moritz, D. and Kaban, I. (**2007**) *International Journal of Materials Research*, **98**, 107.
18 Egry, I., Brillo, J., Holland-Moritz, D. and Plevachuk, Y. *Materials Science and Engineering A*, DOI: 10.1016/J.MSEA.2007.07.104.
19 Brillo, J., Egry, I. and Matsushita, T. (**2006**) *Zeitschrift für Metallkunde*, **97**, 28.
20 Brillo, J. and Egry, I. (**2005**) *Journal of Materials Science*, **40**, 2213.

21 Eichel, R. and Egry, I. (1999) *Zeitschrift fur Metallkunde*, **90**, 5.
22 Plevachuk, Y., Egry, I. and Brillo, J. unpublished data.
23 Brillo, J. and Egry, I. (2007) *International Journal of Thermophysics*, **28**, 1004.
24 Sommer, F. (1982) *Zeitschrift fur Metallkunde*, **73**, 72.
25 Akinlade, O., Singh, R.N. and Sommer, F. (2000) *Journal of Alloys and Compounds*, **299**, 163.
26 Egry, I. (2004) *Journal of Materials Science*, **39**, 6365.
27 Novakovic, R., Ricci, E., Giuranno, D. and Passerone, A. (2005) *Surface Science*, **576**, 175.

5
Measurement of the Solid-Liquid Interface Energy in Ternary Alloy Systems

Annemarie Bulla, Emir Subasic, Ralf Berger, Andreas Bührig-Polaczek, and Andreas Ludwig

5.1
Introduction

The aim of this chapter is to point out a methodology to determine the solid–liquid interface energy in binary and ternary alloy systems. The solid–liquid interface energy, σ_{SL}, is defined as reversible work required for creating a unit area of the interface at constant temperature, volume, and chemical potentials and is of major importance during phase transformation that is phase nucleation, crystal growth, and the final grain structure. In addition to chemical diffusion, this quantity governs the microstructural length scale of solidification morphologies. Therefore, the solid–liquid interface energy is an important parameter in many analytical and numerical models of solidification and ripening. The measurement of the solid–liquid interface energy is very difficult. For binary alloys, the literature provides only limited experimental data, and ternary alloys have been analyzed for the first time in the present study.

One of the most common experimental techniques for measuring the solid–liquid interface energy is the "grain boundary groove in an applied temperature gradient" method [1–18]. To measure the solid–liquid interface energy, a radial heat flow apparatus was constructed and assembled as described by Gündüz [11]. After the equilibration process, the samples were metallographically investigated and the local curvature of the grooves was analyzed. The interface energy was obtained using the Gibbs–Thomson equation, which requires measuring the local curvature of the grain boundary grooves, determining the local undercooling by heat flux simulations, and calculating the entropy of fusion to obtain the solid–liquid interface energy.

In the present chapter, the radial heat flow apparatus was applied to reproduce previous results for a eutectic Al–Cu alloy and to measure the solid–liquid interface energy for an alloy with invariant eutectic composition in the ternary system Al–Cu–Ag. In order to describe the influence of concentration on the solid–liquid interface energy, additional experiments were carried out.

Phase Transformations in Multicomponent Melts. Edited by D. M. Herlach
Copyright © 2008 WILEY-VCH Verlag GmbH & Co. KGaA, Weinheim
ISBN: 978-3-527-31994-7

5.2
Experimental Procedure

5.2.1
The Radial Heat Flow Apparatus

On the basis of the work of Gündüz [11] a radial heat flow apparatus for measuring the solid–liquid interface energy was assembled as shown in Figure 5.1. Because of its axial symmetry the apparatus allows a stable temperature gradient to be maintained, using a single heating wire emitting a power of P over a length of l along the axis of a cylindrical sample, and a water cooled jacket at the outside of the sample.

According to Fourier's law (provided a constant radial heat flux)

$$\dot{Q} = -\lambda 2\pi r l \frac{dT}{dr} \quad \text{with} \quad P = \frac{Q}{t} = \dot{Q} \tag{5.1}$$

where λ (W K^{-1}.m) is the thermal conductivity which is material dependent. The Equation 5.1 may be integrated as follows:

$$\dot{Q} = \frac{-\lambda 2\pi l (T_2 - T_1)}{\ln\left(\frac{r_2}{r_1}\right)} \tag{5.2}$$

where T_1 and T_2 are the temperatures at r_1 and r_2, respectively. The temperature gradient G_S at r required for the measurement of σ_{SL} could be determined by measuring the temperatures T_1 and T_2, respectively.

$$G_S = -\frac{\dot{Q}}{2\pi r l \lambda} \tag{5.3}$$

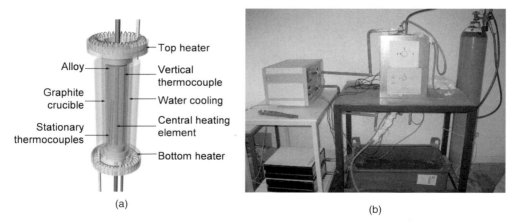

Fig. 5.1 (a) Schematic illustration of the radial heat flow apparatus. (b) The whole assembly of the radial heat flow apparatus.

5.2.2 Equilibration of the Sample

During the equilibration process, the sample was heated at the center using the central heating element and cooled on the outside using the water cooling jacket to establish a stable radial temperature gradient. This was ensured by a temperature controller using calibrated thermocouples. The top and bottom heaters were adjusted so that the vertical isotherms run parallel to the central axis. To prevent a horizontal convectional flow, only a thin liquid layer (1–2 mm thick) was melted along the ceramic tube surrounding the central heating element. The semisolid samples were held in a stable radial temperature gradient for about 2–4 days—depending on the alloy—until a dynamic balance has been reached in which the solid–liquid interface has stabilized in a constant temperature field. During the annealing, the mean temperature deviation was about $\pm 0.02\,°C$ in 1 h and $\pm 0.05\,°C$ in 2 days.

5.2.3 Quenching

The shape of the cusps had to be preserved by rapid quenching. This was realized by turning off the input power to all heaters. Figure 5.2 shows the temperature versus time plot during quenching. At the beginning of the quenching process, the cooling rate at the thermocouples was approximately $40\,°C\,min^{-1}$.

5.3 Evaluation of the Local Curvature of the Grain Boundary Grooves

5.3.1 Preparation of the Sample

The cylindrical samples were cut in a transverse direction in 20 mm slices and metallographically prepared. The microexamination of the samples was done using

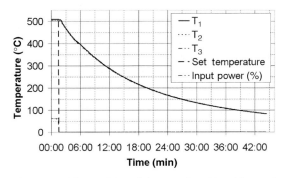

Fig. 5.2 Rapid quenching of the sample achieved by turning off the input power.

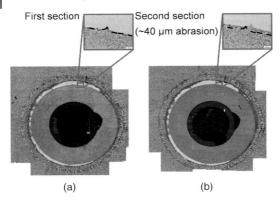

Fig. 5.3 Two cross sections of a specimen with a defined distance of a eutectic Al–Cu–Ag alloy consisting of 50 single pictures magnified 50 times: (a) first plane, (b) second plane.

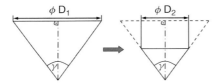

Fig. 5.4 Changes in the diameter of the central holes after grinding. The half angle at the tip of the drill bit, γ, was used to determine the amount of abrasion d, where D is the diameter of the drilled holes at the surface.

an optical light microscope. The grain boundary grooves were photographed at a magnification of 500, to allow an accurate measurement of the local curvature of the grooves. The x and y coordinates of the grain boundary grooves were determined using a computer-aided design (CAD) software. To determine the orientation of the grain boundary grooves relative to the polished surface, two cross-sectional cuts with a defined distance were required (Figure 5.3).

In order to determine the amount of abrasion, d, in the specimen, four opposite center holes were drilled into the polished surface with a drill bit angle of 90°. From the variation in the diameter, the amount of abrasion was calculated (Figure 5.4).

5.3.2
Geometrical Correction of the Groove Coordinates

To calculate the magnitude of the grain, it is assumed that no curvature exists along the direction of the cusp line (z direction) of the grain boundary groove. Owing to the translation invariance in this direction, the 3D geometry of the grain boundary groove can be reduced to a 2D one by a projection to a plane orthogonal to the cusp line. Since the polished surfaces of the specimens are not perpendicular to

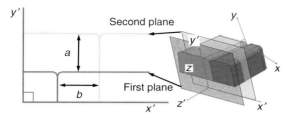

Fig. 5.5 Schematic drawing to show the relation between the coordinate systems of the grain and of the ground cross section [14].

the surfaces of the grain boundary groove (Figure 5.5), a transformation of the x and y coordinates of the grain boundary groove is necessary.

The coordinates of the cusps, x', y', from the metallographic section must be projected into an x-, y-, z-coordinate system aligned with the grain. This new coordinate system is oriented such that the z axis is parallel to the base of the grain boundary groove and the y axis is perpendicular to the macroscopic solid–liquid interface plane. The measured x', y' coordinates are then projected onto the $z = 0$ plane in the coordinate system of the grain. The transformation can be expressed as follows:

$$x = x' \frac{\sqrt{a^2 + d^2}}{\sqrt{a^2 + b^2 + d^2}} + y' \frac{-ab}{\sqrt{a^2 + d^2}\sqrt{a^2 + b^2 + d^2}} \tag{5.4}$$

$$y = y' \frac{d}{\sqrt{a^2 + d^2}} \tag{5.5}$$

and

$$d = \frac{D_1 - D_2}{2 \tan \gamma} \tag{5.6}$$

5.3.3
Determination of the Local Undercooling

If the solid and liquid phases have different thermal conductivities, the isotherms at the grain boundary groove cusps are deformed and the local undercooling has to be determined numerically. Figure 5.6 shows an example of a numerical simulation of the microscopic temperature field, to determine the local deformation of the isotherms at the grain boundary groove.

The shape of the solid–liquid interface was extrapolated using the transformed experimental groove shape. Fixed temperatures have been set at the bottom and the top of the domain, deduced from the macroscopic temperature gradient measured during the equilibration experiment. On either side of the domain the lateral heat flux is zero. The temperature field simulations were carried out using the commercial software FLUENT. To automate the calculation of the local undercooling,

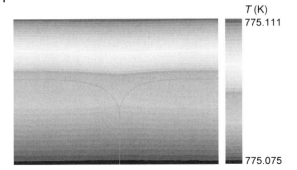

Fig. 5.6 Numerically determined temperature distribution at a grain boundary groove in an Al–Cu–Ag alloy.

a suitable program code was developed that quantifies the undercooling along a grain boundary groove in order to calculate the Gibbs–Thomson coefficient.

5.3.4
Determining the Interface Energy

On the basis of the two-dimensional geometry of the grain boundary groove obtained by the coordinate transformation (Section 3.2) the Gibbs–Thomson equation at any point of the curve of the grain boundary groove could be expressed as follows:

$$\Delta T_r = \frac{\Gamma}{r} \tag{5.7}$$

where r is the radius of the curvature at this point. Since measurement errors could lead to large inaccuracies in the determination of the curvature, the Gibbs–Thomson equation was not evaluated directly but in an integral form as shown below:

$$\int_{y_1}^{y_n} \Delta T_r dy = \Gamma \int_{y_1}^{y_n} \frac{1}{r} dy \tag{5.8}$$

The left hand side of the equation was evaluated numerically, determining the appropriate undercooling, ΔT_r, at a point, y_n, in the simulated temperature field.

$$\int_{y_1}^{y_n} \Delta T_r dy \approx \sum_{i=1}^{n-1} (y_i - y_{i+1}) \left(\frac{\Delta T_i + \Delta T_{i+1}}{2} \right) \tag{5.9}$$

The right-hand side of Equation 5.8 may be evaluated for any shape by setting the length element $ds = r\, d\theta$, where s is the distance along the interface and θ is the angle of a tangent to the interface with the y axis (Figure 5.7). Hence, $dy = \cos(\theta)\, ds = \cos(\theta) r\, d\theta$ can be substituted on the right-hand side of Equation 5.8

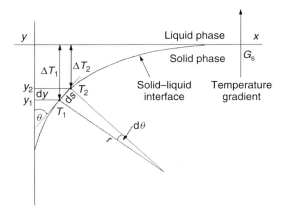

Fig. 5.7 Temperature difference at two different points of the grain boundary groove and description of ds, $d\theta$, and r.

which is as follows for an arbitrary surface:

$$\Gamma \int_{y_1}^{y_n} \frac{1}{r} dy = \Gamma \int_{\theta_1}^{\theta_n} \frac{1}{r} r \cos(\theta) d\theta = \Gamma(1-\sin\theta)\Big|_{\theta_1}^{\theta_n} = \Gamma(\sin\theta_1 - \sin\theta_n) \qquad (5.10)$$

This allows the Gibbs–Thomson coefficient to be calculated by numerically evaluating the right-hand side of Equation 5.8 using the undercooling temperatures from temperature field simulations and measuring the angle θ by constructing a tangent to the surface at y_n. The solid–liquid interface energy is obtained from the definition of the Gibbs–Thomson coefficient:

$$\Gamma = \frac{\sigma_{SL}}{\Delta S^*} \qquad (5.11)$$

ΔS^* is the specific entropy change per unit volume at transformation temperature, which must be known or obtained from other sources. In this study, the entropy has been determined by Thermo-Calc using the database described by Witusiewicz et al. [19].

5.4
Results and Discussion

The described procedure was tested to reproduce previous results obtained by Gündüz [11, 12] and Maraşli [14, 15] for a binary Al–Cu alloy with a eutectic composition. Afterwards, the solid–liquid interface energy was measured for the first time for an alloy with an invariant eutectic composition in the ternary system Al–Cu–Ag. The used alloy had a eutectic composition of 16.86 wt% Cu and 39.97 wt% Ag at a temperature of 502 °C. In order to describe the influence of the concentration on the solid–liquid interface energy, further experiments were

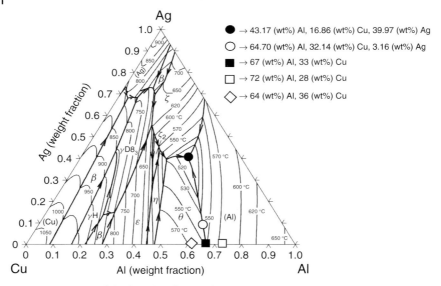

Fig. 5.8 Projection of the liquid surfaces with isothermals plotted in the different phase regions of the Al–Cu–Ag system, with the examined alloys.

carried out for binary Al–Cu alloys adjacent to the Al–Cu alloy with eutectic composition using samples with 28 and 36 wt% Cu, respectively. Furthermore, the solid–liquid interface energy was calculated for an Al–Cu–Ag alloy with a monovariant composition. In Figure 5.8 the examined alloys are displayed.

5.4.1
Al–Cu System with Eutectic Composition

Figure 5.9 shows grain boundary groove shapes of the Al$_{(\alpha)}$ and CuAl$_2$ phase in equilibrium with the quenched liquid and the temperature field simulation of these grains (Figure 5.9a and b). The samples were left in the radial heat flow apparatus for two days until the grain boundary grooves were equilibrated. After quenching, the samples were metallographically prepared and analyzed. An average value of the Gibbs–Thomson coefficient for the solid Al$_{(\alpha)}$–liquid Al–Cu system was found to be $\Gamma = 28 \pm 7 \times 10^{-8}$ K m. With a specific entropy change per unit volume of $\Delta S^* = 6.78 \times 10^5$ J m^{-3} K [14] a solid–liquid interface energy of $\sigma_{SL} = 190 \pm 48$ mJ m^{-2} was calculated. The relative error for the solid–liquid interface energy was about 25%. The small amount of abrasion in this experiment ($d = 6.9$ μm) has led to large inaccuracies for the geometrical transformations and hence to the curvature of the grain boundary groove shapes as well as the simulated temperatures.

The evaluation of the solid CuAl$_2$–liquid Al–Cu phase gave an average value of $\Gamma = 5.0 \pm 0.9 \times 10^{-8}$ K m for the Gibbs–Thomson coefficient. The solid–liquid interface energy, σ_{SL}, was found to be 80 ± 15 mJ m^{-2} with a specific entropy

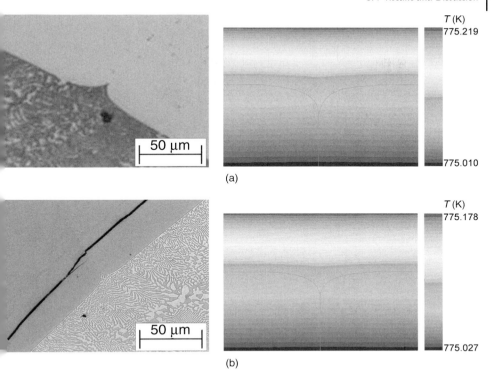

Fig. 5.9 Grain boundary groove shapes of the phases in equilibrium with the quenched liquid. (a) Al$_{(\alpha)}$ phase and (b) CuAl$_2$ phase in equilibrium with the quenched liquid.

change per unit volume of $\Delta S^* = 15.95 \times 10^5$ J m^{-3} K [14]. A relative error of 18% for the solid–liquid interface energy was calculated. The comparison of the average value of the solid–liquid interface energy of the Al$_{(\alpha)}$ and CuAl$_2$ phases shows that within the margin of error, the values of σ_{SL} were in a good accordance with previous results of Gündüz [11, 12] and Maraşli [14, 15].

5.4.2
Al–Cu–Ag System with Invariant Eutectic Composition

Further experiments were carried out to determine the solid–liquid interface energy in the ternary Al–Cu–Ag system. The samples were left in the radial heat flow apparatus for 4 days maintaining a constant temperature gradient until the grain boundary grooves were in local equilibrium (Figure 5.10). The Gibbs–Thomson coefficient was calculated for each phase. For numerical determination of the local undercooling, the thermal conductivity of the solid phase was determined experimentally for each phase by means of the laser-flash method [20–22] and differential scanning calorimetry (DSC). The thermal conductivity of the liquid phase was

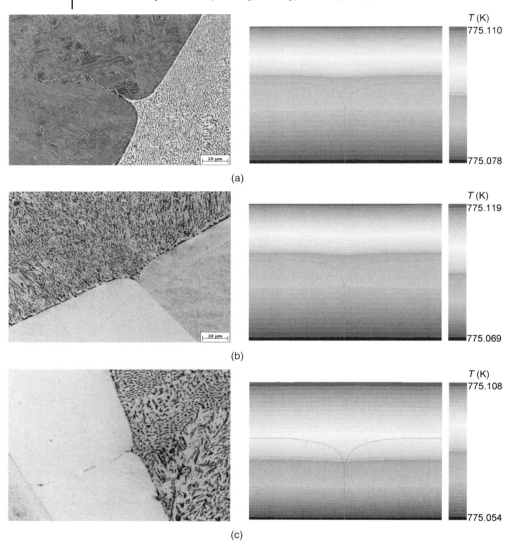

Fig. 5.10 Grain boundary groove shapes of the different phases in equilibrium with the quenched liquid. (a) $Al_{(\alpha)}$ in equilibrium with the quenched liquid; (b) $CuAl_2$ in equilibrium with the quenched liquid; (c) Ag_2Al in equilibrium with the quenched liquid.

determined experimentally by means of a unidirectional growth apparatus that is a Bridgman furnace. The average value of the Gibbs–Thomson coefficient for the solid $Al_{(\alpha)}$–liquid Al–Cu–Ag system was found to be $\Gamma = 4.8 \pm 0.7 \times 10^{-8}$ K m. With a specific entropy change per unit volume of $\Delta S^* = 10.75612 \times 10^5$ J m^{-3} K, the solid–liquid interface energy, σ_{SL}, was found to be 51 ± 8 mJ m^{-2}. The relative error was about 15%.

For the solid CuAl$_2$–liquid Al–Cu–Ag system, an average value of the Gibbs–Thomson coefficient of $\Gamma = 5.9 \pm 0.9 \times 10^{-8}$ K m was determined. With a specific entropy change per unit volume of $\Delta S^* = 18.88085 \times 10^5$ J m^{-3} K, the solid–liquid interface energy, σ_{SL}, was found to be 111 ± 17 mJ m^{-2}. The relative error was about 15%.

For the solid Ag$_2$Al–liquid Al–Cu–Ag system, the average value of the Gibbs–Thomson coefficient was found to be $\Gamma = 5 \pm 0.9 \times 10^{-8}$ K m. The solid–liquid interface energy, σ_{SL}, was found to be 59 ± 11 mJ m^{-2}. A specific entropy change per unit volume of $\Delta S^* = 11.70052 \times 10^5$ J m^{-3} K was calculated. The relative error was about 18%. Comparing the results of the solid–liquid interface energy for the Al$_{(\alpha)}$ phase shows that in the ternary system the value is about 2.5 times smaller than in the binary system, whereas the value for the CuAl$_2$ phase is in the same range.

5.4.3
Concentration Dependence of σ_{SL}

The radial heat flow apparatus was used to measure the solid–liquid interface energy in the binary Al–Cu system with 28 and 36 wt% Cu. In Table 5.1 the average values of the Gibbs–Thomson coefficient and the solid–liquid interface energy for the different phases of the alloys investigated are displayed.

For the sample Al–Cu28, an average value of the solid–liquid interface energy of 117 ± 14 mJ m^{-2} for the Al$_{(\alpha)}$ phase and 83 ± 12 mJ m^{-2} for CuAl$_2$ phase was calculated. For the sample Al–Cu36, the solid–liquid interface energy was found to be 216 ± 41 mJ m^{-2} for the Al$_{(\alpha)}$ phase and 214 ± 13 mJ m^{-2} for CuAl$_2$ phase. Comparing the results of the solid–liquid interface energy for the CuAl$_2$ phase shows that in the Al–Cu system with 28 wt% Cu the value was in good accordance with the results in the Al–Cu system with eutectic composition, whereas in the Al–Cu system with 36 wt% Cu the value was 2.5 times higher. For the determination of the concentration dependency of the solid–liquid interface energy in the Al–Cu–Ag system, a sample with 32.14 wt% Cu and 3.16 wt% Ag was used (Figure 5.8). For the Al$_{(\alpha)}$ phase an average value of the solid–liquid interface

Table 5.1 Results of the Gibbs–Thomson coefficient, Γ, and the solid–liquid interface energy, σ_{SL}, (a) Al—28 wt% Cu; (b) Al—36 wt% Cu; (c) Al—32.14 wt% Cu—3.16 wt% Ag.

Solid phase	$\Gamma \times 10^{-8}$ (K m)	σ_{SL} (mJ m^{-2})
(a) Alpha Al—7.68 wt% Cu	17 ± 2	117 ± 14
(a) Theta Al—55.09 wt% Cu	5.2 ± 0.8	83 ± 12
(b) Alpha Al—7.89 wt% Cu	32 ± 6	216 ± 41
(b) Theta Al—55.13 wt% Cu	13 ± 2	214 ± 13
(c) Alpha Al—7.87 wt% Cu—4.96 wt% Ag	14 ± 3	92 ± 17
(c) Theta Al—56.56 wt% Cu–0.29 wt% Ag	4.5 ± 0.9	84 ± 17

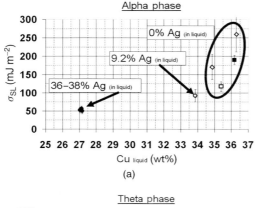

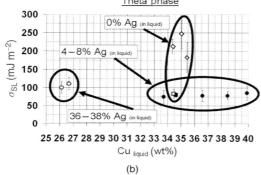

Fig. 5.11 Solid–liquid interface energy as a function of the copper concentration. (a) Alpha phase; (b) theta phase.

energy of 92 ± 17 mJ m^{-2} was calculated and for the CuAl$_2$ phase the solid–liquid interface energy was found to be 84 ± 17 mJ m^{-2}.

In Figure 5.11 the average values of the solid–liquid interface energy of the Al$_{(\alpha)}$ phase and the CuAl$_2$ phase with respect to the Cu concentration in the liquid phase are displayed. The comparison of the average value of the solid–liquid interface energy of the Al$_{(\alpha)}$ phase versus the concentration of Cu in the liquid phase shows that, within the margin of error, the solid–liquid interface energy increases with the concentration of Cu and decreases with increasing Ag concentration. A comparison of the average value of the solid–liquid interface energy of the CuAl$_2$ phase shows no dependency between the Cu concentration and the solid–liquid interface energy.

5.5
Summary and Conclusions

The radial heat flow apparatus in combination with the "grain boundary groove in an applied temperature gradient" method can be applied to measure the

Gibbs–Thomson coefficient, Γ, and the solid–liquid interface energy, σ_{SL}, for grain boundary grooves in alloys where the groove shapes can be investigated after quenching. In this method, the local curvature of the grain boundary grooves and the local undercooling by heat flux simulations must be determined using the Gibbs–Thomson equation. For the simulation of the local undercooling, the temperature gradients in the liquid and the solid phases must be known in addition to the groove shape. The accuracy of the determined solid–liquid interface energy depends on many factors such as the temperature gradient, G_S, the processing time, the purity of the ingot material, the accuracy of the material-specific thermophysical data such as the thermal conductivity and the specific entropy change per unit volume.

Further sources of errors could be from the preparation of the samples, coordinate transformation, geometrical evaluation, and temperature field simulation. Owing to the segregation-induced variation of the concentrations along the height of each sample, grain boundary grooves of the $Al_{(\alpha)}$, $CuAl_2$, and Ag_2Al phase in equilibrium with the liquid phase could be observed. Finally, the average values of the solid–liquid interface energy (millijoule mJ/per square m²eter) and the concentration of Cu (wt%) in the liquid phase were compared for the individual samples. From the obtained results it could be concluded that, within the margin of error, for the $Al_{(\alpha)}$ phase the solid–liquid interface energy decreases with increasing Ag concentration, whereas for the $CuAl_2$ phase no dependency of the solid–liquid interface energy on the Cu concentration could be observed for the investigated range of alloy composition.

Acknowledgments

This work was supported by the DFG (German Research Foundation) within the frame of the SPP1120 "Phase Transformations in Multicomponent Melts". The authors are grateful for the financial support.

References

1 Jones, D.R.H. and Chadwick, G.A. (**1971**) *Journal of Crystal Growth*, **11**, 260.
2 Jones, D.R.H. (**1978**) *Philosophical Magazine*, **27**, 569.
3 Schaefer, R.J., Glicksman, M.E. and Ayers, J.D. (**1975**) *Philosophical Magazine*, **32**, 725.
4 Hardy, S.C. (**1977**) *Philosophical Magazine*, **35**, 471.
5 Nash, G.E. and Glicksman, M.E. (**1977**) *Philosophical Magazine*, **24**, 577.
6 Bolling, G.F. and Tiller, W.A. (**1960**) *Journal of Applied Physics*, **31** (8), 1345.
7 Singh, N.B. and Glicksman, M.E. (**1989**) *Journal of Crystal Growth*, **98**, 573.
8 Bayender, B., Maraşli, N., Cadirli, E., Sisman, H. and Gündüz, M. (**1998**) *Journal of Crystal Growth*, **194** (1), 119.
9 Bayender, B., Maraşli, N., Cadirli, E. and Gündüz, M. (**1999**) *Materials Science And Engineering A*, **270**, 343–48.

10 Maraşli, N., Keşlioğlu, K. and Arslan, B. (2003) "Solid-liquid interface energies in the succinonitrile and succinonitrile-carbon tetrabromide eutectic system". *Journal of Crystal Growth*, **247**, 613–22.

11 Gündüz, M. (1984) *The Measurement of Solid-Liquid Surface Energy*, Ph.D. thesis, University of Oxford.

12 Gündüz, M. and Hunt, J.D. (1985) "The measurement of solid-liquid surface energy in the Al-Cu, Al-Si and Pb-Sn systems". *Acta Metallurgica*, **33** (9), 1651–72.

13 Gündüz, M. and Hunt, J.D. (1989) "Solid-liquid surface energy in the Al-Mg system". *Acta Metallurgica*, **37** (7), 1839.

14 Maraşli, N. (1994) *The Measurement of Solid-Liquid Surface Energy*, Ph.D. thesis, University of Oxford.

15 Maraşli, N. and Hunt, J.D. (1996) "Solid-liquid surface energies in the Al-CuAl$_2$, Al-NiAl$_3$ and Al-Ti systems". *Acta Materialia*, **44**, S1085.

16 Keşlioğlu, K. (2002) The Measurement of Solid-Liquid Surface Energy, Ph.D. Thesis, Erciyes University.

17 Keşlioğlu, K. and Maraşli, N. (2004) "Solid-liquid interfacial energy of the eutectoid b phase in the Al-Zn eutectic system". *Material Science and Engineering A*, **369**, 294–301.

18 Erol, M., Maraşli, N., Keşlioğlu, K. and Gündüz, M. (2004) "Solid-liquid interfacial energy of bismuth in the Bi-Cd eutectic system". *Scripta Materialia*, **51**, 131–36.

19 Witusiewicz, V.T., Hecht, U., Fries, S.G. and Rex, S. (2005) "The Ag-Al-Cu system II. A thermodynamic evaluation of the ternary system". *Journal of Alloys and Compounds*, **387**, 217–27.

20 Parker, W.J., Jenkins, R.J., Butler, C.P. and Abbott, G.L. (1961) "Flash method for determining thermal diffusivity". *Journal of Applied Physics*, **32**, 1679.

21 Bräuer, G., Dusza, L. and Schulz, B. (1992) "The new laser flash equipment LFA-427". *Interceram*, **41**, 7.

22 Dusza, L. (1996) *Wärmetransport–Modelle zur Bestimmung der Temperaturleitfähigkeit von Werkstoffen mit der instationären Laser Flash Methode*, Wissenschaftlicher Bericht FZKA 5820, Forschungszentrum Karlsruhe GmbH.

6
Phase Equilibria of Nanoscale Metals and Alloys

Gerhard Wilde, Peter Bunzel, Harald Rösner and Jörg Weissmüller

6.1
Introduction

Properties of materials are often modified for spatially confined or finite-size systems [1, 2]. Depending on the type of property, this behavior is explained by the crossing of length scales when characteristic interaction lengths or wavelengths become comparable with the system size. This type of argument is usually invoked for explaining the well-known size dependence of, for example optical—or magnetic properties of nanostructured materials. In these cases, the size of the nanoscale structural unit (the nanoparticle or the nanocrystalline grain) becomes equal to or smaller than a characteristic correlation length. Concerning ferromagnetism, the ferromagnetic correlation length $L_0 = \sqrt{A/K_1}$, with the exchange interaction constant, A, and the local magnetocrystalline anisotropy, K_1 [3] becomes similar to or even smaller than the average diameter of the particles or grains if size effects become significant.

A second type of argument concerning the size dependence of properties is related to the presence of interfaces, or—more specifically—the presence of a large fraction of the atoms of the system at or near a surface or an internal interface. In addition, and as shown here, the atomistic details of these interfaces matter as well [2]. Traditionally, the impact of the internal or external interfaces has been implemented into the description of interface-controlled property modifications by describing the interface and the core of the particles or grains in either of two ways: as two separate phases with intrinsically different properties, or as microstructural elements of different dimensionality, where the contribution of the 2D interface component to any extensive property is described by an excess contribution per area. In the simplest case, such models result in a rule of mixture-type behavior. One aspect of this behavior considers that the atoms situated at or near such an interface are energetically in a different state compared to the atoms in the core of a crystallite or a nanoparticle. Transport properties are current examples for property modifications that are discussed by two-phase descriptions. Here again, long-range interactions lead to qualitatively different scenarios. For instance, when

Phase Transformations in Multicomponent Melts. Edited by D. M. Herlach
Copyright © 2008 WILEY-VCH Verlag GmbH & Co. KGaA, Weinheim
ISBN: 978-3-527-31994-7

stress is important for the solubility—as for instance in interstitial alloys—the compositions of interfaces *and* bulk couple via elastic interaction, with significant consequences for the phase equilibrium [4]. This has been found to be relevant in the context of hydrogen storage.

Similar approaches also apply for describing reversible phase transformations between thermodynamically stable phases, which are often modified for spatially confined or finite-size systems. Size-dependent melting phenomena are well-known examples for this aspect of nanostructured materials [5]. In this context, it is still a matter of controversy whether the change in the equilibrium *thermodynamics* of the interface (or of an interface "phase") accounts for the observed behavior or whether the size-dependent differences of the transformation *kinetics* are responsible.

6.2
Phase Stability and Phase Transformations in Nanoscale Systems

6.2.1
Single-Phase Material: External Interfaces

It is one of the earliest findings concerning finite-size effects on materials properties that a decreasing diameter of a particle leads to a shift in the melting temperature, T_m [6]. When the size of a particle is reduced, the excess free energy, the product of the surface area A and of an interfacial free energy density γ, diminishes more slowly than the free energies of the bulk phases and capillary effects will therefore increasingly affect the thermodynamic equilibrium. The atomistic details of this phenomenon are not entirely understood, but in phenomenological theory, the size dependence of T_m of elemental solids is mostly described as a consequence of the change $\Delta\gamma$ in the interfacial free energy upon melting, in other words a free energy change $A\Delta\gamma$ [7, 8].

Formally, it is always possible to decompose the actual values of the thermodynamic potentials per volume of the particle material, g, h, and s, (the free enthalpy, enthalpy, and entropy, respectively) into the values for the bulk material, g_0, h_0, and s_0, and the respective Gibbs excess quantity per area, A, of interface, $\{G\}$, $\{H\}$, and $\{S\}$ as given below:

$$Vg = Vg_0 + A\{G\} \quad (6.1)$$

When melting is treated as a first-order phase transition with $g = h - Ts$, this leads immediately to the following:

$$T_m = \frac{(\Delta h_{m,0} + \alpha \Delta \{H\}_m)}{(\Delta s_{m,0} + \alpha \Delta \{S\}_m)} \quad (6.2)$$

where α denotes the specific interface area, $\alpha = A/V$. Models that ignore the excess entropy, for instance by assuming temperature-independent excess free energies $\{G\}$, and which approximate h and s as temperature independent and the excess enthalpy as size independent, will therefore predict a linear dependence of T_m on α. Since α is inversely proportional to the particle size, this is equivalent to a dependency of the form

$$T_m = T_{m,0} + \frac{c}{D} \tag{6.3}$$

with c as constant. Experimental data for T_m as a function of D were repeatedly analyzed by this law and good agreement was generally found. Classically, this treatment as well as the experimental observations suggested that T_m would always shift to lower temperatures when D is reduced. However, there is yet no single, generally accepted model for the size dependence of T_m and the existing models yield different predictions for the variations of measurable quantities, such as the melting temperature and the melting enthalpy, ΔH_m, on the particle size. Rather recently, it has been realized that the situation is even more complex since an *increase* of T_m at decreased particle sizes is observed for matrix-encased particles [9] and for interlayer material [10] with specific crystallographic orientation relationships and "ordered" interfaces. In terms of model explanations for size-dependent melting it is found in general that models that can reproduce the size-dependent *decrease* of T_m at least mathematically utterly fail in describing the *increase* of T_m of the same material at identical particle sizes. Thus, there is still a discrepancy even concerning the qualitative description of the melting process at small system sizes. In view of the apparent disparities in the description of the melting behavior of nanoscaled particles, the melting behavior of Pb nanocrystals embedded in Al treated as a model system was analyzed in dependence of microstructure and morphology.

The immiscible system Al–Pb has been chosen here since nanocrystalline Pb dispersions prepared by rapid quenching techniques melt at an increased T_m whereas Pb nanocrystals prepared by high-energy ball-milling melt at a decreased T_m [11]. In addition, both components remain mutually immiscible even at nanometer grain/particle size, as confirmed by experiment and modeling [12]. Thus, this immiscible binary alloy system provides a model character to study the impact of the morphology of the surface (or interface) of particles with identical chemistry and identical size on the melting characteristics. The solid dispersions of Pb nanocrystals in a polycrystalline Al matrix were checked to be free from contaminations, especially with respect to the particle–matrix interfaces where electron energy loss spectroscopy (EELS) analysis confirmed the absence of any detectable impurity level. Up to 2 at% Fe was added deliberately in one set of ball-milling experiments, without any detectable effect concerning the melting behavior of the Pb nanoparticles. It was found that the Pb dispersoids display a spheroidal shape without facets after mechanical attrition and that a high density of lattice defects is present in the as-ball-milled material. Yet, the Pb nanocrystals retain a clear cube-on-cube orientation relationship with the polycrystalline Al matrix, as indicated by the Moiré fringes that are parallel for different particles within each Al grain (Figure 6.1a) [13].

Rapid quenching of an Al–Pb melt that has been homogenized in the liquid state at high temperatures also resulted in nanocrystalline Pb dispersions within

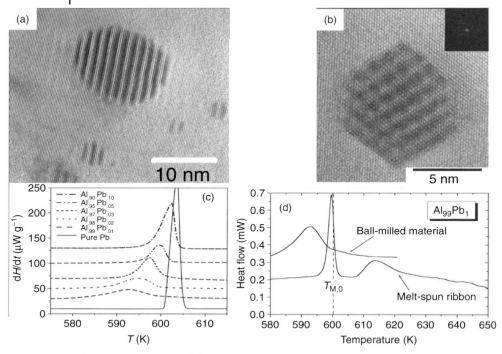

Fig. 6.1 (a) TEM bright field image of ball-milled $Al_{99}Pb_1$. (b) HRTEM image of a faceted Pb nanocrystal in melt-quenched $Al_{99}Pb_1$. The insert shows the selected area electron diffraction image indicating that pure Pb nanocrystals are embedded in a pure Al matrix. (c) Calorimetry results for the melting of ball-milled Al–Pb samples with different nominal compositions, corresponding to different average grain sizes of the Pb nanoparticles (reproduced from 41). (d) Calorimetry result for the melting of melt-quenched $Al_{99}Pb_1$. The maximum at temperatures above the melting point of bulk Pb is due to the smaller, faceted Pb nanoparticles whereas larger nanoparticles with curved interfaces that reside at grain boundaries of the matrix melt at temperatures slightly below the bulk melting temperature. The comparison with the curve obtained on ball-milled material of identical composition indicates the dependence of the melting behavior on the processing treatment.

a polycrystalline Al matrix but, the particle size distribution here is bimodal as confirmed by TEM transmission electron microscopy and SAXS small angle X-ray scattering and the interfaces of particles *within* Al grains are faceted (Figure 6.1b). Thus the different synthesis routes result in particles of identical chemistry and identical orientation relationship with the matrix and similar sizes if only one fraction of the melt-quenched material is considered. Yet, calorimetry experiments show that with decreasing Pb particle size, the position of the endothermic melting peak moves toward lower temperatures for the spheroidal particles of the ball-milled material (Figure 6.1c) whereas the small and faceted particles of the quenched composite melt at a temperature that is considerably higher than the melting temperature of the bulk material (Figure 6.1d) [11].

It is emphasized that both phenomena, the melting point depression in ball-milled Al–Pb as well as the enhanced T_m of the quenched material, are reproduced well during several subsequent melting/resolidification cycles on the same sample. This shows that the phenomena are intrinsic to a given microstructure, and it precludes the possible role of nonequilibrium states of particles or interfaces that arise from the different synthesis routes.

It is also remarkable that each of the two materials can be treated so as to acquire the behavior of the other. Annealing the ball-milled material led to partial faceting of the Pb particles with the evolution of a melting signal above the melting temperature of the bulk material (Figure 6.2a and b) [11, 13]. For the melt-quenched material, plastic deformation by high-pressure torsion straining resulted in the loss of faceting of the interfaces and, concomitantly, in a decrease of the melting temperature to values lower than the respective melting temperature of bulk Pb (Figure 6.2c and d) [11]. These results indicate strongly the importance of the structure of the matrix, and its effect on the interface, that can increase or decrease the melting temperature by several ten degrees.

In order to elucidate the structure of the Pb/Al interface and the difference between material that was synthesized by plastic deformation or rapid melt quenching, an HRTEM high resolution transmission electron microscopy study was conducted and the resulting images were analyzed by Fourier filtering and inverse Fourier retransform construction to remove the Moiré contrast that otherwise prevents the observation of the interface structure with atomic resolution [14]. In general, the shape of the faceted Pb particles was found to be cube-octahedral with larger {111}- facets and smaller {100}- facets and with a cube-on-cube orientation relationship with the Al matrix (Figure 6.3a), in agreement with previous studies in the literature [15, 16]. Fourier filtering was used to enhance the observability of possible defects at the Pb/Al interface: in the corresponding Fourier transform of Figure 6.3a, the Pb reflections were blocked. An identical procedure was applied separately to the Al reflections. The intensities of the Fourier-filtered and enhanced images were then added to obtain a reconstructed image. For easier observation of the misfit dislocations, an additional Fourier filtering of the entire image was applied in such a way that only the Fourier components near the Bragg reflections of Al and Pb were used for computing the inverse Fourier transform, so that noise is suppressed in the real-space image (Figure 6.3b). The various crystallographic directions, which are relevant to the discussion of the defect structure, can be identified by inspection of the appropriate stereographic projection along [011] [17]. For clarity, these are shown in the insert of Figure 6.3b. From Figure 6.3b one can readily identify localized dislocation cores at the (111) and (100) facets of the Al/Pb interface in the form of Al planes ending in a common Pb plane. On both facets, the misfit dislocations appear on every fifth {111} Al habit plane. Moreover, analyses by Burgers circuits at the different facets reveal that specific types of dislocations are present at the respective interfaces, that is full-edge dislocations at the (100) facets and dislocations with edge *and* screw component for the $(\bar{1}1\bar{1})$ facet. This dislocation structure accommodates the lattice mismatch almost exactly while maintaining the bulk lattice parameters of the two phases without long-range strain. It is worth noting that no defects are found within the Pb particle. For

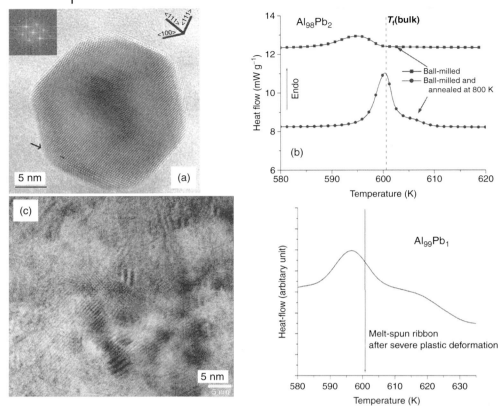

Fig. 6.2 (a) HRTEM image of a Pb nanoparticle in ball-milled material after high-temperature annealing (reproduced from Ref. 42). The insert indicates the electron diffraction pattern obtained by Fourier transformation of the HRTEM image. The arrow on the lower left side of the particle indicates the presence of a dislocation within the particle. (b) Calorimetry results on ball-milled material after high-temperature annealing (reproduced from Ref. 42). The shoulder at temperatures above the bulk melting temperature is indicated by an arrow. (c) HRTEM image of Pb inclusions in melt-quenched material after additional high-pressure torsion straining. (d) Calorimetry results on melt-quenched material after additional high-pressure torsion straining. The maximum at higher temperatures is almost completely removed.

ball-milled material, an identical analysis indicated that only one type of dislocation with mixed character—as present at the {111} facets—were observed.

The interrelation between melting/resolidification and two coupled structural aspects is now studied in more detail; that is, the interface structure and the orientation relationship. Therefore, the case of an "uncovered" Pb inclusion was considered. In contrast to earlier work [18], a Pb nanoparticle was melted under the influence of the electron beam and after resolidification, a nonequilibrium Al–Pb interface was formed at the thin edge of an electron-transparent foil (Figure 6.4a).

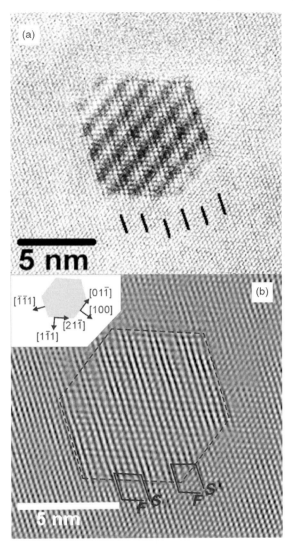

Fig. 6.3 (a) HRTEM image of a faceted Pb nanoparticle embedded in an Al matrix that is almost strain free in the region of the particle (reproduced from Ref. 14). The contrast within the particle results from the Moiré pattern of the overlapping Al and Pb lattices. (b) Image from (a) after Fourier filtering and inverse Fourier transformation without Moiré contrast (reproduced from Ref. 14). The various crystallographic directions are given as inset. Two Burgers circuits have been drawn at the $(1\bar{1}1)$ and (100) facets showing closing failures due to the presence of misfit dislocations. S refers to the starting point and F to the finishing point of the Burgers circuit. The closing failures determine the projected Burgers vectors as $\frac{a_0}{4}[21\bar{1}]$ and $\frac{a_0}{2}[01\bar{1}]$, respectively.

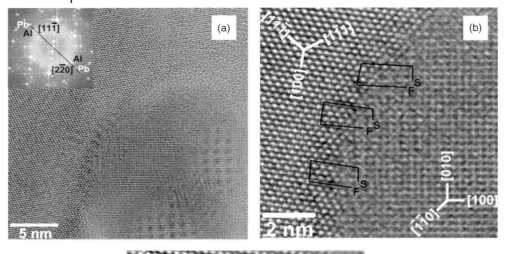
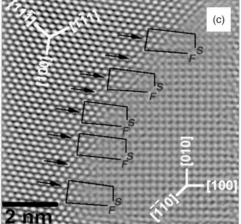

Fig. 6.4 (a) HRTEM micrograph (C_s-corrected): a Pb nanoparticle (right) was melted under the influence of the electron beam and has formed an artificial interface with the Al matrix (left) at the very thin edge of the TEM foil. The inset shows the corresponding Fourier transform revealing the orientation of the Al matrix as $[011]_{Al}$ and the Pb nanoparticle as $[001]_{Pb}$. The black line indicates the orientation relationship between the Pb nanoparticle and the Al matrix in the foil plane to be: $[\bar{1}10]_{Pb} \parallel [11\bar{1}]_{Al}$. Note the sharp bend of the interface. (b) Enlarged detail (raw image) of the Al–Pb interface shown in (a). The left side shows the Al matrix and the right side, a part of the Pb particle, respectively. The corresponding orientations are indicated schematically with respect to their lattice. Three Burgers circuits are drawn revealing $\frac{a_0}{2}[\bar{1}\bar{1}0]$ edge dislocations to accommodate the misfit. (c) Fourier-filtered micrograph of the Al–Pb interface shown in (b) using circular masks with a diameter of $0.54\,\text{nm}^{-1}$ around the Bragg reflections. The left side shows the Al matrix and the right side, a part of the Pb particle, respectively. The corresponding orientations are indicated schematically with respect to their lattice. The orientation relationship is determined to be: $[001]_{Pb} \parallel [011]_{Al}$ and $[\bar{1}10]_{Pb} \parallel [11\bar{1}]_{Al}$. The position of the misfit dislocations is indicated by arrows on the Al side. The Burgers circuits reveal $\frac{a_0}{2}[\bar{1}\bar{1}0]$ edge dislocations to accommodate the misfit.

This situation allows a direct comparison with the atomic structure of "classical" cube-on-cube-oriented Pb inclusions in Al [19].

For this purpose, TEM investigations have been carried out in an FEI Titan 80-300 (field-emission gun, super-twin lens, $C_s = 1.2$ mm) operated at 300 kV and equipped with a CEOS image C_s-corrector. The spherical aberration of the objective lens has been corrected down to the level of a few microns. Since the spherical aberration of the objective lens leads to a delocalization of the information, the compensation of the spherical aberration improves the image quality and enhances the reliability for determining the atomic positions in high-resolution TEM [20, 21].

Figure 6.4a shows a Pb nanoparticle at the very thin edge of the TEM foil that was melted under the influence of the electron beam and which subsequently formed a crystalline interface with the Al matrix during solidification. The corresponding Fourier transform, shown as inset in Figure 6.4a, allows determining the crystallographic orientation relationship between the Pb nanoparticle and the Al matrix. The fact that both lattices are oriented simultaneously in a zone axis reveals already that $[001]_{Pb}$ is parallel to $[011]_{Al}$. A second variant of the orientation relationship is indicated by a straight line in the inset of Figure 6.4a. The analysis shows that $[\bar{1}10]_{Pb}$ is parallel to $[11\bar{1}]_{Al}$. Vector algebra then yields for the third direction that $[1\bar{1}0]_{Pb}$ is parallel to $[\bar{2}1\bar{1}]_{Al}$. This orientation relationship corresponds to a misfit of 33.6% with respect to the Pb lattice and 25% with respect to the Al lattice, that is a significantly larger misfit as compared to the "cube-on-cube" orientation relationship that is commonly found.

The lattice distances of the individual phases have been measured in the Fourier transforms indicating lattice parameters of pure Al and Pb only. Therefore, the formation of new phases, such as oxides, can be excluded.

Figure 6.4b is an enlarged view of the Al–Pb interface shown in (a). Misfit dislocations are noticeable. For a better perception of the entire interface structure, the area shown in (b) has been Fourier-filtered using circular masks around the Bragg reflections. The result after the retransformation shown in Figure 6.4c presents an enhanced visibility of the interface structure. The mismatch is clearly localized in the form of interfacial dislocations that have been indicated by arrows on the Al matrix. Five Burgers circuits are drawn around misfit dislocations to determine their Burgers vectors with reference to the Pb lattice. The analysis reveals $\frac{a_0}{2}[1\bar{1}0]$ edge dislocations to accommodate the misfit.

Moreover, the shape of the Al–Pb interface makes a sharp bend at which the interface normal is changing abruptly from $[1\bar{1}0]$ (top part of the interface) toward $[100]$ with respect to the Pb lattice. However, the orientation relationship between Pb nanoparticle and Al matrix, for example $[\bar{1}10]_{Pb} \parallel [11\bar{1}]_{Al}$, as well as the character of the misfit dislocations always remain the same.

Based on the investigation of an "artificially" formed Al–Pb interface, we can conclude that the "classical" cube-on-cube orientation relationship of Pb inclusions in Al is energetically favorable since less misfit dislocations are needed to accommodate the misfit in that specific configuration. Thus the total excess energy contribution due to the dislocation arrangement is substantially larger for the current configuration as compared to the cube-on-cube orientation relationship.

Table 6.1 The energies $E(\theta)$ for (111), $\frac{1}{2}[\bar{1}10]$ dislocations in Al and Pb have been computed after reference data [39] using the elastic moduli C_{ik} according to published data [40]. The corresponding values are listed in the table

	C_{ik} (GPa)	$E(60°)$ $(10^{-9}$ J m$^{-1})$	$E(90°)$ $(10^{-9}$ J m$^{-1})$	$\dfrac{E(90°) - E(60°)}{E(90°)}$
Al	$C_{11} = 107$ $C_{12} = 60.8$ $C_{44} = 28.3$	1.4474	1.5901	0.0897
Pb	$C_{11} = 49.5$ $C_{12} = 42.3$ $C_{44} = 14.9$	0.5406	0.6026	0.103

Furthermore, the self-energies per unit length of straight dislocation lines have been calculated for two different orientation angles θ between line and Burgers vector. Table 6.1 shows that the energy of a $60°\frac{a_0}{2}\langle 110 \rangle$ dislocation, as found to accommodate the misfit for cube-on-cube-oriented Pb inclusions in Al, is about 10% lower compared to pure edge dislocations as found in this study. Thus, both factors, that is the number of misfit dislocations as well as their specific line energies, indicate that the cube-on-cube orientation relationship presents the energetically favorable configuration for Pb inclusions in Al.

Since the initial mobility during ball milling or during rapid melt quenching is rather high, the cube-on-cube orientation relationship that minimizes the excess energy per unit area of interface for Al–Pb is generally attained for this material, as observed experimentally [13, 22]. The presence of $\frac{a_0}{2}\langle 110 \rangle$ edge dislocations with reference to the Pb lattice in the present case shows that the minimization of the mismatch strain energy proceeds via the formation of localized misfit dislocations and not by attaining specific shapes that might accommodate the misfit elastically [23], even for situations that entail a high-energy configuration.

In addition, the present analysis provides valuable data concerning the observed shape dependence of the melting point shift of Pb nanoparticles that are embedded in Al. If we consider a spherical Pb particle as a close approximation of the shape of the Pb particles obtained after ball-milling, then the radius vector presents all possible directions of the interface normal, that is, the interface direction. Yet, that also entails that interface directions in all directions exist, including energetically unfavorable directions where—as for example in the present case—$[001]_{Pb} \parallel [011]_{Al}$ and $[\bar{1}10]_{Pb} \parallel [11\bar{1}]_{Al}$.

With the current result confirming that even in such high-energy configurations the misfit is localized in the form of interfacial dislocation, it can be concluded that the total excess energy of the interface of a spherical particle (or, more generally, of a particle with curved, noncrystallographic interfaces) is larger than the excess energy of a particle of similar volume but with faceted interfaces and a favorable orientation relationship. The observation that initially curved Pb particles start faceting at higher temperatures further confirms this interpretation [22]. Thus, taking the results on the melting behavior of particles with different morphology into account [13, 22, 24–28], the experimental results indicate that a melting point decrease of small particles is associated with interfaces that entail a high excess free energy. These interfaces are necessarily present with particles with a curved morphology that—in turn—is stabilized by matrix strains [18]. On the other hand, particles that melt at an elevated temperature require faceted interfaces that are oriented in favorable crystallographic orientations such that the interface excess contribution to the total free energy is minimal and even smaller than in the bulk case. Such morphologies are found in the case of an unstrained, relaxed matrix. Furthermore, in view of the role of dislocations as nucleation sites for melting [29], the observation of misfit dislocations at interfaces in a system which is known to exhibit a large increase of T_m at small size provides further support to the notion that the size-dependent shift of the experimental melting temperature is not a kinetic phenomenon, but rather an increase of the equilibrium temperature for melting due to the interface contributions to the total free energy balance. Yet, this argument only holds if the dislocations are not sessile. The mobility of such misfit dislocations is not completely understood and might also depend on the localization of the dislocation core [30], which also needs to be regarded while discussing the observed difference between quenched and deformed material. Thus, more detailed experimental *in situ* studies as well as dedicated molecular dynamics simulations are required to completely unveil the complex relationship between interface structure—especially the defect structure at the interface—and the phase stability and phase transformations. Yet, the present results demonstrate the importance of the atomistic details of the external particle–matrix interface for the stability and the properties of embedded nanoparticles or nanocomposites in general.

6.2.2
Binary Nanoalloys: Internal Heterophase Interfaces

In addition to the energetic contribution of the external surface, a qualitatively similar contribution arises due to the excess energy associated with internal heterophase interfaces in multiphase, multicomponent systems that are necessarily formed due to temperature- or composition-dependent variations of the relative amount of matter per phase. Therefore, the free energy balance must contain terms of the form $\gamma \Delta A$ (specific interface free energy density of the heterophase interface multiplied by the change in area of that interface), on top of the term $A \Delta \gamma$ that is dominant in elemental systems, as indicated in the previous section.

Much less work has been devoted to phase equilibria of nanoscale *alloys*, despite their importance for future nanotechnology devices, which will require the extra degrees of freedom in materials design provided by the use of alloys as opposed to elemental solids. Alloys differ from elemental materials in the fact that constitutional alloy phase diagrams exhibit intervals of temperature and composition in which two (or more) phases coexist at equilibrium. In the following text and without loss of generality, attention will be restricted to binary alloys, where at constant pressure, at the maximum, three phases can coexist in defined points of the phase diagram (zero degrees of freedom for three-phase coexistence, according to Gibbs' phase rule). The central questions in modeling size-dependent alloy phase diagrams are therefore as follows: can two phases coexist in a small particle and what are the conditions for equilibrium? As compared to elemental particles, this question raises a new issue related to the energetics of the internal interface separating the phases within the particle since varying the relative amount of matter in the phases requires the creation or removal of internal interface area. Although it is established that interfacial enrichment or depletion in solute (interfacial segregation) [31] and elastic interactions between the interfaces and the bulk (interface stress) [4] can significantly affect the relative stability of *single-phase* states in nanoscale alloys at constant interfacial area, the consequences of capillarity for the two-phase coexistence within a particle remain widely unexplored. Yet, the capillary energy of the interface between coexisting phases can lead to significant changes in the constitutional phase diagram, that is of the composition–temperature fields in which the different phases represent the thermodynamically stable state, and which may be observable even for sizes as large as 100 nm, that is well above the structure size of next-generation microelectronics devices. These changes are not mere shifts of temperatures or of compositions at equilibrium; instead, several qualitative rules, which are universally obeyed in conventional alloy phase diagrams of macroscopic systems, are no longer applicable at the nanometer scale [33].

In order to analyze the impact of this internal interface to the thermodynamic equilibrium at different particle sizes, an idealized particle embedded in a solid matrix is regarded where the matrix, as in an experiment, serves to prevent coarsening; this implies that the particle shape and consequently (when volume changes during the phase transition can be neglected) the particle–matrix interface area are fixed. The excess free energy due to the outer surface of the particle is then a constant, which can be ignored altogether since it does not affect the phase equilibrium. Thus, in the following equation the notion of an interfacial area A refers exclusively to the internal interfaces between coexisting phases.

The free energy per particle, G, can be related to the molar free energy, g_0, as shown below:

$$G(T, N, x) = N g_0(T, x) + \sum_i \gamma_i(T) A_i(T, N, x) \qquad (6.4)$$

where the subscript labels the possible interfaces. N is the total amount of matter, that is the sum of the amounts of solvent, N_1 and solute, N_2 and x denotes the solute fraction, $x = N_2/N$. At equilibrium, the A_i are not independent state variables, but internal thermodynamic parameters which are functions of T, N, x, determined by the Wulff construction [32]. Generally, the functional dependence of the A_i on N is not linear; this leads to the size dependence of the chemical potentials of single-phase particles embodied in Gibbs–Thompson–Freundlich-type equations.

In cases where the particle contains two phases, φ and β, (e.g. solid, S and liquid, L) that coexist, the Gibbs free energy, $\tilde{G}$, of the two-phase state with arbitrary compositions (that are not necessarily the compositions at equilibrium that minimize the free enthalpy) is given as follows:

$$\tilde{G} = (N^\varphi/N)\, G^\varphi + (1 - (N^\varphi/N))\, G^\beta \tag{6.5}$$

The phase fractions for the macroscopic case are then given by the lever rule.

However, the formation of a new phase necessarily entails changes of the area of interfaces and the creation of new interfaces. Thus the dependence of $\tilde{G}$ on the phase fraction will cease to be linear, contrary to Equation 6.5 and consequently the Gibbs free energy of *two-phase* states (in a nanoparticle) are expressed as [33]:

$$\tilde{G} = (N^\varphi/N)\, G^\varphi + (1 - (N^\varphi/N))\, G^\beta + \Delta G_c \tag{6.6}$$

The term ΔG_c represents the deviation from linearity and becomes equal to zero for single-phase particles. Between two single-phase states, a curved graph as indicated in Figure 6.5a must result. Thus the capillary term, $\partial A/\partial V^\beta$, in general removes

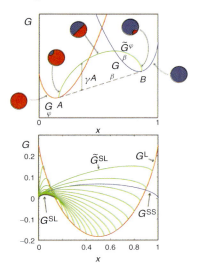

Fig. 6.5 (a) Schematic diagram of molar Gibbs free energies G^φ (red) and G^β (blue) versus solute fraction x for two phases φ and β (reproduced from Ref. 33). Construction of the Gibbs free energy curve for the two-phase coexistence of alloys represented by points A and B. Black dashed: macroscopic system; solid green: finite-size system. Inserts represent cross sections through the particle, illustrating the geometric arrangement of the phases (red, phase φ; blue, phase β) and the maximum of the interfacial area A for equal amounts of phase φ and β. (b) Example of the molar Gibbs free energies (in units of the enthalpy of melting) with various values for x_L used in the computation (reproduced from Ref. 33). Parameters are $T/T_m = 0.75$ and $D = 5$ nm. G^L represents the Gibbs free energy of the single-phase liquid state, G^{SS} represents the two-phase solid state, and $G^{SL}(T,x)$ is the lower envelope of the set of functions $\tilde{G}^{SL}(T, x, x_L)$.

the coincidence of the tangent lines at equilibrium, as indicated in Figure 6.5a. It should be noted that Equation 6.6 holds in general, also for the macroscopic case. However, when the capillary term is negligible, as for macroscopic bulk systems, the condition that the tangents coincide at two-phase coexistence holds with good approximation. The compositions of the coexisting phases in the bulk case are then given by the points of tangency of the common tangent to the free energy functions G^φ and G^β; at each temperature these compositions are constants in the two-phase state independent of the overall composition and the tangent line represents the total free energy.

For small particles, however, the energetic contribution of the internal interface becomes significant and leads to a convex free enthalpy curve as indicated in Figure 6.5a. Different geometric assemblies of the two phases are possible, depending on the relative values of the γ_i, but in general, ΔG_c is a nonlinear function of the phase fraction. The most important consequence of the loss of linearity is that the tangent rule ceases to apply. Instead, the compositions of the two coexisting phases at *equilibrium* are not *a priori* known and is determined by energy minimization: the entire set of functions $\tilde{G}^{SL}$ with the solute fraction of the liquid as a parameter needs to be calculated (Figure 6.5b). The set of curves for $\tilde{G}^{SL}$ also indicates that the composition of the coexisting phases cannot be read from the phase boundaries as in the case of the bulk material. The respective stable states are given by the lower enveloping curve of the Gibbs free energy of all possible phase states and the transition between different phase states that define the minimum of the total Gibbs free energy marks the boundaries of the stability ranges of the different phases.

For the bulk case, the well-known phase diagrams result from an equivalent treatment that minimizes the total Gibbs free energy. However, it is important to note that for nanoscale systems not only is the topology of the phase boundaries changed, but also the way in which the resulting "phase diagrams" are to be used is completely different: the compositions of the coexisting phases are no longer invariant upon isothermal variations of the solute fraction and the composition of the majority phase is no longer continuous across phase boundary lines. Thus, at first sight, a more appropriate term for the resulting phase diagrams would be "*stability*" diagrams since the properties conventionally associated with phase diagrams are no longer applicable for nanoscaled alloys. It should be emphasized however, that these properties merely result from the applicability of linear approximations in the macroscopic world—in principle the results derived here for nanoscale systems are generally applicable—they just become significant at small system size and—this is important to note—the calculated diagrams as well as the construction rules extrapolate to the accepted behavior for large system sizes.

The numerical computation of phase diagrams (or stability diagrams) for different particle sizes requires assumptions on the functional form of the equations of state. A simple case is given by an alloy with no solid solubility and an ideal liquid solution, which for the bulk case results in a simple eutectic-phase diagram that is symmetric concerning the equi-atomic composition. In a reduced representation, the three materials constants that need to be specified (the atomic volume, melting entropy, and interfacial free energy in scaled representations) are similar for most metals. Details of the computation as well as concerning the analytical model are given in [33]. The resulting phase diagrams for the bulk system and for two alloy particles with different sizes are summarized in Figure 6.6. It is seen that, as the particle size is reduced, the phase diagram undergoes several qualitative changes, each of which breaks one of the rules that apply universally to the construction of the phase diagram for macroscopic systems. First, it is observed that the invariance of the solidus temperature is lost in favor of a significant composition dependence. Secondly, as illustrated by the colored lines representing states of identical composition x_L of the liquid phase at equilibrium, the compositions of the constituent phases in two-phase equilibria are no longer invariant at constant temperature. Thirdly, the equi-composition lines lose their continuity at the intersection with the liquidus line. This implies that there is a discrete jump in liquid fraction across the liquidus of the small alloy particles, consistent with the result of numerical modeling matched to Sn–Bi nanoparticles [34], where the ends of the tie lines were found to detach from the phase boundary lines. It should be stressed that relaxing the stringent boundary conditions that were used for constructing a simple model system does not qualitatively change the resulting phase equilibria. In fact, recent calculations based on a model eutectic with finite solubility of the terminal phases have shown that the stability fields of the different one- and two-phase states are shifted and that the two-phase solid–liquid stability fields are detached from the terminal phases [35] as also found by numerical calculations [34].

However, the most fundamental consequence of the finite system size is a topological change in the phase diagram, the degeneration of the eutectic point of the macroscopic system into a line representing an interval of compositions Δx_d (defined in Figure 6.6c) for which the particle undergoes a discontinuous transition between the two-phase solid–solid state and the single-phase liquid state. In the macroscopic system, three phases can coexist at equilibrium at the eutectic point; by contrast, discontinuous melting in this model is a transition between a two-phase equilibrium (solid–solid) and a single-phase state, without three-phase equilibrium (clearly, three phases will coexist during melting, but this is a transient, nonequilibrium configuration). It is because of this loss of three-phase equilibrium in the finite-size system that the transition from a eutectic point to a discontinuous melting line can be reconciled with the phase rule. In fact, recent experimental studies of isothermal composition variation within the electron microscope [36] as well as calorimetric investigations on a Bi–Cd eutectic that closely resembles the assumptions of the simple model eutectic [37, 38] are well consistent with the model results.

In order to verify the theory, experiments were performed on a series of $Al_{98}(Bi_x–Cd_{1-x})_2$ alloys synthesized via melt spinning. The Bi–Cd system presents similar conditions concerning the constitutive behavior as assumed in our simplified theoretical model system, especially concerning the negligible mutual solubility of Bi and Cd in the solid state and the negligible solubility of both components

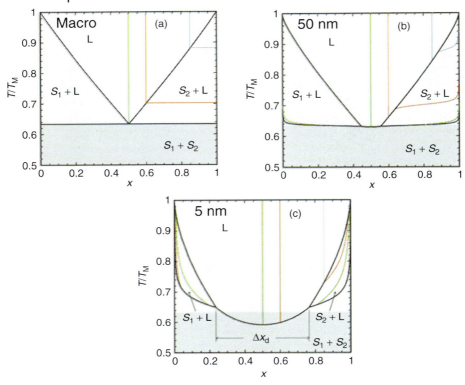

Fig. 6.6 (a) Phase diagram of a bulk eutectic alloy with no solid solubility (reproduced from Ref. [33]). Black: phase coexistence lines; colored: lines of equal solute fraction x_L in the liquid phase for three arbitrarily chosen values of x_L. (b) and (c) (reproduced from Ref. [33]) as in (a), but finite-size systems with particle diameters $D = 50$ and 5 nm, respectively. Δx_d: discontinuous melting interval where a direct transition from a two-phase solid to a single-phase liquid without three-phase coexistence occurs. Capital letters indicate the phases that are stable in the respective regions of temperature/composition space. $S_{1,2}$, solid phases; L, liquid. The gray shades represent the topologic features of the bulk phase diagram for easier comparison.

in solid Al. Thus, eutectic Bi–Cd nanoparticles embedded in an Al matrix were obtained after rapid quenching, as indicated in the TEM bright field image in Figure 6.7a. However, the melt spinning process resulted in a bimodal size distribution of the Bi–Cd particles with larger particles located at grain boundaries of the Al matrix and small particles within the Al grains, as already observed for the Al–Pb alloys. *In situ* melting experiments within the TEM have served to associate the calorimetric melting signals with the respective particle fractions. Thus the melting signal of the different size fractions could be deconvoluted. Quantitative analyses of the size distribution from TEM bright field images have shown that the average sizes of the smaller and the larger particles do not depend on the alloy composition (Figure 6.7b). Figure 6.8a shows the experimental results of calorimetric melting

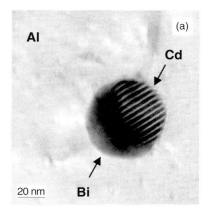

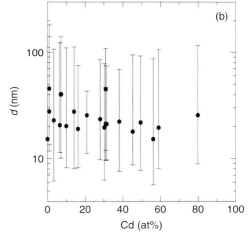

Fig. 6.7 (a) TEM bright field image of a Bi–Cd particle embedded in an Al grain (reproduced from Ref. [38]). The particle belongs to the "smaller" fraction that shows pronounced size dependence of the melting behavior. The two differently appearing parts of the particle (i.e. with or without Moiré effect) are due to the eutectic nature of the particle with one side consisting of Bi and the other of Cd. (b) Average particle diameter as measured from TEM bright field images (reproduced from Ref. [37]). The particle sizes refer to the "smaller" particles that are located within the Al grains. The error bars indicate the 95% Confidence Interval range. The results indicate clearly that the average size of the particles is independent of the alloy composition.

experiments on a series of $Al_{98}(Bi_x-Cd_{1-x})_2$ alloys with $x \in [0, 1]$. The two different peaks that are labeled as peak 1 and peak 2, respectively, refer to the melting signals of the two size fractions. It is clear from the calorimetric results in conjunction with the *in situ* TEM melting experiments, that peak 2 is associated with the melting process of the smaller particles that are located within the Al grains. This peak shows an onset temperature of about 3 °C above the signal maximum due to the larger particles (peak 1) at small Cd concentration but decreases significantly with

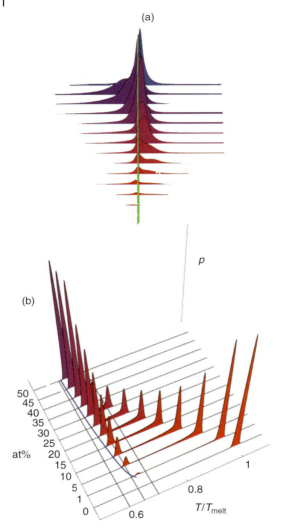

Fig. 6.8 (a) Calorimetric results showing the onset of melting of a series of $Al_{98}(Bi_x-Cd_{1-x})_2$ alloys (reproduced from Ref. [37]). P denotes the calorimetric signal in (watt per gram). The two peaks that are indicated in the figure refer to the fraction of larger (peak 1) and smaller (peak 2) particles. The results indicate that the onset of melting (the "eutectic" temperature) for the smaller particles depends on the alloy composition and is no longer constant as it is for the larger particles or the bulk alloy. (b) Calculated calorimetry curves based on the theoretical model eutectic for a particle diameter of 20 nm (i.e. similar to the average particle size observed experimentally) (reproduced from Ref. [37]). The blue curve outlines the onset of melting for the different alloy compositions. The variation of the melting onset is similar to the experimental results shown in (a).

increasing Cd concentration. At about 60 at% Cd, the onset of melting of the smaller particles is about 6 °C below the onset for the larger particles. Thus the total variation of the onset temperature is about 9 °C. Since the standard deviation for the onset of melting (determined on the same macroscopic alloy system) is only about ±0.8 °C, these results support the theory. In fact, calculating calorimetric curves for the theoretical model system for the same particle size as observed experimentally results in a similar variation of the onset temperature for melting in dependence of the alloy composition (Figure 6.8b). It should be emphasized that observing a dependence of the solidus temperature on the alloy composition is clear evidence for the importance of the interface contributions since for macroscopic systems and according to standard phase diagram construction, the solidus temperature in a eutectic system such as Bi–Cd would strictly remain constant.

6.3
Summary

The experimental results as well as the theory presented here corroborate the importance of internal and external heterophase interfaces for determining the phase equilibrium at any given size of the thermodynamic system and thus for predicting synthesis pathways, microstructures, and the stability of nanostructured multicomponent materials. Moreover, the nature, assembly, and relative amount of phases present in a material—together with the defect structure—generally determine the properties of any given material, including for example mechanical or magnetic properties. Thus, with functional nanostructures or functional nanocrystalline materials, size-dependent and interface-controlled phase equilibria should be considered in general and at the outset of any study, especially when functional properties are concerned.

Acknowledgments

The financial support of the research work by the Deutsche Forschungsgemeinschaft is most gratefully acknowledged. G.W. would like to acknowledge the collaboration with Prof. R. Valiev (Ufa, Russia) and stimulating discussions with Prof. H. Gleiter (Karlsruhe, Germany) on the topic of size-dependent melting.

References

1 Gleiter, H. (2000) *Acta Materialia*, **48**, 1.
2 Gleiter, H. (1995) *Nanostructured Materials*, **6**, 3.
3 Herzer, G. (1990) *IEEE Transactions on Magnetics*, **26**, 1397.
4 Weissmüller, J. and Lemier, C. (1999) *Physical Review Letters*, **82**, 213; (a) Lemier, C. and Weissmüller, J. (2007) *Acta Materialia*, **55**, 1241.
5 Cahn, R.W. (1986) *Nature*, **323**, 668.

6 Pawlow, P. (1909) *Zeitschrift Fur Physikalische Chemie*, **65**, 1.
7 Chattopadhyay, K. and Goswami, R. (1997) *Progress in Materials Science*, **42**, 287.
8 Zhang, D.L. and Cantor, B. (1991) *Acta Metallurgica et Materialia*, **39**, 1595.
9 Dahmen, U., Xiao, S.Q., Paciornik, S., Johnson, E. and Johansen, E.A. (1997) *Physical Review Letters*, **78**, 471.
10 Zhang, L., Jin, Z.H., Zhang, L.H., Sui, M.L. and Lu, K. (2000) *Physical Review Letters*, **85**, 1484.
11 Wilde, G., Bunzel, P., Rösner, H., Valiev, R.Z. and Weissmüller, J. (2004) in *Surfaces and Interfaces of Nanostructured Materials and Trends in LIGA, Miniaturization and Nanoscale Materials*, (eds S.M. Mukhopadhyay, S. Seal, N.B. Dahotre, A. Agarwal, J.E. Smugeresky and N. Moody), TMS (The Minerals, Metals & Materials Society), Warrendale, pp. 61–70.
12 Landa, A., Wynblatt, P., Johnson, E. and Dahmen, U. (2000) *Acta Materialia*, **48**, 2557.
13 Rösner, H., Scheer, P., Weissmüller, J. and Wilde, G. (2003) *Philosophical Magazine Letters*, **83**, 511.
14 Rösner, H., Weissmüller, J. and Wilde, G. (2004) *Philosophical Magazine Letters*, **84**, 673.
15 Thackery, P.A. and Nelson, R.S. (1969) *Philosophical Magazine*, **19**, 169.
16 Moore, K.I., Zhang, D.L. and Cantor, B. (1990) *Acta Metallurgica et Materialia*, **38**, 1327.
17 Rösner, H., Markmann, J. and Weissmüller, J. (2004) *Philosophical Magazine Letters*, **84**, 321.
18 Rösner, H., Weissmüller, J. and Wilde, G. (2006) *Philosophical Magazine Letters*, **86**, 623.
19 Rösner, H., Freitag, B. and Wilde, G. (2007) *Philosophical Magazine Letters*, **87**, 341.
20 Haider, M., Uhlemann, S., Schwan, E., Rose, H., Kabius, B. and Urban, K. (1998) *Nature*, **392**, 768.
21 Rose, H. (1971) *Optik*, **33**, 1.
22 Rösner, H. and Wilde, G. (2006) *Scripta Materialia*, **55**, 119.
23 Dahmen, U., Xiao, S.Q., Paciornik, S., Johnson, E. and Johansen, A. (1997) *Physical Review Letters*, **78**, 471.
24 Moore, K.I., Chattopadhyay, K. and Cantor, B. (1987) *Proceedings of the Royal Society of London*, A **414**, 499.
25 Moore, K.I., Zhang, D.L. and Cantor, B. (1990) *Acta Metallurgica et Materialia*, **38**, 1327.
26 Zhang, D.L. and Cantor, B. (1991) *Acta Metallurgica et Materialia*, **39**, 1595.
27 Sheng, H.W., Ren, G., Peng, L.M., Hu, Z.Q. and Lu, K. (1996) *Philosophical Magazine Letters*, **73**, 179.
28 Dahmen, U., Hagège, S., Faudot, F., Radetic, T. and Johnson, E. (2004) *Philosophical Magazine*, **84**, 2651.
29 Däges, J., Gleiter, H. and Perepezko, J.H. (1986) *Physics Letters A*, **119**, 79.
30 Srinivasan, S.G., Liao, X.Z., Baskes, M.I., McCabe, R.J., Zhao, Y.H. and Zhu, Y.T. (2005) *Physical Review Letters*, **94**, 125502.
31 Weissmüller, J. (1993) *Nanostructured Materials*, **3**, 261.
32 Wulff, G. (1901) *Acta Krystallographica*, **XXXIV**, 449.
33 Weissmüller, J., Bunzel, P. and Wilde, G. (2004) *Scripta Materialia*, **51**, 813.
34 Jesser, W.A., Shiflet, G.J., Allen, G.L. and Crawford, J.L. (1999) *Materials Research Innovations*, **2**, 211.
35 Gährken, M. and Wilde G., to be published.
36 Lee, J.G. and Mori, H. (2004) *Philosophical Magazine*, **84**, 2675.
37 Bunzel, P. (2004) PhD-Thesis, *Phasengleichgewichte in nanoskaligen binären Legierungspartikeln*, Saarland University.
38 Bunzel, P., Wilde, G., Rösner, H. and Weissmüller, J. (2004) in *In Solidification and Crystallization*, (ed. D.M. Herlach), John Wiley & Sons, Weinheim, pp. 157–65.
39 Barnett, D.M., Asaro, R.J., Gavazza, S.D., Bacon, D.J. and Scattergood, R.O. (1972) *Journal of Physics F: Metal Physics*, **2**, 854.

40 Hearmon, R.F.S. (1966) *Landolt-Börnstein*, Springer, Berlin, Vol. II, 1–39.
41 Ehrhardt, H., Weissmüller, J. and Wilde, G. (2001) *Materials Research Society Symposium Proceedings*, **634**, B8.6.
42 Wilde, G., Bunzel, P., Rösner, H. and Weissmüller, J. (2007) *Journal of Alloys and Compounds*, **434**, 286.

Part Two
Microscopic and Macroscopic Dynamics

7
Melt Structure and Atomic Diffusion in Multicomponent Metallic Melts

Dirk Holland-Moritz, Oliver Heinen, Suresh Mavila Chathoth, Anja Ines Pommrich, Sebastian Stüber, Thomas Voigtmann, and Andreas Meyer

7.1
Introduction

Processes on atomic scale determine the mass transport in liquids as well as the crystal growth and the microstructure formation during the solidification of melts. In order to gain a better understanding of such atomic-scale processes, profound information on the atomic structure and the dynamics in liquids is of fundamental importance. For instance, the short-range order of the melt is predicted to influence the energy of the interface between a solid nucleus and the liquid [1–3], while the atomic attachment kinetics at the solid–liquid interface is determined by dynamics on atomic length and time scales [3–5]. Solid–liquid interfacial energy and the diffusion coefficients enter into current models for crystal nucleation and crystal growth [3–5]. Consequently, the knowledge of these parameters is a necessary prerequisite for an understanding of solidification processes.

Atomic-scale processes are also of key importance to understand the glass transition. Here, the nucleation of crystalline phases during the cooling of a melt is avoided so that the melt freezes under formation of an amorphous solid. It was shown in previous investigations that by addition of further alloy components to binary glass-forming alloys, such as Ni–P and Zr–Ni, the melting temperature can be greatly decreased and the glass-forming ability can be significantly improved. The influence of such alloying on the atomic dynamics, however, is not well understood. We will show that for melts of glass-forming Ni–P-based alloys, the addition of further alloy components, such as Pd or Cu, to the binary Ni–P system has negligible impact on the Ni self-diffusivity [6, 7]. On the other hand, Zr–Ti–Cu–Ni–Be melts forming bulk metallic glasses exhibit considerably smaller values of the mean Ni and Ti self-diffusivity as that of Ni in Pd–Ni–Cu–P alloys at same temperatures. These differences may originate from the topological and the chemical short-range structure of the different melts.

Phase Transformations in Multicomponent Melts. Edited by D. M. Herlach
Copyright © 2008 WILEY-VCH Verlag GmbH & Co. KGaA, Weinheim
ISBN: 978-3-527-31994-7

A careful investigation of the short-range order of alloy melts requires the determination of partial pair correlation functions, a challenging task still next to impossible, for multicomponent alloy melts. As will be shown in this work, undercooled binary Zr–Ni melts exhibit similar Ni self-diffusivities as that of Zr–Ti–Cu–Ni–Be melts at same temperatures, suggesting similar mechanisms of atomic diffusion in the binary system as in the complex multicomponent alloys. Moreover, the activation energies for atomic self-diffusion are significantly larger compared to those for the self-diffusion in pure Ni [8] and Al–Ni alloys [9]. In order to obtain further insight into the interplay between short-range order and atomic dynamics, in this chapter we present a combined study of short-range order and atomic dynamics in binary $Zr_{64}Ni_{36}$ alloy melts. Partial structure factors have been determined by neutron scattering using the method of isotopic substitution. The experimental results are analyzed in the framework of mode coupling theory [10, 11], which provides a direct link between the static structure factor of a melt and its atomic dynamics.

From the experimental side, classical diffusion experiments in liquid metals, for example using the long capillary method [12], are hampered by buoyancy-driven convective flow and chemical reactions of the melt with the capillary. As discussed in the work by Griesche *et al.* [13], reliable diffusion data can be obtained if the influence of gravity-driven buoyancy convection is avoided by working under microgravity conditions or is monitored *in situ* by X-ray radiography for the special case of interdiffusion measurements in melts with a large X-ray contrast. Because of these reasons, experimental diffusion data in liquid alloys are rare, more so because the processing temperatures involved are larger. Quasi elastic neutron scattering probes sample dynamics on atomic time and length scales, and therefore the resulting data are not altered by convective flow. This enables us to derive self-diffusion coefficients on an absolute scale for liquids containing an incoherently scattering element [6, 7, 14].

Although special thin-walled SiC (Refs 14–16) and Al_2O_3 (Refs 7, 17, 18) sample crucibles were used, quasi elastic neutron scattering experiments have been limited so far to low melting and/or chemically fairly inert metallic systems and to temperatures slightly above the liquidus temperature. This is the reason why such scattering experiments in a crucible were feasible only for part of the experiments performed in this study. In order to give access to the study of refractory and chemically reactive melts, containerless processing techniques are required. We have developed a specially designed electromagnetic levitation facility in order to perform neutron scattering experiments on containerless processed samples. The absence of a sample holder makes it not only possible to extend the accessible temperature range to temperatures of up to 2300 K but also to the metastable regime of the undercooled liquid several hundreds of Kelvin below the equilibrium melting point, owing to the avoidance of heterogeneous nucleation at the crucible walls. This allows investigation of the temperature dependence of the diffusivity in a broad temperature regime [8, 19, 20].

7.2
Experimental Details

7.2.1
Quasi elastic Neutron Scattering

The atomic dynamics in melts of Ni, $Ni_{80}P_{20}$, $Pd_{40}Ni_{40}P_{20}$, $Pd_{43}Ni_{10}Cu_{27}P_{20}$, $Zr_{64}Ni_{36}$, $Zr_{41.2}Ti_{13.8}Cu_{12.5}Ni_{10}Be_{22.5}$ (V1), and $Zr_{46.8}Ti_{8.2}Cu_{7.5}Ni_{10}Be_{27.5}$ (V4) were studied by quasi elastic neutron scattering using different time-of-flight (TOF) spectrometers. Some of the experiments were performed in a crucible, others by application of the containerless processing technique of electromagnetic levitation.

Melts of Ni, Ni–P, Pd–Ni–P, Pd–Ni–Cu–P, and Zr–Ti–Cu–Ni–Be were investigated in a crucible by quasi-elastic neutron scattering. For the neutron TOF experiments on the Ni-based systems, thin-walled Al_2O_3 containers were used that provide a hollow cylindrical sample geometry with 22 mm in diameter, 40 mm in height, and a wall thickness of 0.6 mm in the case of Ni and Ni–P and of 1.2 mm for Pd–Ni–P and Pd–Ni–Cu–P. In the case of Zr–Ti–Cu–Ni–Be, the glassy samples were sealed in a 0.35 mm thin-walled SiC container giving a 30 × 40 mm flat plate sample geometry with a thickness of 1.5 mm. For the chosen sample geometries and wavelengths of the incoming neutrons, the samples scatter less than 8%. Effects of multiple scattering, which may alter the data especially toward low q, could not be detected.

The Pd–Ni–P melts were investigated on the spectrometer FOCUS at the Paul Scherrer Institut(PSI) in Villigen, Switzerland, whereas melts of Ni, Ni–P, Pd–Ni–Cu–P and Zr–Ni–Ti–Cu–Be were examined using the spectrometer IN6 of the Institut Laue-Langevin (ILL) in Grenoble, France. The wavelength of the incident neutrons (between 5.1 and 5.9 Å) resulted in an accessible q range between 0.4 and 2.1 Å^{-1} and an energy resolution between 92 and 50 µeV both at zero energy transfer. Regarding the scattering cross sections of the individual elements, the incoherent scattering is dominated by the contribution from Ni. In Pd–Ni–Cu–P, incoherent scattering on the Cu atoms contributes about 20% to the signal. Following the arguments in Ref. 6, the diffusion coefficients of the Ni and Cu atoms are quite similar. For Zr–Ni–Ti–Cu–Be, Ti contributes approximately 35%(25%) for V1 (V4), Cu about 6%(4%) and Zr about 6%(9%) to the incoherent scattering.

To perform quasi elastic neutron scattering investigations also for chemically reactive melts and/or in the undercooled regime, we designed a dedicated electromagnetic levitation device, which was combined with the TOF spectrometer, TOFTOF, [21] of the Munich research reactor (FRM II) in Garching, Germany. A new generation of neutron TOF spectrometers provides an increased neutron flux and a signal-to-noise ratio of more than an order of magnitude. By combining with a compact electromagnetic levitation device [22, 23], quasi elastic neutron scattering on small levitated droplets is now feasible. In addition, we succeeded in developing a coil design of the electromagnetic levitator with a large opening (8 mm) between the upper and lower parts of the coil without loss of stability in the sample positioning. This enhanced the visibility of the sample for the incoming neutron beam and enabled a nearly full coverage of the detector banks up to 20° vertical angle and thereby increased the count rate by another factor of 10.

The roughly spherical, electrically conductive samples, 7–8 mm in diameter are levitated within an inhomogeneous electromagnetic radio frequency(RF) field. Because of the RF field, eddy currents are induced in the specimen. On one hand, this leads to an inductive heating of the sample, which allows melting of the specimen. On the other, the interaction of the eddy currents with the inhomogeneous magnetic field of the levitation coil leads to a force in the direction of low magnetic field strength such that gravity is compensated. The convective stirring induced by the inductive currents in combination with the large heat conductivity of the sample results in a homogeneous sample temperature [24], which is measured contact free with a two-color pyrometer. The absolute temperature is derived by gaging the measured melting temperature to the literature values. The temperature of the melt is controlled via the flow of ultrahigh-purity cooling gas (He or, for the experiments on pure Ni, a He/4% H_2 mixture) which is injected by a nozzle that is installed below the sample. The addition of hydrogen gas in some of the experiments is to reduce the oxides on the sample surface. Before the samples are levitated, the sample chamber is evacuated and subsequently filled with the cooling gas.

Using this setup, melts of pure Ni [8] and of $Zr_{64}Ni_{36}$ [19, 25] have been investigated. The data acquisition times ranged between 1 and 4 h. At TOFTOF the used incident neutron wavelength of 5.1 (Zr–Ni) and 5.4 Å (Ni) gives an accessible range of momentum transfer q between 0.4 and 2.2 $Å^{-1}$ at zero energy transfer. At q values below 2 $Å^{-1}$ coherent contributions to the scattered intensity could not be detected and the signal was dominated by incoherent contributions. Therefore, the signal displays the self-motion of the Ni atoms. A measurement of the solid sample at room temperature yields the instrumental energy resolution function that is well described by a Gaussian function with an energy resolution of $\delta E \approx 95\,\mu eV$ ($\lambda = 5.1$Å) and $\delta E \approx 78\,\mu eV$ ($\lambda = 5.4$Å) full width at half-maximum. The absence of a sample container in combination with a shielding of the Cu coils with ^{6}Li-containing rubber results in an excellent signal-to-noise ratio, despite the small sample size and the high temperatures involved.

The scattering law $S(q,\omega)$ is obtained by normalization of the measured quasi elastic neutron scattering spectra to a vanadium standard, correction for self-absorption and container scattering, and interpolation to constant wavenumbers q. Further, $S(q,\omega)$ is symmetrized with respect to the energy transfer $\hbar\omega$ by means of the detailed balance factor. Fourier transformation of $S(q,\omega)$, deconvolution of the instrumental resolution, and normalization with the value at $t = 0$ results in the time correlation function $\Phi(q,t)$. In dense liquids the structural relaxation from a plateau in $\Phi(q,t)$ to zero is usually described by a stretched exponential function, $f_q \exp[(t/\tau_q)^{\beta_q}]$, where τ_q denotes the relaxation time and β_q the stretching exponent [10]. By fitting of the experimentally determined $\Phi(q,t)$, the relaxation times are determined. At small and even intermediate momentum transfer, τ_q

shows a linear q^2 dependence, from which the diffusion coefficient can be inferred by $D = 1/(\tau_q q^2)$ [26].

7.2.2
Elastic Neutron Scattering

The short-range order of melts of $Zr_{64}Ni_{36}$ has been investigated by elastic neutron scattering at the high-intensity two-axis diffractometer D20 of the Institut Laue-Langevin in Grenoble, France, using a wavelength of the incident neutrons of $\lambda = 0.94$Å. To determine partial static structure factors, the technique of isotopic substitution has been employed using samples prepared with ^{60}Ni, ^{58}Ni, and natural Ni. Because of the chemical reactivity of the Zr-based melts, the liquids have been processed without container in an electromagnetic levitation device similar to the one used for the quasi elastic neutron scattering experiments. The experimental setup and the data treatment procedure are described in detail in Ref. 22.

7.3
Results and Discussion

7.3.1
Atomic Dynamics in Liquid Ni

To investigate the influence of alloying on the atomic dynamics in Ni-based glass-forming melts, we first investigated melts of pure Ni as a reference system by quasi elastic neutron scattering. Some of these experiments have been performed at temperatures just above the melting point at the spectrometer IN6 of the ILL in a thin-walled sample geometry [7]; the other experiments [8] were performed in a broad temperature range including the metastable regime of undercooled melts at the TOFTOF spectrometer of the FRM II by application of the electromagnetic levitation technique. Because Ni is a strong incoherent scatterer ($\sigma_i = 5.2$ barn), the comparison of both sets of data allows verification of the levitation approach, and proving that the larger thickness of the levitated samples does not hamper the determination of the Ni self-diffusion coefficients with high precision due to influences of multiple scattering.

Figure 7.1 shows the TOF signal measured for a levitated liquid Ni sample at a temperature of $T = 1546$ K using the spectrometer TOFTOF, together with the background signal of the setup without the specimen. Noteworthy is the excellent signal-to-background ratio achieved with this experimental setup. Here, containerless processing offers the additional advantage that scattering from the sample container that usually contributes to the background signal is avoided. In the spectrum of the liquid sample, the regime of inelastic scattering at TOFs and that of quasi elastic scattering around the elastic line at approximately 1.37 ms m^{-1} TOF are visible. From the TOF signals acquired under different scattering angles, the dynamic structure factors $S(q, \omega)$ and finally the density correlation functions $\Phi(q, t)$ are derived.

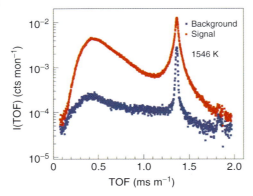

Fig. 7.1 Time-of-flight (TOF) signal measured for an electromagnetically levitated liquid Ni sample at $T = 1546$ K using the spectrometer TOFTOF at FRM II and corresponding background measured without specimen.

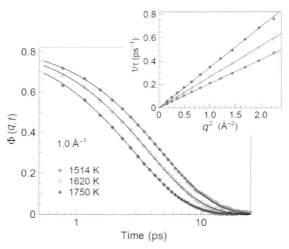

Fig. 7.2 Density correlation functions $\Phi(q,t)$ in the structural relaxation regime for liquid Ni, at different temperatures at a momentum transfer of $q = 1\text{Å}^{-1}$. The solid lines represent fits of $\Phi(q,t)$ under assumption of an exponential decay ($\beta_q = 1$). The inset shows the resulting inverse relaxation times as a function of q^2, which show a linear q^2 dependence.

Figure 7.2 shows the density correlation functions $\Phi(q,t)$ in the structural relaxation regime for liquid Ni, at different temperatures for a fixed momentum transfer of $q = 1\text{Å}^{-1}$. The solid lines represent fits of $\Phi(q,t)$ under assumption of an exponential decay ($\beta_q = 1$). The inverse relaxation times, $1/\tau_q$, inferred from such fits are plotted as a function of q^2 in the inset of Figure 7.2. $1/\tau_q$ shows a linear q^2 dependence. From the slope of $1/\tau_q$ as function of q^2, the self-diffusion coefficient D is derived for each temperature on an absolute scale with high precision. The

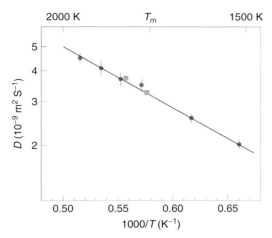

Fig. 7.3 Ni self-diffusion coefficient D as a function of inverse temperature in a range spanning more than 200 K above to more than 200 K below the melting point. The circles are the data points determined by the levitation experiments [8], and the squares are those from the experiments in a crucible [7]. The line is an Arrhenius fit of the experimental data, resulting in an activation energy of $E_A = 0.47$ eV.

temperature dependence of the Ni self-diffusion coefficients is plotted in Figure 7.3. Two data points shown in the figure have been measured in a special crucible that gives a sample geometry of a hollow cylinder with 0.6 mm wall thickness [7]. These data points are in excellent agreement with the results of the levitation experiments. This shows that the increase in sample thickness and the corresponding increase in multiple scattering events do not alter the determination of the diffusion coefficients. While the quasi elastic neutron scattering measurements in the crucible can be performed only in a narrow regime of temperatures above the melting point, the application of the electromagnetic levitation technique gives access to a significantly wider temperature range, including also the metastable regime of an undercooled liquid. Over this large temperature range—from more than 200 K above to more than 200 K below the melting point of 1726 K—the temperature dependence of the diffusion coefficients can be described by an Arrhenius law, $D = D_0 \exp(-E_A/k_B T)$, with an activation energy $E_A = 0.47 \pm 0.03$ eV and a prefactor $D_0 = 77 \pm 8 \times 10^{-9}$ m^2 s^{-1}. This also shows that the change from a liquid in thermodynamic equilibrium to a metastable liquid at temperatures below the melting point is not reflected in its atomic dynamics.

The Arrhenius behavior observed here for liquid Ni differs from the T^2 behavior reported for metallic melts with a lower density of packing such as Sn, Pb, In, and Sb [27, 28]. There, long capillary experiments were performed in space under microgravity conditions, where the absence of buoyancy-driven convective flow resulted in the required precision of the diffusion coefficients. As a consequence, a T^2 behavior of the temperature dependence of the self-diffusion coefficient does not hold as a general rule in one-component liquid metals around the melting point. In dense-packed liquids, mode coupling theory predicts a temperature dependence of diffusion of $(T - T_c)^\gamma$ close to a critical temperature T_c [10]. We were not able to undercool liquid Ni to temperatures at which such a behavior is expected.

7.3.2
Atomic Dynamics in Ni–P-based Glass-forming Alloy Melts

To investigate the influence of alloying in Ni–P-based glass-forming melts on the atomic dynamics, quasi elastic neutron scattering experiments were performed on melts of $Ni_{80}P_{20}$, $Pd_{40}Ni_{40}P_{20}$, and $Pd_{43}Ni_{10}Cu_{27}P_{20}$. Since for these samples the quasi elastic signal is dominated by the incoherent scattering of Ni, the data analysis gives the self-diffusion coefficient of the Ni atoms. Figure 7.4 displays self-diffusion coefficients determined for these melts, together with the results for pure Ni (see Section 7.3.1)

Over the accessible temperature range, there is an overall agreement in the Ni self-diffusion coefficients of the glass-forming Ni–P-based alloys within ±10%. The agreement is even better for temperatures above 1390 K. There, diffusion coefficients of Ni–P and Pd–Ni–Cu–P are equal within the errors bars. Absolute values in Ni–P and Pd–Ni–Cu–P, respectively, range from $1.96 \pm 0.06 \times 10^{-9}$ m^2 s^{-1} and $2.09 \pm 0.11 \times 10^{-9}$ m^2 s^{-1} at 1390 K to $4.65 \pm 0.05 \times 10^{-9}$ m^2 s^{-1} and $4.74 \pm 0.18 \times 10^{-9}$ m^2 s^{-1} at 1795 K. Replacing more than 85% of the Ni atoms by Pd and Cu has, at least in this temperature range, no resolvable effect on the diffusion of the Ni atoms. At temperatures around 1000 K, the data for $Pd_{40}Ni_{40}P_{20}$ and for $Pd_{40}Ni_{10}Cu_{30}P_{20}$ [16] overlap: differences in the diffusion coefficients of the two alloys are not larger than the error bars.

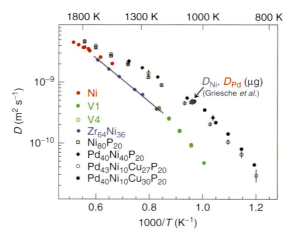

Fig. 7.4 Temperature dependence of the mean Ni self-diffusivities in liquid Ni, Ni–P, Pd–Ni–P, Pd–Cu–Ni–P, and $Zr_{64}Ni_{36}$ and of the mean Ni and Ti diffusivities in liquid Zr–Ti–Ni–Cu–Be [6–8, 14, 16]. The figure contains also two data points for the Ni and Pd self-diffusivity in $Pd_{40}Cu_{30}Ni_{10}P_{20}$ measured by Griesche et al. [29] under microgravity (μg).

Even in pure Ni, the value of $3.80 \pm 0.06 \times 10^{-9}$ m^2 s^{-1} at 1795 K is only 20% below the diffusion coefficients of the Ni–P and Pd–Ni–Cu–P alloys. Around the liquidus temperature of 865 K, atomic dynamics in the Pd$_{43}$Ni$_{10}$Cu$_{27}$P$_{20}$ alloy are in accordance with the mode coupling theory of the liquid to glass transition, and Ni self-diffusion coefficients are proportional to $(T - T_c)^\gamma$, with $\gamma = 2.7$ and $T_c = 710$ K [6]. At higher temperatures, the Ni self-diffusion coefficient of the Pd–Ni–Cu–P alloy [7] becomes as large as the measured diffusion coefficient for pure liquid Ni and exhibits also a temperature dependence that is well described with an Arrhenius law.

Figure 7.4 also shows the diffusion coefficients for the Ni and Pd self-diffusion in Pd$_{40}$Cu$_{30}$Ni$_{10}$P$_{20}$ measured under microgravity conditions by Griesche et al. [29]. The Ni self-diffusion coefficient measured under reduced gravity is in excellent agreement with the data determined in this work by quasi elastic neutron scattering. Essentially the same diffusion coefficient was measured for the Pd self-diffusion in this alloy, showing that neither the mass nor the size of the atoms in a multicomponent alloy has a significant effect on the atomic diffusion in densely packed glass-forming melts.

7.3.3
Atomic Dynamics in Zr–Ti–Ni–Cu–Be and Zr$_{64}$Ni$_{36}$ Alloy Melts

The atomic dynamics in Zr-based glass-forming alloys have been studied by quasi elastic neutron scattering for melts of Zr$_{41.2}$Ti$_{13.8}$Cu$_{12.5}$Ni$_{10}$Be$_{22.5}$ (V1) and Zr$_{46.8}$Ti$_{8.2}$Cu$_{7.5}$Ni$_{10}$Be$_{27.5}$ (V4). The temperature dependence of the mean Ni and Ti diffusivities in V1 and V4 are also shown in Figure 7.4, which allows a comparison with the Ni self-diffusivities determined for the Ni-based systems. The data for the Zr-based melts are about 1 order of magnitude smaller than self-diffusivities in the other liquids at similar temperatures. The absolute value of the diffusivity is barely affected by the change in the alloy's composition. In the idealized version of the mode coupling theory, which does not account for hopping processes, $D \sim (T - T_c)^\gamma$ for temperatures above T_c. Indeed, the temperature dependence of the mean self-diffusivity of the Ni and Ti atoms in the Zr–Ti–Cu–Bi–Be melts is described well by this scaling law. The best fit to the V1 data gives $\gamma = 2.5$ and $T_c = 850$ K. This compares well with $\gamma \approx 2.65$ and $T_c \approx 875$ K obtained from the mode coupling analysis of the fast relaxation process in V4 [15].

It may be speculated that the differences in the atomic motion between Zr–Ti–Ni–Cu–Be and the Ni–P-based melts visible in Figure 7.4 originate from the short-range order in the liquids. Therefore, investigations on the short-range order are of special interest. For complex multicomponent alloys, such as Zr–Ti–Ni–Cu–Be, the determination of partial structure factors is practically impossible. Nevertheless, we have found for Ni–P-based glass-forming alloys that the addition of further alloy components such as Pd or Cu has only a negligible influence on the Ni self-diffusivity at a given temperature. If it can be shown that binary Zr–Ni melts show a similar diffusion behavior as the complex bulk glass-forming Zr–Ti–Ni–Cu–Be alloys, structural investigations on the less complex binary Zr–Ni system may allow establishing a link between short-range order and atomic dynamics. Therefore, we have performed quasi elastic neutron scattering experiments on binary $Zr_{64}Ni_{36}$ melts. The liquids have been investigated at six different temperatures ($T = 1210, 1290, 1345, 1455, 1545$, and 1650 K) around the liquidus temperature of $T_L = 1283$ K. The results are plotted in Figure 7.4. In the investigated temperature regime, $D(T)$ of the $Zr_{64}Ni_{36}$ melts shows essentially an Arrhenius-type behavior (Figure 7.4), from which an activation energy for the Ni self- diffusion of $E_A(Zr - Ni) = 0.64$ eV is inferred. This value is large compared to activation energies for Ni self-diffusion observed in other melts. For instance, our experiments on pure Ni gave $E_A(Ni) = 0.47$ eV (compare Section 7.3.1). For liquid $Al_{80}Ni_{20}$, a value of $E_A(Al - Ni) = 0.36$ eV is reported [9]. Moreover, it is evident that the data points for Zr–Ti–Ni–Cu–Be are on the same $D(T)$ curve, suggesting a similar Ni diffusion mechanism in both systems.

7.3.4
The Short-Range Order of Liquid $Zr_{64}Ni_{36}$

In order to find the reasons for the peculiar atomic dynamics in the Zr–Ni-based melts, we have investigated three $Zr_{64}Ni_{36}$ melts prepared with natural Ni, ^{58}Ni, and ^{60}Ni at a temperature of $T = 1375$ K by elastic neutron scattering. Figure 7.5 shows the resulting three total static structure factors $S(q)$. The marked differences between these structure factors are obvious. From the three total structure factors, partial structure factors were calculated within the Faber–Ziman [30] and the Bhatia–Thornton [31] formalisms.

The static Faber–Ziman structure factors $S_{ZrZr}(q)$, $S_{ZrNi}(q)$, and $S_{NiNi}(q)$ describe the contributions to the total structure factor $S(q)$, which result from the three different types of atomic pairs (Zr–Zr, Zr–Ni, and Ni–Ni). Within the Bhatia–Thornton

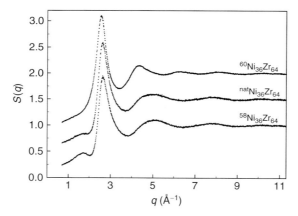

Fig. 7.5 Total structure factors measured by neutron scattering for $Zr_{64}^{58}Ni_{36}$, $Zr_{64}^{60}Ni_{36}$ and $Zr_{64}^{nat}Ni_{36}$ at $T = 1375$ K. The curves are shifted by multiples of 0.5 along the vertical axis [19].

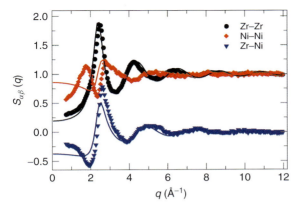

Fig. 7.6 Static partial Faber–Ziman structure factors $S_{\alpha\beta}(q)$ (α,β = Ni, Zr) of liquid $Zr_{64}Ni_{36}$ at $T = 1375$ K. The solid lines are modeled structure factors within a hard sphere model based on the Percus–Yevick approximation.

formalism, the static partial structure factor $S_{NN}(q)$ describes solely the topological short-range order of the system, $S_{CC}(q)$, the chemical short-range order, and $S_{NC}(q)$ the correlation of number density and chemical composition. Figure 7.6 shows the Faber–Ziman structure factors determined from the neutron scattering experiments for liquid $Ni_{36}Zr_{64}$ at a temperature of $T \approx 1375$ K. The pair correlation functions, $g_{NN}(r)$, $g_{CC}(r)g_{NC}(r)$, $g_{ZrZr}(r)$, $g_{NiNi}(r)$, and $g_{ZrNi}(r)$ calculated by Fourier transformation from the static Bhatia–Thornton and Faber–Ziman structure factors are depicted in Figure 7.7.

The Bhatia–Thornton pair correlation function g_{CC} is characterized by a remarkable minimum at $r \approx 2.7$Å. This minimum is a signature of a chemical short-range order prevailing in the liquid, which is the result of an affinity for the formation of Ni–Zr nearest neighbors. The same conclusion can be drawn from the fact that the first maximum of the Faber–Ziman pair correlation function $g_{NiZr}(r)$ is significantly larger than the first maxima of $g_{NiNi}(r)$ and $g_{ZrZr}(r)$.

The different nearest neighbor coordination numbers, $Z_{\alpha\beta}$ (α,β = N, Zr, Ni), were determined by integrating the partial radial distribution function $4\pi c_\beta \rho r^2 g_{\alpha\beta}(r)$ over its first maximum (with the first and second minimum as integration boundaries). c_β denotes the composition of the component β and ρ the atomic density. The value of $\rho = 0.052$ at/Å^3 is used, which is inferred from measurements of the density of $Zr_{64}Ni_{36}$ melts employing electromagnetic levitation. There are other methods to determine Z_{ij} [32] that may lead to a spread of the absolute values of $Z_{\alpha\beta}$ by about 10%. The calculations gave $Z_{NN} = 13.8 \pm 0.5$, $Z_{NiNi} = 2.5 \pm 0.5$, $Z_{ZrZr} = 10.4 \pm 0.5$, and $Z_{NiZr} = 7.4 \pm 0.5$ [19]. The coordination number, $Z_{NN} \approx 13.8$, is slightly higher than the typical values of $Z_{NN} \approx 12$ determined for most metallic melts [23, 32]. This may indicate a comparatively high local density of packing in molten $Zr_{64}Ni_{36}$.

For a further analysis of the topological short-range order of liquid $Zr_{64}Ni_{36}$, $S_{NN}(q)$ was modeled in the regime of large q vectors by assuming that the short-range

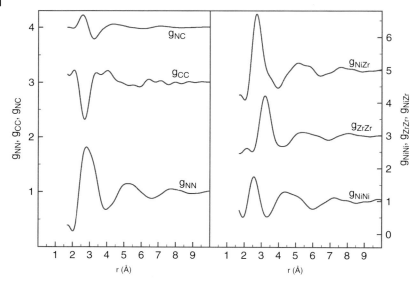

Fig. 7.7 Partial Bhatia–Thornton (left side) and Faber–Ziman (right side) pair correlation functions for liquid $Zr_{64}Ni_{36}$ at $T = 1375$ K. The curves are shifted by integers along the vertical axis [19].

order is dominated by one type of structural units. The following structures were investigated: icosahedral, dodecahedral, face centered cubic (fcc), hexagonal close packed (hcp), and body centered cubic (bcc). The used simulation method [33, 34] has the advantage that it depends on three free parameters only. These are the shortest mean distance, $\langle r_0 \rangle$, of atoms within the unit; its mean thermal variation $\langle \delta r_0^2 \rangle$ that determines the Debye–Waller factor, $\exp(-2q^2 \langle \delta r_0^2 \rangle /3)$; and the concentration, X, of atoms belonging to the aggregates that make up the short-range order in the liquid. The parameters are adjusted such that the best fit of the experimentally determined $S_{NN}(q)$ is obtained especially at large q vectors. This regime of large momentum transfer is mainly determined by the short-range order because the contributions from the less tightly bound intercluster distances are damped out by thermal motions. Effects of long-range correlations that affect $S_{NN}(q)$ mainly at small q vectors are neglected in this simple approach.

The result of these simulations for icosahedral short-range order is shown in Figure 7.8, together with the experimentally determined $S_{NN}(q)$ at large momentum transfer. In contrast to all other melts of pure metals and metallic alloys with a small difference of the atomic radii of the components investigated so far [23, 35–43], for liquid $Zr_{64}Ni_{36}$ the measured $S_{NN}(q)$ is not well reproduced under the assumption of an icosahedral short-range order. This may be a result of the comparatively large difference of the atomic radii of Zr and Ni atoms, which may promote the formation of alternative structures [44].

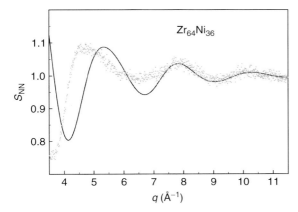

Fig. 7.8 Measured (gray dots) and simulated (solid line) $S_{NN}(q)$ of liquid $Zr_{64}Ni_{36}$ at large momentum transfer. For the simulation, an icosahedral short-range order is assumed to prevail in the melt [19].

We also modeled the structure factors under assumption of a trigonal prismatic short-range order as suggested by molecular dynamics investigations for amorphous Zr–Ni–Al alloys [45]. For this type of short-range order, a good description of the measured structure factor was obtained only if atomic distances were assumed that are in contradiction with the nearest neighbor distances inferred from the first maxima of the Faber–Ziman pair correlation functions (Fig. 7.7), such as a Zr–Ni distance of approximately 3.2 Å.

Moreover, we modeled the structure factor of Zr–Ni using a simple hard sphere model based on the Percus–Yevick approximation matching composition and (covalent) radii to those of the experimental system: $R_{Ni} = 1.15$ Å and $R_{Zr} = 1.45$ Å [46]. The results are also shown in Figure 7.6. Marked differences are visible. Most strikingly, the hard-sphere structure factor does not reproduce the experimentally observed pronounced prepeak visible in $S_{NiNi}(q)$ at $q \approx 1.9$ Å^{-1}. This is due to the fact that the hard sphere model does not account for any chemical order in the melt.

The activation energy we have determined for the Ni self-diffusion in liquid $Zr_{64}Ni_{36}$ (see Section 7.3.4) is significantly higher than that in liquid $Al_{80}Ni_{20}$ of 0.36 eV [9]. Partial structure factors were calculated by molecular dynamics simulations for $Al_{80}Ni_{20}$ [9]. The simulations are indicative of a chemical short-range order such that Al–Ni nearest neighbors are preferred. The Bhatia–Thornton structure factor S_{NN} calculated for liquid $Al_{80}Ni_{20}$ develops a shoulder on the second maximum if the temperature is decreased. Such a shoulder is considered as a first indication of an icosahedral short-range order prevailing in the melt [47]. Our diffraction studies have revealed that the topological short-range order of $Zr_{64}Ni_{36}$ melts is different from an icosahedral one. It may be speculated whether the marked differences in the activation energies for Ni self-diffusion observed between liquid $Zr_{64}Ni_{36}$ and $Al_{80}Ni_{20}$ are a result of the different topological short-range order. Especially, the comparatively large coordination number observed for $Zr_{64}Ni_{36}$

melts suggests a large local density of packing possibly resulting in the high activation energy for atomic diffusion.

7.3.5
Analysis Within Mode Coupling Theory

Using the measured partial structure factors, calculations in the framework of mode coupling theory were performed [25]. Figure 7.9 shows the results from mode coupling theory for the self-diffusion of Ni and Zr in liquid $Zr_{64}Ni_{36}$. We find that the values of D_{Ni} and D_{Zr} inferred from mode coupling theory are nearly identical. This qualitatively agrees with recent experimental results for glass-forming Pd–Ni–Cu–P melts [29], where also Pd and Ni were found to have equal self-diffusion coefficients. As in our case, this holds despite a notable difference in size and mass of the different atom types. This differs from the mode coupling results obtained for the hard sphere model, which suggest a 2 times smaller self-diffusion coefficient for the large atoms as compared to that of small atoms [25]. These findings also contrast with molecular dynamics simulation results modeling a deeply quenched $Zr_{50}Ni_{50}$ system [48], where Ni diffusion was found to be significantly faster than Zr diffusion. For our Zr–Ni results, the error bar in Figure 7.9 indicates systematic errors due to uncertainties in the input quantities $S(q)$ and ρ. Small changes there significantly affect the absolute values of the transport coefficients predicted by mode coupling theory. But this systematic error cancels in relative values, and we find $D_{Zr}/D_{Ni} = 1.02 \pm 0.02$. Figure 7.9 also shows the Ni self-diffusion coefficients determined experimentally

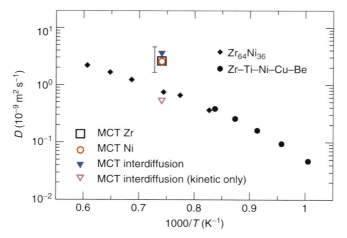

Fig. 7.9 Diffusion and interdiffusion coefficients for Zr–Ni as a function of inverse temperature. Diamonds and circles are results from quasi elastic neutron scattering for $Zr_{64}Ni_{36}$ and $Zr_{41.2}Ti_{13.8}Cu_{12.5}Ni_{10}Be_{22.5}$, respectively. Open squares (circles) show mode coupling results for Zr (Ni); inverted triangles are interdiffusion coefficients (filled: including thermodynamic contributions, open: without). The error bar indicates the systematic error for D_{Zr}. Other errors are smaller than the symbol size [25].

by quasi elastic neutron scattering for melts of $Zr_{64}Ni_{36}$ and of the five-component $Zr_{41.2}Ti_{13.8}Ni_{10}Cu_{12.5}Be_{22.5}$ alloy. The absolute value of D_{Ni} from mode coupling theory is about 2–3 times larger than the experimental value, which is within the usual quality of the mode coupling approximation.

The mode coupling theory results allow us to calculate further quantities that have not yet been experimentally measured for $Zr_{64}Ni_{36}$. Of particular interest is the interdiffusion coefficient D_{CC}, which consists of a kinetic contribution (the Onsager coefficient) and a prefactor $1/S_{CC}(q=0)$ that is purely thermodynamic in origin. Here, $S_{CC}(q)$ denotes the q-dependent generalized susceptibility with respect to concentration fluctuations [49]. Both the kinetic-only and the full contribution to D_{CC} are shown in Figure 7.9 as inverted open and filled triangles, respectively. The Onsager coefficient for the interdiffusion process is about a factor 5 smaller than both D_{Zr} and D_{Ni}. Only the large thermodynamic prefactor raises the measured interdiffusion coefficient above both self-diffusivities. Since the estimation of $S_{CC}(q)$ from the experimental $S(q)$ data bears a relatively large uncertainty, we cannot comment on the numerical value of D_{CC}/D_{Ni}, but we expect it to be slightly larger than unity.

Also for the hard sphere model leading to the structure factors shown in Figure 7.6, mode coupling theory predicts the kinetic part of the interdiffusion coefficient to be much slower than either of the self-diffusion coefficients [25]. Only if one takes into account the thermodynamic contribution, $D_{CC} \approx D_{small}$ results. A similar relation between the diffusion coefficients has also been reported in earlier mode coupling calculations of an equimolar hard sphere mixture [50].

Recent simulations and experiments on atomic diffusion in $Al_{80}Ni_{20}$ [9, 51] reveal that diffusion of the individual components is strongly coupled by chemical short-range order in this system also [52]. However, interdiffusion in this case was found to be well predicted by the Darken equation [53], in contrast to our results. This difference may be related to the higher density of packing of the $Zr_{64}Ni_{36}$ melt.

7.4
Conclusions

In conclusion, we measured Ni self-diffusion coefficients in melts of Ni, Ni–P, Pd–Ni–P, Pd–Ni–Cu–P, and Zr–Ni and the mean Ni and Ti diffusion coefficients in liquid Zr–Ti–Ni–Cu–Be alloys by quasi elastic neutron scattering. Part of the studies has been performed using a newly developed experimental setup in which the containerless processing technique of electromagnetic levitation is combined with quasi-elastic neutron scattering at the TOF spectrometer TOFTOF of the FRM II. This allowed us to determine self-diffusion coefficients with high precision even for chemically reactive melts. Moreover, by avoidance of heterogeneous nucleation at crucible walls, the melts can be undercooled below the equilibrium melting temperature such that diffusion data are available also in the metastable regime of an undercooled liquid.

In the multicomponent glass-forming liquids, an increasing number of components leads to a continuous decrease of the liquidus temperature, the critical cooling rate for glass formation, and the atomic mobility at the liquidus temperature as compared to the binary Ni–P and Zr–Ni systems. The diffusion coefficient at the corresponding liquidus temperature is nearly 2 orders of magnitude smaller in Pd–Ni–Cu–P as compared to pure Ni. At a constant temperature, the replacement of more than 85% of the Ni atoms in Ni–P by Pd and Cu has no resolvable effect on the Ni self-diffusion coefficient. In these dense liquids, atomic transport is controlled by the packing fraction, which is very similar in these alloys. For the Ni–P-based glass-forming alloys, the viscosity is coupled to the diffusivity by the Stokes–Einstein relation. On the other hand, Zr–Ti–Cu–Ni–Be melts forming bulk metallic glasses are characterized by considerably smaller values of the mean Ni and Ti diffusivity as determined for the Ni–P-based glass-forming melt at similar temperatures. Moreover, viscosity and diffusivity are not coupled. We have shown that undercooled binary Zr–Ni melts exhibit the same Ni self-diffusivities as the mean Ni and Ti self-diffusivities in Zr–Ti–Cu–Ni–Be melts at same temperature, suggesting that the binary system shows similar mechanisms of atomic diffusion as the complex multicomponent alloys.

To investigate the influence of short-range order on the atomic dynamics in the Zr-based alloy melts, we have determined partial structure factors of liquid $Zr_{64}Ni_{36}$ by neutron scattering using the technique of isotopic substituion. The $Zr_{64}Ni_{36}$ melts exhibit a pronounced chemical short-range order such that the Zr–Ni nearest neighbors are preferred. Different from most other metallic melts investigated so far, we found no indications for the existence of an icosahedral short-range order in stable $Zr_{64}Ni_{36}$ melts. This may be a result of the comparatively large difference of the atomic radii of Ni and Zr ($R_{Zr}/R_{Ni} = 1.29$).

Combining the neutron scattering data on the partial static structure factors and mode coupling theory of the glass transition, we have calculated self-diffusion and interdiffusion coefficients for molten $Zr_{64}N_{36}$. Our calculations within mode coupling theory predict that Zr and Ni diffusion proceeds with almost equal rates, and therefore is more strongly coupled than one would expect from the analogy with a binary hard sphere mixture, where only the difference in covalent radii between the two elements is accounted for, but no chemical ordering effects. The chemical ordering in Zr–Ni is connected with a prepeak in the static structure factors. Mode coupling theory predicts the dynamics of the system using solely this static equilibrium input, establishing a firm relation between the structure and dynamical transport in such melts. Interdiffusion is found to be only slightly faster than self-diffusion. We show that this is due to a large thermodynamic prefactor, whereas the Onsager coefficient is approximately a factor of 5 smaller than the self-diffusion coefficients.

Acknowledgments

The authors thank R. Bellissent, T. Hansen, H. Hartmann, D.M. Herlach, S. Janssen, M. Koza, T. Mehaddene, D. Menke, W. Petry, V. Simonet, T. Unruh, T.

Volkmann, and F. Yang for fruitful discussions and/or their support during the preparation and performance of the experiments. We thank J. Brillo for providing the density measurements. We thank the Institut Laue-Langevin for beam time and support at the instruments D20 and IN6, the FRM II for beam time and support at the instrument TOFTOF, and the Paul Scherrer Institut for beam time and support at the instrument FOCUS. Financial support was provided by the Deutsche Forschungsgemeinschaft (DFG) within the priority program "Phasenumwandlungen in Mehrkomponentigen Schmelzen" under Grants No. Me1958/2-1, 2-2, 2-3 and No. Ho1942/6-3.

References

1 Nelson, D.R. and Spaepen, F. (**1989**) in *Solid State Phys*, Vol. **42**, (eds H. Ehrenreich, F. Seitz and D. Turnbull), Academic Press, New York, p. 1.
2 Holland-Moritz, D. (**1998**) *International Journal of Non-Equilibrium Processing*, **11**, 169–99.
3 Herlach, D.M., Galenko, P. and Holland-Moritz, D. (**2007**) *Metastable Solids from Undercooled Melts*, Pergamon Materials Series, (ed R.W. Cahn), Elsevier, Oxford.
4 Rappaz, M. and Boettinger, W.J. (**1999**) *Acta Materialia*, **47**, 3205–19.
5 Boettinger, W.J., Warren, J.A., Beckermann, C. and Karma, A. (**2002**) *Annual Review of Materials Research*, **32**, 163–94.
6 Meyer, A. (**2002**) *Physical Review B*, **66**, 134202-1–134202-9.
7 Mavila Chathoth, S., Meyer, A., Koza, M.M. and Juranyi, F. (**2004**) *Applied Physics Letters*, **85**, 4881–83.
8 Meyer, A., Stüber, S., Holland-Moritz, D., Heinen, O. and Unruh, T. (**2008**) *Physical Review B*, **77**, 092201-1–092201-4.
9 Horbach, J., Das, S.K., Griesche, A., Macht, M.-P., Frohberg, G. and Meyer, A. (**2007**) *Physical Review B*, **75**, 174304-1–174304-8.
10 Götze, W. and Sjögren, L. (**1992**) *Reports on Progress in Physics*, **55**, 241.
11 Götze, W. and Voigtmann, TH. (**2003**) *Physical Review E*, **67**, 021502-1–021501-14.
12 Griesche, A., Macht, M.-P. and Frohberg, G. (**2007**) *Journal of Non-Crystalline Solids*, **353**, 3305–09.
13 Griesche, A., Macht, M.-P. and Frohberg, G. (**2008**) *Diffusion in Multicomponent Metallic Melts Near the Melting Temperature*, this issue.
14 Meyer, A., Petry, W., Koza, M. and Macht, M.-P. (**2003**) *Applied Physics Letters*, **83**, 3894–96.
15 Meyer, A., Wuttke, J., Petry, W., Randl, O.G. and Schober, H. (**1998**) *Physical Review Letters*, **80**, 4454–57.
16 Meyer, A., Busch, R. and Schober, H. (**1999**) *Physical Review Letters*, **83**, 5027–29.
17 Dahlborg, U., Besser, M., Calvo-Dahlborg, M., Janssen, S., Juranyi, F., Kramer, M.J., Morris, J.R., Sordelet, D.J. and Non-Cryst, J. (**2007**) *Solids*, **353**, 3295–99.
18 Ruiz-Martín, M.D., Jiménez-Ruiz, M., Plazanet, M., Bermejo, F.J., Fernández-Perea, R. and Cabrillo, C. (**2007**) *Physical Review B*, **75**, 224202-1–224202-11.
19 Pommrich, A.I., Meyer, A., Holland-Moritz, D. and Unruh, T. (**2008**) *Applied Physics Letters*, **92**, 241922-1–241922-3.
20 Stüber, S., Meyer, A., Holland-Moritz, D. and Unruh, T. in preparation.
21 Unruh, T., Neuhaus, J. and Petry, W. (**2007**) *Nuclear Instruments and Methods A*, **580**, 1414–22.
22 Holland-Moritz, D., Schenk, T., Convert, P., Hansen, T. and Herlach, D.M. (**2005**) *Measurement Science and Technology*, **16**, 372–80.
23 Schenk, T., Holland-Moritz, D., Simonet, V., Bellissent, R. and

Herlach, D.M. (**2002**) *Physical Review Letters*, **89**, 075507-1–075507-4.
24 El-Kaddah, N. and Szekely, J. (**1984**) *Metallurgical Transactions B*, **15**, 183.
25 Voigtmann, Th., Meyer, A., Holland-Moritz, D., Stüber, S., Hansen, T. and Unruh, T. (**2008**) *Europhysics Letters*, **82**, 66001-p1–66001-p6.
26 Boon, J.P. and Yip, S. (**1980**) *Molecular Hydrodynamics*, McGraw-Hill, New York.
27 Mathiak, G., Griesche, A., Kraatz, K.H. and Frohberg, G. (**1996**) *Journal of Non-Crystalline Solids*, **205–207**, 412–16.
28 Itami, T., Masaki, T., Aoki, H., Munejiri, S., Uchida, M., Masumoto, S., Kamiyama, K. and Hoshino, K. (**2002**) *Journal of Non-Crystalline Solids*, **312–314**, 177–81.
29 Griesche, A., Macht, M.P., Suzuki, S., Kraatz, K.-H. and Frohberg, G. (**2007**) *Scripta Materialia*, **57**, 477–80.
30 Faber, T.E. and Ziman, J.M. (**1965**) *Philosophical Magazine*, **11**, 153.
31 Bhatia, A.B. and Thornton, D.E. (**1970**) *Physical Review B*, **2**, 3004–12.
32 Waseda, Y. (**1980**) *The Structure of Non-Crystalline Materials*, McGraw-Hill, New York.
33 Simonet, V., Hippert, F., Klein, H., Audier, M., Bellissent, R., Fischer, H., Murani, A.P. and Boursier, D. (**1998**) *Physical Review B*, **58**, 6273–86.
34 Simonet, V., Hippert, F., Audier, M. and Bellissent, R. (**2001**) *Physical Review B*, **65**, 024203-1–024203-11.
35 Lee, G.W., Gangopadhyay, A.K., Kelton, K.F., Hyers, R.W., Rathz, T.J., Rogers, J.R. and Robinson, D.S. (**2004**) *Physical Review Letters*, **93**, 037802-1–037802-4.
36 Holland-Moritz, D., Schenk, T., Bellissent, R., Simonet, V., Funakoshi, F., Merino, J.M., Buslaps, T. and Reutzel, S. (**2002**) *Journal of Non-Crystalline Solids*, **312–314**, 47–51.
37 Holland-Moritz, D., Heinen, O., Bellissent, R. and Schenk, T. (**2007**) *Materials Science and Engineering A*, **449–451**, 42–45.
38 Schenk, T., Simonet, V., Holland-Moritz, D., Bellissent, R., Hansen, T., Convert, P. and Herlach, D.M. (**2004**) *Europhysics Letters*, **65**, 34–40.
39 Holland-Moritz, D., Schenk, T., Simonet, V. and Bellissent, R. (**2006**) *Philosophical Magazine*, **86**, 255–62.
40 Holland-Moritz, D., Schenk, T., Simonet, V., Bellissent, R., Convert, P., Hansen, T. and Herlach, D.M. (**2004**) *Materials Science and Engineering A*, **375–377**, 98–103.
41 Holland-Moritz, D., Schenk, T., Simonet, V., Bellissent, R., Convert, P. and Hansen, T. (**2002**) *Journal of Alloys and Compounds*, **342**, 77–81.
42 Holland-Moritz, D., Schenk, T., Simonet, V., Bellissent, R. and Herlach, D.M. (**2005**) *Journal of Metastable and Nanocrystalline Materials*, **24–25**, 305–10.
43 Holland-Moritz, D., Heinen, O., Bellissent, R., Schenk, T. and Herlach, D.M. (**2006**) *International Journal of Materials Research*, **97**, 948–53.
44 Ronchetti, M. and Cozzini, S. (**1994**) *Materials Science and Engineering A*, **178**, 19–22.
45 Guerdane, M. and Teichler, H. (**2002**) *Physical Review B*, **65**, 142031-1–142031-10.
46 Pauling, L.J. (**1947**) *American Chemical Society*, **69**, 542.
47 Steinhardt, P.J., Nelson, D.R. and Ronchetti, M. (**1983**) *Physical Review B*, **28**, 784–805.
48 Teichler, H. and Non-Cryst, J. (**2001**) *Solids*, **293–295**, 339–44.
49 Balucani, U. and Zoppi, M. (**1995**) *Dynamics of the Liquid State*, Oxford University Press.
50 Barrat, J.-L. and Latz, A. (**1990**) *Journal of Physics: Condensed Matter*, **2**, 4289–95.
51 Das, S.K., Horbach, J., Koza, M.M., Mavila Chatoth, S. and Meyer, A.

(**2005**) *Applied Physics Letters*, **86**, 011918-1–011918-3.

52 Maret, M., Pomme, T., Pasturel, A. and Chieux, P. (**1990**) *Physical Review B*, **42**, 1598–604.

53 Das, S.K., Kerrache, A., Horbach, J. and Binder, K. (**2008**) *Phase Behavior and Microscopic Transport Processes in Binary Metallic Alloys: Computer Simulation Studies*, this book.

8
Diffusion in Multicomponent Metallic Melts Near the Melting Temperature

Axel Griesche, Michael-Peter Macht, and Günter Frohberg

8.1
Introduction

The decomposition of a multicomponent melt in front of the liquid–solid phase boundary during solidification is controlled by diffusion. The related volume fluxes of the alloy constituents can be described by Fick's laws if convection effects are neglected. These fluxes of the different alloy constituents are named chemical diffusion or interdiffusion (ID). They depend on the concentration differences in the given examples that are determined mainly by the partition coefficients of the alloy constituents.

In the general case of solidification, that is the crystallization of multicomponent melts, the concentration differences between solid and liquid are high, and the concentration dependence of diffusion cannot be neglected. Furthermore, the large concentration gradients are related to gradients of the chemical potentials of the components, which, in addition, govern diffusion. In the theory of irreversible thermodynamics the latter concept is described well in Ref. [1]. Here any force can cause a flux (e.g. heat flux, mass flux, etc.) and can be described by a linear relation. Onsager [2] showed for diffusive mass flux that the proportionality constant due to its symmetry properties is the same as the diffusion constant in the first Fickian law. This allows the calculation of the chemical diffusion if the mobility of the species and the driving force are known. This force is usually described by the dimensionless thermodynamic factor [3].

Experimentally, for chemical diffusion this concept has been proved by Darken [4] in solid state diffusion experiments. Manning [5] revised the Darken equation introducing a correction factor that accounts for a vacancy-assisted diffusion mechanism in solids. In liquids, up to now, only our investigation that exists within this priority program shows the validity of this concept [6] and the applicability of the Darken–Manning equation.

In this chapter we present two examples that show how chemical diffusion fluxes can be governed by thermodynamic forces. One example shows the enhancement of ID compared to self-diffusion (SD) by 300–400% in a wide temperature region from

Phase Transformations in Multicomponent Melts. Edited by D. M. Herlach
Copyright © 2008 WILEY-VCH Verlag GmbH & Co. KGaA, Weinheim
ISBN: 978-3-527-31994-7

the equilibrium melt to the supercooled state (see Section 8.3.1). The other example demonstrates the demixing of a multicomponent melt near the liquidus temperature due to thermodynamic forces (see Section 8.3.2). Our diffusion data fit well with the results of the groups of Meyer and Horbach (see Ref. [6] and other chapters in this book) and are discussed there in the respective context in more detail.

In order to obtain diffusion coefficients with minimal errors by direct measurement, our investigations require sophisticated experimental techniques. Therefore, we report here mostly about the methods for direct measurements and in particular about a new methodological development. This is the combination of X-ray radiography (XRR) with the classical capillary diffusion technique (see Section 8.2.2).

8.2
Experimental Diffusion Techniques

Capillary techniques are the most universal methods to measure both, ID and SD [7]. The long-capillary (LC) method [8], which will be described briefly in Section 8.2.1, and the shear cell (SC) method [9] are well established. These techniques suffer from buoyancy-driven convection that contributes to the total mass transport superimposing the diffusive transport. This problem can partly be overcome by switching off gravity in space experiments [10]. Since most experiments are performed in the lab on ground, any improvement on the used lab techniques allows for experiments with higher precision. Up to now the *post-mortem* analysis of the diffusion profiles allows not to separate diffusive transport and convective transport. Thus, visual *in situ* information would help identify disturbances and exclude them from the experiment analysis. A novel XRR system for the LC technique is described in Section 8.2.2. Other techniques used in our joint collaboration are quasi-elastic neutron scattering (QNS) [11] and molecular dynamic (MD) simulation [6].

8.2.1
Long-Capillary Method

The classical LC technique [8] is shown in Figure 8.1.

In a combined diffusion experiment, both ID and SD can be measured simultaneously. The diffusion couple consists of two cylinders with a different chemical composition; in the given example $Al_{85}Ni_{15}$ at the top and $Al_{75}Ni_{25}$ at the bottom. In between, a small disc of $Al_{80}Ni_{20}$ is placed, containing the enriched stable isotope ^{62}Ni (Figure 8.1a). In order to neglect the concentration dependence of diffusion, the concentration difference in a diffusion couple is minimized but still large enough to achieve a good resolution of the concentration profile. The diffusion couple is placed inside a graphite diffusion capillary.

Prior to total mixing of the diffusion sample, the concentration profiles obtained in the liquid state are frozen in by quenching (Figure 8.1b). Then the chemical concentration profiles are measured by means of atomic absorption spectroscopy

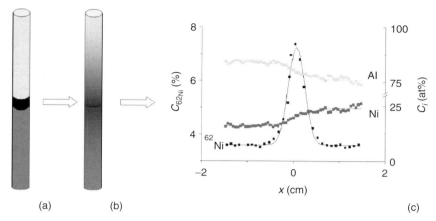

Fig. 8.1 Principle sketch of the long-capillary technique. (a) Initial diffusion couple configuration, (b) after diffusion annealing, (c) *post-mortem* measured concentration profiles.

(AAS) or energy-dispersive X-ray spectrometry (EDX) and the tracer concentration profile is measured by means of an inductively coupled plasma mass spectrometry (ICP-MS) (Figure 8.1c). For AAS and ICP-MS the diffusion sample rod is cut perpendicular to the axis in thin slices, which are dissolved in a solvent and analyzed subsequently for their composition. For EDX spectrometry, the diffusion sample is embedded in plastics, ground, and polished to half of the diameter over the entire length. Then each measurement point is obtained by scanning the electron beam over a certain area depending on the microstructure. This procedure allows to integrate over a sufficient large number of phase constituents, which might have different compositions, to get a mean value of concentration. Figure 8.2 shows the principle of such analyzing method and in Figure 8.6 a real example is given.

The diffusion coefficients, which are assumed to be constant, are obtained by fitting the appropriate solution of Fick's second equation to the given one-dimensional diffusion problem (solid lines in Figure 8.1). In the case of ID the solution equation reads

$$c(x,t) = c_1 + \frac{c_0 - c_1}{2} \operatorname{erfc}\left(\frac{x}{\sqrt{4\tilde{D}t}}\right) \tag{8.1}$$

with c_0, c_1 the initial concentrations of the diffusion couple, the ID coefficient $\tilde{D}$, the space coordinate x, and the diffusion time t. The boundary conditions for this solution are $c = c_0$ for $x < 0$ and $c = c_1$ for $x > 0$ at $t = 0$. In the case of SD, the solution equation (*thick-film* solution) reads

$$c(x,t) = \frac{1}{2}c_0 \left(\operatorname{erf}\frac{h-x}{\sqrt{4D^*t}} + \operatorname{erf}\frac{h+x}{\sqrt{4D^*t}}\right) \tag{8.2}$$

with c_0 the initial concentration of the tracer, the thickness h of the tracer slice, the SD coefficient D^*, the space coordinate x, and the diffusion time t. The boundary

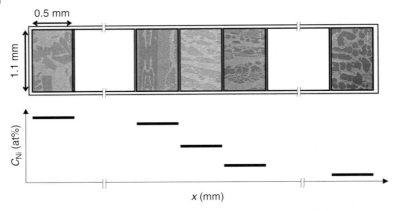

Fig. 8.2 Sketch of the concentration analysis of a long capillary diffusion sample by EDX spectrometry in the system Pd–Cu–Ni–P. The microstructure pictures are obtained using electron microscopy in the element contrast modus and show the area of concentration averaging. The graph concentration c versus distance x below shows the averaged values of Ni concentration for the given areas.

conditions for this solution are $c = c_0$ for $-h/2 < x < h/2$ at $t = 0$. Then the only fitting parameters are the ID coefficient $\tilde{D}$ and the SD coefficient D^* of Ni.

The LC technique requires principally two corrections. First the sample length of the *post-mortem* analysis has to be corrected for the sample length in the liquid state. Second the additional time between reaching the annealing temperature after melting and during cooling prior to solidification, respectively, has to be taken into account. The LC technique and its corrections and the concentrations analysis procedures are described in detail in Ref. 8. Results of the measurements in the system Al–Ni are given in Section 8.3.1.

8.2.2
Long-Capillary Method Combined with X-ray Radiography

In the case of chemical diffusion, when the diffusion couple is transparent for X-rays and the electron density difference between both partners of the couple produces a sufficiently high gray value contrast, a series of radiograms of an LC diffusion experiment can be recorded. In order to make such a movie, the capillary setup is placed between a micro focus X-ray source and a flat panel detector. Then, every two seconds a radiogram can be taken. Figure 8.3 shows an example of a radiogram.

The markers and references allow for both, a concentration calibration and a distance calibration for each radiogram (picture) of the movie, ruling out any problems related to intensity drift. Thus, reading out the gray values along the capillary axis for any picture of the movie yields not just a concentration profile equal to that of the *post-mortem* analysis with EDS shown in Figure 8.1 but now directly for the liquid state. The pressure device allows reducing the free volume between melt and capillary wall in the case of nonwetting melts and thereby reduces Marangoni-driven convection [12].

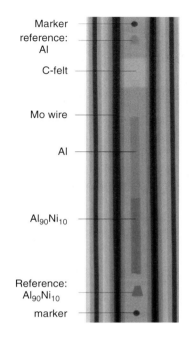

Fig. 8.3 Radiogram of a long capillary with a diffusion couple Al versus Al$_{90}$Ni$_{10}$ at% before melting. The position markers are ZrO$_2$ precision spheres, the C-felt is a carbon spring-like device to apply pressure to the liquid, Mo wires serve as electrical resistance heater, and the reference material is the original sample material.

The result of a typical gray value reading with related calibration is given in Figure 8.4.

The ID coefficient can be calculated in the same way as described in Section 8.2.1. One advantage of the *in situ* technique is that a correction in the length of the sample is not necessary. Furthermore, the time span in which the diffusive transport is evaluated can be chosen freely. It is not necessary to follow the mean square penetration from the very beginning of the experiment because it is sufficient to look for the relative change of penetration depth in an arbitrary time interval. This allows for a check of the time dependence of diffusion, for the *in situ* recognition of convective mass transport contributions, and for measuring ID at different temperatures during one experiment (if the temperature–time profile is controlled accordingly). In time regions where $\tilde{D}$ is constant, $\tilde{D}$ can be determined as an average over all $\tilde{D}$-values from quite a number of radiograms, by which the accuracy of the result is drastically increased. The method is published in detail in Refs. 13–15 and the first result is shown in Section 8.3.1.

8.3
Influence of Thermodynamic Forces on Diffusion

The phenomenological Darken–Manning relation, which connects thermodynamic forces with diffusion, reads for binary systems

$$\tilde{D} = (N_A D_B^* + N_B D_A^*)\Phi S \tag{8.3}$$

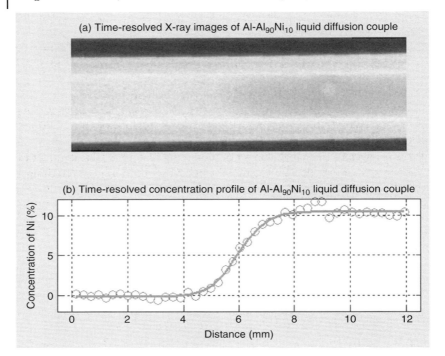

Fig. 8.4 Radiogram (a) and corresponding concentration profile (b) at an arbitrary time during an interdiffusion experiment Al versus Al$_{90}$Ni$_{10}$ at% in the liquid state.

with $\tilde{D}$ the ID coefficient, D_A^*, D_B^* the SD coefficients in the alloy, the mole fractions N_A, N_B, the thermodynamic factor Φ, and the Manning factor S. Section 8.3.1 shows an example in which all quantities of Equation 8.3 were determined. This allowed a check of the validity of the formula. Section 8.3.2 gives an impressive example of uphill diffusion that shows the control of chemical diffusion by thermodynamic forces.

8.3.1
Systems with Mixing Tendency: Al–Ni

Al–Ni has a pronounced chemical short range order that enhances the SD in the Al-rich composition region by a factor of two [16]. ID in Al$_{90}$Ni$_{10}$ and Al$_{80}$Ni$_{20}$ is also enhanced, but due to thermodynamic forces again by a factor of about 2–3. The latter behavior can be described quantitatively by the Darken–Manning relation (Equation 8.3). We could show for the first time that the Manning factor is only weakly temperature dependent but the physical meaning of S is still an open question. Moreover, the value of S is close to 1, which means that dynamic cross correlations of S are here almost negligible. Therefore one may conclude that in the given example the temperature dependence of Φ influences significantly the temperature behavior of $\tilde{D}$. A detailed analysis is given in Ref. 6 and in the related contribution of Das *et al.* in this book.

A comparison of measured and calculated diffusion coefficients shows that the MD simulation fits well to the real dynamics in the equilibrium melt above T_{liq}. Both in experiment and in simulation, the ID coefficient is higher than the SD coefficients. This is valid in the whole temperature range considered, that is in the normal liquid state as well as in the undercooled regime. In the latter regime, which is only accessible by the simulation, the difference between ID coefficient and SD coefficients increases with decreasing temperature.

These results demonstrate the power of combining experimental and numerical methods. Furthermore, we can state from Figure 8.5 that the improved LC technique with XRR shows less scattering and a smaller relative error of $\tilde{D}$ data than the conventional LC technique, which proves the benefit of the *in situ* diagnostic.

8.3.2
Systems with Demixing Tendency: Pd–Cu–Ni–P

Another example for the influence of thermodynamic forces on chemical diffusion is the observed uphill diffusion in an ID experiment in liquid Pd–Cu–Ni–P. Figure 8.6 shows the concentration profiles $c(x)$ obtained in two independent measurements—with and without buoyancy-driven convection.

Figure 8.6 depicts the demixing of Pd and P in presence of the Cu and Ni concentration gradients, respectively, caused by thermodynamic forces. This diffusive flux in opposite direction to a concentration gradient is named "uphill"

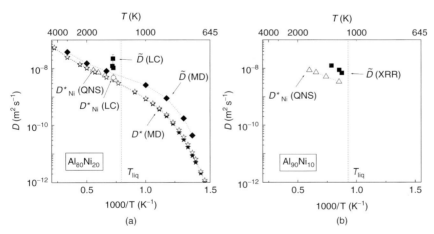

Fig. 8.5 Arrhenius plots of self-diffusion and interdiffusion coefficients in two different Al–Ni alloys as a function of inverse temperature. The abbreviations in parentheses denote the used method for determining the respective diffusion coefficient. All data in $Al_{80}Ni_{20}$ are taken from [6]. Self-diffusion data of $Al_{90}Ni_{10}$ are taken from [17]. The dashed lines are a guide to the eye. Error bars are given or are in the order of the symbol's size.

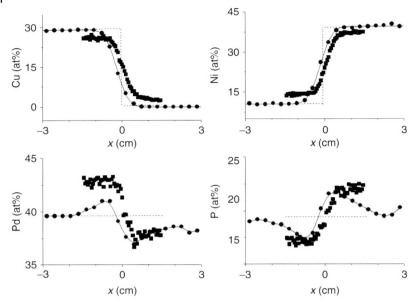

Fig. 8.6 Concentration profiles of two different interdiffusion experiments $Pd_{40}Cu_{30}Ni_{10}P_{20}$ versus $Pd_{40}Ni_{40}P_{20}$ at 1073 K. square denote all concentration profiles $c(x)$ of a long-capillary (LC) experiment in the lab measured by energy-dispersive X-ray spectrometry (EDS) [18]. Points denote the $c(x)$ of a shear cell (SC) experiment in space onboard FOTON M2 measured by inductively coupled plasma mass spectrometry (ICP-MS). The dashed lines denote the initial $c(x)$ of the experiments. The solid lines are a guide to the eye.

diffusion. For the first time such demixing by uphill diffusion in a multicomponent system was found by Darken in a solid state ID experiment between two steel alloys with different compositions [4]. Although no thermodynamic data for the quaternary PdCuNiP-system exist, the direction of the diffusive fluxes can be correlated to different heats of mixing of the binary subsystems [19]. The different $c(x)$ shapes of Pd and P under gravity and μ-gravity conditions are a result of buoyancy-driven convection. The space experiment reveals the true $c(x)$ that equals a snapshot of the time-dependent $c(x)$ in another experiment from literature (see Figure 8.7). The uphill diffusive flux of Pd and P yields a non-monotonic density distribution profile in the liquid column. This produces a destabilisation of the density layering in the vertical capillary by gravity. Thus, gravity-driven buoyancy convection is system immanent and hence correct diffusion coefficients can be obtained in general only under μ-gravity conditions.

Figure 8.7 shows an example of solid state uphill diffusion of Al during interdiffusion between two different NiCrAl alloys. It is quite obvious that in multicomponent systems a prediction of the diffusive fluxes from SD alone without knowledge of thermodynamics can be misleading. However, the phenomenological Darken–Manning approach seems to be a good approximation for the prediction of ID, if SD coefficients, thermodynamic data, and S values are available.

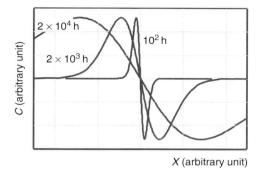

Fig. 8.7 Concentration profiles of Al as function of annealing time in the solid state interdiffusion experiment $Ni_{78}Cr_{17}Al_5$ versus $Ni_{87}Cr_8Al_5$ at 1373 K (in [20]).

Acknowledgments

Financial support by the German Science Foundation (DFG) within the Priority Program 1120 *Phase transformations in multicomponent melts* under grant no. Ma1832/3 and from DFG grant no. Gr2714/2 is gratefully acknowledged. We also would like to thank R. Schmid-Fetzer, J. Brillo, J. Horbach and A. Meyer for the fruitful collaboration that lead to results presented in this contribution.

References

1. Kirkaldy, J.S. and Young D.J. (**1987**) *Diffusion in the Condensed State*, The Institute of Metals, London, UK, pp. 127–41.
2. Onsager, L. (**1931**) *Physical Review*, **37**, 405.
3. Haasen, P. (**1984**) *Physikalische Metallkunde*, Springer Verlag, Berlin, Germany, pp. 154–59.
4. Darken, L.S. (**1948**) *Transactions of the AIME*, **180**, 430.
5. Manning, J.R. (**1961**) *Physical Review*, **124**, 470–82.
6. Horbach, J., Das, S.K., Griesche, A., Macht, M.-P., Frohberg, G. and Meyer, A. (**2007**) *Physical Review B*, **75**, 174304.
7. Shimoji, M. and Itami, T. (**1986**) *Atomic Transport in Liquid Metals*, Trans Tech Publications Ltd, Aedermannsdorf, Switzerland, pp. 154–62.
8. Griesche, A., Macht, M.-P., Garandet, J.-P. and Frohberg, G. (**2004**) *Journal of Non-Crystalline Solids*, **336/3**, 173–78.
9. Griesche, A., Kraatz, K.-H. and Frohberg, G. (**1998**) *Review of Scientific Instruments*, **69**, 315.
10. Malmejac, Y. and Frohberg, G. (**1987**) Mass transport by diffusion, in *Fluid Sciences and Materials Science in Space*, (ed H. U. Walter), Springer Verlag, Heidelberg, Germany, pp. 161–74.
11. Meyer, A. (**2002**) *Physical Review B*, **66**, 134205.
12. Suzuki, S., Kraatz, K.-H., Frohberg, G., Roşu-Pflumm, R. and Müller-Vogt, G. (**2007**) *Journal of Non-Crystalline Solids*, **353**, 3300–4.
13. Griesche, A., Garcia-Moreno, F., Macht, M.-P. and Frohberg, G. (**2006**) *Materials Science Forum*, **508**, 567–72.

14 Griesche, A., Macht, M.-P. and Frohberg, G. (2007) *Defect Diffusion Forum*, **266**, 101–8.
15 Zhang, B., Solorzano, E., Garcia-Moreno, F. and Griesche, A. (2008) *Review of Scientific Instruments*, in preparation.
16 Das, S.K., Horbach, J., Koza, M.M., Mavila Chatoth, S. and Meyer, A. (2005) *Applied Physics Letters*, **86**, 011918.
17 Griesche, A., S.M. Chathoth, M.-P. Macht, G. Frohberg, M.M. Koza, A. Meyer, (2008) *High Temperatures—High Pressures*, **37**, 153–162.
18 Griesche, A., Zumkley, Th., Macht, M.-P., Suzuki, S. and Frohberg, G. (2004) *Materials Science and Engineering A*, **375-377C**, 285–87.
19 Griesche, A., Macht, M.-P. and Frohberg, G. (2005) *Scripta Materialia*, **53**, 1395–1400.
20 Glicksman, M.E. (2000) *Diffusion in Solids*, John Wiley & Sons, New York, pp. 391–405.

9
Phase Behavior and Microscopic Transport Processes in Binary Metallic Alloys: Computer Simulation Studies

Subir K. Das[1], Ali Kerrache[2], Jürgen Horbach and Kurt Binder

9.1
Introduction

In a binary liquid mixture, different kinds of phase transitions can occur that are associated with various mass transport phenomena in the liquid. First, there is the possibility that the liquid undergoes a liquid–liquid demixing transition [1]. Near the critical point of this transition, a slowing down of dynamic properties is observed which is characterized, for example, by a vanishing interdiffusion coefficient at the critical point [2, 3]. Another possible phase transition is a first-order transition of the liquid into a crystalline structure. In this case, crystal nucleation and growth are limited by the diffusive transport in the liquid [1, 4]. In a binary liquid, crystal nucleation processes can be so slow that crystallization processes are inhibited on an experimental time scale. Then, the liquid solidifies into a metastable glass state, where one finds a structure that is very similar to that of the liquid. Toward the glass transition, transport properties such as diffusion coefficients or the shear viscosity exhibit a drastic change by many orders of magnitude in a relatively small temperature range. The detailed understanding of the glass transition is still one of the challenging problems in condensed matter physics [5].

The molecular dynamics (MD) simulation method [6, 7] is an ideal tool to study the interplay between transport processes and the aforementioned phase transitions. In the MD simulation method, Newton's equations of motion are solved for an interacting many-particle system, yielding the trajectories of all the particles, that is, the positions and velocities of the particles as a function of time. From the trajectories, any quantity of interest can be computed, in particular static and dynamic correlation functions, as well as transport coefficients. This provides insight

---

[1] Jawaharlal Nehru Center for Advanced Scientific Research, Jakkur, Bangalore 560064, India
[2] Université de Montréal, Département de Physique, C.P. 6128, succ. Centre-ville, Montréal (Québec) H3C 3J7, Canada

Phase Transformations in Multicomponent Melts. Edited by D. M. Herlach
Copyright © 2008 WILEY-VCH Verlag GmbH & Co. KGaA, Weinheim
ISBN: 978-3-527-31994-7

into mechanisms of structure formation, transport, and phase behavior on a microscopic scale. In this work, MD simulations of different model systems are used to elucidate the role of mass transport with respect to critical slowing down, glassy dynamics, and crystallization from the melt.

The dynamics in the vicinity of the critical point of a liquid–liquid demixing transition is studied for a simple model system, a symmetric Lennard–Jones (LJ) mixture. Monte Carlo simulations in the semi-grandcanonical ensemble (SGMC) [8] are used to compute the phase diagram; in particular the critical temperature is located accurately. Then, structural quantities and transport coefficients such as the self-diffusion and interdiffusion coefficients are determined by MD simulation.

Detailed predictions for the critical dynamics are provided by mode-coupling theories [9] and renormalization group theories [2, 10], giving universal exponents and amplitude ratios for systems that belong to the same universality class (see below). In a recent simulation work [11], it was claimed that the vanishing of the interdiffusion coefficient D_{AB} is not correctly described by the latter theories. In contrast to this, we show that one has to take into account "background contributions" due to the regular liquid diffusion dynamics when analyzing the behavior of D_{AB} near the critical point. Then, the simulation results for D_{AB} are fully consistent with the theoretical predictions.

To study the diffusion dynamics in glass-forming melts, a model of Al–Ni is considered. We aim at understanding the relation between chemical ordering and diffusion in this system. Simulations toward the deeply undercooled liquid are performed for $Al_{80}Ni_{20}$. Here, the fundamental question about the relation between one-particle transport and collective transport properties arises. We address this issue for the relation between self-diffusion and interdiffusion. The temperature dependence of self-diffusion and interdiffusion coefficients is discussed and a simple phenomenological relation between them, the so-called Darken equation [12], is checked.

Finally, we consider the crystal growth kinetics in $Al_{50}Ni_{50}$. The experimental melting temperature for this system is at 1920 K where it exhibits a first-order phase transition from a liquid to an intermetallic B2 phase. It is known from recent experiments that the growth kinetics in $Al_{50}Ni_{50}$ is about two orders of magnitude slower than for pure metals, indicating that $Al_{50}Ni_{50}$ may be the prototype of a system with a diffusion-limited growth mechanism. The classical model for the description of the growth of a planar crystal–liquid interface is the one by Wilson and Frenkel [13]. It considers crystal growth as an activated process, controlled by the diffusion in the liquid. We demonstrate that the Wilson–Frenkel model is not able to describe the growth kinetics of $Al_{50}Ni_{50}$ on a quantitative level.

The rest of the chapter is organized as follows: In the next section, we give the basic formulas for diffusion coefficients that are used to compute them from the MD simulation. Then, the results for a demixing binary LJ fluid (Section 9.3) and Al–Ni alloys (Section 9.4) are presented. Finally, we briefly summarize the results.

9.2
Transport Coefficients

The transport coefficients for a liquid can be calculated from the fluctuations at equilibrium. To this end, the trajectories of the particles, as provided by the MD computer simulation, are required. In this section, we report the formulas that were used to determine diffusion coefficients from equilibrium MD simulations.

Consider a three-dimensional, binary AB fluid of $N = N_A + N_B$ particles (with N_A and N_B the number of A and B particles, respectively). The self-diffusion constant $D_{s,\alpha}$ ($\alpha = A, B$) describes the tagged particle motion of species α on hydrodynamic scales. It can be calculated from the long-time limit of the mean-squared displacement [14] as shown below:

$$D_{s,\alpha} = \lim_{t\to\infty} \frac{1}{N_\alpha} \sum_{j=1}^{N_\alpha} \frac{\left\langle \left[\mathbf{r}_j^{(\alpha)}(t) - \mathbf{r}_j^{(\alpha)}(0) \right]^2 \right\rangle}{6t} \tag{9.1}$$

where $\mathbf{r}_j^{(\alpha)}(t)$ is the position of particle j of species α at time t.

The interdiffusion constant D_{AB} is related to the collective transport of mass driven by concentration gradients. It can be written as [14]

$$D_{AB} = \frac{\Lambda}{k_B T \chi} \tag{9.2}$$

with k_B the Boltzmann constant, T the temperature, χ the static concentration susceptibility, and Λ the Onsager coefficient for interdiffusion. The susceptibility χ can be expressed by the concentration structure factor $S_{cc}(q)$ at zero wavenumber, $q = 0$, as

$$\chi = \frac{S_{cc}(q=0)}{k_B T} \tag{9.3}$$

The structure factor $S_{cc}(q)$ can be determined from the partial static structure factors,

$$S_{\alpha\beta}(q) = \frac{1}{N} \sum_{k=1}^{N_\alpha} \sum_{l=1}^{N_\beta} \left\langle \exp\left[i\mathbf{q}\cdot(\mathbf{r}_k - \mathbf{r}_l)\right]\right\rangle, \quad \alpha\beta = AA, AB, BB, \tag{9.4}$$

with $c_A = N_A/N$ and $c_B = N_B/N$ the total concentration of A and B particles, respectively, via

$$S_{cc}(q) = \frac{c_B}{c_A} S_{AA}(q) + \frac{c_A}{c_B} S_{BB}(q) - 2S_{AB}(q) \tag{9.5}$$

With the definition 9.5, the function $S_{cc}(q)$ approaches 1 for $q \to \infty$.

The Onsager coefficient Λ describes the kinetic contribution to the interdiffusion coefficient. It can be expressed by a generalized mean-squared displacement [14]:

$$\Lambda = \lim_{t\to\infty} \left(1 + \frac{m_A c_A}{m_B c_B}\right)^2 N c_A c_B \frac{\left\langle \left[\mathbf{R}_A(t) - \mathbf{R}_A(0)\right]^2\right\rangle}{6t} \tag{9.6}$$

with m_A and m_B the masses of A and B particles, respectively. In Equation 9.6, $\mathbf{R}_A(t) - \mathbf{R}_A(0)$ is the displacement of the center of mass of the A Particles and

$$\mathbf{R}_A(t) - \mathbf{R}_A(0) = \int_0^t \mathbf{V}_A(t')\,dt', \qquad \mathbf{V}_A(t) = \frac{1}{N_A} \sum_{j=1}^{N_A} \mathbf{v}_j^{(A)}(t) \qquad (9.7)$$

where $\mathbf{v}_j^{(A)}(t)$ is the velocity of particle j of species A at time t. Note that the formula 9.6 is applicable in the microcanonical ensemble with zero total momentum which is the natural ensemble for MD simulations.

Formally, the Onsager coefficient can be written as a linear combination of the self-diffusion constants:

$$\Lambda = S\left(c_A D_{s,B} + c_B D_{s,A}\right) \qquad (9.8)$$

where the so-called Manning factor [15] S is defined by

$$S = 1 + \frac{\Lambda_d}{c_A D_{s,B} + c_B D_{s,A}} \qquad (9.9)$$

Λ_d comprises all the correlations between distinct particles in Equation 9.6. If $S = 1$, then the interdiffusion constant follows the Darken equation [12]:

$$D_{AB} = \left(c_A D_{s,B} + c_B D_{s,A}\right)/(k_B T \chi) \qquad (9.10)$$

Note that this equation cannot be correct in the vicinity of a critical point of a demixing transition since then the Manning factor S is expected to diverge when approaching the critical point (see below).

We note that the transport coefficients can be also expressed by the so-called Green–Kubo relations. These expressions are equivalent to the mean squared displacement formulas, presented in this section. For details on Green–Kubo relations, the reader is referred to standard textbooks [5, 14].

9.3
A Symmetric LJ Mixture with a Liquid–Liquid Demixing Transition

We consider an LJ potential for a binary AB fluid,

$$u_{LJ,\alpha\beta}(r) = 4\epsilon_{\alpha\beta}\left[\left(\frac{\sigma_{\alpha\beta}}{r}\right)^{12} - \left(\frac{\sigma_{\alpha\beta}}{r}\right)^{6}\right], \qquad \alpha, \beta = A, B \qquad (9.11)$$

which is truncated at $r = r_{cut}$ and zero for $r \geq r_{cut}$ such that the final form of the potential is

$$u_{\alpha\beta}(r) = u_{LJ,\alpha\beta}(r) - u_{LJ,\alpha\beta}(r_{cut}) - (r - r_{cut})\left(\frac{du_{LJ,\alpha\beta}(r)}{dr}\right)\bigg|_{r=r_{cut}} \qquad (9.12)$$

Here, the last term ensures that the potential and its first derivative are continuous everywhere. For the parameters in Equation 9.11, we have chosen $\sigma_{AA} = \sigma_{BB} = \sigma_{AB} = \sigma$ and $\epsilon_{AA} = \epsilon_{BB} = 2\epsilon_{AB} = \epsilon_0$. The reduced temperature is given by $T^* = k_B T/\epsilon_0$. The cut-off has been set to $r_{cut} = 2.5\sigma$. The simulations were done at a fixed reduced density $\rho^* = \rho\sigma^3 = N\sigma^3/V = 1$, considering N particles in a cubic simulation box of volume $V = L^3$ with periodic boundary conditions. At the chosen density, $\rho^* = 1$, a dense liquid is formed (rather than a gas) and, at the temperatures of interest, crystallization is not a problem. For more details on the model and its properties, we refer to the original publications [16–19].

The LJ mixture, defined by Equations 9.11 and 9.12, exhibits a liquid–liquid demixing transition. In order to determine the critical point and the binodal for this transition, Monte Carlo simulations in the SGMC [8] were performed. In this ensemble the volume V, the temperature T, and the total number of particles N are fixed, whereas N_A and N_B, respectively the number of A and B particles, are allowed to fluctuate, keeping the chemical potential difference between A and B particles, $\Delta\mu$, fixed. Before an SGMC run was started at a given temperature, the sample was equilibrated in the canonical ensemble by carrying out displacement moves of the particles using the standard Metropolis method. The SGMC run was done as follows: After 10 displacement steps per particle, $N/10$ particles were randomly chosen successively and an identity switch was attempted for these particles, controlled by the energy change caused by the switch and the chemical potential difference $\Delta\mu$. Owing to the symmetry of our model, we chose $\Delta\mu = 0$. This ensures that above the critical temperature, $T > T_c$, states with an average concentration $\langle c_A \rangle = \langle c_B \rangle = 0.5$ are sampled, while for $T < T_c$ states along the A-rich and B-rich branch of the coexistence curve in the (T, c_A) plane are obtained.

As a result of the SGMC runs, the probability distribution $P(c_A)$ of the concentration variable c_A is obtained. Examples of $P(c_A)$ for states above and below T_c are shown in Figure 9.1. The distributions are symmetric around $c_A = 0.5$ and bimodal below T_c, where the location of the two maxima corresponds to the concentrations of the coexisting A-rich and B-rich states.

Interesting information can be extracted from the moments $\langle c_A^k \rangle$ of $P(c_A)$, given by

$$\langle c_A^k \rangle = 2 \int_{0.5}^{1} c_A^k P(c_A) \, dc_A \qquad (9.13)$$

From the second moment, the concentration susceptibility for $T > T_c$ can be determined by,

$$k_B T \chi = N \left(\langle c_A^2 \rangle - 0.25 \right) \qquad (9.14)$$

Note the alternative expression for χ, as given by Equation 9.3.

Another important quantity that is related to the moments of $P(x_A)$ is the so-called fourth-order cumulant U_L, defined by

$$U_L = \frac{\langle (c_A - 0.5)^4 \rangle}{\left[\langle (c_A - 0.5)^2 \rangle^2 \right]} \qquad (9.15)$$

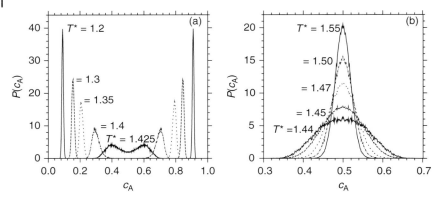

Fig. 9.1 Probability distribution $P(c_A)$ of the relative concentration c_A of A particles for $N = 6400$ and $\Delta\mu = 0$ at several temperatures, (a) below T_c and (b) above T_c [18].

This quantity can be used to locate the critical temperature [20]. In Figure 9.2a, U_L is shown for different system sizes N. In the finite-size scaling limit, the cumulants for the different N should intersect at T_c at a value that depends essentially on the universality class [20]. Indeed, the data shown in Figure 9.2a are consistent with a unique intersection point at $T_c^* = 1.4230 \pm 0.0005$ and $U_L(T_c^*) = 0.6236$. The latter value corresponds to the predicted one for the 3D Ising universality class [21]. Also the static susceptibility $\chi(T)$ follows the expected 3D Ising behavior, $\chi \propto [(T - T_c)/T_c]^{-\gamma} \equiv \epsilon^{-\gamma}$ with $\gamma \approx 1.239$, as indicated by Figure 9.2b. The correlation length ξ was estimated from the concentration structure factor via fits to the

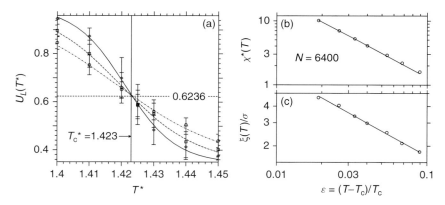

Fig. 9.2 (a) The fourth-order cumulant $U_L(T)$ plotted versus T^* for $N = 1600$ (circles), 3200 (squares) and 6400 (diamonds). The vertical straight line indicates the resulting estimate for T_c^*, while the horizontal broken straight line indicates the theoretical value that $U_L(T = T_c^*)$ should take for the 3D Ising universality class. (b) Log–log plot of the reduced susceptibility $\chi^* = \epsilon_0 \chi$. (c) Log–log plot of the reduced correlation length $\xi(T)/\sigma$ versus ϵ. The lines in (b) and (c) represent fits using the anticipated Ising exponents. The data in (b) and (c) refer to systems of $N = 6400$ particles. From Das et al. [18].

Ornstein–Zernicke form [14] $S_{cc}(q) = k_B T \chi / [1 + q^2 \xi^2 + \cdots]$ (using for χ the values estimated from the probability distributions $P(c_A)$). The Ising prediction $\xi \propto \epsilon^{-\nu}$ with $\nu \approx 0.629$ is obviously consistent with the simulation data (Figure 9.2c).

Self-diffusion and interdiffusion coefficients were determined by means of MD simulations in the microcanonical ensemble. Identical masses for particles were chosen, $m_A = m_B = m$. The velocity form of the Verlet algorithm was applied to integrate the equations of motion, using a time step $\delta t^* = 0.01/\sqrt{48}$ (with $t^* = t/t_0$ and $t_0 = (m\sigma^2/\epsilon_0)^{1/2}$). As starting configurations for the MD runs, well-equilibrated configurations as generated by SGMC were used since SGMC allows for a faster equilibration in the vicinity of the critical point (see Ref. [18, 19]). For $T^* \geq 1.5$, runs over 10^6 MD steps were performed. In the temperature range $1.5 > T^* > T_c^* = 1.423$, the MD runs lasted over 2.8×10^6 MD steps.

For $T > T_c$ and $\Delta \mu = 0$, the average concentration is $\langle c_A \rangle = \langle c_B \rangle = 0.5$. Thus, for the determination of the diffusion constants, we did runs for systems with $c_A = c_B = 0.5$ approaching the critical point from above. For this case, the symmetry of our model results in $D_{s,A}^* = D_{s,B}^* = D_s^*$ for the self-diffusion coefficient.

As can be inferred from Figure 9.3a, the self-diffusion constant exhibits a very weak temperature dependence, showing no anomalous critical behavior when approaching the critical temperature. The interdiffusion constant shows a much stronger dependence on temperature. The theoretical prediction [2, 3] for the asymptotic behavior of the interdiffusion coefficient (i.e. for $\epsilon \to 0$) is a power law, $D_{AB} \propto \xi^{-x_D} \propto \epsilon^{\nu x_D}$ with $\nu x_D \approx 0.672$. However, from the fit to D_{AB}, shown in Figure 9.3a, an effective exponent $\nu x_{eff} \approx 1.0$ is obtained. A similar observation has been made in a recent simulation study [11]. At first glance, this is surprising since the theoretical prediction is consistent with accurate experimental data [22].

As we have seen in Figure 9.2b, the susceptibility is consistent with 3D Ising criticality. Thus, the Onsager coefficient $\Lambda(T)$ has to be considered to resolve the

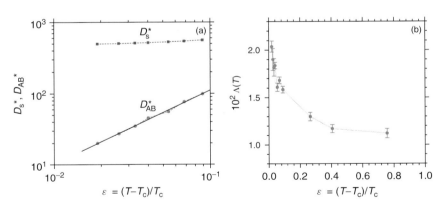

Fig. 9.3 (a) Self-diffusion and interdiffusion coefficient of the LJ mixture as a function of reduced temperature. (b) Onsager coefficient $\Lambda(T)$ for the interdiffusion plotted versus temperature. In (a) and (b), data for $N = 6400$ with $c_A = c_B = 1/2$ is used. From Das et al. [18].

origin of the apparent disagreement between theory and simulation. As Figure 9.3b shows, $\Lambda(T)$ exhibits a very weak temperature dependence far above T_c, whereas in the vicinity of T_c, a sharp rise of Λ is observed. This indicates that there is a noncritical background contribution to Λ, as has been already shown much earlier to reconcile experimental data for D_{AB} with theory [23]. Indeed, as can be formally inferred from Equation 9.8, there are contributions to Λ from the self-diffusion constant D_s that do not display any sign of a critical singularity. Thus, the Onsager coefficient can be decomposed into a singular part $\Delta\Lambda(T)$ and a noncritical background contribution $\Lambda_b(T)$,

$$\Lambda(T) = \Lambda_b(T) + \Delta\Lambda(T), \quad \Delta\Lambda(T) = L_0 \epsilon^{-\nu_\lambda} \tag{9.16}$$

with L_0 a critical amplitude. The prediction for the critical exponent is $\nu_\lambda = 0.567$. By doing a careful finite-size scaling analysis using the ansatz 9.16, one can show that the simulation data is fully consistent with the theoretical predictions [18, 19, 24].

9.4
Structure, Transport, and Crystallization in Al–Ni Alloys

Now we present the results of extensive MD simulations on Al–Ni mixtures. To model the interactions between the atoms in Al–Ni at various compositions, an embedded atom potential proposed by Mishin et al. [25] was used. All the simulations were done at zero pressure to allow direct comparison with experiments. For more details on the simulations discussed in this section, we refer to the original publications [26–29].

A useful quantity to characterize the short-range order in a liquid as well as the chemical ordering in a liquid mixture is the pair correlation function. For a binary AB mixture, three partial pair correlation functions $g_{\alpha\beta}(r)$ ($\alpha, \beta = $ A,B) are defined as [5, 14]

$$g_{\alpha\beta}(r) = \frac{N}{\rho N_\alpha (N_\beta - \delta_{\alpha\beta})} \left\langle \sum_{i=1}^{N_\alpha} \sum_{j=1}^{N_\beta} {}'\delta(r - r_{ij}) \right\rangle \tag{9.17}$$

where the prime in the second sum means that $i \neq j$ if $\alpha = \beta$; $\delta_{\alpha\beta}$ is the Kronecker delta. Physically $g_{\alpha\beta}(r)$ is proportional to the conditional probability of finding a particle of species β at a distance r from a particle of species α fixed at the center. In case of an ideal gas, there are no correlations between the particles for all distances r. Then, the functions $g_{\alpha\beta}(r)$ are equal to 1. For a liquid, the limit $g_{\alpha\beta}(r) \to 1$ is approached for $r \to \infty$.

Figure 9.4a–c displays the functions $g_{\alpha\beta}(r)$ for three different Al–Ni compositions at $T = 1525$ K. The location of the first peak in the $g_{\alpha\beta}(r)$ is essentially independent of composition, indicating that the distance between neighboring particles does not change. Thereby, the distance between NiNi and AlNi nearest neighbors is smaller

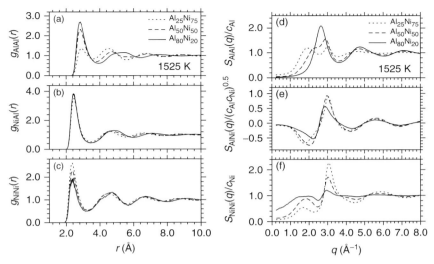

Fig. 9.4 Pair correlation functions at $T = 1525$ K for the AlAl (a), NiAl (b), and NiNi (c) correlations at different compositions, as indicated. Panels (d–f) show corresponding results for the partial static structure factors.

than that of AlAl neighbors ($r_{NiNi} \approx r_{AlNi} \approx 2.4$ Å for nearest NiNi and AlNi pairs, respectively, and $r_{AlAl} \approx 2.7$ Å for nearest AlAl pairs). The functions $g_{AlNi}(r)$ and $g_{NiNi}(r)$ exhibit only minor changes with composition for all distances, whereas in the function $g_{AlAl}(r)$ the second and third peaks shift to smaller distances with increasing Ni concentration. The latter behavior of $g_{AlAl}(r)$ is a consequence of $r_{NiNi} \approx r_{AlNi}$ being smaller than r_{AlAl}. Thus, at high Ni concentration the Al atoms are more efficiently packed into the structure, leading to a smaller periodicity of $g_{AlAl}(r)$.

Structural features on intermediate length scales can be well identified by the partial static structure factors $S_{\alpha\beta}(q)$, defined by Equation 9.4. These functions are essentially the Fourier transforms of the partial pair correlation functions. As can be inferred from Figure 9.4f, a peculiar feature of $S_{NiNi}(q)$ is the emergence of a prepeak around $q = 1.8$ Å^{-1} for all considered compositions. Such a feature is also present in $S_{AlAl}(q)$ and $S_{AlNi}(q)$ at high Ni concentration, albeit in the latter function with a negative amplitude. The occurrence of the prepeak is due to an inhomogeneous distribution of Ni atoms on intermediate length scales (for a detailed discussion see Ref. 26). In the Ni-rich Systems, a prepeak occurs in the AlAl correlations also, because then the minority species Al is built into the Ni structure.

The chemical ordering in the different Al–Ni systems leads to a nonlinear dependence of the atomic volume V on composition (Figure 9.5a). The nonlinear behavior of V is due to the more efficient packing of Al atoms in Ni-rich than in Al-rich systems. As a matter of fact, the packing of the particles is reflected in the compositional dependence of the self-diffusion constants (Figure 9.5b). Whereas

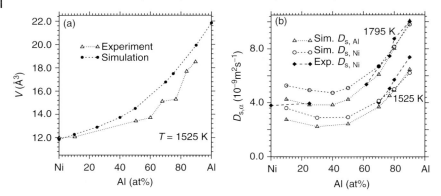

Fig. 9.5 (a) Atomic volume from the simulation, compared to experimental data from Ref. 30 as a function of composition. (b) Diffusion coefficients at 1795 and 1525 K as a function of the aluminum concentration from simulation and neutron scattering. The broken lines are guides to the eye. [26].

on the Al-rich side the diffusion coefficients show a pronounced increase with increasing Al content, on the Ni-rich side, substitution of 25% Ni by Al does not affect the diffusion coefficients significantly. This connection between the packing of the particles and their diffusional motion suggests that one can understand the essential features of mass transport in the liquid by means of hard sphere models. Also included in Figure 9.5b are results for $D_{s,Ni}$, as measured by quasi-elastic neutron scattering (QNS) experiments. Very good agreement is found between simulation and experiment.

Having discussed the dependence of the self-diffusion constants on composition, we turn now our attention to their dependence on temperature, considering the mixture $Al_{80}Ni_{20}$ in its deeply undercooled state. Here, our main aim is to understand the relation between self-diffusion and interdiffusion. In order to check the validity of the Darken Equation 9.10, we have determined the Manning factor S and the concentration susceptibility χ from extensive MD simulations [27]. S and $1/(k_B T \chi)$ are shown in Figure 9.6a. A completely different behavior as for the critical dynamics of the LJ mixture of the previous section is revealed. In Al–Ni system S is almost constant over the whole temperature range, but it diverges for the LJ system. Furthermore, the value of S is close to 1 and thus one can conclude that the Darken equation is a good approximation for $Al_{80}Ni_{20}$. On the other hand, $1/(k_B T \chi)$ for the LJ system goes to zero at the critical point, thus causing the vanishing of D_{AB} at the critical temperature. In the case of $Al_{80}Ni_{20}$, the function $1/(k_B T \chi)$ increases substantially with decreasing temperature, and therefore the interdiffusion constant D_{AB} is significantly larger than the self-diffusion constants $D_{s,Al}$ and $D_{s,Ni}$, the latter being very similar over the whole temperature range (Figure 9.6b). Again the simulation is in very good agreement with experimental data, as measured by QNS and the long capillary (LC) technique [31]. For a detailed discussion of temperature dependence of diffusion coefficients, shear viscosity, and

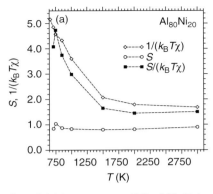
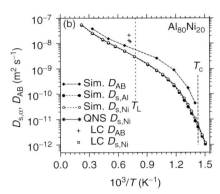

Fig. 9.6 (a) Inverse susceptibility $1/(k_B T \chi)$, "Manning" factor $S(T)$, and the product of both, as obtained from the simulation. (b) Arrhenius plot of interdiffusion and self-diffusion constants, as obtained from experiment and simulation, as indicated. The vertical dotted line in (a) marks the location of the experimental liquidus temperature, $T_L \approx 1280\,\text{K}$. The vertical dashed line is at the location of the critical temperature of mode-coupling theory, $T_c \approx 700\,\text{K}$ [27].

density correlation functions as well as their analysis in terms of the mode-coupling theory (MCT) of the glass transition we refer to Refs. 27, 28. Note that the critical temperature of MCT is around 700 K for our model of $Al_{80}Ni_{20}$, as indicated in Figure 9.6b.

The last issue in this section is on the relation between diffusion dynamics and crystal growth kinetics. To this end, we study the crystallization of $Al_{50}Ni_{50}$ from the melt, considering the temperature range $1600\,\text{K} \geq T \geq 1200\,\text{K}$. At each temperature, 12 independent samples with solid–liquid interfaces were prepared. For this, the B2 phase of $Al_{50}Ni_{50}$ was equilibrated at the target temperature for 1 ns. The simulations were done for a system of $N = 3072$ particles ($N_{Al} = N_{Ni} = 1536$) in an elongated simulation box of size $L \times L \times D$ (with $D = 3 \times L$), considering the (100) direction of the crystal. Periodic boundary conditions were employed in all three spatial directions. Having relaxed the crystal sample, one-third of the particles in the middle of the box were fixed and the rest of the system was melted during 500 ps at $T = 3000\,\text{K}$. Then, the whole system was annealed at the target temperature for another 500 ps, before we started the production runs over 1 ns in the NpT ensemble.

The behavior of samples as the one shown in Figure 9.7 depends strongly on the temperature at which they are simulated. Although, below the melting temperature T_m, the crystal will grow, it will melt above T_m. From the simulation, the velocity v_I with which the crystal–liquid interface moves can be determined. At $T = T_m$, the interface velocity v_I vanishes. Thus, by the extrapolation $v_I \rightarrow 0$, the melting temperature T_m can be estimated. In the following text, we show that this procedure yields a rather accurate estimate of T_m.

Figure 9.7 displays the partial number density profiles $\rho(z)$ of Al and Ni at $T = 1460\,\text{K}$ along the z direction, that is, perpendicular to the solid–liquid interfaces. The lower profiles in Figure 9.7 correspond to the starting configuration,

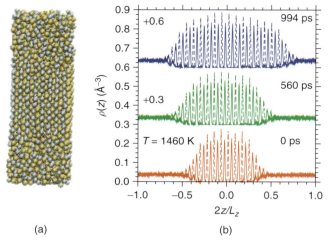

Fig. 9.7 (a) Snapshot of a simulated configuration with two crystal–melt interfaces of the system $Al_{50}Ni_{50}$. (b) Number density profiles for the system $Al_{50}Ni_{50}$ during crystal growth at $T = 1460$ K for Al (solid lines) and Ni (dashed lines). The profiles corresponding to $t = 560$ ps and $t = 994$ ps are shifted with respect to the $t = 0$ ps profiles by 0.3 and 0.6 Å, as indicated [29].

while the second and the third ones correspond to $t = 560$ ps and 994 ps. Note that in Figure 9.7 the z coordinate is scaled by the factor $2/D$, placing $z = 0$ in the middle of the simulations box. Because the crystal structure leads to pronounced peaks in $\rho(z)$, a constant density is observed for the liquid regions along the z direction, as expected. We can also infer from Figure 9.7 that the intermetallic B2 phase (here in (100) orientation) exhibits a pronounced chemical ordering, characterized by the alternate sequence of Al and Ni layers. This indicates that, different from one-component metals, the crystal growth kinetics relies on local rearrangements in the liquid structure. Thus, one may expect that diffusive transport is required to bring the atoms of each species to a suitable site in the B2 crystal. As one can further see in Figure 9.7, the crystal is growing at $T = 1460$ K. Thus, this temperature is below the melting temperature of our $Al_{50}Ni_{50}$ model.

Since the density of the crystalline B2 phase is higher than that of the liquid phase, the total volume of the system decreases at temperatures $T < T_m$ whereas it increases above T_m. Figure 9.8a shows the time dependence of the volume per particle, V_p, for different temperatures between 1400 and 1600 K. From this plot, one can infer that the melting temperature is between 1500 and 1530 K. Also shown in Figure 9.8a are examples of linear fits of the form $f(t) = A - \dot{V}_p t$. We use these fits to determine the change of the volume $\dot{V}$ per unit time.

From the volume change $\dot{V}_p$, the velocity v_I, with which the crystal–liquid interfaces move, can be estimated as $v_I = [\dot{V}_p d]/[2N_l(V_c - V_l)]$. In this formula, the product $N_l(V_c - V_l)$ quantifies the increase of the volume caused by the addition of a crystalline layer (with N_l the average number of particles in a layer, and V_c and V_l

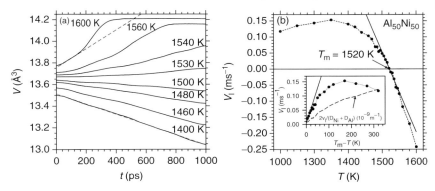

Fig. 9.8 (a) Atomic volume V for the system $Al_{50}Ni_{50}$ as a function of time for different temperatures, as indicated. The bold dashed lines are examples of linear fits from which the volume velocity V is determined. (b) Interface velocity as a function of temperature (filled circles, the dashed line is a guide to the eye). The bold line is a linear fit, $v_I = k(T_m - T)$, yielding the kinetic growth coefficient $k = 0.0025$ m s^{-1} K^{-1} and the melting temperature $T_m = 1520$ K. The inset shows the interface velocity as a function of undercooling $T_m - T$ and the interface velocity divided by the averaged self-diffusion coefficient (dashed line) [29].

the specific volumes of the crystal and the liquid phase, respectively). The length d is the spacing between crystalline layers.

Figure 9.8b displays the interface velocity v_I as a function of temperature. We see that v_I vanishes around 1520 K and thus this temperature is the estimate for the melting temperature T_m of our simulation model. Note that the experimental value for T_m is around 1920 K and so our simulation underestimates the experimental value by about 20%. We attribute this discrepancy to the fact that the effective potential [25] is not accurate enough to predict the differences in the thermodynamic potentials of molten and crystalline $Al_{50}Ni_{50}$ correctly. Around T_m, the simulation data for v_I can be fitted by the linear law $V_I = k(T_m - T)$ where the fit parameter k is the so-called kinetic coefficient. The fit, that is shown in Figure 9.8b, yields the value $k = 0.0025$ m s^{-1} K^{-1}. This value is about two orders of magnitude smaller than the typical values for kinetic coefficients that have been found in simulations of one-component metals [4, 32, 33].

The inset in Figure 9.8b shows the interface velocity as a function of undercooling $\Delta T = T_m - T$. We see that v_I increases linearly up to an undercooling of about 30 K. At $\Delta T \approx 180$ K, the interface velocity reaches a maximum value of about 0.15 m s^{-1}. Note that our simulation data are in good agreement with recent experimental data on $Al_{50}Ni_{50}$, measured under reduced gravity conditions during a parabolic flight campaign [34].

The very low value of the kinetic coefficient indicates that the crystal growth in $Al_{50}Ni_{50}$ is limited by mass diffusion in the liquid, as proposed by the classical model by Wilson and Frenkel [13]. However, we show in Ref. 29 that, on the basis of sensible assumptions of various parameters, even the Wilson–Frenkel model overestimates the kinetic coefficient for $Al_{50}Ni_{50}$ by at least one order of

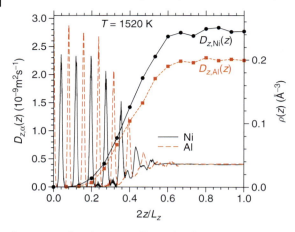

Fig. 9.9 Number density profiles and diffusion profiles for Ni and Al in $Al_{50}Ni_{50}$ at coexistence, as indicated [29].

magnitude. The reason for this can be elucidated by the simulation, which reveals that mass diffusion in the crystal–liquid interface controls crystal growth. Since the diffusion in the interface region is much slower than in the bulk liquid (Figure 9.9), the kinetic coefficient is much lower than expected from the Wilson–Frenkel model.

9.5
Summary

Computer simulations have been used to study the interplay between structure, transport, and phase behavior of binary alloys. We have demonstrated how the detailed information, as provided by the simulation, gives insight into the microscopic mechanisms of mass transport. The critical dynamics of a phase-separating LJ system as well as the diffusion dynamics of glass-forming Al–Ni melts and the crystal growth kinetics of B2-$Al_{50}Ni_{50}$ have been considered. In all these cases, different types of diffusive motion of the particles play a crucial role. In the LJ system, there is no sign of a critical anomaly in the self-diffusion although the interdiffusion constant D_{AB} vanishes owing to the critical slowing down. However, when analyzing D_{AB} with respect to the theoretically predicted critical power laws, noncritical fluctuations, as the ones reflected in the self-diffusion constants, have to be taken into account. In the Al–Ni systems, the differences between self-diffusion and interdiffusion are mainly due to the static susceptibility that appears in the expression for D_{AB}. In this case, the Darken equation is thus a good approximation. As the crystal growth in $Al_{50}Ni_{50}$ is limited by the diffusion in the melt and in the crystal–liquid interface, it cannot be described by the classical Wilson–Frenkel theory at a quantitative level. New microscopic approaches are required for a deeper understanding of crystallization processes from the melt.

Acknowledgments

We thank M. E. Fisher, A. Griesche, S. Mavila Chathoth, A. Meyer, and J. V. Sengers for fruitful collaborations that lad to part of the results presented in this review. We gratefully acknowledge the financial support within the SPP 1120 of the Deutsche Forschungsgemeinschaft (DFG) under Grant No. Bi314/18 and the computing time on the JUMP at the NIC Jülich.

References

1 Kostorz, G. (ed) (2001) *Phase Transformations in Materials*, Wiley-VCH Verlag GmbH, Berlin.
2 Hohenberg, P.C. and Halperin, B.I. (1977) *Reviews of Modern Physics*, 49, 435.
3 Onuki, A. (2002) *Phase Transition Dynamics*, Cambridge University Press, Cambridge.
4 Herlach, D., Galenko, P. and Holland-Moritz, D. (2007) *Metastable Solids from Undercooled Melts*, Elsevier, Amsterdam.
5 Binder, K. and Kob, W. (2007) *Glassy Materials and Disordered Solids: An Introduction to Their Statistical Mechanics*, World Scientific, London.
6 Allen, M.P. and Tildesley, D.J. (1987) *Computer Simulation of Liquids*, Clarendon Press, Oxford.
7 Binder, K., Horbach, J., Kob, W., Paul, W. and Varnik, F. (2004) *Journal of Physics: Condensed Matter*, 16, S429.
8 Landau, D.P. and Binder, K. (2005) *A Guide to Monte Carlo Simulations in Statistical Physics*, 2nd edn, Cambridge University Press, Cambridge.
9 Kadanoff, L.P., Swift, J. (1968) *Physical Review A*, 166, 89   (a) Kawasaki, K. (1970) *Physical Review B*, 1, 1750   (b) Mistura, L. (1975) *Journal of Chemical Physics*, 62, 4571   (c) Luettmer-Strathmann, J., Sengers, J.V., Olchowy, G.A. (1995) *Journal of Chemical Physics*, 103, 7482   (d) Onuki, A. (1997) *Physical Review E*, 55, 403.
10 Siggia, E.D., Halperin, B.I., Hohenberg, P.C. (1976) *Physical Review B* 13, 2110;   (a) Folk, R. and Moser, G. (1995) *Physical Review Letters*, 75, 2706.
11 Jagannathan, K. and Yethiraj, A. (2004) *Physical Review Letters*, 93, 015701.
12 Darken, L.S. (1949) *Transactions of AIME*, 180, 430.
13 Wilson, H.A. (1900) *Philosophical Magazine* 50, 238   (a) Frenkel, J. (1932) *Physikalische Zeitschrift der Sowjetunion*, 1, 498.
14 Hansen, J.-P. and McDonald, I.R. (1986) *Theory of Simple Liquids*, Academic Press, London.
15 Manning, J.R. (1961) *Physical Review*, 124, 470.
16 Das, S.K., Horbach, J. and Binder, K. (2003) *Journal of Chemical Physics*, 119, 1547.
17 Das, S.K., Horbach, J. and Binder, K. (2004) *Phase Transitions*, 77, 823.
18 Das, S.K., Fisher, M.E., Sengers, J.V., Horbach, J. and Binder, K. (2006) *Physical Review Letters*, 97, 025702.
19 Das, S.K., Horbach, J., Binder, K., Fisher, M.E. and Sengers, J.V. (2006) *Journal of Chemical Physics*, 125, 024506.
20 Binder, K. (1981) *Zeitschrift für Physik B*, 43, 119.
21 Wilding, N.B. (1997) *Journal of Physics: Condensed Matter*, 9, 585.
22 Burstyn, H.C. and Sengers, J.V. (1982) *Physical Review A*, 26, 448.
23 Sengers, J.V., Keyes, P.H. (1971) *Physical Review Letters*, 26, 70; Swinney, H.L., Henry, D.L. (1973) *Physical Review A*, 8, 2586.
24 Das, S.K., Sengers, J.V. and Fisher, M.E. (2007) *Journal of Chemical Physics*, 127, 144506.

25 Mishin, Y., Mehl, M.J. and Papaconstantopoulos, D.A. (**2002**) *Physical Review B*, **65**, 224114.
26 Das, S.K., Horbach, J., Koza, M.M., Mavila Chathoth, S. and Meyer, A. (**2005**) *Applied Physics Letters*, **86**, 011918.
27 Horbach, J., Das, S.K., Griesche, A., Macht, M.-P., Frohberg, G. and Meyer, A. (**2007**) *Physical Review B*, **75**, 174304.
28 Das, S.K., Horbach, J., *Physical Review B*, submitted.
29 Kerrache, A., Horbach, J. and Binder, K. (**2008**) *Europhysics Letters*, **81**, 58001.
30 Ayushina, G.D., Levin, E.S. and Geld, P.V. (**1969**) *Russian Journal of Physical Chemistry*, **43**, 2756.
31 For a Detailed Discussion of Experimental Results of the Diffusion Constants, as Obtained by QNS and the LC Technique, we refer to chapter 7 (Holland-Moritz *et al.* and chapter 8 (Friesche *et al.*) in this book.
32 Broughton, J.Q. and Gilmer, G.H. (**1986**) *Journal of Chemical Physics*, **84**, 5759.
33 Jackson, K.A. (**2002**) *Interface Science*, **10**, 159.
34 Reutzel, S., Hartmann, H., Galenko, P.K., Schneider, S. and Herlach, D.M. (**2007**) *Applied Physics Letters*, **91**, 041913.

10
Molecular Dynamics Modeling of Atomic Processes During Crystal Growth in Metallic Alloy Melts

Helmar Teichler and Mohamed Guerdane

10.1
Introduction

The project was aimed to provide, by molecular dynamics (MD) simulations, the characteristic parameters necessary to describe the processes during the crystallization of multicomponent metallic melts. Recent improvements in modeling atomic interactions and the enormous increase in computer capacity have made MD simulations a powerful tool to evaluate material properties at the mesoscale beyond the size limitations of full atomistic first-principles calculations. For the ergodic metallic melts, isothermal MD simulations allow determination of various kinds of quantities, such as equilibrium averages of the enthalpy or transport parameters like diffusion coefficients, from modeled atomic trajectories, provided the simulation time is long enough to approximate the thermodynamic ensemble average by the time average of the system.

In addition to these parameters, understanding the crystallization process of alloy melts needs knowledge of the thermodynamic potentials of the involved phases and of the corresponding chemical potentials, as they are the driving forces for the phase transition. This means, in essence, a need for knowing the entropies of crystalline and liquid alloy phases.

The particular importance of the thermodynamic potentials for modeling the crystallization process is best demonstrated by the phase field (PF) theories in the present research program, which deduce the microstructure of the solidified system, which means for example dendritic growth and lamellar or globular structure, from competition between the thermodynamic potentials of the involved phases and their growth parameters during crystallization. At present, the PF theories use either entropy data from CALPHAD analysis of experimental parameters or related methods, or they rely on phenomenological model assumptions for the free enthalpy (FE). MD modeling of entropy and FE of the liquid phase may broaden the basis for estimating these quantities. It may also help to gain a deeper understanding of these quantities from the atomistic point of view, which may support also our understanding of the liquid-related part of multicomponent phase diagrams.

Phase Transformations in Multicomponent Melts. Edited by D. M. Herlach
Copyright © 2008 WILEY-VCH Verlag GmbH & Co. KGaA, Weinheim
ISBN: 978-3-527-31994-7

Realizing the need for having suitable tools to determine the FEs of melts, we laid our main emphasis in the present project on implementing methods for evaluating these quantities from MD simulations. Two such methods were developed and they are addressed below. The first one, limited to binary alloy systems, was formulated in the initial period of the project. It was applied in the second period to the Zr-rich NiZr system. In this context, we demonstrated that the estimated entropy and FE of the liquid in competition with that of crystalline phases describes well the Zr-rich part of the experimental phase diagram.

During the second period of the project, for the NiZr system the propagation of a crystalline Zr/liquid-Ni_xZr_{1-x} solidification front was simulated by MD and the parameters were estimated for PF modeling this process. The related PF calculations were carried out by Nestler and her group. The comparison between PF results and the MD data for the propagation of the solidification front and the spatial distribution of the composition of the melt are sketched in the following text.

The third period of the project, which covered only one year of funding, was devoted to implementing an alternative method for MD estimating the entropy of alloy melts, applicable to ternary and higher component liquids. Results of this method for the $Al_xNi_{40-x}Zr_{60}$ system also will be discussed.

10.2
Entropy and Free Enthalpy of Zr-rich Ni_xZr_{1-x} Melts from MD Simulations and Their Application to the Thermodynamics of Crystallization

10.2.1
Survey

The investigation addressed here was aimed at developing a preliminary method for evaluating entropy and FE of liquid binary metallic alloys, where the particular model of Zr-rich Ni_xZr_{1-x} melts was considered. The study can be subdivided into three parts:

(a) development of the method
(b) test of the internal reliability of the method
(c) application to the Zr-rich part of the (x, T) phase diagram for NiZr.

In the following sections, the results of these three parts are sketched. The details of the investigation are documented in the Thesis of Küchemann [1].

10.2.2
Results and Their Meaning

10.2.2.1 The Method that Works
The method consists of two steps: (i) evaluation of entropy and FE from MD simulations at compositions x_α with congruent melting point, such as $Ni_{0.5}Zr_{0.5}$,

10.2 Entropy and Free Enthalpy of Zr-rich Ni_xZr_{1-x} Melts from MD Simulations

$Ni_{0.33}Zr_{0.67}$, or the monatomic cases $x = 0$ and 1; (ii) evaluating from MD data the entropy and FE for general composition x using the results of (i).

For ordered crystalline phases with negligible defect density, at sufficient low temperature T_0, the entropy can be well approximated by the vibration entropy, $S_{vib}(T_0)$. $S_{vib}(T_0)$ requires the knowledge of the vibration spectrum, a quantity easily accessible to MD simulations from the velocity correlation function. At composition x_α with congruent melting point, the entropy of the melt above the melting temperature T_m there from follows by integrating $dS(T) = c_p(T)/T dT$ from T_0 to $T > T_m$, using $S_{vib}(T_0)$ as the initial condition. $c_p(T)$ means the specific heat of the crystal and melt, which for both phases results as $\partial_T H(T)$ from the MD enthalpy data, $H(T)$. In addition, the singular melting entropy, $\Delta S(T_m) = [H_{liq}(T_m) - H_{cry}(T_m)]T_m$, has to be added.

From the entropy $S(T, x_\alpha)$, FE values in the liquid at compositions x_α, $g(T, x_\alpha)$, follow by

$$g(T, x_\alpha) = H(T, x_\alpha) - TS(T, x_\alpha) \tag{10.1}$$

A more elaborate treatment is necessary for deducing entropy and FE in the melt for general compositions x, outside those with congruent melting reactions. In this case, we determine the change of the chemical potential with composition x, $\partial_x[\mu_{Ni}(T, x) - \mu_{Zr}(T, x)]$, from analyzing the spatial change of the Ni concentration, $x(r)$, under the influence of a spatially varying auxiliary potential applied to the Ni atoms [1]. Use of the equation

$$\partial_x^2 g(T, x) = \partial_x[\mu_{Ni}(T, x) - \mu_{Zr}(T, x)] \tag{10.2}$$

yields $g(T, x)$ by integrating $\partial_x[\mu_{Ni}(T, x) - \mu_{Zr}(T, x)]$ twice with respect to x. (Here and in the following text, all extensive thermodynamic quantities refer to the density per atom.) The approach needs two integration constants, which we determine from the FEs of the congruently melting compositions $x = 0.33$ and 0.5, evaluated by Equation 10.1. This completes our method for evaluating the entropy and FE of the binary metallic melt.

10.2.2.2 Free Enthalpy Results for Zr-rich Ni_xZr_{1-x} Melts

Figure 10.1 shows the FE of Zr-rich Ni_xZr_{1-x} binary melts at 1600 K deduced by our approach. In addition, the FE of the high-temperature bcc Zr crystal is included and a "double-tangent" that determines the equilibrium concentration $x_L(T = 1600\,K)$ between crystalline Zr and liquid Ni_xZr_{1-x}. The FE of Zr crystal is determined from the MD-simulated data according to Equation 10.1, with the modification that T remains below T_m.

From the values at one fixed T, the FEs of melt, $g(T, x)$, and Zr crystal, $g_c(T)$, can be evaluated for any arbitrary T by use of the specific heat of melt, $c_{p,liq}(T, x)$, and Zr crystal, both accessible to MD simulations from the T dependence of the enthalpies. Therefrom one obtains $x_L(T)$ by the double-tangent construction for an

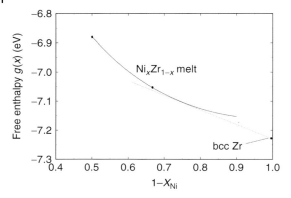

Fig. 10.1 Free enthalpy at 1600 K for Zr-rich Ni_xZr_{1-x} melts and bcc Zr from MD simulations [1]. The gray line presents the "double-tangent" that allows determination of the liquidus composition $x_L(1600\ K)$.

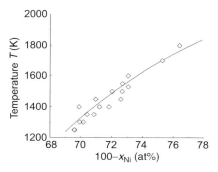

Fig. 10.2 Liquidus line $x_L(T)$ between Ni_xZr_{1-x} melt and bcc Zr from free enthalpy estimates (solid line) and equilibrium compositions from modeling $[Ni_xZr_{1-x}]_{liquid}$ and $Zr_{crystal}$ layer structures (diamonds) [1].

arbitrary T. Figure 10.2 displays the resulting $x_L(T)$ that means the liquidus curve between the melt and crystalline Zr.

Independent of the estimated thermodynamic potentials, the liquidus line is accessible to isothermal MD simulations by modeling layered structures of Zr crystal and Ni_xZr_{1-x} liquid parts as shown in Figure 10.3. At constant T, the composition of the melt in this layered system will change by growth or shrinkage of the crystalline layer thickness until the equilibrium concentration $x_L(T)^*$ in the melt is reached. The $x_L(T)^*$ values determined by this approach are included in Figure 10.2.

The excellent agreement between $x_L(T)$ and $x_L(T)^*$ can be interpreted in two ways: on one hand, it proves that isothermal simulations do reproduce the thermodynamic equilibrium between the phases as predicted by the thermodynamic relations. On

Fig. 10.3 Equilibrated [Ni$_x$Zr$_{1-x}$]$_{liquid}$ and Zr$_{crystal}$ two-phase layer structure from MD simulation showing two crystal–liquid interfaces.

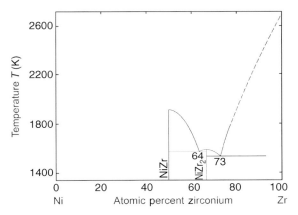

Fig. 10.4 Zr-rich part of the Ni$_x$Zr$_{1-x}$ phase diagram constructed by use of the free enthalpy estimates from MD analysis [1].

the other, it shows that the estimated FEs are the quantities governing the simulated phase equilibrium.

10.2.2.3 Zr-rich Part of the (x, T) Phase Diagram

Figure 10.4 displays the Zr-rich part of the (x, T) phase diagram, determined from simulated FEs of the melt and of the crystalline NiZr and NiZr$_2$ phases by use of the double-tangent construction. As a by-product, the construction yields the eutectic temperatures and the eutectic compositions, which complete the Zr-rich part of the phase diagram.

For the MD-simulated part of the diagram, there is encouraging agreement with the experimental one [2], in particular regarding the eutectic compositions $x^E_{Zr} = 64$ and 73 at% Zr (experimentally, $x^E_{Zr} = 64$ and 76 at% Zr). However, it needs reducing the temperatures from MD modeling by about 20% to bring the simulated T_m, eutectic temperatures and the liquidus lines in acceptable accordance with the experimental values (up to T_m of NiZr$_2$, where a larger deviation is found). The difference in temperature scale was already observed in Ref. 3 regarding T_m of NiZr. There we emphasized that the MD data rely on atomic interaction potentials adapted to the first-principles theoretical results of Hausleitner and Hafner [4] for binary transition metal liquids, which apparently imply too large temperatures.

One result of the present analysis is the explicit demonstration that there are no particular features in the FE of the melt around the eutectic concentrations and temperature. This disproves a long-standing conjecture [5], according to which distinguished spatial structures characterize the eutectic concentrations: in order to be of effect in thermodynamics, such structures have to be visible in the FE. Our calculations definitely show that the eutectic points are due to a competition of crystal FEs and a *smooth* FE of the melt.

10.3
Bridging the Gap between Phase Field Modeling and Molecular Dynamics Simulations. Dynamics of the Planar [Ni$_x$Zr$_{1-x}$]$_{liquid}$ − Zr$_{crystal}$ Crystallization Front

10.3.1
Survey

It is a particular challenge of present material sciences to provide a link between atomistic descriptions of matter, such as *ab initio* calculations or MD simulations, and meso- and macroscale approaches, such as PF modeling or elasticity theory. Regarding this, we have carried out, in cooperation with Prof. B. Nestler and her group, a combined MD and PF investigation on the propagation of a planar [Ni$_x$Zr$_{1-x}$]$_{liquid}$ − Zr$_{crystal}$ crystallization front under nonequilibrium conditions. The details of the investigation are compiled in a manuscript submitted for publication [6].

In detail, we consider a two-phase crystal–liquid sample equilibrated at temperature T brought into a nonequilibrium state by an abrupt temperature drop ΔT. The subsequent propagation of the [Ni$_x$Zr$_{1-x}$]$_{liquid}$ − Zr$_{crystal}$ crystallization front at $T - \Delta T$ is considered by MD and PF modeling and the changes of the concentration profiles are investigated, which take place while the liquid relaxes into the new equilibrium. The particular questions here are whether PF modeling is applicable down to the range of the atomistic structure and whether isothermal nonequilibrium MD simulations are capable of treating properly relaxation dynamics driven by thermodynamic forces.

The MD simulations are carried out with the Ni$_x$Zr$_{1-x}$ model used in the previous section. The PF modeling relies on the formulation of alloy solidification recently proposed for a general class of multicomponent, multiphase systems [7].

10.3.2
Results and Their Meaning

10.3.2.1   MD-Generated Input Parameter for PF Modeling
The PF modeling needs parameters from outside, such as the liquidus line $c_L(x)$, the interface thickness ε, the diffusion coefficients D_S and D_L of the solute in solid and liquid phase, the equilibrium partition coefficient k_E, and the interface mobility ν.

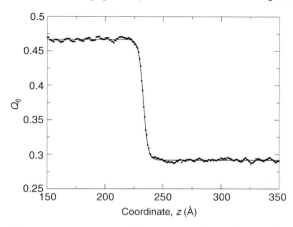

Fig. 10.5 $Q_6(z)$ parameter from MD simulation (dots) describing the transition from liquid Ni_xZr_{1-x} (Q_6^L) to crystalline Zr (Q_6^S) layer structure and the adapted analytical expression $Q_6(z) = Q_6^L + (Q_6^S - Q_6^L)\varphi(z)$ for the static interface in the PF approach [6].

The liquidus line is estimated as described in Section 10.2.2.2. Further data, up to v, follow from isothermal MD simulations of a crystal–liquid sandwich under equilibrium conditions. Estimating v needs information from (nonequilibrium) interface dynamics. Figure 10.5 presents the results at 1900 K for Steinhardt's Q_6 parameter [8] used to discriminate the crystalline ($Q_6^S = 0.467$) and the molten ($Q_6^L = 0.292$) part and the smooth transition in between. The solid line presents the analytical expressions $Q_6(z) = Q_6^L + (Q_6^S - Q_6^L)\varphi(z)$. Here, $\varphi(z) = [1 - \text{th}(3z/2\varepsilon)]/2$ is the static kink solution of the φ^4 potential used in PF theory to model the liquid ($\varphi = 0$) to solid ($\varphi = 1$) transition. The Q_6 data at 1700 K and similar data for the diffusion coefficients allow deducing the coefficients D^S, D^L, and the thickness parameter ε (= 0.613 nm). Moreover, the ratio of the Ni concentration in the solid and melt gives a partition coefficient $k = 0.036$.

The nearly perfect match between the simulated profiles of $Q_6(z)$ and the theoretical curve of the PF approach means that the mathematical features of the latter describe well the MD-simulated interface structure down to the atomistic scale.

10.3.2.2 Comparison of MD and PF Results for the Concentration Profiles and Propagation of the Crystallization Front

The MD simulation of propagation of the crystallization front makes use of an orthorhombic basic volume of about 178 nm length and 3.8×3.8 nm^2 cross section, repeated periodically in all three dimensions. The basic volume contains two crystal–liquid interfaces. The structure is equilibrated at 1900 K and quenched into a nonequilibrium state at 1700 K. In the subsequent equilibration process, interface propagation is recorded, caused by crystallization of the melt. In order to model electronic heat transport in the metallic system, heat production from

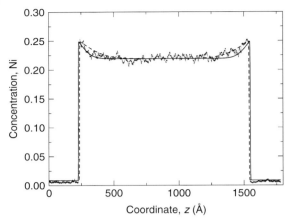

Fig. 10.6 Ni concentration profile in a $[Ni_xZr_{1-x}]_{liquid}$ and $Zr_{crystal}$ layer package after 20 ns of annealing at 1700 K with nonequilibrium Ni enhancement in the melt in front of the interfaces according to MD simulation (dots) and PF modeling (dashed line: $D_L = 6.47$ nm² ns⁻¹; solid line: $D_L = 1.64$ nm² ns⁻¹) [6].

crystallization is compensated by keeping the interfaces and their surroundings at constant temperature, in addition to the global temperature control of the system by a Berendsen thermostat. This keeps the temperature spatially constant over the whole structure up to temporary local fluctuations of less than 1%, even after more than 50 ns of crystallization. The PF modeling uses the MD structure at 1900 K as the initial configuration and studies its evolution at 1700 K.

Figure 10.6 presents the simulated Ni concentration profile after 20 ns of annealing. Between 220 and 1560 Å there is the molten region with nonequilibrium Ni enhancement in front of the interface. By crystallization, the interface moves with time, as displayed by Figure 10.7. The figure indicates that the crystal growth follows closely the parabolic law. In both figures, the dots mean the MD data, while the smooth solid, the dashed, and the dash-dotted lines are the results of PF modeling. In Figure 10.7, the dashed line is obtained from PF calculations using the MD-simulated diffusion coefficients. The solid line results from enhancing them by a factor 4. Obviously, the latter fit well the MD data, whereas in the former case there are significant deviations. Moreover, from comparing the PF and MD data, the interface mobility v can be estimated. In case of the original MD values for D_S and D_L, the deviations between PF and MD crystallization velocity data and between the Ni profiles cannot be brought to match by varying v (see, e.g., the dash-dotted line in Figure 10.7). Agreement between the data needs the mentioned increase of the diffusion coefficients by a factor 4. Then, a value of $v = 10^{-7}$ m² s⁻¹ follows.

The present observations demonstrate that mesoscale PF modeling of the crystallization dynamics of a nonequilibrium [100]Zr crystal–Ni_xZr_{1-x} liquid ($x \approx$

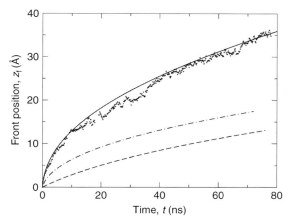

Fig. 10.7 Change of interface position during annealing time t from MD simulation (dots) PF modeling using different data sets (solid line: $D_L = 6.47$ nm^2 ns^{-1}, $\nu = 100$ nm^2 ns^{-1}; dash-dotted line: $D_L = 1.64$ nm^2 ns^{-1}, $\nu = 100$ nm^2 ns^{-1}; dashed line: $D_L = 1.64$ nm^2 ns^{-1}, $\nu = 250$ nm^2 ns^{-1}) [6].

0.22) interface yields results comparable to those of atomistic MD treatments, provided the diffusion coefficients are suitably rescaled. At present, one can only speculate about the reason for the discrepancy. There remains the question of so-far-hidden differences in the processes described by the MD and PF approach or, for example the role of activity coefficients in the chemical potential of the liquid and crystal phases in PF modeling.

10.4
Entropy and Free Enthalpy in Ternary $Al_y Ni_{0.4-y} Zr_{0.6}$ Alloy Melts

10.4.1
Survey

At the beginning of the last funding period, which covered only one year for the present project, we found a way to calculate by MD simulations the entropy and FE of multicomponent alloy melts, not amenable to treatment by our initial approach given in Section 10.2. Considering its importance in, for example modeling ternary and higher-components phase diagrams or providing the input FEs for a PF description of microstructure evolution under crystallization in such systems, we concentrated in the last year of funding on this topic. A manuscript for publication of the material is in preparation (authors H. Teichler and M. Guerdane).

Below we sketch some basic elements of this approach and we provide as special results a test of the numerical reliability of the method as well as some findings about the entropy in ternary $Al_y Ni_{0.4-y} Zr_{0.6}$ alloy melts.

The analysis of $Al_yNi_{0.4-y}Zr_{0.6}$ relies on our earlier MD model [9]. In the binary $Ni_{0.4}Zr_{0.6}$ alloy melt, the Ni atoms, owing to size mismatch, reside in tri-capped trigonal-prismatic holes (and derivatives thereof) within the Zr matrix, while the binary $Al_{0.4}Zr_{0.6}$ melt is close to a polytetrahedral structure with icosahedral basic elements. Intermediate compositions have to combine both structure types. Our previous MD study [9] of $Al_{0.15}Ni_{0.25}Zr_{0.6}$ melts shows that the Al atoms reside in icosahedral 12-neighbor holes that tend to form linear arrangements, while the Ni atoms are hosted in 9-neighbor trigonal prisms dispersed in between. Regarding this, there are the obvious questions of the ratio between the entropies of the binary border composition melts, $Ni_{0.4}Zr_{0.6}$ and $Al_{0.4}Zr_{0.6}$, and for the entropy in the mixed $Al_yNi_{0.4-y}Zr_{0.6}$ melts.

10.4.2
Results and Their Meaning

10.4.2.1 The Method: Test of Its Numerical Reliability

The method determines absolute values of entropy for alloy melts at compositions with congruent melting point, as described in Section 10.2.2.1. In our treatment of $Al_yNi_{0.4-y}Zr_{0.6}$ melts, we use, for example $Ni_{0.5}Zr_{0.5}$ as the reference system.

For evaluating by MD simulations the entropy at arbitrary compositions, we make use of Kirkwood's coupling parameter approach [10], which allows evaluation of isothermal FE differences, Δg_{fi}, between a final (f) and an initial (i) composition of the melt at a fixed T. The method introduces auxiliary intermediate systems, which smoothly couple the Hamilton function of i to that of f. Evaluation of Δg_{fi} needs MD simulations for the intermediate systems. From Δg_{fi}, the entropy change follows according to

$$\Delta S_{fi}(T) = \frac{[\Delta H_{fi}(T) - \Delta g_{fi}(T)]}{T} \quad (10.3)$$

with $\Delta H_{fi}(T)$ the enthalpy difference between f and i at T.

As an example, Figure 10.8 displays data of the entropy change in $Al_yNi_{0.4-y}Zr_{0.6}$ melts with varying Al concentration at 1700 K. For each concentration y, MD simulations are carried out for a set of 11 auxiliary intermediate systems coupling smoothly $Al_yNi_{0.4-y}Zr_{0.6}$ at fixed y to the system at $y = 0$.

In order to test the numerical reliability of the method, we have carried out a set of simulations along the closed path $Ni_{0.4}Zr_{0.6} \rightarrow Al_{0.4}Zr_{0.6} \rightarrow Zr \rightarrow Ni_{0.4}Zr_{0.6}$ at $T = 2400$ K. The arrows indicate pairs of configurations for which the entropy differences are evaluated. For these differences we find 0.0404, 0.3611, and $-0.3935 k_B$/atom. The closure failure, that means the sum of the three contributions, turns out as 0.008 k_B. For $T = 2800$ K a similar analysis yields a failure of 0.009 k_B. The small values of the closure failure can be used as an indication that there is high numerical precision in the calculated entropy differences, with uncertainty of the order of 0.01 k_B or even better.

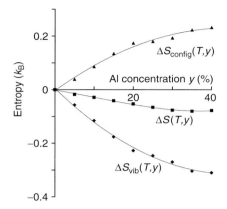

Fig. 10.8 MD results for the change of entropy in $Al_y Ni_{0.4-y} Zr_{0.6}$ against $Ni_{0.4} Zr_{0.6}$ showing the total structural entropy $\Delta S(T; y)$, the vibrational contribution $\Delta S_{vib}(T; y)$, and the configuration part $\Delta S_{config}(T; y)$ at $T = 1700$ K.

10.4.2.2 Results for the Entropy Change in the $Al_y Ni_{0.4-y} Zr_{0.6}$ Melt Series at 1700 K

The results in Figure 10.8 for the entropy change $\Delta S(T = 1700\,\text{K}; y)$ in the $Al_y Ni_{0.4-y} Zr_{0.6}$ melt series indicate a slight negative entropy of mixing, which means that the ternary mixtures have lower entropy than predicted by linear interpolation between $Ni_{0.4} Zr_{0.6}$ and $Al_{0.4} Zr_{0.6}$. Figure 10.8, in addition, predicts that at $T = 1700$ K the trigonal-prismatic $Ni_{0.4} Zr_{0.6}$ melt has a larger entropy per atom than the nearly polytetrahedral $Al_{0.4} Zr_{0.6}$.

The latter finding sounds strange since trigonal-prismatic melts mean a more special atomic arrangement than a nearly polytetrahedral distribution. Therefore, a larger number of realizations should exist for the former and therefore larger configuration entropy. This raises the question of properly deducing the configuration entropy S_{config} from the total structural entropy considered so far.

Around 1100 K, the configuration dynamics in the undercooled, highly viscous melts is slow and permits discrimination between atomic vibrations around temporary equilibrium positions and slow changes of the pattern of the equilibrium positions. Therefore, at, for example $T_1 = 1100$ K, the vibration entropy, $S_{vib}(T_1)$, can be estimated from the MD mean vibrational spectrum in the viscous melt. The concept of inherent structure [11] allows determination of the specific heat of thermal vibrations, $c_{vib}(T) = \partial_T[H(T) - H^{is}(T)]$, from MD modeling. Here $H^{is}(T)$ means the enthalpy of the inherent structure, and $H(T)$ the total enthalpy. This argument allows deducing $S_{vib}(T)$ from $S_{vib}(T_1)$ by integrating $dS_{vib} = (c_{vib}/T)dT$ from T_1 to T. Therefrom, one obtains

$$S_{config}(T) = S(T) - S_{vib}(T) \tag{10.4}$$

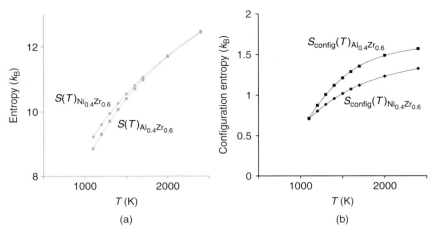

Fig. 10.9 MD results for (a) total structural entropy and (b) configuration entropy of $Ni_{0.4}Zr_{0.6}$ and $Al_{0.4}Zr_{0.6}$ as function of temperature.

and $\Delta S_{config}(T; y)$, the change of configuration entropy between $Al_y Ni_{0.4-y} Zr_{0.6}$ and $Ni_{0.4}Zr_{0.6}$ melt at any arbitrary temperature.

Besides $\Delta S(T; y)$, Figure 10.8 displays $\Delta S_{vib}(T; y)$ and $\Delta S_{config}(T; y)$ at $T = 1700$ K determined in this way. There is a strong composition dependence of $\Delta S_{vib}(T; y)$, having major influence on the y-dependence of $\Delta S_{config}(T; y)$. Owing to this, the trigonal-prismatic $Ni_{0.4}Zr_{0.6}$ melt at 1700 K has a lower configuration entropy per atom than the nearly polytetrahedral $Al_{0.4}Zr_{0.6}$ alloy melt, as expected. Moreover, the figure shows a slight positive mixing effect in $\Delta S_{config}(T; y)$, which means a gain in configuration entropy by ternary mixing, which is, however, much smaller than an ideal mixture would predict.

As further results, we present in Figure 10.9 the entropy data $S(T; y)$ and $S_{config}(T; y)$ for $Al_{0.4}Zr_{0.6}$ ($y = 0.4$) and $Ni_{0.4}Zr_{0.6}$ ($y = 0$) as functions of temperature, obtained by the specific heat values from MD simulations. For constructing the figures, we determined $S(1700 \text{ K}; Ni_{0.4}Zr_{0.6})$ from the entropy value of the congruently melting $Ni_{0.5}Zr_{0.5}$ by an additional coupling parameter calculation.

Figure 10.9 indicates for $S(T; y)$ and $S_{config}(T; y)$ a change of order between $Ni_{0.4}Zr_{0.6}$ and $Al_{0.4}Zr_{0.6}$ with temperature. These changes can be explained by assuming for the $Al_{0.4}Zr_{0.6}$ melt a large number of accessible configurations with a small low-enthalpy tail and for $Ni_{0.4}Zr_{0.6}$ a smaller number of configurations without a marked low-enthalpy tail.

The details presented here make obvious that understanding of the composition and temperature dependence of structural entropy in complex melts needs the understanding of various aspects, such as separation of configuration and vibrational entropy, nonlinear mixing effects in the former, and composition dependence of the latter, which are all accessible to MD modeling, as demonstrated.

10.5
Concluding Remarks

The results of the project cover a rather broad range of examples in which our methods can be applied with success. These are, among others, the smooth composition dependence of the FE across the eutectic concentrations, the construction of the (T, x) phase diagram for the Zr-rich Ni_xZr_{1-x} melts, or the explanation of the change of order between total structural entropy and the configuration part of $Ni_{0.4}Zr_{0.6}$ and $Al_{0.4}Zr_{0.6}$. Beyond these examples, it is left to future research to explore the full spectrum of questions where the method can be applied.

Another type of open question concerns, for example testing the broad variety of interatomic potentials in the literature regarding their ability to reproduce experimental thermodynamic parameters of liquid alloys, or establishing methods to combine the elaborate approaches for calculating the entropy in ordered crystals, with the present treatment for the melt. We also shall mention the discrepancy in the crystallization front propagation velocity from PF and isothermal MD modeling, which indicates so-far-hidden differences in the processes described by the present isothermal MD approach and the used PF description and needs full solution.

Finally, we shall address that during the course of the project some related studies on estimating entropy of melts from atomistic simulations have been presented in the literature: for example the two-phases model [12], or an elaborated inherent structure approach [13]. The former constructs the total structural entropy by combining the vibrational part with a hard sphere gas term aimed at modeling particles in a mobile phase. It is an open question whether such a treatment is suited for describing dense, complex metallic alloy melts. Ref. 13 (and references therein) deals with the entropy and free energy in liquid and amorphous silica and uses the inherent structure concept for extracting the configuration part entropy. With regard to our analysis, there is the essential difference that these approaches do not consider the question of how to relate the entropy or thermodynamic potential of two liquids with different compositions to each other. In our treatment, this is the central point that allows estimating the composition dependence of the entropy and FE of multicomponent melts with acceptable numerical reliability.

Acknowledgments

The authors gratefully acknowledge financial support by the DFG in the priority program SPP 1120 "Phase Transformations in Multicomponent Melts". They acknowledge the kind and very fruitful collaboration with B. Nestler and D. Danilov, Karlsruhe, within the SPP 1120 regarding combined PF and MD modeling of the crystallization dynamics in $[Ni_xZr_{1-x}]_{liquid} - Zr_{crystal}$ layers. The numerical computations concerning the thermodynamic potential in binary Ni_xZr_{1-x} melts have been carried out at the parallel computer system of the ZAM, Juelich, and those concerning combined MD and PF modeling at the High Performance Computer Center Hannover, HLRN.

References

1 Küchemann, K.-B. (2004) Freie Enthalpie binärer metallischer Legierungsschmelzen: Molekulardynamik Simulationen für Ni_xZr_{1-x}, Thesis, Universität Göttingen, http://webdoc.sub.gwdg.de/diss/2005/kuechemann/kuechemann.pdf.
2 Massalski T. (ed) (1990) *Binary Alloy Phase Diagrams*, Vol. 3, 2nd ed, ASM International, Ohio.
3 Teichler, H. (1999) "Melting transition in molecular-dynamics simulations of the $Ni_{0.5}Zr_{0.5}$ intermetallic compound". *Physical Review B*, 59, 8473–80.
4 Hausleitner, Ch. and Hafner, J. (1992) "Hybridized nearly-free-electron tight-binding-bond approach to interatomic forces in disordered transition-metal alloys. II. Modelling of metallic glasses". *Physical Review B*, 45, 128–42.
5 Yavari, A.R. (2005) "Solving the puzzle of eutectic compositions with 'Miracle glasses'", *Nature Materials*, 4, 2–3.; (a) Hume-Rothery, W. and Anderson, E. (1960) "Eutectic compositions and liquid immiscibility in certain binary alloy", *Philosophical Magazine*, 5, 383–405.
6 Danilov, D., Nestler, B., Guerdane, M. and Teichler, H., *Bridging the gap between molecular dynamics simulations and phase field modeling: dynamics of a $[Ni_xZr_{1-x}]_{liquid} - Zr_{crystal}$ solidification front*, submitted for publication.
7 Garcke, H., Nestler, B. and Stinner, B. (2004) "A diffuse interface model for alloys with multiple components and phases". *SIAM Journal on Applied Mathematics*, 64, 775–99.
8 Steinhardt, P.J., Nelson, D.R. and Ronchetti, M. (1983) "Bond-orientational order in liquids and glasses". *Physical Review B*, 28, 784–805.
9 Guerdane, M. and Teichler, H. (2001) "Structure of the amorphous, massive-metallic-glass forming $Ni_{25}Zr_{60}Al_{15}$ alloy from molecular dynamics simulations". *Physical Review B*, 65, 014203.
10 Kirkwood, J.G. (1935) "Statistical mechanics of fluid mixtures". *Journal of Chemical Physics*, 3, 300–13.
11 Stillinger, F.H. and Weber, T.A. (1984) "Packing structures and transitions in liquids and solids". *Science*, 225, 983–89.
12 Lin, S.-T., Blanco, M. and Goddard, W.A. III (2003) "The two-phase model for calculating thermodynamic properties of liquids from molecular dynamics: validation for the phase diagram of Lannard-Jones fluids". *Journal of Chemical Physics*, 119, 11792–805.
13 Saika-Voivod, I., Sciortino, F. and Poole, P.H. (2004) "Free energy and configurational entropy of liquid silica: fragile-to-strong crossover and polyamorphism". *Physical Review E*, 69, 041503.

11
Computational Optimization of Multicomponent Bernal's Liquids

Helmut Hermann, Antje Elsner, and Valentin Kokotin

11.1
Introduction

Bulk metallic glasses (BMGs) have been prepared from the liquid state and investigated with respect to their structure and properties for nearly two decades [1–3]. The ability of multicomponent metallic melts to form a glass during solidification depends upon whether or not the glassy state can be obtained at a given cooling rate. It depends on a series of parameters that have been extracted from numerous experiments. Inoue [4] has summarized three empirical rules describing the conditions for good glass-forming ability of a broad class of multicomponent metallic melts:
1. Multicomponent systems consisting of more than three elements are required.
2. The difference in atomic size among the three main constituent elements should be above 12%.
3. Negative heats of mixing among the three main constituent elements are necessary.

Egami [5] has extracted four conditions that would favor bulk metallic glass formation from liquid alloys:
1. Increase the atomic size ratio of the constituent elements.
2. Increase the number of elements involved.
3. Increase the interaction between the small and large atoms.
4. Introduce repulsive interactions between small atoms.

Despite the large body of experimental work, the intrinsic properties of the multicomponent metallic melts, however, remain nebulous. Even in the case of monoatomic metallic melts, the progress from the first paper on the structure of a liquid metal (mercury) [6] to recent reports on advanced experimental studies of liquid Fe, Co, Ni, and Zr [7–9] is moderate. Icosahedral clusters have been discussed as possible structural elements in liquid metals on the basis of the

Phase Transformations in Multicomponent Melts. Edited by D. M. Herlach
Copyright © 2008 WILEY-VCH Verlag GmbH & Co. KGaA, Weinheim
ISBN: 978-3-527-31994-7

knowledge that the binding energy of an isolated 13-atom icosahedral cluster of argon atoms is 8.4% greater than for hexagonal and face centered cubic (fcc) structures [10]. This discussion has been reopened by the experimental observation of fivefold local symmetry in liquid lead at a solid–liquid interface [11, 12]. However, it is not clear whether the fivefold symmetry observed is a property of the solid–liquid interface or an intrinsic feature of the liquid state. It is obvious that the experimental investigation of multicomponent melts cannot achieve detailed knowledge about their structure. Nevertheless, it is possible to observe phenomena such as phase separation in Ni–Nb–Y melts [13, 14] or continuous development of structural changes in the liquid state with temperature [15].

Owing to the fundamental problems in the structural research of liquids and taking into account the theoretical difficulties in describing the liquid state [16, 17], computer simulations have become very important and are successful. *Ab initio* molecular dynamics simulations are suitable to simulate specific small systems consisting of about 10^2 atoms [18, 19]. Larger systems can be described by model interaction potentials or a parameterized empirical potential for realistic liquid metals and alloys; see, for example [20–27].

In the case of BMGs, for the relationship among the multicomponent liquid state, the glass-forming ability of the melt, and the properties of the solidified material, there is, at the moment, no fundamental theory available [5]. Moreover, computer simulation of multicomponent metallic melts is an intricate problem since, even by restricting to pairwise interaction potentials, the determination of the numerous adjustable parameters of the potentials is scarcely possible. Therefore, in this chapter, the simplest approach of hard sphere potentials is chosen. This approach might be called the generalized Bernal's model linking to the first successful description of the structure of liquids [28, 29].

11.2
Methods

11.2.1
Force Biased Algorithm

The force biased algorithm belongs to the class of collective rearrangement methods. The version used here as fundamental for the simulation procedure is based on classical papers [30, 31]. Improvements have been included in implementing the force biased algorithm as described in [32, 33].

The initial configuration of spheres is characterized by a random distribution of the sphere centers in which no correlation between their coordinates is present. This corresponds to a Poisson point field (see, e.g. [34]). By giving each sphere a finite diameter, overlaps between spheres will appear. Then, the algorithm tries to reduce the overlaps by shifting the spheres but keeping the number of spheres constant. A repulsive force, f_{ij}, between overlapping spheres i and j is introduced that controls the extent of the shifts. Apart from the shifting, the diameters of the spheres are changed. The procedure is continued until all overlaps are removed. Subsequently, all radii are scaled up by a certain amount, and the step of removing overlaps is repeated. The process is terminated when a predefined packing fraction is achieved or if no further improvement to the packing is possible. The procedure is carried out using periodic boundary conditions. It should be emphasized that any arbitrary size distribution can be chosen.

The force biased algorithm has been successfully applied to many problems in recent years; see for example [35–40].

11.2.2
The Nelder–Mead Algorithm

The Nelder–Mead simplex algorithm [41] is a numerical method to find extreme values of functions, which is particularly suitable for complicated functions. It starts from a set of $(f + 1)$ sampling points, where f is the number of degrees of freedom. Considering a system with discrete size distribution of spheres and n different species, the number of degrees of freedom is $f = 2(n - 1)$. Starting from a given set of sampling points, the algorithm tries to find better values of the packing fraction by varying the parameters according to a specific procedure. The procedure is illustrated in Figure 11.1 for the case of two adjustable parameters. Here, a sampling point P_k stands for two parameters, for example concentrations x and y of a ternary system $A_x B_y C_{1-x-y}$ with fixed radii of the components A, B, C. The sampling points P_1, P_2, and P_3 are ordered with respect to the packing fraction, $g(P)$, achieved for the sampling point P, $g(P_1) \geq g(P_2) \geq g(P_3)$, forming a simplex. Then, a new simplex is created adjacent to the former one but closer to the (local) maximum of the function $g(P)$. This is done by replacing P_3 by P_R. If $g(P_R) > g(P_3)$, the search is extended to P_E. If $g(P_E) > g(P_R)$, P_E replaces the sampling point of the previous simplex with the lowest packing fraction. If $g(P_E) \leq g(P_R)$, P_R is used for this replacement. For other relations between $g(P_1)$, $g(P_2)$, $g(P_3)$, $g(P_E)$, $g(P_R)$, similar procedures are applied. The simplex algorithm can be refined by so-called threshold-accepting [42, 43] in order to avoid trapping in weak local minima. This option is implemented in the present study.

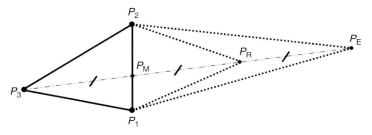

Fig. 11.1 The Nelder–Mead simplex for the special case of the two-dimensional problem.

11.2.3
Voronoi Tessellation

The method of Voronoi tessellation is used in a generalized form (S-Voronoi or hyperbolic Voronoi tessellation) to analyze the local structure of the hard sphere models in terms of local density. The local packing fraction, p_k, of a sphere k with radius R_k is defined by

$$p_k = \frac{\frac{4}{3}\pi R_k^3}{V_k} \tag{11.1}$$

where V_k is the volume of the corresponding S-Voronoi cell. The S-Voronoi cell of sphere k is defined by the set of all random test points that are closer to the surface of this sphere than to any other sphere [44]. Using this method one avoids allotting shares of the volume of a large sphere to the volume occupied by a small sphere. This definition is identical with the conventional Voronoi tessellation if all spheres have the same diameter. Figure 11.2 illustrates both methods. The S-Voronoi cell volume is calculated by statistical integration in such a way that the statistical error of the estimated local packing fractions is less than 0.3%.

One of the interesting items in models for liquids is the search for atoms with icosahedral environment because this type of short-range order is often considered to be an essential feature regarding the stability against crystallization [10]. The local packing fraction analysis provides a criterion to identify icosahedral order because the local packing fraction of the central atom takes the value of 0.76, which cannot be reached by other arrangements. However, the fluctuations typical for

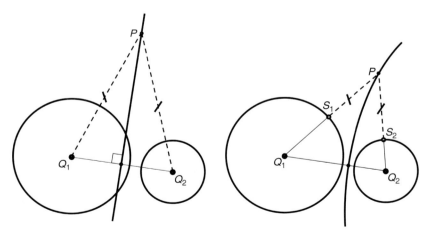

Fig. 11.2 Definition of Voronoi (left) and hyperbolic or S-Voronoi (right) tessellation. The point P is situated at the boundary separating the adjacent Voronoi and S-Voronoi cells if, respectively, $PQ_1 = PQ_2$ and $PS_1 = PS_2$, where AB is the distance between points A and B.

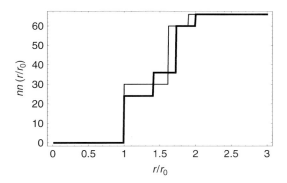

Fig. 11.3 Distance distribution of next-nearest neighbors, nn, of an atom with icosahedral (gray line) and fcc (bold line) short-range order.

a liquid (and models for liquids) may cause the icosahedral clusters, if present, to be distorted to some degree, which would result in a reduced local packing fraction. Therefore, an atom with a local packing fraction of, for example 0.74 is not unambiguously characterized as fcc but might also represent a distorted icosahedral cluster. We used an additional criterion to discriminate between fcc and distorted icosahedral clusters. The distance distribution of the neighbors of an atom in question differs essentially for these two types of short-range order. Figure 11.3 shows the corresponding cumulative distributions. The distribution for fcc has a step at $r/r_0 = 1.414$, while it is constant in the interval $1 < r/r_0 < 1.628$ for the icosahedral arrangement. Denoting the number of pairs of atoms in the neighborhood of a given atom having a distance in the range $1 < r/r_0 < 1.628$ by N^*, the criterion is $N^* = 0$ for icosahedral order and $N^* = 12$ for fcc-like order. This criterion will be used in the subsequent text.

11.3
Results and Discussion

11.3.1
Monoatomic Liquids

The Bernal's model of simple liquids, which is the model of dense random packing of equal hard spheres, has been studied extensively during the last decades. We used this classical model to test our improved algorithms. Additionally, we considered the question of what happens if the global packing density, g, of the model is enhanced beyond the value of 0.636, which is characteristic of the classical Bernal's model. It was possible to increase the packing up to about 0.70 by means of the force biased algorithm. The concept of local packing was used for a first-order analysis of the generated models. The frequency distribution of the local packing fraction for the Bernal's model shows, in good approximation, a Gaussian shape

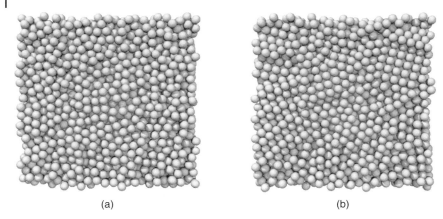

Fig. 11.4 (a) Liquid system with fcc nuclei ($g = 0.648$) and (b) partially crystallized arrangement of hard spheres ($g = 0.664$) obtained by densification of random initial structures.

with mean value of $g = \langle p_i \rangle = 0.636$. With increasing global packing fraction, an additional peak is observed at the local packing fraction of 0.74, which points to the formation of face centered nanocrystals [39]. Corresponding systems were carefully analyzed in [35, 37, 39]. In several papers, the appearance of local packing fraction in the range of 0.74 was discussed as the beginning of the formation of icosahedral short-range order [45, 46] with $p = 0.76$ for the atom situated at the center of the icosahedron. In order to discriminate between fcc short-range order ($p = 0.74$) and (distorted) icosahedral arrangement, we defined the parameter N^* as the number of distances between the next-nearest neighbors surrounding the central atom of a 13-atom cluster in the distance range $1 < r/r_0 < 1.618$. It takes the values $N^* = 0$ and 12 for icosahedral and fcc order, respectively (see Figure 11.3). This parameter is useful since, in contrast to nanometer-scale crystallization (Figure 11.4), icosahedral clusters are difficult to observe by visual inspection.

Defining an ideal (or distorted) icosahedral cluster by a local packing fraction of at least 0.74 and $N^* = 0$ (or 1), the number of clusters in systems with a global packing fraction of 0.636 or higher fitting these conditions was always of the order of about 1 icosahedral cluster per 10 000 atoms. In order to study the behavior of icosahedral clusters in hard sphere packing, we generated initial systems with low global packing fraction containing preformed icosahedral (Mackay) clusters consisting of 13 and 55 atoms. The density of the systems was enhanced by applying the force biased algorithm, and the change of the number of clusters with icosahedral short-range order was determined. Figure 11.5 shows the results. Obviously, transitions occur from low-density systems with icosahedral clusters to high-density ones without icosahedral short-range order. The density, g_c, where this transition takes place is higher for the 55-atom Mackay cluster ($g_c = 0.52$) than for the simple 13-atom icosahedral arrangements ($g_c = 0.43$).

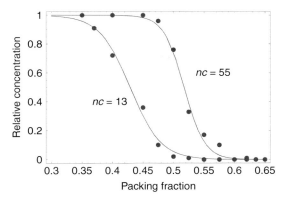

Fig. 11.5 Dependence of relative concentration of icosahedral clusters consisting of 13 and 55 atoms, respectively, versus global packing fraction.

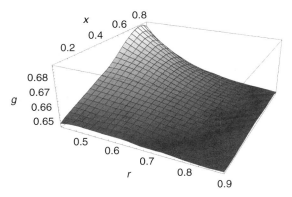

Fig. 11.6 Maximum packing fraction, g, of binary random close packing models versus size ratio of spheres, r, and percentage, x, of small spheres.

11.3.2
Multicomponent Liquids

At first, the methods described in Section 11.2 are tested for two-component systems. Figure 11.6 shows the results of the force biased algorithm for a systematic simulation in the parameter range $0.45 \leq r \leq 0.95$, $0 \leq x \leq 1.0$, where r is the size ratio of the spheres and x is the concentration of the small spheres. The results are in quantitative agreement with previous simulations [47]. Furthermore, the data show that the variation of the packing fraction with the parameters of the system is smooth, which is important for the convergence of the Nelder–Mead algorithm for systems with many degrees of freedom.

In a second test we consider the Cu–Zr amorphous alloy, which has been experimentally studied in great detail; see [48] for example. Assuming atomic

Table 11.1 Partial coordination numbers, N_{ij}, for the $Cu_{50}Zr_{50}$ BMG.

Method	N_{CuCu}	N_{CuZr}	N_{ZrZr}
Present work	4.27	5.56	7.15
MD simulation [49]	4.44	6.24	7.59
EXAFS [50]	4.1	5.4	7.4
XRD [51]	4.6	4.8	7.8

Table 11.2 Sphere radius and the related atomic radius of selected metals [52]).

Component	Normalized sphere radius	Atomic radius of related elements (nm)	Related elements
A	0.625	0.124–0.125	Fe, Ni, Cr, Co
B	0.725	0.143–0.146	Nb, Ta, Al, Au, Ag, Ti
C	0.800	0.158–0.164	Hf, Mg, Zr, Pm, Nd
D	0.900	0.178–0.181	Dy, Tb, Ho, Gd, Y, Sm
E	1.000	0.198–0.215	Ca, Eu, Sr, Ba

diameters of 0.256 and 0.320 nm for Cu and Zr, respectively, and searching for the composition at which the maximum packing fraction can be achieved in the liquid state, we find $Cu_{66.4}Zr_{33.6}$. This is in agreement with [48], where $Cu_{65}Zr_{35}$ was obtained as the composition with maximum packing fraction. The partial coordination numbers obtained with the present simulation are collected in Table 11.1 and compared with the results of a molecular dynamics study [49] as well as with experimental data [50, 51]. Obviously, the present simulation results correspond to the independently obtained theoretical and experimental data [49–51]. The additivity of atomic radii, that is $2R_{CuZr} = R_{Cu} + R_{Zr}$, required for the validity of the generalized Bernal's model is realized with an accuracy of 1.5 and 4.5% according to extended X-ray absorption fine structure (EXAFS) [50] and X-ray diffraction (XRD) [51] measurements, respectively.

Now we consider a five-component system with fixed values of the radii. The chosen radii and metals with similar atomic radii are given in Table 11.2 (values taken from [52]). Each system has four degrees of freedom given by the concentrations c_i, $i = 1, \ldots, 5$, taking into account the condition $c_1 + c_2 + \cdots + c_5 = 1$. Two different initial conditions are chosen (gray bars in Figure 11.7). The results of the combined force biased and Nelder–Mead procedures are given by black bars. The improvement of the packing fraction is moderate (from 0.655 to 0.657) for the system starting with the convex initial size distribution but it is remarkable (from 0.650 to 0.660) for the initial set with concave distribution. In both cases the resulting distributions are quite similar to those corresponding to Zr-based bulk amorphous metallic alloys [53]. Figure 11.8 shows the results for systems with fixed relative size ratio of largest and smallest spheres but variable remaining

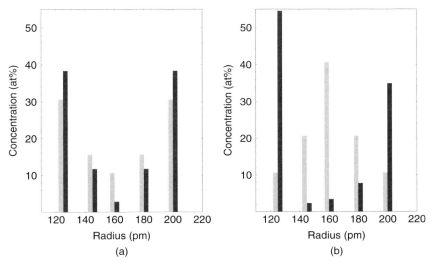

Fig. 11.7 Atomic size distribution plots for five-component systems with fixed radii and variable concentration. Convex (a) and concave (b) initial distributions. Gray and black bars correspond to initial and final distribution.

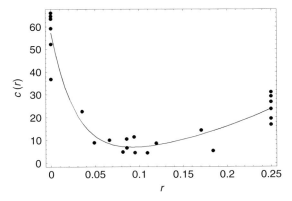

Fig. 11.8 Atomic size distribution plot of optimized three-, four-, and five-component systems. A constant relative difference (25%) of the largest and smallest spheres is assumed with variable concentration and radii; c is the concentration of species with relative size difference r to the smallest spheres.

radii and variable concentrations. Again, the resulting atomic size distribution plot resembles those determined from experimental data [52, 53].

The simulations show that the maximum packing fraction that can be achieved for multicomponent, random, close-packed arrangements of spheres depends on the size distribution of the species. It is possible to optimize size distributions with respect to maximum packing fraction using the force biased algorithm for the packing procedure and the Nelder–Mead simplex method to search for maxima by varying the size distribution. The combination of these methods is important for systems with more than three degrees of freedom because there no systematic exploration is possible.

11.4
Conclusion

The optimization of multicomponent Bernal's liquids regarding global packing fraction is a high-dimensional problem which cannot be solved by systematic raster-scan methods. For analyzing systems consisting of about 10^4 spheres it takes a computer time of the order of 10^3 s for generating one dense-packed model with given radii distribution using an advanced PC (state of the art in the year 2008). For an N-component system, the number of degrees of freedom is $2(N-1)$. Choosing 10 sampling points for each free parameter, the computer time required for a systematic scan would amount to 10^{2N+1} s. Starting a three-year research program for the systematic computational analysis of the five-component system, the parallel work of 10^3 PCs would be required. This example demonstrates the value of the combined force biased/Nelder–Mead algorithm, which helps determine optimal atomic size distributions for generalized Bernal's liquids. In terms of physics, the contribution of the packing effect to the glass-forming ability of multicomponent metallic melts can be separated from other parameters such as, for instance, mixing enthalpies using the methods presented in this chapter.

Acknowledgments

The authors are very thankful to Kristin Lochmann, Anett Wagner, and Dietrich Stoyan for their permanent and fruitful cooperation. Financial support by the Deutsche Forschungsgemeinschaft is gratefully acknowledged.

References

1 Inoue, A., Zhang, T. and Masumoto, T. (**1990**) "Production of amorphous cylinder and sheet of $La_{55}Al_{25}Ni_{20}$ alloy by a metallic mold casting method". *Materials Transactions, JIM*, **31**, 425–28.

2 Zhang, T., Inoue, A. and Masumoto, T. (**1991**) "Amorphous Zr-Al-TM (TM = Co, Ni, Cu) alloys with significant supercooled liquid region of over 100K". *Materials Transactions, JIM*, **32**, 1005–10.

3 Peker, A. and Johnson, W.L. (**1993**) "Highly processable metallic glass $Zr41.2Ti13.8Cu12.5Ni10.0Be22.5$". *Applied Physics Letters*, **63**, 2342–44.

4 Inoue, A. (**2000**) *Acta Materialia*, **48**, 279–306.

5 Egami, T. (**2003**) *Journal of Non-Crystalline Solids*, **317**, 30–33.

6 Debye, P. and Menke, H. (**1930**) "Bestimmung der inneren Struktur von Flüssigkeiten mit Röntgen-

strahlen". *Physik Zeitschrift*, **31**, 797–98.

7   Jakse, N. and Bretonnet, J.L. (**1995**) "Structure and thermodynamics of liquid transition metals: integral equation study of Fe, Co and Ni". *Journal of Physics: Condensed Matter*, **7**, 3803–15.

8   Schenk, T., Holland-Moritz, D., Simonet, V., Bellisent, R. and Herlach, D.M. (**2002**) "Icosahedral short-range order in deeply undercooled metallic melts". *Physical Review Letters*, **89**, 075507.

9   Holland-Moritz, D., Schenk, T., Bellisent, R., Simonet, V., Funakoshi, K., Merino, J.M., Buslaps, T. and Reutzel, S. (**2002**) "Short-range odder in undercooled Co metals". *Journal of Non-Crystalline Solids*, **312-414**, 47–51.

10   Frank, F.C. (**1952**) "Supercooling of liquids". *Proceedings of the Royal Society of London, Series A*, **215**, 43–46.

11   Reichert, H., Klein, O., Dosch, H., Denk, M., Honkimäki, V., Lippman, T. and Reiter, G. (**2000**) "Observation of five-fold local symmetry in liquid lead". *Nature*, **408**, 839–41.

12   Spaepen, F. (**2000**) "Five-fold symmetry in liquids". *Nature*, **408**, 781–82.

13   Mattern, N. (**2007**) "Structure formation in liquid and amorphous metallic alloys". *Journal of Non-Crystalline Solids*, **353**, 1723–32.

14   Mattern, N., Kühn, U., Gebert, A., Schöps, A., Gemming, T. and Schultz, L. (**2007**) "Phase separation in liquid and amorphous Ni-Nb-Y alloys". *Materials Science and Engineering A*, **449-451**, 207–10.

15   Mattern, N., Kühn, U. and Eckert, J. (**2007**) "Structural behavior of amorphous and liquid metallic alloys at elevated temperatures". *Journal of Non-Crystalline Solids*, **353**, 3327–31.

16   Percus, J.K. and Yevick, G.J. (**1958**) "Analysis of classical statistical mechanics by means of collective coordinates". *Physical Review*, **110**, 1–13.

17   March, N.H. and Tosi, M.P. (**2002**) *Introduction to Liquid state Physics*, World Scientific.

18   Alfè, D., Kresse, G. and Gillan, M.J. (**2000**) "Structure and dynamics of liquid iron under Earth's core conditions". *Physical Review B*, **61**, 132–42.

19   Sheng, H.W., Luo, W.K., Alamgir, F.M., Bai, J.M. and Ma, E. (**2006**) "Atomic packing and short-to-medium-range order in metallic glasses". *Nature*, **439**, 419–25.

20   Kob, W., Donati, C., Plimpton, S.J., Poole, P.H. and Glotzer, S.C. (**1997**) "Dynamical heterogeneities in a supercooled Lennard-Jones liquid". *Physical Review Letters*, **79**, 2827–30.

21   Donati, C., Glotzer, S.C., Poole, P.H., Kob, W. and Plimpton, S.J. (**1999**) "Spatial correlations of mobility and immobility in a glass-forming Lennard-Jones liquid". *Physical Review E*, **60**, 3107–19.

22   Sumi, T., Miyoshi, E. and Tanaka, K. (**1999**) "Molecular- dynamics study of liquid mercury in the density region between metal and nonmetal". *Physical Review B*, **59**, 6153–58.

23   Sadigh, B., Dsugutov, M. and Elliott, S.R. (**1999**) "Vacancy ordering and medium-range structure in a simple monatomic liquid". *Physical Review B*, **59**, 1–4.

24   Guerdane, M. and Teichler, H. (**2001**) "Structure of the amorphous, massive-metallic-glass forming Ni25Zr60Al15 alloy from molecular dynamics simulations". *Physical Review B*, **65**, 014203.

25   Mossa, S. and Tarjus, G. (**2003**) "Locally preferred structure in simple atomic liquids". *Journal of Chemical Physics*, **119**, 8069–8074.

26   Gebremichael, Y., Vogel, M. and Glotzer, S.C. (**2004**) "Particle dynamics and the development of string-like motion in a simulated monoatomic liquid". *Journal of Chemical Physics*, **120**, 4415–27.

27   Duan, G., Xu, D., Zhang, Q., Zhang, G., Cagin, T., Johnson, W.L. and Goddard, W.A. (**2005**) "Molecular dynamics study of Cu46Zr54 metallic glass motivated by experiments: glass formation and atomic-level

28 Bernal, J.D. (**1960**) "Geometry of the structure of monatomic liquids". *Nature*, **185**, 68–70.
29 Bernal, J.D. (**1964**) "The structure of liquids". *Proceedings of the Royal Society of London, Series A*, **280**, 299–322.
30 Jodrey, W.S. and Tory, E.M. (**1979**) "Simulation of random packing of spheres". *Journal of Simulation*, **32**, 1–12.
31 Jodrey, W.S. and Tory, E.M. (**1985**) "Computer simulation of random packing of spheres". *Physical Review A*, **32**, 2347–51.
32 Mosciniski, J. and Bargiel, M. (**1991**) "C-language program for irregular packing of hard spheres". *Computer Physics Communications*, **64**, 183–92.
33 Bezrukov, A., Bargiel, M. and Stoyan, D. (**2002**) "Statistical analysis of simulated random packing of spheres". *Particle and Particle Systems Characterization*, **19**, 111–18.
34 Stoyan, D., Kendall, W.S. and Mecke, J. (**1987**) *Stochastic Geometry and Its Applications*, John Wiley & Sons, Ltd, Chichester.
35 Hermann, H., Elsner, A. and Gemming, T. (**2005**) "Influence of the packing effect on stability and transformation of nanoparticles embedded in random matrices". *Materials Science Poland*, **23**, 541–49.
36 Hermann, H., Elsner, A., Hecker, M. and Stoyan, D. (**2005**) "Computer simulated dense-random packing models as approach to the structure of porous low-k dielectrics". *Microelectronic Engineering*, **81**, 535–43.
37 Lochmann, K., Anikeenko, A., Elsner, A., Medvedev, N. and Stoyan, D. (**2006**) "Statistical verification of crystallization in hard-sphere packings under densification". *European Physical Journal B*, **53**, 67–76.
38 Hermann, H., Elsner, A., Lochmann, K. and Stoyan, D. (**2007**) "Optimisation of multi-component hard sphere liquids with respect to dense packing". *Materials Science and Engineering A*, **449-451**, 666–70.
39 Hermann, H., Elsner, A. and Stoyan, D. (**2007**) "Behaviour of icosahedral clusters in computer simulated hard sphere systems". *Journal of Non-Crystalline Solids*, **353**, 3693–97.
40 Elsner, A. and Hermann, H. (**2007**) "Computer simulation and optimization of properties of porous low-k dielectrics". *Materials Science Poland*, **25**, 1193–202.
41 Nelder, J.A. and Mead, R. (**1965**) "A simplex method for function minimization". *The Computer Journal*, **7**, 308–13.
42 Winkler, P. (**2001**) *Optimization Heuristics in Econometrics*, John Wiley & Sons, Ltd, Chichester.
43 Gilli, M. and Winkler, P. (**2003**) "A global optimization heuristic for estimating agent based models". *Computational Statistics and Data Analysis*, **42**, 299–312.
44 Medvedev, N.N. (**2000**) *The Voronoi-Delaunay Method in the Investigation of the Structure of Noncrystalline Systems*, Izdatelstvo SO RAN, Novosibirsk.
45 Clarke, A. and Jónsson, H. (**1993**) "Structural changes accompanying densification of random hard-sphere packings". *Physical Review E*, **47**, 3975–84.
46 Jullien, R., Jund, P., Caprion, D. and Quitmann, D. (**1996**) "Computer investigation of long-range correlations and local order in random packings of spheres". *Physical Review E*, **54**, 6035–41.
47 Clarke, A.S. and Wiley, J.D. (**1987**) "Numerical simulation of the dense random packing of a binary mixture of hard spheres: Amorphous metals". *Physical Review B*, **35**, 7350–56.
48 Park, K., Jang, J., Wakada, M., Shibutani, Y. and Lee, J. (**2007**) "Atomic packing density and its influence on the properties of Cu-Zr amorphous alloys". *Scripta Materialia*, **57**, 805–8.
49 Păduraru, A., Kenoufi, A., Bailey, N.P. and Schiøtz, J. (**2007**) "An interatomic potential for studying CuZr bulk metallic glasses". *Advanced Engineering Materials*, **9**, 505–8.

50 Bubanov, Y.A., Schvetsov, V.R. and Sidorenko, A.F. (**1995**) "Atomic structure of binary amorphous alloys by combined EXAFS and X-ray scattering". *Physica B*, **208-209**, 375–76.

51 Bionducci, M., Buffa, F., Licheri, G., Navarra, G., Bouchet-Fabre, B. and Tonnerre, J.M. (**1996**) "Determination of the partial structure factors of amorphous CuZr by anomalous X-ray scattering and reverse Monte Carlo". *Zeitschrift fur Naturforschung*, **51a**, 71–82.

52 Senkov, O.N. and Miracle, D.B. (**2001**) "Effect of the atomic size distribution on glass forming ability of amorphous metallic alloys". *Materials Research Bulletin*, **36**, 2183–98.

53 Senkov, O.N. and Miracle, D.B. (**2003**) "A topological model for metallic glass formation". *Journal of Non-Crystalline Solids*, **317**, 34–39.

12
Solidification Experiments in Single-Component and Binary Colloidal Melts

Thomas Palberg, Nina Lorenz, Hans Joachim Schöpe, Patrick Wette, Ina Klassen, Dirk Holland-Moritz, and Dieter M. Herlach

12.1
Introduction

Investigations of colloidal suspensions as model systems for the questions in focus entered the SPP1120 during the last funding period. Previous studies established single-component suspensions as models for equilibrium issues of solid state physics and statistical mechanics [1–6]. Recently, however, interest has shifted from equilibrium situations to phenomena and processes occurring far from equilibrium, including the influence of external fields [7] and the kinetics of phase transitions [8, 9]. Many strong analogies to atomic systems have been discovered, but some colloid-specific effects have also been investigated, for example the influence of polydispersity on phase behavior and nucleation kinetics [10–12]. Increasingly, more investigations have addressed binary systems starting from hard sphere-like mixtures mimicking packing-dominated atomic systems [13–20], proceeding to more metal-like systems comprising charged spheres [21–33] and finally to electrosterically stabilized particles of opposite charges, which provided access to a large variety of salt structures [34–36].

With the focus of SPP1120 on metal solidification, the present contribution reports some systematic investigations concerning the solidification behavior and properties of the resulting solids in suspensions of charged spheres. As an important prerequisite for systematic investigations on solidification, we need precise interaction control. Particles interact via a screened Coulomb potential, and therefore the range and strength of repulsion can be experimentally tailored via the particle's effective charge, the concentration of the screening electrolyte, and the number density of the particles. In mixtures, the composition is an additional parameter. Adjustment and characterization of interactions is performed using advanced conditioning methods in combination with static light scattering, elasticity measurements, and conductometry. A major part of the remaining paper is therefore devoted to a demonstration of the possibilities of interaction adjustment and control and the determination of phase behavior as a function of the interaction parameters. We then apply the full range of experimental methods established for

Phase Transformations in Multicomponent Melts. Edited by D. M. Herlach
Copyright © 2008 WILEY-VCH Verlag GmbH & Co. KGaA, Weinheim
ISBN: 978-3-527-31994-7

solidification experiments on fixed-charge single-component systems [8, 37], to the charge variable single-component silica suspensions, and to two-component systems of different fixed sizes and charge ratios. In particular, we use (time-resolved) static light scattering and microscopy for structural investigations at moderate particle concentrations. To access larger particle concentrations without multiple scattering effects and also to obtain a larger range of scattering vectors to static light scattering and microscopy to determine the melt structure, we complement these measurements by ultra small-angle X-ray scattering (USAXS) [38]. We conclude with a short discussion of the scope and validity of our approach, with particular emphasis on the similarities and differences between colloidal and metallic melts, solids, and the phase transition kinetics. The emerging questions from this conclusion will give direction to future investigations. We start, however, with a short overview on the tunable interactions in colloidal suspensions.

12.2
Experimental Procedure

12.2.1
Tunable Interactions in Charged Colloidal Suspensions

All particles used in this investigation interact via a screened Coulomb or Debye–Hückel potential, which is equivalent to a Yukawa potential often used in the context of metals. The main difference is the use of a correction for the finite size of the particles and the different, much higher effective charge Z_{eff}. Of all (titratable) surface groups N of a given particle, only Z are dissociated (bare charge). Most of these are condensed on the surface of the particles, the remaining Z^* being freely diffusing counterions taking the role of valence electrons. This charge, also known as *renormalized charge* [39–41], is equal to Z at low Z but levels off and finally saturates at large Z. Like in metals, also in colloidal solids many-body effects further lead to an anisotropy of interactions and a metal-like elasticity behavior [42]. In isotropic samples (fluid or polycrystalline), however, this effect can be mapped back onto a still lower effective charge (valence) and the convenient mean-field treatment be retained [43–45]. We therefore write the pair interaction energy for our samples as:

$$V(r) = \frac{(Z_{\text{eff}} e)^2}{4\pi \varepsilon \varepsilon_0} \left(\frac{\exp(\kappa a)}{1 + \kappa a} \right)^2 \frac{\exp(-\kappa r)}{r} \tag{12.1}$$

where a is the particle radius, $\varepsilon \varepsilon_0$ is the dielectric permittivity, and $k_B T$ is the thermal energy. The screening parameter contains the contributions of particle counterions nZ_{eff} (valence electrons), added salt (molar concentration c), and self-dissociation of the solvent:

$$\kappa = \sqrt{\frac{e^2}{\varepsilon \varepsilon_0 k_B T}(nZ_{\text{eff}} + 2 \times 10^3 N_A (c + 2 \times 10^{-7}))} \tag{12.2}$$

Here e is the elementary charge. The number of freely diffusing microions, Z^*, is accessible in conductivity measurements calibrated by static light scattering to obtain the particle density n. The effective charge accounting for counterion condensation and many-body effects, Z_{eff}, is obtained from measurements of elasticity in polycrystalline samples [44, 46] or the evolution of the static structure factor [47]. As expected theoretically, it was observed experimentally that for sufficiently large bare charge Z, the effective charge Z_{eff} is proportional to the renormalized charge Z^* and also shows saturation behavior.

These excursions have been instructive to illustrate the charging process of colloidal particles and its mean-field description. Which of these charges, however, is useful for our purposes? For sufficiently strong and long-ranged interactions, a colloidal suspension will undergo a transition from the fluid to the crystalline state. From Equations 12.1 and 12.2 it is seen that this can be achieved in one-component systems by increasing the number density n of the particles, reducing the amount of screening electrolyte, and increasing the effective charge. In fact, by use of the effective charge from shear modulus measurements Z_{eff} and the universal melting line derived from extensive computer simulations for Yukawa systems [48], a surprisingly good and consistent description of the phase behavior is obtained [49].

For particles of a binary mixture, additional parameters determine the sample properties. First, we have the composition, which, in analogy to atomic systems, is taken as the molar fraction of the smaller species: $p = p_{\text{small}} = n_{\text{small}}/n_{\text{total}}$. Second, we can vary the size ratio $\Gamma = a_{\text{small}}/a_{\text{large}}$. In most cases of highly charged particles, the effective charge ratio $\Lambda = Z_{\text{eff,small}}/Z_{\text{eff,large}}$ follows strictly the size ratio. For small charges, however, differences may occur. In the following text, we investigate a mixture in which the large spheres carry a comparatively low charge such that $\Gamma = 0.68$ but $\Lambda = 1$. Lindsay and Chaikin [21, 50] proposed an extension of Equations 12.1 and 12.2 for the case of binary, charged sphere suspensions in the melt state or in a chemically disordered solid (solid solution), which is based on number-weighted averages:

$$U(r) = \frac{e^2}{4\pi \varepsilon \varepsilon_0} \left(p_A^2 \tilde{Z}_A^2 + 2 p_A p_B \tilde{Z}_A \tilde{Z}_B + p_B^2 \tilde{Z}_B^2 \right) \frac{\exp(-\kappa r)}{r} \tag{12.3}$$

where we used the abbreviation:

$$\tilde{Z}_j = Z_{\text{eff},j} \frac{\exp(\kappa a_j)}{1 + \kappa a_j} \tag{12.4}$$

and the screening parameter reads:

$$\kappa = \sqrt{\frac{e^2}{\varepsilon \varepsilon_0 k_B T} \left(2 \times 10^3 N_A (c + 2 \times 10^{-7})\right) + n \left(p_A Z_{\text{eff,A}} + (1 - p_a) Z_{\text{eff,B}}\right)} \tag{12.5}$$

Results from the literature [21, 50] and our previous studies have shown that this description is well suited for use with solid solutions [27, 29, 30]. In particular, shear moduli were measured extensively, which are related to the interaction potential via:

$$G_{bcc} = f_A \frac{4}{9} n V(d_{NN}) \kappa^2 d_{NN}^2 \tag{12.6}$$

where $d_{NN,bcc} = \sqrt{3}/\sqrt[3]{4n}$ is the nearest-neighbor distance in a bcc lattice and f_A is a known geometrical factor accounting for the orientational averaging in polycrystalline samples. An excellent agreement between measured shear moduli and the expectations of the combined Equations 12.3–12.6 was observed for solid solutions of size and charge ratios close to unity [30]. Also, for conductivity number-weighted averages apply [29]. The description is thus confirmed as the reference case for further investigations. Deviations from number-weighted averages of crystal properties therefore, in general, may indicate deviations from the case of solid solutions.

Table 12.1 presents the used particle species. We employed several species of commercial polystyrene spheres and an industrial batch of poly(n-butylacrylamide) copolymer particles (a kind gift of BASF) (lab codes PSXX and PnBAPSXX, where XX denotes the diameter in nm). For USAXS measurements, particles of high electron contrast in aqueous dispersions were chosen. To be specific, we choose silica (SiO_2) particles with a diameter of 84 nm (Si84). They are suspended in ultrapure water at a concentration of $n = 113\,\mu m^{-3}$ in the stock suspension, corresponding to a volume fraction of $\Phi = 0.0355$. The density of the silica particles is on the order of $\rho = 1.8\,g\,cm^{-3}$. Although as compared to polystyrene spheres ($\rho = 1.05\,g/cm^{-1}$) sedimentation effects have a larger influence, they are still negligible over the duration of solidification experiments (a few minutes) for our sufficiently small particles. Si84 particles carry weakly acidic silanol surface groups (Si–OH). The originally low bare charge under deionized conditions may be considerably increased using an ion-exchange technique. Sodium hydroxide (NaOH) is added to a salt-free suspension triggering the reaction: $SiOH + NaOH \rightarrow SiO^- + Na^+ + H_2O$ [51, 52]. Thus the system stays deionized but the counterion species changes and Z increases. With the increasing Z, also the effective charge increases and saturates. Past the equivalence point, all silanol groups are used up and the electrolyte concentration increases because of excess NaOH. This induces a strong screening, and the interaction decreases again. By repeating the experiment at different particle concentrations, silica particles thus allow independent access to all three interaction-determining parameters.

Sample conditioning proceeds in a closed Teflon tubing system [29]. The tubing connects different components. These comprise a mixed-bed ion-exchange column; a reservoir to add solvent, NaOH, or further particles; the sample cell (for microscopy, light scattering or USAXS measurements); and a conductivity experiment to control the state of deionization or charging at constant n. The suspension is driven through the preparation circuit by a peristaltic pump under an inert argon atmosphere to avoid contamination with airborne CO_2. A computer controls the addition of materials via a dosimeter (Titronic Universal, Schott AG, Germany) and operates the electromagnetic valves to stop the flow instantaneously after

Table 12.1 Compilation of single-particle properties. Sample name; nominal diameter $2a$; effective charges $Z_{eff,G}$ and Z_σ^* from shear modulus and conductivity measurements; polydispersity determined via ultracentrifugation (UZ), static light scattering in combination with Mie theory (Mie), dynamic light scattering (DLS), or transmission electron microscopy (TEM). The last column gives corresponding references.

Sample	Manufacturer and lot no.	$2a$ (nm)	Polydispersity(%)	$Z_{eff,G}$	Z_σ^*	Reference
PnBAPS68	BASF ZK2168/7387 GK0748/9378	68	5 (UZ)	331 ± 3	450 ± 16	53
Si84	—	84	8 (Mie)	Variable $253 \pm 15 - 340 \pm 20$	n/a	n/a
PS85	IDC 767.1	85	7 (TEM)	350 ± 20	530 ± 32	[24–26, 53]
PS90	Bangs 3012	90	2.5 (DLS)	315 ± 8	504 ± 35	53
PS91	Dow chemicals D1C27/Serva 4197	91	7 (TEM)	n/a	190 ($\Phi = 10^{-6}$) 680 (PBC)	[24–26, 45]
PS100A	Bangs 3512	100	6.6 (TEM)	349 ± 10	527 ± 30	53
PS100B	Bangs 3067	100	2.7 (DLS)	327 ± 10	530 ± 38	53
Si103	CS81	103	13.2 (TEM)	n/a	n/a	[24–26]
PS109	Dow chemicals D1B76/Seradyn 2010M9R	109	3 (TEM)	n/a	450 ± 30, 450 (PBC)	[8, 24, 25, 54]
PS120	IDC 10-202-66.3, 10-202-66.4	115	1.6 (Mie)	474 ± 10	685 ± 10	53
PnBAPS122	BASF 2035/7348	122	2 (UZ)	582 ± 18	743 ± 40	31
Si136	CS121	136	8 (TEM)	n/a	n/a	[25, 26]
PS156	IDC 2-179-4	156	n/a	615 ± 50	945 ± 70	53

shut down of the pump. Stop of flow thus marks the start of a crystallization experiment. Automatic control is particularly useful for USAXS, where the setup is inaccessible inside the experimental hutch, but also facilitates convenient performance of time-extensive systematic light scattering and microscopic experiments. In general, this method has the advantage of being very fast, accurate, and reproducible. Its particular advantage for the present measurements of different properties is that it guarantees identical interaction conditions at the different experiments connected by the tubing system. It thus circumvents the necessity to transfer the fragile samples between setups, which could easily alter the sample morphology or even shear melt them.

For the determination of phase diagrams of mixtures, an alternative, material-saving procedure was employed. First, high-concentration stock suspensions were prepared by adding an ion exchanger left in contact for more than two months. From these, the actual samples were prepared by mixing and appropriate dilution with milli-Q water. Fresh ion-exchange resin was added and the samples were left to deionize again for another month. Here the residual ion concentration is well below $10^{-7}\,\mathrm{mol\,l^{-1}}$. They were then shaken up again to homogenize them, and the phase behavior was inspected by microscopy and/or light scattering.

12.2.2
Instrumentation for Time-Resolved Static Light Scattering

Samples at low to moderate particle concentrations can conveniently be investigated by optical methods. For our investigations on concentrated Si84 samples, USAXS was employed. This has the advantage of rendering the data free of the influence of multiple scattering, and in addition giving access to an extended range of scattering vectors $q = (4\pi/\lambda)\sin(\Theta/2)$ needed for the determination of the melt structure. (Here λ denotes the wave length of the radiation in the sample and Θ is the scattering angle.) We first briefly sketch the setup used for light scattering, and then explain the USAXS measurements in some more detail.

Our multipurpose light scattering instrument was constructed following closely an earlier version of the machine [55], but with a few interesting improvements. The goniometer comprises two antiparallelly aligned illumination beams combined with a servomotor-driven double arm, such that both arms can detect scattered light under the same scattering vector but originating from the two different illuminations. It facilitates quasi-simultaneous static light scattering, time-resolved static light scattering, and elasticity measurements on the very same sample of well-controlled interaction. For the details of elasticity measurements and conductometry, the interested reader is referred to previous papers [46, 53, 56, 57].

Standard detection optical components are mounted on one goniometer arm. They allow stepwise increase of Θ and are optimally suited for structure identification. Statistical accuracy is enhanced by rotating the sample about its vertical axis. In addition, a large CCD array was implemented on the opposite goniometer arm to detect a broader sector simultaneously. It is therefore suitable even for systems at high particle number densities where crystallization kinetics is fast. This is exemplified in Figure 12.1 for a sample of pure PnBAPS68 at $n = 76\,\mathrm{\mu m^{-3}}$ suspended in a refractive-index-matching solvent. Microscopy complements the scattering experiments, as it significantly extends the range for kinetic measurements toward low undercoolings and allows precise determination of the position of phase boundaries [58].

The USAXS measurements were performed at the soft matter beamline (BW4) of HASYLAB, Hamburg. BW4 is a wiggler beamline designed as a high-flux materials research setup. X-rays coming off the wiggler pass a double crystal monochromator and have a standard wavelength of $\lambda = 0.138\,\mathrm{nm}$. The beam is focused onto the detector by means of horizontally and vertically mounted mirrors. A commercial

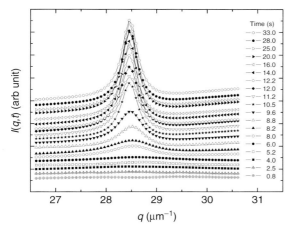

Fig. 12.1 Exemplary time-resolved and background-corrected static light scattering signal from the 110- reflection during crystallization of a sample of PnBAPS68 suspended at $n = 76\,\mu m^{-3}$ in a water/dimethylsulfoxide solvent. Intensity curves are shifted for clarity.

CCD detector (MARCCD165) with an active diameter of 165 mm and a resolution of 2048 × 2048 (pixel size 79.1 μm) was used. At the sample, the diameter of the X-ray beam was 0.4 mm with a Gaussian beam profile. Since particle distances of between 260 and 360 nm were expected, the largest possible sample–detector distance was chosen and calibrated with a collagen sample to be 13.3 m. With these parameters, the accessible range of scattering vectors q was $0.8 < q < 320\,\mu m^{-1}$ [59]. Samples are investigated using the flow-through cell that can be connected to the computer-controlled preparation circuit. A Kapton film of thickness 25 μm and diameter of 4 mm is used as window material, ensuring good transmission at $\lambda = 0.138$ nm. The sample volume is about 150 mm^3 and the wall-to-wall distance 3 mm.

Time-resolved diffraction patterns $I(q, t) = I_0 n P(q) S(q, t)$ of the shear-molten and recrystallizing colloidal silica suspension were taken at $40 \leq n \leq 113\,\mu m^{-3}$. The lower bound is due to the beam stop in front of the detector, and the upper bound is given by the concentration of the stock suspension. The investigated range shows a considerable overlap with that from light scattering. In addition, the intensity at lower n and with a large amount of NaOH added was recorded. In this case, the interactions are suppressed and $S(q) = 1$, and hence the particle form factor $P(q)$ is directly accessible. The X-ray intensity I_0 is measured with a standard Lupolen sample [60]. Thus, $S(q)$ can be calculated in a straightforward manner.

Measurements carefully obeyed the following protocol. After conditioning, circulation was abruptly stopped by the electromagnetic valves 0.5 s before the shutter of the detector opened. This ensured that the suspension was at rest when the data collection started. For each set of interaction parameters, 10 diffraction patterns with an accumulation time of 3–5 s depending on the scattering intensity and the present accelerator current were taken. With a detector readout time of 3.5 s between each collected diffraction pattern, a total accumulation time of 1 min for this sequence resulted. A further diffraction pattern of longer accumulation time of 15–30 s was made 30 s later. Figure 12.2 shows two sets of raw data taken during the crystallization of Si84. To the left we show a sequence taken at the maximum particle number density of $n = 113\,\mu m^{-3}$ and a sodium hydroxide concentration of $c_{NaOH} = 1.1 \times 10^{-3} mol\,l^{-1}$ which is close to maximum interaction. The first diffraction pattern shows the undercooled melt state. After the third picture, the Debye–Scherrer rings become narrower and additional rings appear, indicating the transition to a polycrystalline bcc state. In addition, from the beginning of crystallization a weakly developed fourfold point pattern was visible, which can be ascribed to an oriented, wall-based single crystal created via heterogeneous nucleation [61]. The effect becomes more pronounced on approaching the phase boundary, as the interaction between the particles is decreased. This is shown in the right part of Figure 12.2 for a sequence taken at a small amount of added NaOH ($c_{NaOH} = 4 \times 10^{-4} mol\,l^{-1}$) corresponding to the same n but a much lower charge. Here the appearance of heterogeneous nucleation has a much larger influence on the diffraction pattern.

Evaluation of the diffraction pattern includes normalization of the scattered intensities, the background, and the transmission correction and follows a standard procedure which is described in detail elsewhere [62]. For further data evaluation the program "FIT2D" is used, which is offered by the European synchrotron radiation facility (ESRF) as freeware. To give an idea on its performance, we exemplarily evaluate the left sequence in Figure 12.2. First, an azimuthal integration of the detected intensity of each single diffraction pattern (the so-called 2θ integration)

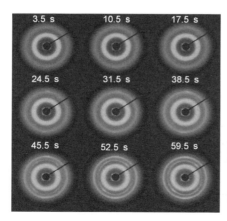
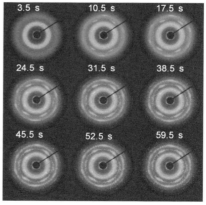

Fig. 12.2 The left picture shows a sequence of nine diffraction patterns of the colloidal silica suspensions with particle diameter of 84 nm taken within 1 min at a number density $n = 113\,\mu m^{-3}$ and sodium hydroxide concentration $c_{NaOH} = 1.1 \times 10^{-3} mol\,l^{-1}$ for example at strongest particle interaction. The right picture shows a sequence of diffraction patterns at identical particle number density and identical accumulation time but at a lower NaOH concentration of $c_{NaOH} = 4 \times 10^{-4} mol\,l^{-1}$.

was carried out. For samples showing contributions of wall crystal scattering, the corresponding regions were excluded from the averaging procedure. The resulting 1D intensity distribution $I(q)$ was divided by the particle form factor $P(q)$. The resulting structure factors $S(q)$ of the sequence at different times are shown in Figure 12.3.

12.3 Results

12.3.1 The Full Phase Diagram of Charge Variable Systems

Early measurements of the phase behavior of charged suspensions employed light scattering and microscopy [13, 63, 64]. A complete phase diagram in terms of salt concentration and particle density was first measured by Sirota et al. [65] using small-angle X-ray scattering, enhancing the concentration range, and accessing both face centered cubic (fcc) phases and a glass phase. Using a combination of light scattering and USAXS on Si84, we here determined the full phase behavior as a function of particle number densities n, effective charge, and concentration of excess electrolyte (NaOH) [66]. As seen from Figure 12.3, the suspension stayed bcc up to the largest n. The full phase diagram is shown in Figure 12.4 as a function of n and the amount of added NaOH.

In contrast to systems of fixed charge, a reentrant crystallization is observed as a function of both c_{NaOH} and n. Under completely deionized conditions, the charge is

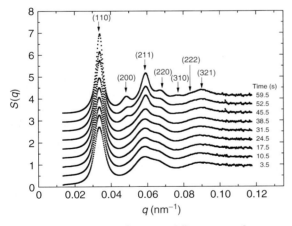

Fig. 12.3 Static structure factors at different times for a crystallizing shear melt of Si84. A particle number density of $n = 113\,\mu m^{-3}$ at maximum interaction ($c_{NaOH} = 1.1 \times 10^{-3}\,mol\,l^{-1}$). $S(q)$ are shifted for clarity, time increases from bottom to top. Miller indices are indicated, clearly identifying the bcc structure of the polycrystalline solid formed.

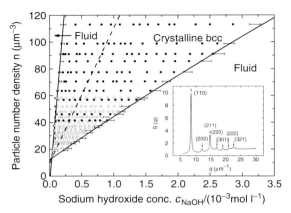

Fig. 12.4 Full phase diagram of Si84 as a function of the particle number density and the sodium hydroxide concentration as determined by USAXS (filled circles) and light scattering (open circles). The coexistence regions on both sides are marked by horizontal bars. Note that their extension in c_{NaOH} direction increases with increasing n. The solid lines are guides to the eyes dividing the crystalline bcc from the fluid phases. The dashed line denotes the maximum interaction where all silanol groups on the particle surface are used up for the charging-up reaction. The insert shows an example of a structure factor obtainable by light scattering.

too small to cause crystalline order. Increasing the charge induces an enhancement of interactions and therefore crystallization. At constant c_{NaOH}, an increase of n reduces the amount of NaOH available per silanol group and therefore first causes a decrease of screening but then a decrease of the charge density. Thus the reentrant behavior is given in terms of two variables. This is a novel observation and different from the reentrant behavior previously observed for polymer particles of constant bare charge. There, upon increasing the particle concentration at deionized conditions, the interaction first increased, then decreased, and then increased again. The decrease was attributed to an intermediate increase of screening due to the added particle counterions outweighing the decrease of interparticle distance [52, 67].

To pursue this behavior further, we performed measurements of the shear modulus G using the multipurpose light scattering instrument. G data were collected in the range of $n = 28\,\mu m^{-3}$ up to $n = 60\,\mu m^{-3}$ in which homogeneous nucleation was the dominant nucleation process and polycrystalline samples were obtained. At a concentration below $n = 28\,\mu m^{-3}$, heterogeneous nucleation dominates, leading to complex morphologies not covered by the presently used evaluation algorithm [68].

Figure 12.5a shows that with increasing NaOH concentration the shear modulus first increases approximately linearly owing to the charging-up reaction of the silica particles. The maximum shear modulus indicates the maximum interaction between the particles. Further addition of excess NaOH linearly decreases the shear modulus owing to the screening of the particles. The maximum values of the shear moduli at different concentration can be simply described by a straight

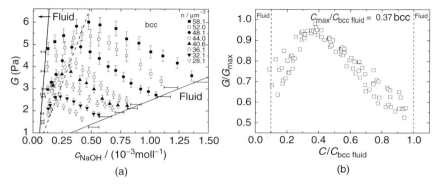

Fig. 12.5 Shear modulus behavior of the colloidal silica crystals in dependence on the concentration of added NaOH. (a) The shear modulus first increases up to a maximum value, then decreases again. The maximum values at each particle number density can be described by a straight dashed line representing the maximum interaction between the particles. (b) Shear moduli normalized by the maximum shear modulus versus the NaOH concentration normalized by the concentration at the bcc fluid phase boundary.

line (see Figure 12.5a). Further, all curves appear to have a similar shape. This in fact is expected if (i) the charging procedure scales with both n and c_{NaOH}, and (ii) a polycrystalline morphology is retained throughout. To check the latter condition we scaled the shear moduli by their maximum value and the NaOH concentration by its value at the upper phase boundary. All data collapse on a single master curve, indicating the self-similarity of the charging process during this shear modulus titration experiment. Moreover, $c_{max}/c_{bcc\,fluid} = 0.37$ holds irrespective of n, indicating that, indeed, both the charging and the following screening process scale with particle concentration (Figure 12.5b).

We, therefore, may identify the maximum interaction with the equivalence point, and the maximum surface charge density σ_a can be calculated from the added amount of base c_{NaOH} (in mol l^{-1}) as:

$$\sigma_a = \frac{10^{-3}}{3} N_A e a \frac{c_{NaOH}}{\Phi} \tag{12.7}$$

where N_A is Avogadro's number, e the elementary charge (in C), a the particle radius (in cm), and $\Phi = n/(4/3)\pi a^3$ the volume fraction of the silica. The conversion of the charge density values at each concentration results in a constant mean number of silanol groups on each particle surface of $N = 4520 \pm 130$.

The corresponding effective charge can be determined from the shear moduli if a polycrystalline morphology is observed. The maximum shear moduli are shown in Figure 12.6 with a fit of Equation 12.6 yielding a constant $Z_{eff} = 340 \pm 20 \ll N$. As our method of effective charge determination can be carried out also for shear moduli off the maximum interaction, we currently check for the development of effective charges also along both phase boundaries. Preliminary results further indicate that the phase behavior can again be excellently described using the elasticity charge in combination with the universal melting line of Robbins et al. [48].

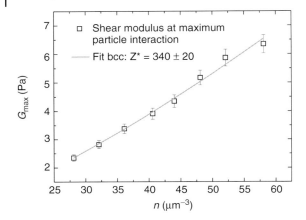

Fig. 12.6 Maximum values of the shear modulus in dependence on the particle number density. The solid curve is the best one parameter least square fit yielding a constant $Z_{\text{eff}} = 340 \pm 20$.

12.3.2
Shapes of Phase Diagrams of Charged Sphere Mixtures

For charged sphere mixtures, two additional parameters govern the shape of phase diagrams in the concentration–composition plane: the size ratio Γ and the charge ratio Λ, although in previous investigations only the former was varied. Spindle-like, lower azeotrope and eutectic-like phase diagrams were obtained [22]. In most cases solid solutions were found [27], but also compounds [13, 14] and glasses [22, 24] have been observed. Both ratios can be controlled by synthesis and, as shown above, the latter also via the conditioning procedure. So far we only have made use of the former approach and mixed several species of fixed effective charged spheres (Table 12.1). In the case of effective charge saturation for both particles, the effective charge scales with the radius. In this case, Γ and Λ are varied in concert. By contrast, the use of a weak acid group for one species will lower its bare charge and in turn will decrease its effective charge. Then Γ and Λ can also be varied independently.

Table 12.2 summarizes the findings of the different mixing experiments. At large size ratios and charge ratios we observe a spindle-like phase diagram, typical for indifferent solubility. Light scattering reveals the formation of chemically disordered bcc crystals in practically all cases. To facilitate comparison with metal systems, we follow earlier work [22] and plot our data in terms of the inverse particle number density $1/n$ and the composition $p = n_{\text{small}}/n_{\text{total}}$. Different to the metal case in this representation, coexistence regions appear already for the pure components, owing to the differences in coexisting density for the solid and melt phases.

This effect is well known from single-component systems and simply transfers to our mixtures. Figure 12.7 shows exemplarily the phase diagram of PS90/PS100B,

Table 12.2 Compilation of properties of investigated mixtures. The last three mixtures are included from the literature [22] for comparison. Size ratio Γ is as determined from the nominal diameter of the single components; effective charge ratio Λ is determined from effective shear modulus charges $Z_{eff,G}$ (*charge ratio determined by conductivity charges); type of the phase diagram observed is a function of the composition and the occurrence of a glass transition. The last column gives the corresponding references.

Samples	Size ratio Γ	Effective charge ratio Λ	Type of phase diagram	Glass	Reference
PnBAPS68/PS100B	0.68	1.01	Upper azeotropic	—	[31, 32]
PnBAPS68/PnBAPS122	0.56	0.57	Eutectic	—	31
PS85/PS100A	0.85	1.00	Spindle	—	[27, 29]
PS90/PS100B	0.9	0.96	Spindle	—	29
PS120/PS156	0.74	0.77	Spindle	—	[28, 29]
PS85/PS109	0.78	n/a	Lower azeotropic or eutectic?	+	24
PS85/PS91	0.93	n/a	Spindle	—	24
Si103/Si136	0.76	n/a	Spindle	+	24
Stavans 1 (PS86/PS99)	0.87	0.78*	Spindle	—	22
Stavans 2 (PS67/PS86)	0.78	0.67*	Lower azeotropic	+	22
Stavans 3 (PS86/PS160)	0.54	0.36*	Eutectic	+	22

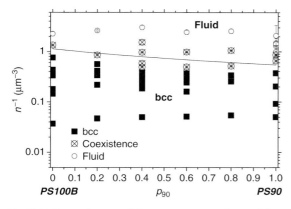

Fig. 12.7 Phase diagram of the PS90/PS100B in the inverse particle number density–composition plane. Data were obtained from visual inspection and static light scattering using the batch deionization technique. The residual uncertainty in n is given by the exemplary error bars; for p it is in the order of 0.05. Symbols denote fully crystalline samples (filled squares), the coexistence region (crossed diamonds), and the fluid phase (open circles). The solid line gives the expectation for melting of an ideal solid solution, as interpreted from the pure sample melting points using number-weighted averages.

for which we have also determined other properties, such as conductivity, shear modulus, growth velocities, and nucleation rate densities. Shear moduli and other properties show a monotonous decrease or increase with composition [27, 30]. Phase behavior and all other properties of this solid solution can be described well using the effective potential suggested by Lindsay and Chaikin, Equation 12.3 [21], which comprises a number-weighted average of particle properties for each mixture. We therefore conclude that at sufficient similarity of particle properties, colloidal mixtures show a behavior of indifferent solubility and solidify as solid solutions.

This is not the case if the size and therefore charge difference is increased. Then a lower azeotrope or even a eutectic behavior is observed [31]. The mixture of PnBAP68/PnBAPS122 was investigated in more detail. At a certain composition of $p = 0.8$ a significantly decreased stability of the crystal phase is found. For $p > 0.8$ and at coexistence, the majority species (PnBAPS68) is enriched at the cell top by differential sedimentation and there forms nearly pure bcc crystals ("head crystal"). The lower part of the cell remains molten and is enriched with the minority component (PnBAPS122). At particle concentrations above coexistence, we found the solidification process of this mixture to be considerably slowed. While for pure samples and mixtures at low p we observe a solidification within minutes, the mixture of PnBAPS68/PnBAPS122 at $p = 0.9$ solidifies on the time scale of days. This was monitored by following the evolution of the shear modulus. The results are shown in Figure 12.8. Interestingly, also a subsample drawn from the head crystal grown in a large reservoir container that was shaken and recrystallized in a separate vial showed a rapid solidification, indicating the separation of species occurring

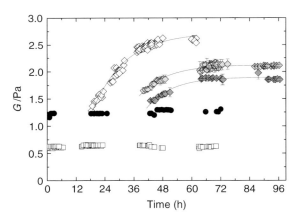

Fig. 12.8 Time evolution of the shear modulus of the pure component PnBAPS68 (filled circles) and for the mixture with PnBAPS122 at $p = 0.9$ at particle densities of 32 μm^{-3} (open diamonds), 29 μm^{-3} (light gray diamonds) and 26 μm^{-3} (dark gray diamonds). Note that all latter samples need a finite time to reach their final rigidity. Lines are fits of an exponential $G = G_0(1 - exp(-t/\tau))$ with a structural relaxation times of 13.8, 11 , and 10.2 h. By contrast the lowest data set (open squares) was measured on a sample taken from the head crystal of the reservoir at $n = 13$ μm^{-3}. Here no slowing of solidification is present.

during the first solidification. Investigations with microscopy in addition showed the formation of two species of crystals of different compositions. We, therefore, conclude that we are observing a eutectic mixture in this case. The investigations on PS85B/PS91 at present still lack the demonstration of two different crystal species forming. So, there the question of whether a lower azeotrope or a eutectic is formed remains open.

A different scenario is encountered if we work at moderate size difference but employ low-charged large particles. For the mixture of PnBAPS68 and PS100B, the size ratio is $\Gamma = 0.68$, whereas the effective charge ratio $\Lambda \approx 1$. In this case, the system assumes a solubility much better than indifferent. At a composition of $p = 0.25$, we observe a strongly enhanced crystal stability, an increased limiting growth velocity, and a rather small increase of difference in chemical potential between the melt and solid as we increase the particle concentration. All these findings indicate an excellent solubility of the larger but equally charged particle in the majority of smaller ones. Consequently, we interpret our findings as an upper azeotrope. We thus have access to a rich variety of phase diagrams, all also known from metal systems.

12.3.3
Growth of Binary Colloidal Crystals

Growth velocities are accessible either from microscopy from direct observations of the thickness of a wall-nucleated crystal as a function of time [54] or from time-resolved light scattering experiments (Figure 12.1). There the crystallite size is obtained from the full width at half-height Δq of the Bragg reflections. To be specific, the average side length of an equivalent crystal of cube shape $\langle L(t) \rangle$ is obtained from $\langle L(t) \rangle = 2\pi K / \Delta q(t)$, where the Scherrer constant K is of the order of 1 and the pointed brackets indicate the ensemble average; $d\langle L(t)\rangle/dt$ then yields the growth velocity. This is shown in Figure 12.9 as an example of pure PnBAPS68 in an index-matching solvent. For Si84, we could perform a systematic study on the growth velocities as a function of all interaction parameters in a range of particle number densities between $n = 18$ and $n = 37\,\mu m^{-3}$ using microscopy. Measurements were taken at fixed n as a function of the added amount of NaOH. This was repeated for different particle number densities. In all cases a linear increase of the crystal dimension was observed, indicating reaction-controlled growth.

The result is given in Figure 12.10a. Different to the shear modulus measurements, the maximum appears flatter and can be better described by a convex parabolic curve. But again, a good scaling is observed in a plot of the velocity normalized by the maximum velocity and the base concentration by that at the upper phase boundary. The ratio $c_{max}/c_{bcc\,fluid} = 0.42$ appears only slightly above the corresponding value extracted from the shear modulus measurements.

The flattening is understandable considering that reaction-controlled growth possesses a limiting velocity. In Figure 12.10b, the growth velocities along the

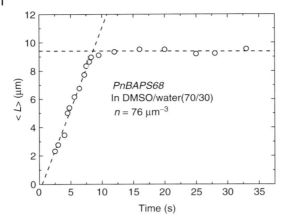

Fig. 12.9 Exemplary measurement of growth velocities of homogeneously nucleated samples at large undercooling evaluated from the data in Figure 12.1. Crystallite sizes first increase linearly then saturate upon intersection. A least square fit to the early data points yields the constant growth velocity as $v = 1.2\,\mu ms^{-1}$.

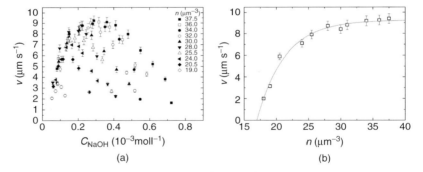

Fig. 12.10 (a) Measured growth velocities of Si84 versus amount of added base for different particle densities n. At small NaOH-concentrations the velocities increase, then go through a flat maximum and decrease again for large base concentrations. (b) Growth velocities vs. particle number density at maximum interaction. The solid curve is a fit of a WF law yielding $v_\infty = 9.25 \pm 0.25\,\mu ms^{-1}$ and $B = 3.91 \pm 0{,}60\,k_B T$.

line of strongest interactions are shown with a fit of a Wilson–Frenkel (WF) law [69, 70]:

$$v = v_\infty \left[1 - \exp\left(-\frac{\Delta\mu}{k_B T}\right)\right] \tag{12.8}$$

where $\Delta\mu$ is the chemical potential difference between the fluid phase and the crystalline phase, which is a measure for the undercooling of the system, and v_∞ is the limiting velocity at infinite $\Delta\mu$. To adapt this expression to colloids, a simple approximation was first suggested by Aastuen et al. [71]: $\Delta\mu = B(n - n_F)/n_F$. Here n_F

denotes the particle number density at freezing, which in our case is $n_F = 17\,\mu m^{-3}$, and the undercooling parameter B is a fit parameter to be determined from the experiment. Later work of Würth et al. [54] included the density dependence of the potential $V(r)$ (Equation 12.4 and 12.5) and approximate $\Delta\mu = B\Pi^*$ with the reduced energy density difference $\Pi^* = (\Pi - \Pi_F)/\Pi_F$ between the melt and the fluid at freezing. Here the energy density $\Pi = \beta n V(d_{NN})$, with β being the coordination number in the short-range ordered state. A least square two parameter version of Equation 12.8 using Ackerson's approximation to the velocities as measured along the line of maximum interaction (Z_{eff} and c are constant) yields $v_\infty = 9.25 \pm 0.25\,\mu m/s$ and $B = 3.91 \pm 0.60\,k_B T$. Similar fits can be made at constant n along increasing charge or salt concentration using Würth's approximation. With the limiting growth velocity and the undercooling parameters, a major prerequisite is gained for the extraction of interfacial tensions from the nucleation experiments within the framework of classical nucleation theory (CNT) as attempted in the next chapter.

The growth behavior was studied also in several mixtures. Again, in all cases, growth was linear above coexistence. Undercooling parameters could conveniently be extracted from WF law fits to the measured data of solid solutions and also for binary mixtures showing spindle-type or upper azeotrope phase behavior. However, two interesting observations could be made on the velocity as a function of composition, both presumably connected to a change of interfacial width between melt and solid.

The mixture of PS120 with PS156 shows a regular solid solution phase behavior and obeys a WF growth law as a function of particle concentration. The composition dependence of the growth velocity was investigated using Bragg microscopy at $n = 0.47\,\mu m^{-3}$, which is a compromise between the requirement of sufficient undercooling to be close to the limiting velocity and the tendency of homogeneous nucleation at larger n [33]. Interestingly, we found a significant deviation from the expected behavior. This is plotted in Figure 12.11. For compositions larger than $p = 0.15$, the data follow the expectation based on the assumption of a monolayer interface. With decreasing p, the velocity increases by roughly a factor of 2 to approach the prediction for an extended interface approximately two layers thick. Layering has been observed before in computer simulations on Lennard-Jones spheres and binary metal melts [72, 73] and is known from the initial stages of heterogeneous nucleation of charged sphere suspensions at charged walls [74]. Würth had discussed it as a possible explanation of the larger-than-expected growth velocities in one-component colloidal melts [54]. The present data then would indicate that one-component samples of low polydispersity could form a layered interface, while the decrease of the velocity would correspond to a decrease of the amount of layering in front of the growing crystals as the impurity concentration and therefore the effective polydispersity was increased.

A complementary behavior was observed in the data for the PnBAPS68/PS100B mixture shown in Figure 12.12 [32]. At large n, a pronounced increase of the limiting velocity is visible for $p_{small} > 0.9$, in line with the above argument. A second feature, however, is seen for $p \approx 0.2$, where the maximum of the azeotrope is observed. The very good mutual solubility of both species aids the formation of

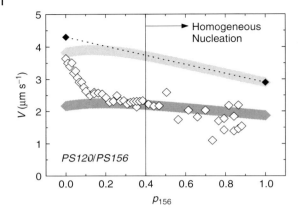

Fig. 12.11 Growth velocities for the mixture PS120/PS156 (open diamonds). The scatter of data for $p > 0.4$ is caused by interference with homogeneous nucleation. The dotted line connects the limiting growth velocities of the pure components (closed diamonds). The light gray (dark gray) line gives the predictions of numerical calculations for WF growth with extended (monolayer) interfacial region. Between $p = 0$ and $p = 0.15$ a transition between these cases is observed.

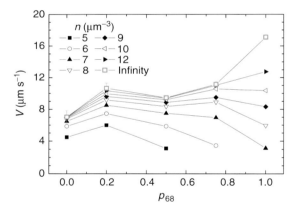

Fig. 12.12 Growth velocities for the mixture PnBAPS68/PS100B. Particle concentrations increase from bottom to top. At large particle concentrations $n > 18\,\mu m^{-3}$ (infinity) the limiting growth velocities are almost reached here denoted as open squares. Note the strong increase at large p. The corresponding decrease at small p is masked by the larger limiting velocity observed in the region where the onset of compositional order sets in.

layering and thus an extended interface. In turn, the maximum growth velocity is increased for this composition, where the structural similarity between the melt and solid is most pronounced. The second fit parameter of the WF evaluation, the undercooling parameter B, also shows an anomaly at this composition. Although at all other compositions it is in the order of 2, at $p = 0.2$ it is just below 1. This corresponds to the structural similarity of the melt and the solid.

12.3.4
Quantitative Determination of Nucleation Kinetics and Extraction of Key Parameters

Homogeneous nucleation is a thermally activated process. Nucleation rate densities increase exponentially with undercooling or the difference in chemical potential between the melt and solid. Thus, nucleation rapidly accelerates with increasing particle number density (with Z_{eff} and c kept constant for most experiments). At low concentrations, nucleation rate densities are directly accessible from microscopic counting experiments. At larger concentrations, time-resolved static structure factors or, if these are not available, the static structure factors after complete solidification are evaluated [58]. In the first case, we follow the procedures outlined in [75]. The amount of crystalline material $X(t)$ is obtained from the integrated intensity of a Bragg reflection normalized to its value after completion of solidification. From the average crystallite sizes from the peak widths, the crystallite density $\rho(t) = 1/\langle L(t) \rangle^3$ is calculated. Division by the relative free-melt volume $V_{\text{free}}(t)/V_{\text{Total}} = 1 - X(t)$ yields the nucleation rate density $J(t)$. For the quickly solidifying samples, the final crystallite density is employed to infer the average nucleation rate density during crystallization using the Kolmogorov–Johnson–Mehl–Avrami Model (KJMA) [76]. Therefore we assume that the nucleation rate density is not a strong function of time.

Figure 12.13a and b show the nucleation rate densities for several species of charged sphere suspensions and a mixture. In all cases the rate density initially increases rapidly with increased n, and then levels off. The mixture

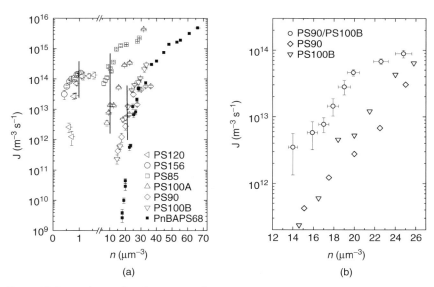

Fig. 12.13 Dependence of nucleation rate densities on particle concentration (a) for several single-component suspensions as indicated, (b) for *PS90/PS100B* as single-component species and as 50 : 50 mixture.

shows slightly larger rate densities than for the pure components. This dependence is parameterized using a version of CNT adapted to the diffusive dynamics of colloidal dispersions to obtain barrier heights and dimensionless interfacial energies [77]. To exemplify this procedure we use the data for pure PnBAPS68. As the freezing concentration is $n_0 = 6.1\,\mu m^{-3}$, experiments were performed at large-scale undercoolings of $(\Delta T)/T_0 \propto (1/n_0 - 1/n)/(1/n_0)$ between 0.666 and 0.909. In CNT, the nucleation rate density reads $J = J_0 \exp(-\Delta G^*/k_B T)$ were the barrier height is given by $\Delta G^* = (16\pi/3)\gamma^3/(n\Delta\mu)^2$ as the ratio between the interfacial energy γ^3 to the undercooling squared. The prefactor reads $J_0 = A(n) \cdot n^{\frac{3}{4}} \sqrt{\gamma(n)} \cdot D(n)$, with $A(n)$ being a numeric factor for the efficiency of barrier-crossing which CNT assumes to be constant at a value of $A = 1.55 \times 10^{11}$.

As the diffusion coefficient $D(n)$ is not known, we may either make an estimate from the literature and calculate $J_0(n)$, or perform a two-parameter fit of the CNT expression with $A(n)$ fixed and both $D(n)$ and $\gamma(n)$ as free parameters, or use an Arrhenius plot (Figure 12.14a) to determine the interfacial tension from the local slope (tangent method) and the kinetic prefactor from extrapolation to infinite undercooling. The resulting kinetic prefactors are shown in Figure 12.14b together with the experimental data for PnBAPS68, and a three-parameter fit of the CNT expression. J_0 from explicit calculation are way above and those from the Arrhenius evaluation rather close to the experimental data. In turn, the interfacial tensions from the explicit calculation are large, while the ones from the Arrhenius evaluation are lower and show a lower slope. This is shown in Figure 12.14c, where we, in addition, use the reduced interfacial energies per area occupied by a particle in the surface $\sigma = \gamma d_{NN}^2$. In both cases the data from the two-parameter fit assume an intermediate position and show a low slope.

At present we cannot yet decide which procedure yields the more realistic values for the interfacial energy. The Arrhenius procedure has the advantage that it does not rely on any assumption about the nucleation dynamics (e.g. addition of individual particles to the nucleus). It only assumes an activated process with competing bulk and surface terms. Hence, at present we may safely state only that these are larger by at least a factor of 2 than measured for hard spheres. There values of $\sigma = 0.51 \pm 0.03\,k_B T$ were obtained [8], which are in good agreement with other experiments on similar spheres [75] and some of the theoretical expectations [78]. In addition, irrespective of the evaluation method, charged sphere surface energies show an increase with increasing particle concentration. This is in line with recent computer simulations [79]. It is further interesting to note that the interfacial energies increase linearly, as does the undercooling. If we plot the reduced interfacial energy per particle $\sigma = \gamma d_{NN}^2$ versus the chemical potential difference, we obtain a plot analogous to the famous one by Turnbull recently supplemented by new experimental and theoretical data [80, 81]. Two differences are observed in comparison to actual versions of this plot, also containing computer simulation results. First, our data lie parallel to the expectations for bcc crystals from simulation and have a lower slope than found for fcc metals. Second, we observe an offset at small $\Delta\mu$. This can already be anticipated in Figure 12.14c,

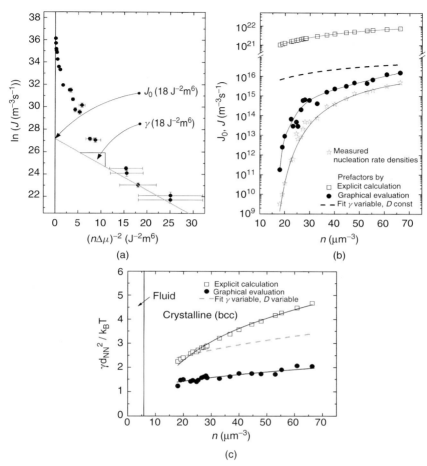

Fig. 12.14 (a) Arrhenius plot of the measured nucleation rate densities illustrating the tangent method to obtain the interfacial energy from the local slope and the kinetic prefactor from extrapolation. (b) Comparison of the derived kinetic prefactors with the experimental data and a full three-parameter fit of the CNT expression (solid line through the data). The solid lines through the kinetic prefactor data are guides for the eye. (c) Reduced interfacial energies per particle in the interface (symbols as before). For all three approaches the derived values are larger than the values for hard spheres and show an increase with increased n.

as $\Delta\mu$ increases linearly with n. Without additional data taken closer to the phase boundary, we have to refrain from further speculation. The present experiments were performed under the variation of n, which has a complex effect on the interaction, as it decreases the particle distance but simultaneously increases the screening. We anticipate that the Arrhenius approach can be rigorously tested for a case in which n, c, and Z can be varied independently. This is realized in the Si84 system.

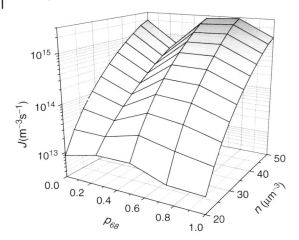

Fig. 12.15 Nucleation rate densities obtained for a deionized mixture of *PnBAPS68* and *PS100B* in water. Note the anomaly at $p = 0.2$ which is induced by a small undercooling factor B.

The nucleation kinetics as a function of composition was studied in the system PnBAPS68/PS100B [32]. There, an interesting decrease of nucleation rate densities close to the azeotropic condition was observed, which is correlated to a minimum of the undercooling factor B. Thus the minimum of rate densities deepened with increasing n. This is shown in Figure 12.15. In all studied cases we obtained the same flattened curves for the nucleation rate densities as a function of particle concentration. The flattening in CNT is mainly due to interfacial tension increase, while the Arrhenius evaluation in addition suggests some slowed increase of the kinetic prefactor. The increase of interfacial energies could be related to a changed melt structure. It, therefore, is instructive to finally also discuss the structure of the undercooled melts accessible from the USAXS measurements.

12.3.5
Investigations on the Structure of Undercooled Melts

$S(q)$ of the fluids and undercooled melts was measurable up to the fourth oscillation. In Figure 12.16a we show the evolution of the structure factor with increased n measured at maximum interaction and with no excess salt. From the growth behavior at maximum interaction (see Figure 12.10b), the corresponding chemical potential differences as a measure for the undercooling were extracted by means of a two-parameter WF fit to the data following the procedure of Würth et al. [54], which was described in Section 12.3.3. With increasing undercooling, a shoulder evolves in the second peak, which is absent in the equilibrium fluid. The feature prevails with large n also when the charge is decreased or the salt concentration is increased. Such a feature has been observed previously for binary colloidal

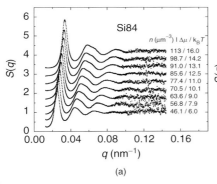 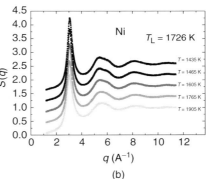

Fig. 12.16 Comparison of structure factors of equilibrium liquids and undercooled melts taken at different undercoolings. (a) USAXS structure factors of the Si84 suspension all taken in crystallizing samples at maximum interaction, (b) neutron scattering structure factors of Ni; the lower two curves were recorded above the liquidus temperature. Note the development of a shoulder in both cases which becomes more pronounced with increasing undercooling, but is practically absent in the equilibrium systems.

suspensions only and was ascribed to a glassy structure, which can be safely excluded in our case [82]. The evolution is compared to structure factors measured by incoherent neutron scattering on Ni melts shown in Figure 12.16b [83]. The evolution of the shoulder feature looks amazingly similar for both systems. We therefore conclude that a change of short-range order is a common feature for both metals and colloids.

For the metal system this observation was interpreted as an increasingly icosahedral short-range order within the single-component melt. First attempts to fit the corresponding theoretical expressions to our data give similar results, indicating that about 95% of the melt could be of an icosahedral or dodecahedral order while some 5% are bcc. The occurrence of crystals with cubic symmetry from such a melt seems difficult. Recent experiments on hard spheres had shown an increased delay of nucleation with increasing concentration, but also evidenced the early formation of crystal-like clusters, which later convert to fcc and hcp structures [84]. Recent computer simulations on hard spheres suggest that nuclei may in some cases be formed with an icosahedral symmetry of fcc slices [85], whereas simulations on single-component Yukawa systems have shown that nucleation proceeds via bcc nuclei, and that if fcc is the stable phase, this polymorph is chosen at the growth stage [86]. A full understanding of the melt structure could help in resolving these puzzles and this is currently under way. In any case, however, the observed increase of interfacial tensions could be attributed to an increasing structural difference between the nuclei and the surrounding melt, which is much less pronounced at low n where the melt structure is more bcc like [87]. And similar concepts, that the underlying phase behavior in contrast or in line with the melt structure will have strongest influence on the nucleation kinetics, are very likely to apply also for the case of mixtures [88].

12.4
Conclusions

We have shown the possibilities of tuning the interactions in colloidal charged sphere suspensions and performed a large variety of systematic measurements on the crystallization kinetics as well as on other properties of melts and resulting solids. We have seen that in many cases a large analogy exists between colloidal and metal systems. The phase behavior is quite similar, both concerning pure samples showing bcc and fcc structures, and for mixtures revealing upper and lower azeotropes and even eutectic behavior. Most often, spindle-like phase diagrams are observed for not too large differences in size and charge of the particles. In these cases, the solid state properties could be described well using number–densty-weighted averages. Reduction of the differences results in the formation of an upper azeotrope. Increasing the differences results in a reduced mutual solubility and an approach of eutectic behavior. It is a fascinating open question how the solid properties react on the corresponding local separation. A second open point is the evolution of microstructures in the eutectic mixtures.

Time-resolved measurements using CCD techniques were employed both with light scattering and USAXS. This allowed quantitative access to the nucleation kinetics, and compiled a comprehensive database covering several important situations. Both the reaction-controlled growth after heterogeneous nucleation and the activated process of homogeneous nucleation could be monitored with high accuracy and the data could well be parameterized using classical phenomenological theories. Owing to the relatively convenient experimental accessibility of colloidal melts, our data are of large statistical accuracy, they are free of heterogeneous nucleation contributions, the interaction potential is well known, and we even have access to the melt structure. Together, this allowed a critical assessment of classical nucleation theory. We find that CNT with the key parameters of interfacial energy and kinetic prefactor indeed provides a useful parameterization. Still, a large number of challenges remain, for example to describe the relation of nucleation rate densities and melt structure beyond qualitative analogies, or to address the case of heterogeneous nucleation which is so important in metal systems. It often has been pointed out that we are still missing a microscopic theory of crystallization. Perhaps an analytic approach is still too difficult. Here, colloids and simulations may play important roles. Microscopic experiments with single-particle resolution have just begun to reveal the details of the nucleation mechanism. If such experiments are cross-checked with macroscopic ones, such as those presented here, and combined with careful computer simulation, a full description of nucleation processes in simple systems will be within reach soon.

Acknowledgments

It is a pleasure to thank J. Horbach for stimulating discussions and the BW4 Team (A. Timmann and S.V. Roth) at HASYLAB, Hamburg, for experimental support.

Financial support of the DFG (SPP1120, He1601/17, Pa459/12,13,14), the Material Science Research Center in Mainz (MWFZ), and the DLR, Cologne, is gratefully acknowledged.

References

1 Pusey, P.N. (1991) In *Liquids, Freezing and Glass Transition, 51st Summer School in Theoretical Physics, Les Houches (F) 1989* (eds J.P. Hansen, D. Levesque and J. Zinn-Justin), Elsevier, Amsterdam, p. 763.
2 Sood, A.K. (1991) *Solid State Physics*, **45**, 1–74.
3 Löwen, H. (1994) *Physics Reports*, **237**, 249–324.
4 Löwen, H., Denton, A.R. and Dhont, J.K.G. (1999) *Journal of Physics: Condensed Matter*, **11**, 10047–198.
5 Tata, B.V.R. and Jena, S.S. (2006) *Solid State Communications*, **139**, 562–80.
6 Yethiraj, A. (2007) *Soft Matter*, **3**, 1099–115.
7 Löwen, H. (2001) *Journal of Physics: Condensed Matter*, **13**, R415–32.
8 Palberg, T. (1999) *Journal of Physics: Condensed Matter*, **11**, R323–60.
9 Anderson, V.J. and Lekkerkerker, H.N.W. (2002) *Nature*, **416**, 811.
10 Bolhuis, P.G. and Kofke, D.A. (1994) *Physical Review E*, **54**, 634–43.
11 Schöpe, H.J., Bryant, G. and van Megen, W. (2006) *Physical Review Letters*, **96**, 175701.
12 Sear, R.P. (2007) *Journal of Physics: Condensed Matter*, **19**, 033101.
13 Hachisu, S. and Yoshimura, S. (1980) *Nature*, **283**, 188.
14 Sanders, J.V. and Murray, M.J. (1978) *Nature*, **275**, 201–3.
15 Bartlett, P., Ottewill, R.H. and Pusey, P.N. (1990) *Journal of Chemical Physics*, **93**, 1299.
16 Bartlett, P., Ottewill, R.H. and Pusey, P.N. (1992) *Physical Review Letters*, **68**, 3801.
17 Hunt, W.J. and Zukoski, C.F. (1999) *Journal of Colloid and Interface Science*, **210**, 332–43.
18 Hunt, W.J., Jardine, R. and Bartlett, P. (2000) *Physical Review E*, **62**, 900.
19 Schofield, A.B. (2001) *Physical Review E*, **64**, 051403.
20 Schofield, A.B., Pusey, P.N. and Radcliffe, P. (2005) *Physical Review E*, **72**, 031407.
21 Lindsay, H.M. and Chaikin, P.M. (1982) *Journal of Chemical Physics*, **76**, 3774.
22 Meller, A. and Stavans, J. (1992) *Physical Review Letters*, **68**, 3646–49.
23 Kaplan, P.D., Rouke, J.L., Yodh, A.G. and Pine, D.J. (1994) *Physical Review Letters*, **72**, 582–85.
24 Okubo, T. and Fujita, H. (1996) *Colloid and Polymer Science*, **274**, 368.
25 Okubo, T., Tsuchida, A., Takahashi, S., Taguchi, K. and Ishikawa, M. (2000) *Colloid and Polymer Science*, **278**, 202.
26 Okubo, T. and Fujita, H. (2001) *Colloid and Polymer Science*, **279**, 571.
27 Wette, P., Schöpe, H.J. and Palberg, T. (2001) *Progress in Colloid and Polymer Science*, **118**, 260.
28 Liu, J., Stipp, A. and Palberg, T. (2001) *Progress in Colloid and Polymer Science*, **118**, 91.
29 Wette, P., Schöpe, H.-J., Biehl, R. and Palberg, T. (2001) *Journal of Chemical Physics*, **114**, 7556.
30 Wette, P., Schöpe, H.J. and Palberg, T. (2005) *Journal of Chemical Physics*, **122**, 144901.
31 Lorenz, N., Liu, J. and Palberg, T. (2007) "Phase behaviour of binary mixtures of colloidal charged spheres". *Colloids and Surfaces A*, **319**, 109–115.
32 Wette, P., Schöpe, H.J. and Palberg, T. "Colloidal crystals formed in binary mixtures of spheres with differing size but equal effective charge". *Journal of Chemical Physics* (submitted).
33 Stipp, A. and Palberg, T. (2007) *Philosophical Magazine Letters*, **87**, 899–908.

34 Redl, F.X., Cho, K.S., Murray, C.B. and O'Brien, S. (2003) *Nature*, **423**, 968.
35 Shevchenko, E.V., Talapin, D.V., O'Brien, S. and Murray, C.B. (2005) *Journal of the American Chemical Society*, **127**, 8741.
36 Leunissen, M.E. et al. (2005) *Nature*, **437**, 235.
37 Ackerson, B.J. (1990) *Phase Transitions*, **21** (2–4), 73–249.
38 Matsuoka, H., Yamamoto, T., Harada, T. and Ikeda, T. (2005) *Langmuir*, **21**, 7105.
39 Alexander, S., Chaikin, P.M., Grant, P., Morales, G.J., Pincus, P. and Hone, D. (1984) *Journal of Chemical Physics*, **80**, 5776.
40 Belloni, L. (1998) *Colloids and Surfaces A*, **140**, 227.
41 Levin, Y. (2002) *Reports on Progress in Physics*, **65**, 1577.
42 Reinke, D., Stark, H., von Grünberg, H.-H., Schofield, A.B., Maret, G. and Gasser, U. (2007) *Physical Review Letters*, **98**, 0380301.
43 Shapran, L., Medebach, M., Wette, P., Schöpe, H.J., Palberg, T., Horbach, J., Kreer, T. and Chaterji, A. (2005) *Colloids and Surfaces A*, **270**, 220.
44 Shapran, L., Schöpe, H.J. and Palberg, T. (2006) *Journal of Chemical Physics*, **125**, 194714.
45 Garbow, N., Evers, M., Palberg, T. and Okubo, T. (2004) *Journal of Physics: Condensed Matter*, **16**, 3835–42.
46 Wette, P., Schöpe, H.-J. and Palberg, T. (2003) *Colloids and Surfaces A*, **222**, 311–21.
47 Bucci, S., Fagotti, C., Degiorgio, V. and Piazza, R. (1991) *Langmuir*, **7**, 824.
48 Robbins, M.O., Kremer, K. and Grest, G.S. (1988) *Journal of Chemical Physics*, **88**, 3286.
49 Wette, P. and Schöpe, H.J. (2006) *Progress in Colloid and Polymer Science*, **133**, 88.
50 Chaikin, P.M., di Meglio, J.M., Dozier, W., Lindsay, H.M. and Weitz, D.A. (1987) In *Physics of Complex and Supramolecular Fluids* (eds S.A. Safran and N.A. Clark), Wiley-Interscience, New York, p. 65.
51 Gisler, T. et al. (1994) *Journal of Chemical Physics*, **101**, 9924.
52 Yamanaka, J., Hayashi, Y., Ise, N. and Yamaguchi, T. (1997) *Physical Review E*, **55**, 3028.
53 Wette, P., Schöpe, H.J. and Palberg, T. (2002) *Journal of Chemical Physics*, **116**, 10981–88.
54 Würth, M., Schwarz, J., Culis, F., Leiderer, P. and Palberg, T. (1995) *Physical Review E*, **52**, 6415–23.
55 Schöpe, H.J. and Palberg, T. (2001) *Journal of Colloid and Interface Science*, **234**, 149.
56 Schöpe, H.J., Decker, T. and Palberg, T. (1998) *Journal of Chemical Physics*, **109**, 10068–74.
57 Hessinger, D., Evers, M. and Palberg, T. (2000) *Physical Review E*, **61**, 5493–506.
58 Wette, P., Schöpe, H.J. and Palberg, T. (2005) *Journal of Chemical Physics*, **123**, 174902.
59 Roth, S.V. et al. (2006) *Review of Scientific Instruments*, **77**, 085106.
60 Endres, A. et al. (1997) *Review of Scientific Instruments*, **68**, 4009–13.
61 Heymann, A., Stipp, A., Sinn, C. and Palberg, T. (1998) *Journal of Colloid and Interface Science*, **207**, 119.
62 Stribeck, N. (2007) *X-Ray Scattering of Soft Matter*, Springer, Berlin, Heidelberg.
63 Luck, W., Klier, M. and Wesslau, H. (1963) *Berichte der Bunsen-Gesellschaft fur Physikalische Chemie*, **67**, 75; (a) Luck, W., Klier, M. and Wesslau, H. (1963) *Berichte der Bunsen-Gesellschaft fur Physikalische Chemie*, **67**, 84.
64 Monovoukas, Y. and Gast, A.P. (1989) *Journal of Colloid and Interface Science*, **128**, 533–48.
65 Sirota, E.B., Ou-Yang, H.D., Sinha, S.K., Chaikin, P.M., Axe, J.D. and Fujii, Y. (1989) *Physical Review Letters*, **62**, 1524.
66 Wette, P. et al. *Full Phase Diagram of Charge Variable Colloidal Spheres*, to be published.
67 Rojas, L.F., Urban, C., Schurtenberger, P., Gisler, T.

and von Grünberg, H.H. (**2002**) *Europhysics Letters*, **60**, 802.
68 Palberg, T. *et al.* (**1994**) *Journal de Physique III (France)*, **4**, 31.
69 Wilson, H.A. (**1900**) *Philosophical Magazine*, **50**, 238.
70 Frenkel, J. (**1932**) *Physikalische Zeitschrift der Sowjetunion*, **1**, 498.
71 Aastuen, D.J.W., Clark, N.A., Cotter, L.K. and Ackerson, B.J. (**1986**) *Physical Review Letters*, **57**, 1733; (**1986**) *Physical Review Letters*, **57**, 2772 (Erratum).
72 Burke, E., Broughton, J.Q. and Gilmer, G.H. (**1988**) *Journal of Chemical Physics*, **89**, 1030.
73 Kerrache, A., Horbach, J. and Binder, K. (**2008**) *Europhysics Letters*, **81**, 58001.
74 Murray, C.A. and Grier, D.A. (**1996**) *Annual Review of Physical Chemistry*, **47**, 421–62.
75 Harland, J.L. and van Megen, W. (**1996**) *Physical Review E*, **55**, 3054–67.
76 Van Siclen, C.De.W. (**1996**) *Physical Review B*, **54**, 11856.
77 Wette, P. and Schöpe, H.J. (**2007**) *Physical Review E*, **75**, 051405.
78 Marr, D.W.M. and Gast, A.P. (**1994**) *Langmuir*, **10**, 1348–50.
79 Auer, S. and Frenkel, D. (**2002**) *Journal of Physics: Condensed Matter*, **14**, 7667.
80 Kelton, K.F. (**1993**) *Solid State Physics*, **45**, 75.
81 Asta, M., Spaepen, F., van der Veen, J.F. (eds.) (**2004**) "Solid-liquid interfaces: molecular structure, thermodynamics and crystallization". *MRS Bulletin*, **29** (12), 920–62.
82 Kesavamoorthy, R., Sood, A.K., Tata, B.V.R. and Arora, A.K. (**1988**) *Journal of Physics C: Solid State Physics*, **21**, 4737.
83 Schenk, T., Holland-Moritz, D., Simonet, V., Bellissent, R. and Herlach, D.M. (**2002**) *Physical Review Letters*, **89**, 075507.
84 Schöpe, H.J., Bryant, G. and van Megen, W. (**2007**) *Journal of Physical Chemistry*, **127**, 084505.
85 O'Malley, B. and Snook, I. (**2003**) *Physical Review Letters*, **90**, 085702.
86 Desgranges, C. and Delhomelle, J. (**2007**) *Journal of Chemical Physics*, **126**, 054501.
87 Cotter, L.K. and Clark, N.A. (**1987**) *Journal of Chemical Physics*, **86**, 6616.
88 Punnathannam, S. and Monson, P.A. (**2006**) *Journal of Chemical Physics*, **125**, 024508.

Part Three
Nd–Fe based Alloys

13
Phase-Field Simulations of Nd–Fe–B: Nucleation and Growth Kinetics During Peritectic Growth

Ricardo Siquieri and Heike Emmerich

13.1
Introduction

In modeling nucleation it is essential to realize that the solid–liquid interface is known to extend to several molecular layers. This has successively been indicated by experiments [1], computer simulations [2], and statistical mechanical treatments based on the density functional theory [3]. The need to pay particular attention to this diffuse interface results from the fact that for nucleation the typical size of critical fluctuations is comparable to the physical thickness of the interface. The success of such careful treatment can be seen in modern nucleation theories for homogeneous nucleation, which do consider the molecular scale diffuseness of the interface. These theories could remove the difference of many orders of magnitude between nucleation rates from the classical sharp-interface approach and experiment [4].

In heterogeneous nucleation we face an even more complex situation, since the main degrees of freedom of the process are larger than in homogeneous nucleation: first of all each phase can nucleate separately. Moreover, several phases can nucleate jointly, that is approximately at the same space and time. Finally one phase can nucleate on top of the other.

Here we are particularly interested in peritectic material systems. Even though many industrially important metallic alloy systems such as, Nd–Fe–B, as well as ceramics are peritectics, much less is known about microstructural pattern formation in peritectic growth [5] than, for example in eutectic growth. Similar to a eutectic system the phase diagram of a peritectic system contains a point—the peritectic point with peritectic temperature T_p—at which two different solid phases, the parent (primary) and peritectic (secondary) phases, coexist with a liquid of higher composition than either solid phase. Above T_p, the parent phase is stable and the peritectic phase is metastable, whereas below T_p, the opposite is true. In the following text, we consider C to be the concentration of the impurity and T_m the melting point of the pure phase. For a figure displaying a respective schematic phase diagram of a peritectic material system the reader is referred to, for example [6].

Phase Transformations in Multicomponent Melts. Edited by D. M. Herlach
Copyright © 2008 WILEY-VCH Verlag GmbH & Co. KGaA, Weinheim
ISBN: 978-3-527-31994-7

In such peritectic material systems it is particularly relevant to understand the nucleation of the peritectic phase on top of the properitectic phase in detail, since this is the nucleation process yielding the stationary growth morphology. For this specific nucleation process the precise configuration of the properitectic phase, that is its free energy on the one hand and its morphology on the other [7], should contribute to the precise nucleation rate.

Nevertheless, the well-established spherical cap model for the nucleation of a new phase β on a planar front of initial phase α predicts the following nucleation rate:

$$I = I_0 e^{-\Delta F^*/k_B T} \tag{13.1}$$

where I_0 is a constant factor (with dimension equal to the number of nucleations per unit volume and unit time) and ΔF^* is the activation energy for heterogeneous nucleation. Assuming the critical nucleus of phase β to be spherical (see Figure 13.1), the interfacial tensions $\gamma_{\alpha L}$, $\gamma_{\alpha \beta}$, and $\gamma_{\beta L}$ balance each other enclosing a contact angle ϑ if the following condition is fulfilled:

$$\gamma_{\alpha L} = \gamma_{\alpha \beta} + \gamma_{\beta L} \cos \theta \tag{13.2}$$

ΔF^* is then given, respectively, in two and three dimensions by

$$\Delta F^* = \begin{cases} \dfrac{\gamma_{\beta L}^2}{\Delta F_B} \times \dfrac{\theta^2}{\theta - (1/2)\sin 2\theta}, & 2D \\ \dfrac{\gamma_{\beta L}^3}{\Delta F_B^2} \times \dfrac{16\pi(2+\cos\theta)(1-\cos\theta)^2}{12}, & 3D \end{cases} \tag{13.3}$$

Here ΔF_B is the difference between the bulk free energy of the peritectic phase and of the liquid phase.

Equation 13.3 determines the classical local nucleation rate and hence the probability per unit time of a nucleus forming as a function of the local temperature at the solid–liquid interface. Thus morphological and energetical contributions to Equation 13.3 resulting from the properitectic microstructure as discussed in [7] are neglected classically.

In the following we derive a phase field model approach for peritectic growth taking into account hydrodynamics in the molten phase, which is capable of treating

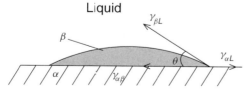

Fig. 13.1 Heterogeneous nucleation of a "spherical cap" shaped second phase β on a planar initial phase α according to the spherical cap model [8].

this open issue. The way we proceed here is different from the further scientific advance of the authors of [9–11] in the sense, that we analyze the nucleation rate belonging to a heterogeneous nucleation event for a peritectic system. In contrast, the authors of [9] extended their own work to investigate several stochastically initialized homogeneous nucleation events of different phases and their subsequent growth in multiphase systems [12]. More recently, Gránásy et al. [11] and Castro [10] investigated heterogeneous nucleation using a phase-field approach for a pure liquid crystal.

Here we describe our new approach to investigate the rate of a nucleation event of the second phase on top of the first one in detail in Section 13.2. Also, in Sections 13.2 and 13.3 we investigate the microstructure growth in a peritectic system under the influence of hydrodynamic convection in the melt. We then report on numerical investigations of the nucleation kinetics in such peritectic material systems, in particular on a morphological contribution from the properitectic phase to the activation energy, in Section 13.3. Moreover, we discuss the relation of our results to classical nucleation theory in Section 13.3. Finally, we conclude with a discussion of the general impact of our new approach for peritectic materials under the influence of convection.

13.2
Phase-Field Model with Hydrodynamic Convection

In the following text, we extend the phase-field model described in [13] to simulate peritectic growth including hydrodynamic convection in the molten phase. The free-energy functional of a representative volume of the investigated material system is given by

$$\mathcal{F} = \int_V f \, dV \tag{13.4}$$

with the free-energy density defined as

$$f = \frac{W(\theta)^2}{2} \sum_i (\nabla p_i)^2 + \sum_i p_i^2 (1 - p_i)^2$$

$$+ \tilde{\lambda} \left[\frac{1}{2} [c - \sum_i A_i(T) g_i(\vec{p})]^2 + \sum_i B_i(T) g_i(\vec{p}) \right] \tag{13.5}$$

where $W(\theta) = W_0(1 + \epsilon_{p_i} \cos 4\theta)$ depends on the orientation of the interface [14], with $\theta = \arctan \partial_y p_i / \partial_x p_i$, ϵ_{p_i} being the measure of the anisotropy (here we consider ϵ_{p_2} always equal to zero) and $\tilde{\lambda}$ being a constant. The function g_i couples the phase field to the concentration and the temperature. The coefficients $A_i(T)$ and $B_i(T)$ define the equilibrium phase diagram [15],

$$A_i(T) = c_i \mp (k_i - 1) U, \quad A_L = 0,$$
$$B_i(T) = \mp A_i U, \quad B_L = 0,$$

where $U = (T_p - T)/(|m_i|\Delta C)$ is the dimensionless undercooling, k_i are the partition coefficients, A_L, B_L are the corresponding liquid coefficients, m_i are the liquidus slope, and $c_i = (C_i - C_p)/\Delta C$ is the scaled concentration, where C_p is the liquidus concentration at T_p.

We use three phase fields $p_i \in [0,1]$, where $\sum_{i=1}^{3} p_i = 1$. The p_i labels the properitectic, the peritectic, and the liquid phase, respectively, that is $i = \alpha$ (for the properitectic phase), $i = \beta$ (for the peritectic phase) and $i = L$ (for the liquid phase) denote $\vec{p} \equiv (p_\alpha, p_\beta, p_L)$.

Their dynamics is derived from the free-energy functional $\mathcal{F}$

$$\frac{\partial p_i}{\partial t} = \frac{1}{\tau}\frac{\delta \mathcal{F}}{\delta p_i}$$

where τ is a relaxation time.

The concentration field is given by:

$$\frac{\partial c}{\partial t} + p_L \mathbf{v} \cdot \vec{\nabla} c - \vec{\nabla} \cdot \left(M(p_i) \vec{\nabla} \frac{\delta \mathcal{F}}{\delta c} - \vec{J}_{AT} \right) = 0 \tag{13.6}$$

where $M(p_i)$ is a mobility and $\vec{J}_{AT}$ is the antitrapping term.

The flow field is modeled by coupling a modified Navier–Stokes equation with the phase field [15],

$$\frac{\partial p_L \mathbf{v}}{\partial t} = -p_L \mathbf{v} \cdot \vec{\nabla} \mathbf{v} - p_L \vec{\nabla} p + \frac{1}{Re}\nabla^2 p_L \mathbf{v} + M_1^2 \tag{13.7}$$

where $Re = \frac{\rho U}{\nu}$ and M_1^2 is a dissipative interfacial force per unit volume [16].

This model allows us to investigate the peritectic transformation under the influence of fluid flow quantitatively. A representative evolution is shown in Figure 13.2, where time runs from the left picture to the right. The row of pictures on the top represents the case where flow in the melt phase is taken into account, on the bottom the case without flow. The pictures (with and without flow) are depicted at the same time step for the same process parameters. The dark circle indicates the properitectic phase, the light structure the peritectic phase, which is nucleating on top of the properitectic one. Comparing peritectic growth with and without convection, we find that hydrodynamic transport in the melt enhances the growth process considerably. To stress this effect we investigated a scale relation that describes the influence of the inflow velocity on the solid volume fraction of the microstructure evolution for a fixed time instant. In Figure 13.3 we resume this finding by plotting the logarithm of the solid volume fraction against the logarithm of the inflow velocity. We can see that for a larger inflow velocity a linear regime was found. More details about these findings are addressed elsewhere in [17]. These results are in qualitative agreement with the experimental investigation of the peritectic material system Nd–Fe–B in [18].

13.2 Phase-Field Model with Hydrodynamic Convection | 219

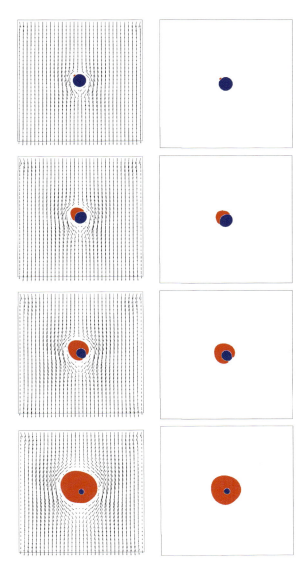

Fig. 13.2 Numerical simulations of the peritectic transformation under the influence of convection (on the left) and without convection (on the right). The dark circle indicates the properitectic phase, the light structure the peritectic phase. Arrows are vectors indicating the velocity of the hydrodynamic field in the molten phase.

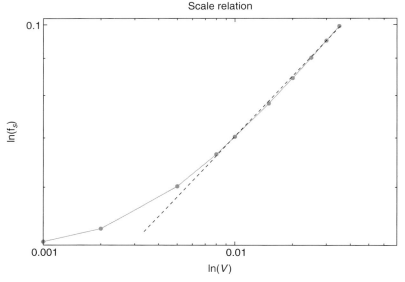

Fig. 13.3 Plot of the logarithm of the solid volume fraction against the logarithm of the inflow velocity.

13.3
Investigating Heterogeneous Nucleation in Peritectic Materials via the Phase-Field Method

In solidification experiments the final microstructure is determined by both the peritectic growth dynamics as well as the microstructure growth kinetics. Therefore, for a full quantitative comparison with experiments, it is essential to analyze the heterogeneous nucleation kinetics of the above peritectic material system, as well. For such a system a nucleation event arises as a critical fluctuation, which is a nontrivial time-independent solution of the governing equations we can derive from the underlying free energy functional. Our derivation follows the standard variational procedure of phase-field theory (for a review see e.g. [19–21]). By solving the Equations 13.4–13.7 numerically under boundary conditions that prescribe bulk liquid properties far from the fluctuations ($p_L \to 1$, and $c \to c_\infty$ at the outer domain boundaries) and zero field-gradients at the center of the respective phases, one obtains the free energy of the nucleation event as

$$\Delta F^* = F - F_0 \qquad (13.8)$$

Here F is obtained by numerically evaluating the integration over F after having the time-independent solutions inserted, while F_0 is the free energy of the initial liquid. The zero field-gradients arise naturally because of the stationarity of the problem if the "seed" phase is chosen large enough.[1] On the basis of Equation 13.8, the homogeneous nucleation rate is calculated as

$$I = I_0 \exp\{-\Delta F^*/kT\} \qquad (13.9)$$

where the nucleation factor I_0 of the classical kinetic approach is used, which proved consistent with experiments [8, 22].

As introduced in Section 13.1, in a peritectic material sample it is particularly relevant to understand the nucleation of the peritectic phase on top of the properitectic phase in detail, since this is the nucleation process yielding the stationary growth morphology. As demonstrated previously via analytical predictions and Monte Carlo studies (see e.g. [7, 23]), for this specific nucleation process the precise configuration of the properitectic phase, that is its free energy on the one hand and its morphology on the other, should contribute to the precise nucleation rate. This, as well as the experimental evidence for deviations from classical nucleation theory in the system Nd–Fe–B [24], motivated us to study the effect of two morphological features of the properitectic phase on the nucleation rate of the peritectic one, namely (i) the effect of facettes and (ii) the effect of its radius. In this context the underlying facetted shape of the properitectic phase is initialized as "seed" for the peritectic phase to nucleate upon as depicted in Figure 13.4. Figure 13.4 reveals, too, that in our investigations the peritectic phase is nucleating at the corner of the properitectic phase. The anisotropic form $W(\theta) = W_0(1 + \epsilon_{p_i} \cos 4\theta)$ for

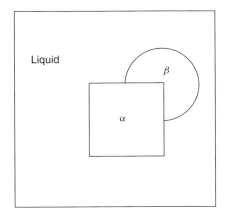

Fig. 13.4 Schematic sketch to elucidate the initialization of the relaxation procedure and subsequent calculation to determine the heterogeneous nucleation rate via the phase-field method for the case of an anisotropic "seed".

---

[1] Thermodynamically this is always possible. The functioning of the underneath relaxation procedure does not depend on the volume of the properitectic phase as such, but on the relative volume of the properitectic phase to the volume which we chose as initialization for the peritectic phase. This has to be tuned close to a ratio to be expected from the position in the phase diagram to ensure convergence within the limit of a reasonable number of variations.

$W(\theta)$ allows us to obtain this state for the anisotropic case. However, we are aware that for a simulation of the full dynamic microstructure evolution into a facetted shape more elaborate anisotropic forms of $W(\theta)$ are required, as for example given in [25]. To calculate the nucleation rate of the peritectic phase on top of this facetted seed, it is then—as described above—essential to determine the corresponding time-independent configuration, at which neither of the two phases grow, and at which also all diffuse fields are fully relaxed, that is stationary. To find this state we vary the radius of the properitectic phase systematically, keeping the position of its center relative to the peritectic phase constant. For each variation we carry out the relaxation procedure. There is exactly one radius where stationarity can be achieved, namely the radius of the critical nucleus. The precise morphology of the critical "two-phase" nucleus, in particular the ratio of the volume of the two phases, depends—as indicated above—on the precise thermodynamic state of the system under investigation, as can easily be understood from the phase diagram.

In Figure 13.5 and Figure 13.6 we summarize our results. As can be taken from Figure 13.5, the less facetted the properitectic phase, the larger the nucleation probability for a peritectic nucleation on top of it. For the contribution resulting from the radius of the properitectic phase a similar relation is true: the larger the radius of the properitectic phase, the larger the probability of a peritectic nucleation on top of it. Both findings are in qualitative agreement with the following atomistic picture: Unfaceted nuclei offer a great number of surface kinks for nucleation. This holds for nuclei of small radii, as well. However, small radius nuclei are also subject to large surface diffusion owing to kink flow [26]. This overrides the first effect such that the overall nucleation rate turns out to be smaller for smaller

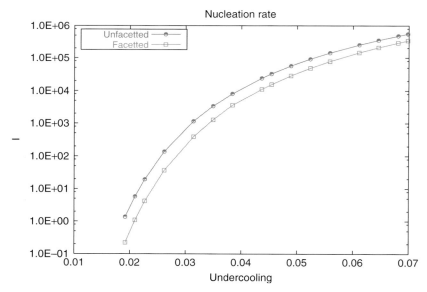

Fig. 13.5 Comparison of the nucleation rate on top of a facetted nucleus to the one on top of an unfaceted nucleus.

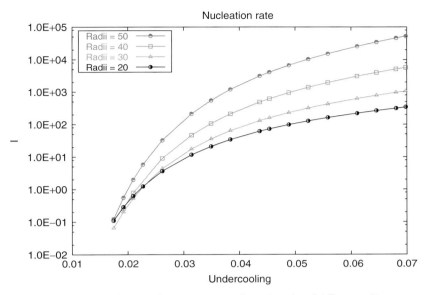

Fig. 13.6 Comparison of the nucleation rate on unfaceted nuclei of different radii.

radii. Moreover, these findings are in qualitative agreement with [7] and thus provide a first qualitative validation for our new approach toward heterogeneous nucleation. However, it should be noted that the atomistic picture is just given for a common sense estimation of what our model should do. In the continuum picture underlying our investigations, the differences of the different curves arise owing to the fact that the total surface energy tied to the diffuse surface area of the properitectic nucleus depends on its morphology. Thus the latter naturally has an impact on the nucleation rate just as indicated experimentally. This can be analyzed in more detail making use of the phase-field profiles at the stationary point [17]. The benefit of using our continuum model approach rather than atomistic models for the proposed studies is twofold: (i) The approach is computationally considerably more efficient, that is it easily allows for subsequent 3D simulation and simulations of several nuclei competing in the course of initial growth. This also implies that the time scales that can be accessed are larger than for atomistic simulations. Only because of this does simulation of nucleation as well as initial growth become possible. (ii) Moreover, it can easily be extended to additional physical mechanisms influencing the nucleation process such as, for example elastic ones [21] or anisotropies of the solid–liquid interfacial free energy [17].

13.4
Conclusion

To summarize, in this chapter we have introduced a new phase-field modeling approach for peritectic growth taking into account hydrodynamic transport in the molten phase. We applied this approach successfully to investigate the influence of melt flow on the peritectic transformation. Moreover, we employed our model to identify the precise mechanisms of the heterogeneous nucleation kinetics in a peritectic system, that is essentially mechanisms beyond the classical nucleation theory. In this context it is important to notice that the new features of our approach to heterogeneous nucleation inherently included are (i) the notion of a diffuse interface as well as (ii) long-range interaction effects due to our continuum field approach toward the problem. On the basis of these features our model can explain the differences between the classical nucleation theory and experiments as morphological contributions to the nucleation rate. Moreover, it compares well to careful statistical studies of the effects of long-range interactions. In this sense it poses a valuable new approach toward heterogeneous nucleation in general, taking into account kinetic, thermodynamic, as well as long-range interaction effects, which still has to be developed further.

References

1 Huisman, W.J., Peters, J.F., Zwanenburg, M.J., de Vries, S.A., Derry, T.E., Albernathy, D. and van der Veen, J.F. (**1997**) *Nature*, **390**, 379.
2 Davidchack, R.L. and Laird, B.B. (**1998**) *Journal of Chemical Physics*, **108**, 9452.
3 Ohnesorge, R., Löwen, H. and Wagner, H. (**1994**) *Physical Review E*, **50**, 4801.
4 Gránásy, L. and Iglói, F. (**1997**) *Journal of Chemical Physics*, **107**, 3634.
5 Boettinger W.J., Coriell, S.R., Greer, A.L., Karma, A., Kurz, W., Rappaz, M. and Trivedi, R. (**2000**) *Acta Materialia*, **48**, 43.
6 Lo, T.S., Karma, A. and Plapp, M. (**2001**) *Physical Review E*, **63**, 031504.
7 Jackson, K A. (**1958**) in *Growth and Perfection of Crystals* (eds R.H. Doremus, B.W. Roberts and D. Turnbull), John Wiley & Sons, Ltd, New York, p. 319.
8 Porter, D.A. and Easterling, K.E. (**1981**) *Phase Transformations in Metals and Alloys*, 2nd edn, Van Nostrand Reinhold, New York.
9 Gránásy, L., Pusztai, T. and Börzsönyi, T. (**2003**) in *Interface and Transport Dynamics, Lecture Notes in Computational Science and Engineering*, Vol. 32 (eds H. Emmerich, B. Nestler and M. Schreckenberg), Springer-Verlag, Berlin, p. 190.
10 Castro, M. (**2003**) *Physical Review B* **67**, 035412.
11 Gránásy, L., Pusztai, T., Saylor, D. and Warren, J.A. (**2007**) *Physical Review Letters*, **98**, 035703.
12 Gránásy, L., Pusztai, T., Tegze, G., Warren, J.A. and Douglas, J.F. (**2005**) *Physical Review E*, **72**, 011605; (a) Gránásy, L., Pusztai, T., Börzsönyi, T., Warren, J.A. and Douglas, F. (**2004**) *Nature Materials*, **3**, 645.
13 Folch, R. and Plapp, M. (**2005**) *Physical Review E*, **72**, 011602.
14 Karma, A. and Rappel, W.-J. (**1996**) *Physical Review E*, **53**, R3017.
15 Emmerich, H. and Siquieri, R. (**2006**) *Journal of Physics: Condensed Matter*, **49**, 11121.
16 Beckermann, C., Diepers, H.-J., Steinbach, I., Karma, A. and Tong, X. (**1999**) *Journal of Computational Physics*, **154**, 468.
17 Emmerich, H. and Siquieri, R. in preparation
18 Filip, O., Hermann, R. and Schultz, L. (**2004**) *Journal of Materials Science and Engineering A*, **375**, 1044.
19 Warren, J.A. and Boettinger, W.J. (**1995**) *Acta Metallurgica et Materialia*, **43**, 689.

20 Boettinger, W.J. and Warren, J.A. (1996) *Metallurgical and Materials Transactions A*, **27**, 657.
21 Emmerich, H. (2003) *The Diffuse Interface Approach in Material Science—Thermodynamic Concepts and Applications of Phase-Field Models, Lecture Notes in Physics LNPm 73*, Springer-Verlag, Berlin.
22 Kelton, K.F. (1991) *Solid State Physics*, **45**, 75.
23 Shneidman, V.A., Jackson, K.A. and Beatty, K.M. (1999) *Physical Review B*, **59**, 3579.
24 Strohmenger, J., Volkmann, T., Gao, J. and Herlach, D.M. (2004) *Journal of Materials Science and Engineering A*, **375**, 561; (a) Volkmann, T., Strohmenger, J., Gao, J. and Herlach, D.M. (2004) *Applied Physics Letters*, **85**, 2232.
25 Debierre, J.M., Karma, A., Celestini, F. and Guerin, R. (2003) *Physical Review E*, **68**, 041604.
26 Pierre-Louis, O., D'Orsogna, M.R. and Einstein, T.L. (1999) *Physical Review Letters*, **82**, 3661.

Further Reading

Emmerich, H. (2008) *Advances of and by phase-field modeling in condensed matter physics, Advances in Physics*, **57**, 1–87.

14
Investigations of Phase Selection in Undercooled Melts of Nd–Fe–B Alloys Using Synchrotron Radiation

Thomas Volkmann, Jörn Strohmenger and Dieter M. Herlach

14.1
Introduction

This chapter concentrates on investigations of phase selection processes in melts of Nd–Fe–B alloys with special emphasis to the crystallization of the hard magnetic $Nd_2Fe_{14}B_1$ phase (Φ phase) with its extraordinary hard magnetic properties [1–3]. Under near equilibrium conditions, the alloys of stoichiometric composition $Nd_2Fe_{14}B_1$ solidify in the γ phase from which Φ phase is formed by a peritectic reaction. Undercooling of the melt prior to solidification opens up the possibility of various solidification pathways into different solid phases [4]. This study concentrates on the investigations of phase formation as a function of undercooling, in particular to determine the criteria for primary crystallization of Φ phase from the undercooled melt and circumventing the crystallization of γ phase.

Figure 14.1 shows the quasi-binary phase diagram of Nd–Fe–B with fixed concentration ratio Nd : B = 2 : 1 with the equilibrium liquidus temperature T_L (solid lines) and the extensions of the liquidus temperature of γ phase and Φ phase into the regime of the undercooled melt (dotted lines). It is obvious that only a moderate undercooling is sufficient to generate a thermodynamic driving force for primary crystallization of hard magnetic Φ phase in the regime of Fe concentrations of 83–76 at%. In addition to γ phase and Φ phase, a metastable χ phase is found by Schneider *et al.* [5] of composition $Nd_2Fe_{17}B_x$ (x ≈ 1), which is formed by a peritectic reaction of γ-Fe at temperatures of about 50 K below the peritectic temperature of Φ phase. This metastable χ phase decays into γ-Fe and Φ phase rapidly after its formation. Therefore, no investigations of the structure of this χ-phase exist so far. In this study, the sequence of phase formation as a function of undercooling is *in situ* observed by energy dispersive X-ray (EDX) diffraction on levitation-processed samples. The high intensity synchrotron radiation allows for recording full diffraction spectra within a time period of less than 1 s. This makes it possible to detect any phase transformation during undercooling and to determine unambiguously the structures of the various phases formed including that of the metastable χ phase

Phase Transformations in Multicomponent Melts. Edited by D. M. Herlach
Copyright © 2008 WILEY-VCH Verlag GmbH & Co. KGaA, Weinheim
ISBN: 978-3-527-31994-7

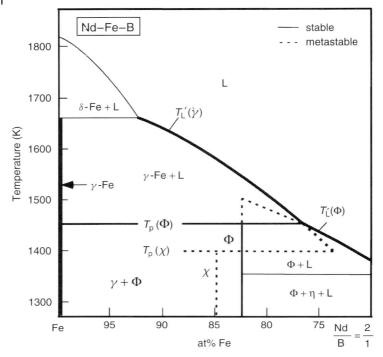

Fig. 14.1 Quasi-binary phase diagram of Nd–Fe–B at fixed concentration ratio Nd : B = 2 : 1 with varying Fe concentration. The solid lines show the equilibrium liquidus temperature while the dotted lines correspond to the extensions of the liquidus temperatures in the regime of the undercooled melt. Also, the peritectic reaction of the formation of metastable χ phase is schematically indicated.

It will be demonstrated that primary crystallization of γ-Fe can be suppressed if the melt is undercooled prior to solidification [6–12]. By investigating the phase formation as a function of undercooling and Fe concentration at fixed Nd : B = 2 : 1 composition, a phase selection diagram is constructed for the quasi-binary alloy with undercooling as an additional process parameter to the concentration.

14.2
Description of the Investigations

For conducting investigations of phase selection as a function of undercooling, an electromagnetic levitation facility was integrated in beam line ID15A of the European Synchrotron Radiation Facility (ESRF) in Grenoble. By applying EDX diffraction (white X-ray beam, fixed diffraction angle) complete diffraction spectra were recorded during subsequent time periods of cooling and undercooling the liquid sample within a time interval of about 1 s [13, 14]. All as solidified samples were investigated with respect to phase coexistence at ambient temperatures by conventional X-ray diffraction and scanning electron microscopy including EDX to check whether any primary formed crystalline phase is transformed during postrecalescence and subsequent cooling (cooling rates 10–100 K s^{-1}) of the as solidified sample to room temperature.

The experimental results are analyzed within nucleation theory and models of dendrite growth. Since no thermodynamic data are available for the metastable χ phase, the analysis is limited to γ phase and Φ-phase. The phase diagram including also the metastable extensions in the undercooled melt regime was computed using a model by Hallemans et al. [15] that gives an expression for the dependence of the free enthalpy of both γ and Φ phase as a function of temperature and concentration.

The activation energy ΔG^* to form a critical nucleus is given by the following equation:

$$\Delta G^* = \frac{16 \pi}{3} \cdot \frac{\sigma^3}{\Delta G_V^2} \cdot f(\theta) \tag{14.1}$$

with σ the interfacial energy between solid nucleus and liquid, ΔG_v the difference of free enthalpy between solid and liquid phase and $f(\theta)$ the catalytic potency factor of heterogeneous nucleation. The interfacial energy σ was determined according to the negentropic model by Spaepen [16] as shown below:

$$\sigma = \alpha \cdot \frac{\Delta S_m \cdot T_L}{N_L^{1/3} \cdot V_m^{2/3}} \cdot T \tag{14.2}$$

ΔS_f denotes the entropy of fusion, T_L the melting temperature, N_L Avogadro's number, V_m the molar volume, and T the temperature. The factor α depends on the crystalline structure of the nucleus and is analytically estimated as $\alpha_{fcc} = 0.86$ for the fcc structure of γ-Fe [17]. The interfacial energy of a complex phase as intermetallic Φ-phase can only be determined by numerical calculations, which result in $\alpha_\Phi = 0.37$ for the (001) plane of the tetragonal unit cell of Φ phase [18]. Apparently, $\alpha_\Phi = 0.37$ is much smaller than $\alpha_{fcc} = 0.86$ and consequently, the interfacial energy of Φ phase, σ_Φ is smaller than that of fcc phase, σ_{fcc} [19]. Therefore, according to Equation 14.1 the nucleation of Φ phase should be energetically favored if a positive driving force is generated by sufficient undercooling of the melt. The change of the difference of free enthalpy ΔG_v between solid and liquid is estimated by the thermodynamic functions given by Hallemanns et al. [15]. To do so the different concentrations in the nucleus and the undercooled liquid are taken into account. The composition of nucleus and liquid is different in particular for the γ phase in alloys with high Nd and B content. The γ phase shows a very small solubility of up to 1 at% Nd and 0.02 at% B, respectively. The concentration in the solid nucleus was determined by minimizing its free energy with respect to the concentration of the nucleus.

The growth of γ phase and Φ phase takes place via dendritic growth. To describe the growth behavior of both phases, we apply the sharp interface model by Lipton, Trivedi and Kurz [20] and its extension to ternary alloys by Bobadilla et al. [21].

According to these models the bulk undercooling ΔT consists of four different contributions:

$$\Delta T = \Delta T_\mathrm{T} + \Delta T_\mathrm{c} + \Delta T_\mathrm{R} + \Delta T_\mathrm{k} \tag{14.3}$$

ΔT_T is the thermal undercooling, ΔT_c is the constitutional undercooling and the curvature undercooling ΔT_R describes the decrease of the melting temperature at a strongly curved interface due to the Gibbs–Thomson effect. The kinetic undercooling ΔT_k gives the deviation from local equilibrium at a rapidly propagating solid–liquid interface and is related to the interface mobility μ, also called the kinetic coefficient, by the expression

$$v = \mu \Delta T_\mathrm{k} \tag{14.4}$$

here, v is the normal velocity of the advancing solid–liquid interface, and T_i is the temperature at the dendrite tip. According to Coriell and Turnbull [22] the kinetic coefficient is as follows:

$$\mu = \frac{v_o \Delta H_\mathrm{f}}{N_a k_\mathrm{B} T_\mathrm{L}^2} \tag{14.5}$$

where ΔH_f is the latent heat of fusion, N_A is Avogadro's number, k_B is Boltzmann constant, T_L is the melting temperature, and v_0 is related to the atomic attachment kinetics at the interface. In the case of collision-limited growth, the atomic attachment kinetics is controlled by the vibration frequency in the order of 10^{13} Hz. Its value of v_0 equals the speed of sound being of the order of 4000 m s^{-1} in liquid metals. The growth of intermetallic Φ phase is expected to be diffusion limited [23] in contrast to the solid solution of γ-Fe. Therefore, a sluggish growth of the chemically ordered Φ phase is taken into consideration by a parameter v_0 that corresponds to the speed of diffusion being of the order of 10 m s^{-1} in computing growth velocities as a function of undercooling. The redistribution of solute at the interface of Φ phase and undercooled melt is described by referring to the model of Trivedi and Kurz developed for the growth of intermetallic phases in concentrated alloys [24]. In this model it is taken into account that for intermetallic phases the partition coefficient $k = c_\mathrm{S}{}^*/c_\mathrm{L}{}^*$, in which $c_\mathrm{S}{}^*$ is the concentration of the dendrite tip and $c_\mathrm{L}{}^*$ is the concentration in the liquid at the interface, which is not constant but varying with the interface temperature. Solute redistribution at the advancing solid–liquid interface is expressed by a modified parameter k^* that is related to partition coefficient k and its derivative with composition. For stoichiometric phases such as the Φ phase, the modified partition coefficient is $k^* = 0$. The thermophysical data used for calculation of dendrite growth velocities such as the latent heat of fusion and the partitioning coefficient are derived from the thermodynamic functions given by Hallemans [15]. Umeda *et al.* analyzed the dendrite growth of γ-Fe and Φ phase for a pseudobinary alloy of Nd–Fe–B with a diffusion coefficient of $D = 5 \times 10^{-9}$ m^2 s^{-1} [25]. In the present analysis for the ternary system, the diffusion coefficient for Nd is assumed to be smaller

Table 14.1 Thermophysical properties. Parameters used for calculation of dendrite growth velocities of γ-Fe and Φ phase in $Nd_{14}Fe_{79}B_7$ alloy melts (see text for details)

Parameter	γ	Φ
Latent heat of fusion ΔH_f (J mol^{-1})	14 470	19 890
Entropy of fusion ΔS_f (J molK^{-1})	8.12	13.80
Molar volume of the solid V_m (m^3 mol^{-1})	6.72×10^{-6}	8.36×10^{-6}
Interfacial energy σ (J m^{-2})	0.385	0.21
Partition coefficient k for Nd	0.06	$K^* = 0$
Partition coefficient k for B	0.002	$K^* = 0$
Slope of liquidus surface m_L(Nd) (K/at%)	-16.6	-2.0
Slope of liquidus surface m_L(B) (K/at%)	-15.0	-3.0
Kinetic growth coefficient μ (m s^{-1} · K)	3.1	0.1
Heat capacity of the liquid c_p^L (J mol^{-1} · K)	42.0	42.0
Diffusion coefficient of Nd in the liquid D_L(Nd) (m^2 s^{-1})	2×10^{-9}	2×10^{-9}
Diffusion coefficient of B in the liquid D_L(B) (m^2 s^{-1})	5×10^{-9}	5×10^{-9}
Thermal diffusivity of the liquid a (m^2 s^{-1})	4×10^{-6}	4×10^{-6}

than that for the element B. All material parameters used for the calculations are summarized in Table 14.1.

In analyzing the growth kinetics in electromagnetically levitated samples a strong stirring of the melt by the strong alternating electromagnetic fields has to be considered that leads to fluid flow with velocities ranging up to $u = 0.3 \, \text{m s}^{-1}$ [26]. Fluid flow effects by convection during solidification of undercooled melts of Nd–Fe–B alloys are treated by R. Hermann *et al.* in another chapter of this book. As a consequence, the calculation of dendrite growth velocities as a function of undercooling for electromagnetically processed melts has to be taken into account the influence of forced convection in the heat and mass transport. This is realized by a model, in which changes in heat and mass transport by forced convection in alternating electromagnetic fields is taken into consideration. On the basis of a Navier–Stokes equation, thermal and chemical Peclet numbers for heat and mass transport by convection are introduced. The fluid flow velocity induced by forced convection is computed by considering an energetic balance between gravity-driven convection, electromagnetically induced convection, and viscous dissipation in the melt [27, 28]. We are using this model to compute the dendrite growth velocities of γ and ϕ phase as a function of undercooling, fluid flow velocity u in the melt and the concentration of the alloy.

14.3
Experimental Results and Discussion

It is supposed that the metastable χ phase of composition $Nd_2Fe_{17}B_x$ ($x \approx 1$) is a ternary extension of the binary Nd_2Fe_{17} phase. It is assumed that the B atoms occupy the interstitial lattice sides of the ordered Nd_2Fe_{17} crystal similar to the C atoms in the $Nd_2Fe_{17}C_x$ phase (with $x = 0–0.6$) that is stable in the ternary Nd–Fe–C alloy [29]. To investigate the relation of binary Nd_2Fe_{17} phase and ternary $Nd_2Fe_{17}B_x$ phase (χ) we changed systematically the B content but keeping the ratio Nd:Fe $= 1 : 7$ unchanged. Figure 14.2 shows the diffraction pattern recorded during solidification of $Nd_{12.3}Fe_{86.3}B_{1.4}$ and $Nd_{12}Fe_{84}B_4$ melts together with the corresponding temperature time profiles. Owing to the addition of 1.4 at% B, two peritectic phases are formed in direct sequence as can be inferred from Figure 14.2a. In ternary $Nd_{12.3}Fe_{86.3}B_{1.4}$ alloy, a phase is solidified that can be indexed in the diffraction pattern like the binary Nd_2Fe_{17} phase assuming the same rhombohedral crystal structure (prototype Zn_2Th_{17} [30]). Since B atoms are solved likely in the rhombohedral phase, it is denoted as χ $Nd_2Fe_{17}B_x$ in Figure 14.2. In a second reaction step Φ-phase crystallizes whereby χ-phase is conserved. The corresponding

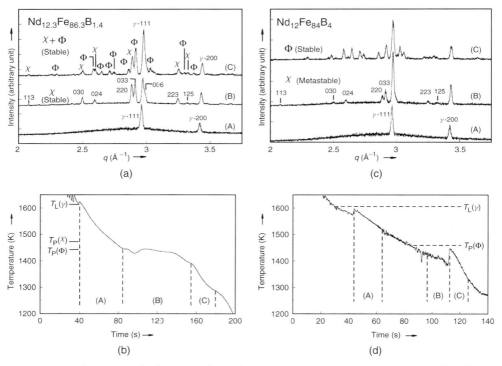

Fig. 14.2 (a, b) Sequence of crystallizing phases in undercooled melt of $Nd_{12.3}Fe_{86.3}B_{1.4}$ alloy with the corresponding temperature–time profile (A). γ-Fe, χ-$Nd_2Fe_{17}B_x$ (stable), and Φ-$Nd_2Fe_{14}B_1$ phase are found in the X-ray spectrum. The individual spectra were recorded within an integration time of 5 s within the three experiment intervals denoted by (A), (B), and (C) in the temperature–time profile. The peritectic temperatures were determined by the plateaus in the temperature–time profiles during remelting of the sample. (c, d) Sequence of crystallizing phases in $Nd_{12}Fe_{84}B_4$ melts and the corresponding temperature–time profile: γ-Fe, χ-$Nd_2Fe_{17}B_x$ (metastable), and Φ-$Nd_2Fe_{14}B_1$.

Bragg interferences in the diffraction pattern disappear during reheating the sample. χ-phase remelts before Φ-phase. Apparently, both phases are stable.

The as solidified sample consists of a microstructure of α-Fe dendrites embedded in the peritectic χ-phase. The latter one is surrounded by Φ-phase (Figure 14.3). The χ-phase was identified by its composition ratio Nd:Fe = 2 : 17. Since B cannot be detected by EDX backscattering in the electron microscope, its concentration remains unknown.

With increasing B concentration the χ phase becomes metastable. This transition occurs at B concentration between 1.4 and 4 at% B if the concentration of Nd : Fe = 1 : 7 is kept constant. In $Nd_{12}Fe_{84}B_4$ alloys χ phase can only be formed after primary crystallization of γ-Fe if solidification of peritectic Φ phase is being delayed, that is if the remaining liquid is undercooled below the peritectic temperature $T_P(\Phi)$ (see Figure 14.2 c, d). Subsequently, after formation of χ phase, Φ phase crystallizes and χ phase decays into γ-Fe and Φ phase as indicated by the disappearance of the Bragg reflexes of χ phase in the respective diffraction pattern. The metastable χ phase as formed in levitation undercooled Nd–Fe–B samples differs from a metastable phase found in melt-spun ribbons [31]. In the rapidly quenched Nd–Fe–B alloy, the metastable phase could be retained to ambient temperatures. It is hexagonal of $TbCu_7$ type. This structure is a disordered modification of Zn_2Th_{17}-type structure. The ordered crystal is identified in the diffraction pattern by the superlattice reflex (024) that disappears in the diffraction spectrum of disordered $TbCu_7$-type phase [32].

The metastable peritectic χ phase of the structural type of Zn_2Th_{17} phase was also observed in alloys of higher B concentration as $Nd_{11.8}Fe_{82.3}B_{5.9}$ and $Nd_{14}Fe_{79}B_7$ [12, 33]. It has been shown that the metastable χ phase is primarily formed at largest undercoolings. The primary solidification of γ-Fe is increasingly

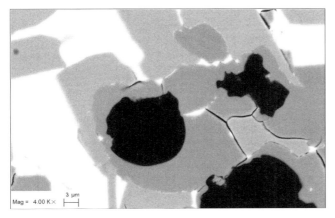

Fig. 14.3 Microstructure observed by scanning electron microscopy (backscattering mode) of $Nd_{12.3}Fe_{86.3}B_{1.4}$ sample corresponding to the temperature–time profile shown in Figure 14.2 (a, b). A–Fe (black), χ-$Nd_2Fe_{17}B_x$ (dark gray), Φ-$Nd_2Fe_{14}B_1$ (light gray), and Nd-rich phase (white).

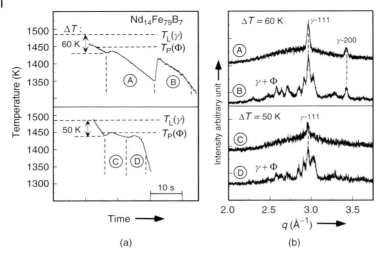

Fig. 14.4 Temperature–time profiles (a) and diffraction spectra (b) during solidification of $Nd_{14}Fe_{79}B_7$ melts which have been undercooled below the peritectic temperature $T_P(\Phi)$ prior to solidification. The diffraction patterns were recorded during an integration time of 5 s.

depressed with growing undercooling in favor of χ phase. Figure 14.4 shows temperature–time profiles measured on $Nd_{14}Fe_{79}B_7$ alloy during containerless cooling and solidification. Solidification starts at a temperature below the peritectic temperature $T_P(\Phi)$ corresponding to an undercooling of about 60 K. In this temperature range γ-Fe is metastable. Nevertheless γ-Fe primarily crystallizes at this temperature and Φ phase forms in a second reaction step at a temperature $T \approx$ 1350 K ($\Delta T \approx 160$ K). The fraction of primary formed γ-Fe depends on the time interval between two crystallization events. If this time interval becomes shorter, the volume fraction of γ-Fe decreases and that of Φ phase increases as expected. This can easily be observed from the comparison of intensities of the Bragg interferences of γ-Fe phase in the spectra of Figure 14.4(A) and (C) or (B) and (D), respectively.

Figure 14.5 exhibits an example of a solidification event in which crystallization of γ-Fe and Φ phase sets in at the same time so that only one recalescence event is observed. The diffraction spectra confirm that both phases are formed but it cannot be decided which phase is primarily crystallized. In some cases, also Φ phase solidifies exclusively without solidification of γ-Fe as illustrated in the lower part of Figure 14.4.

The diffraction experiments on levitation undercooled samples of Nd–Fe–B alloys reveal that γ-Fe crystallizes primarily in the state of undercooled liquid. However, the volume fraction of Φ phase increases at the expense of γ-Fe as soon as the secondary crystallization of Φ phase sets in. This observation hints on a growth dynamics of Φ phase being faster than that of γ phase.

With increasing undercooling, a transition from primary crystallization of the properitectic γ-Fe phase to primary crystallization of metastable χ phase is

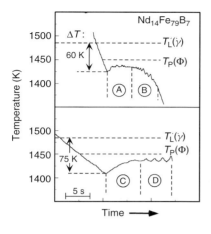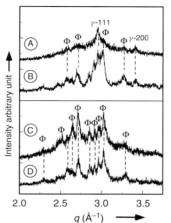

Fig. 14.5 Temperature–time profiles and diffraction spectra measured on $Nd_{14}Fe_{79}B_7$ melts during a single step solidification process at undercoolings in the range $\Delta T > T_L(\gamma) - T_P(\Phi)$ during which it cannot be discriminated between primary crystallization of γ-Fe and Φ phase and vice versa (upper part) and primary crystallization of Φ phase without detecting any Bragg interferences of γ-Fe (lower part).

observed. As can be seen in the temperature–time profile of Figure 14.6 the temperature–time characteristics shows a slight change in its slope indicating a crystallization event before the rapid temperature rise of the recalescence sets in at a temperature of 1350 K. Referring to the corresponding diffraction pattern the primary formation of a crystalline phase is indicated by the appearance of three Bragg interferences, which can be attributed to solidification of a phase whose structure is similar to that of the binary rhombohedral Nd_2Fe_{17} phase [12]. After the primary solidification of metastable χ phase in the deeply undercooled melt, the hard magnetic Φ phase crystallizes in a second reaction step during the time period at which the steep rise of temperature during the recalescence event at $T \approx 1350$ K is observed in the respective temperature–time profile. In the last diffraction spectrum of Figure 14.6 the Bragg reflexes of metastable χ phase disappear and only interferences are recorded which are attributed to γ-Fe and Φ phase. From this finding it is concluded that the metastable χ phase decays in γ-Fe and Φ phase when cooling the as solidified sample to ambient temperatures.

The results with respect to the different sequences of phase formation observed in the experimental investigations are summarized in Figure 14.7 that shows the quasi-binary phase diagram of Fe–Ni–B at fixed concentration ratio Nd : B = 2 : 1. At the stoichiometric composition $Nd_2Fe_{14}B_1$ and also at higher concentrations of Nd and B in the hyperperitectic region primary crystallization of γ-Fe is observed at undercoolings up to $\Delta T \approx 100$ K. That corresponds to a temperature being 60 K below $T_P(\Phi)$ of $Nd_{14}Fe_{79}B_7$ alloy. At undercoolings below the peritectic temperature also single step crystallization events are observed in melts of $Nd_{14}Fe_{79}B_9$ alloy, at which γ Fe and Φ phase are formed simultaneously or exclusively Φ phase is crystallizing. However, also in these experiments it is supposed that primary

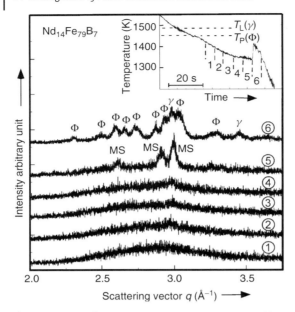

Fig. 14.6 Energy dispersive X-ray spectra measured during containerless undercooling and solidification at an undercooling of $\Delta T = 130$ K of a $Nd_{14}Fe_{79}B_7$ sample. Solidification starts with primary formation of metastable χ phase (MS). The insert gives the temperature–time profile in which the times are marked by an arrow at which the X-ray spectra counted by (1)–(6) are recorded.

crystallization of γ-Fe is setting in slightly before the formation of Φ phase, but the different events cannot be separated because of the very short time interval between solidification of γ-Fe and Φ phase.

At the highest undercoolings achieved for Nd–Fe–B alloys always metastable χ phase crystallizes primarily. The hard magnetic phase is formed in most cases in a second reaction step. Apparently, the crystallization of metastable χ phase is energetically favored at the highest undercoolings. In fact, primary crystallization of χ phase is also observed in the nonperitectic $Nd_{18}Fe_{73}B_9$ alloy if undercooling exceeds $\Delta T = 60$ K [34].

14.4
Analysis Within Models of Nucleation and Dendrite Growth

A primary crystallization of γ-Fe at temperatures below the peritectic temperature is understood if an energetically favored nucleation of γ-Fe compared with Φ phase is assumed. According to Equation 14.1 the activation energy to form a critical nucleus, ΔG^*, depends on interfacial energy σ, the difference of free enthalpy ΔG_v between solid and liquid phase and in case of heterogeneous nucleation on the catalytic potency factor $f(\theta)$. The energy gain during formation of a nucleus

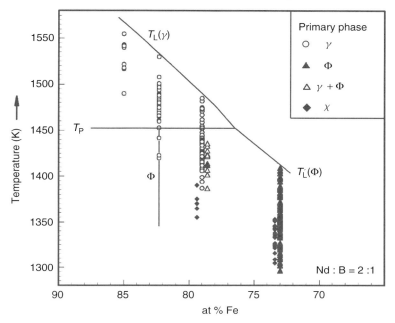

Fig. 14.7 Phase selection diagram of quasi-binary alloy of Nd–Fe–B at fixed concentration ratio Nd:B = 2 : 1. The solid lines represent the equilibrium liquidus temperatures and the peritectic temperature T_P, while the symbols denote the undercooling and concentration range of primary crystallization of γ-Fe (open circles), Φ phase (closed triangles), simultaneous solidification of γ-Fe and Φ phase (open triangles), and metastable χ phase (closed diamonds).

of γ-Fe is computed as a function of undercooling. Through undercooling the melt below the peritectic temperature $T_P(\Phi)$ the decrease of the free energy is more pronounced if a nucleus of γ-Fe instead of Φ phase is formed even though γ-Fe is not thermodynamically stable in this regime. This is due the composition dependence of ΔG_v. The composition in the remaining melt changes during crystallization due to segregation. Therefore, the assumption of a preferred nucleation of γ-Fe is only valid during the very first stage of solidification. The decrease of free energy during the formation of a nucleus of Φ phase exceeds that of a γ-Fe nucleus if temperature becomes less than $T_P(\Phi)$ that is 70 K.

The activation energy ΔG^* for the formation of a nucleus of Φ phase is smaller compared with the formation of a γ-Fe nucleus if the temperature falls below a critical temperature $T_{\Delta G^*}$ (cf. Figure 14.8). With increasing undercooling, a transition occurs at a critical undercooling $\Delta T_{\Delta G^*} = T_L(\gamma) - T_{\Delta G^*}$ at which the nucleation rate of Φ phase equals that of γ phase and with further increase in undercooling, exceeds that of γ-Fe. The nucleation rate is given as follows:

$$I = I_0 \cdot \exp\left(-\frac{\Delta G^*}{k_B \cdot T}\right) \tag{14.6}$$

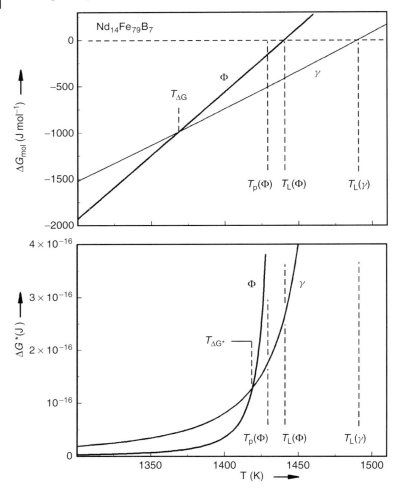

Fig. 14.8 Computed change of molar free energy, ΔG_{mol}, and computed activation energy for the formation of a critical nucleus, ΔG^*, in melts of $Nd_{14}Fe_{79}B_7$ alloy as a function of temperature. At temperatures below the transition temperatures, $T_{\Delta G}$, and $T_{\Delta G^*}$, the energy gain $|\Delta G_{mol}|$ is greater and the activation energy ΔG^* smaller if a nucleus of Φ phase is formed instead of a nucleus of γ-Fe. The dashed vertical lines represent the computed liquidus temperatures of γ-Fe and Φ phase, $T_L(\gamma)$ and $T_L(\Phi)$, as well as the peritectic temperature $T_P(\Phi)$.

Since the interfacial energy σ of Φ phase is smaller than that of γ phase, it holds $T_{\Delta G^*} > T_{\Delta G}$. The critical temperature $T_{\Delta G^*}$ is located below the peritectic temperature $T_P(\Phi)$. Despite the larger interfacial energy of γ phase nucleation of γ-Fe is still energetically preferred even at temperatures below the peritectic temperature since the larger driving force ΔG_v of γ-Fe overcompensates the effect due to its larger interfacial energy as long as the temperature does not fall below $T_{\Delta G^*}$.

Figure 14.9 shows the computed phase diagram including the metastable extension of the liquidus line of Φ phase, $T_L(\Phi)$, together with the transition temperatures $T_{\Delta G}$ and $T_{\Delta G^*}$ as a function of concentration. For comparison, the symbols denote the results of experimental investigations with respect to phase selection as a function of undercooling and concentration. The calculated liquidus lines deviate by about 10–20 K from the experimentally determined phase diagram [5]. To compare the critical undercoolings for changes in phase selection behavior with those computed from thermodynamics and nucleation theory, the critical temperatures for changes in phase formation were determined from the measured undercoolings and the computed liquidus temperature of γ phase. It is obvious from Figure 14.9 that the gain of free enthalpy during nucleation of Φ phase becomes greater than that of nucleation of γ phase only at very large undercoolings below the peritectic temperature. The transition to preferred nucleation of Φ phase takes place at smaller undercoolings according to the calculation of the activation energies of the various phases; however, the critical temperature $T_{\Delta G^*}$ is located always below the peritectic temperature $T_P(\Phi)$. According to the experimental results primary crystallization of γ phase is sometimes found not only at temperatures $T < T_P(\Phi)$, but at temperatures that are located below the calculated transition temperatures

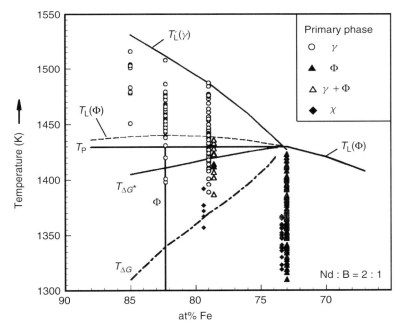

Fig. 14.9 Computed phase diagram of quasi-binary Fe–Ni–B alloy at fixed composition ratio Nd : B = 2 : 1 with varying Fe concentration with the metastable extensions of liquidus line of Φ phase. Also shown are the concentration dependence of the critical temperatures $T_{\Delta G^*}$ and $T_{\Delta G}$. The temperatures of the measured data were inferred from the experimentally determined undercoolings and the computed liquidus line of γ phase.

$T_{\Delta G^*}$. Therefore, the analysis of the results on phase selection within classical nucleation theory gives a qualitative understanding of primary crystallization of γ phase in undercooled melts of Nd–Fe–B alloy, but a quantitative description of the phase selection diagram (Figure 14.7) is not possible.

The experimental observation that at simultaneous solidification of γ and Φ phase the volume fraction of γ phase is essentially reduced can be understood if a more sluggish growth kinetics of γ phase is assumed compared with the growth kinetics of Φ phase. To investigate the growth kinetics we refer to an extended sharp interface model of dendrite growth (see Chapter 4) as presented elsewhere in more detail in this book [35]. Figure 14.10 illustrates the variation of the dendrite growth velocities of γ and Φ phase with undercooling as calculated within extended sharp interface theory of dendrite growth. For the computations it is assumed that the fluid flow velocity u induced by forced convection in electromagnetic levitation experiments is in the order of $u \approx 30\,\text{cm}\,\text{s}^{-1}$ (solid curve in Figure 14.10). Dendrite growth velocities of maximum $10\,\text{cm}\,\text{s}^{-1}$ at the undercooling of $100\,\text{K}$ of both phases are very small compared with the growth velocities up to several 1–$70\,\text{m}\,\text{s}^{-1}$ of pure metals and solid solutions [35]. Small growth velocities of the intermetallic Φ phase are expected since the crystallization of a superlattice structure as always present in intermetallic phases requires short-range atomic diffusion so that the growth kinetics is diffusion limited [36]. But on the first sight the sluggish growth of solid solution of γ-Fe phase is surprising. It can be understood if the very small solubility of Nd and B in γ-Fe is taken into account. As previously shown by measurements on dilute Ni–B [37] and Ni–Zr [38] even small amounts of elements with a small equilibrium partition coefficient in a solvent metal require a strong redistribution of the solute atoms at the solid/liquid interface during solidification that reduces drastically the growth velocity of chemical dendrites. Apparently, this effect is very pronounced so that even at small undercoolings below the liquidus temperature $T_L(\Phi)$ of Φ phase the growth velocity of γ-Fe is smaller than the growth velocity of Φ phase. The difference between the growth velocities of Φ and γ phase can be even as large as one order of magnitude if influences of convection are neglected so that the growth kinetics is exclusively controlled by conduction of heat and atomic diffusion (cf. dashed lines in Figure 14.10).

14.5
Summary and Conclusion

Electromagnetic levitation technique was successfully applied for containerless undercooling and solidification of liquid drops of Nd–Fe–B alloys. Large undercoolings were achieved up to $150\,\text{K}$. In Nd–Fe–B alloys various crystalline phases competes during solidification. The electromagnetic levitation was combined with the diagnostic means at the ESRF in Grenoble. The synchrotron radiation allows for EDX with high time resolution. Complete diffraction spectra were recorded within time intervals of a few seconds. It is demonstrated that with this diagnostics, the sequence of phase formation during rapid solidification of undercooled melts

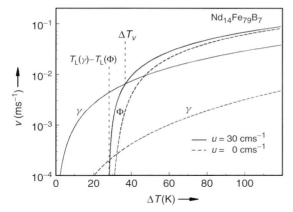

Fig. 14.10 Dendrite growth velocity V of γ-Fe and Φ phase as a function of undercooling ΔT in melts of $Nd_{14}Fe_{79}B_7$ alloys without convection (dashed lines) and taking into account a fluid flow velocity $u = 30\,\mathrm{cm\,s^{-1}}$ (solid lines) induced by forced convection in electromagnetically levitated drops. A critical undercooling $\Delta T_v \approx 35$ K is found at which the growth velocity of γ-Fe becomes equal to the growth velocity of Φ phase.

is observed *in situ*. In this way, a phase selection diagram was determined for quasi-binary Nd–Fe–B alloy of fixed composition ratio Nd : B = 2 : 1, which shows the sequence of phase formation as a function of Fe concentration and the undercooling prior to solidification. An important result found is that the volume fraction of soft magnetic γ Fe decreases with increasing undercooling in favor of hard magnetic Φ phase. The experimental results were analyzed within models of nucleation and dendrite growth. For doing so, the equilibrium phase diagram with the metastable extensions of γ-Fe and hard magnetic Φ phase of the quasi-binary Nd–Fe–B alloy was computed on the basis of the concentration-dependent free enthalpy taken from literature. The selection mode of phase formation as a function of undercooling can be understood within classical nucleation theory in a qualitative manner. Calculations of the dendrite growth velocities of γ and Φ phase show a sluggish growth dynamics where the growth velocities of the solid solution of γ-Fe are even smaller than the diffusion-controlled growth velocities of superlattice structure of intermetallic Φ phase at undercoolings $\Delta T \approx 35$ K. This surprising result is understood by the fact that Nd and B show a very small solubility that is partition coefficient in Fe as solvent material. The investigations clearly reveal the importance of undercooling in phase selection during nonequilibrium solidification of undercooled melts and demonstrate that undercooling can be used as control parameter in materials production from the liquid state.

Acknowledgments

We thank D. Holland-Moritz and O. Heinen for their support to the experiments and fruitful discussions. Their help in performing the energy dispersive X-ray scattering experiments at ESRF Grenoble is appreciated. We are grateful to the Deutsche Forschungsgemeinschaft for the financial support within the Priority Programme SPP 1120 under grant no. HE 1601/14-1,2.

References

1 Sagawa, M., Fujimura, S., Togawa, N., Yamamoto, H. and Matsuura, Y. (**1984**) "New material for permanent magnets on a base of Nd and Fe (invited)". *Journal of Applied Physics*, **55**, 2083–87.

2 Croat, J.J. Herbst, J.F., Lee, R.W. and Pinkerton, F.E. (**1984**) "Pr-Fe and Nd-Fe-based materials: a new class of high performance permanent magnets". *Journal of Applied Physics*, **55**, 2078–82.

3 Herbst, J.F. (**1991**) "$R_2Fe_{14}B$ materials: intrinsic properties and technological aspects". *Reviews of Modern Physics*, **63**, 819–98.

4 Herlach, D.M., Galenko, P., Holland-Moritz, D. (**2007**) Metastable solids from undercooled melts, in *Pergamon Materials Series*, (ed. R. Cahn), Elsevier, ISBN 978-0-08-043638-8.

5 Schneider, G., Henig, E.-T., Petzow, G. and Stadelmeier, H.H. (**1986**) "Phase relations in the system Fe-Nd-B". *Zeitschrift Fur Metallkunde*, **77**, 755–61.

6 Herlach, D.M. and Volkmann, T. and Gao, J. (**2002**) *Verfahren zur Herstellung von Nd-Fe-B-Basislegierungen*, German Patent DE 10 106 217.

7 Volkmann, T., Gao, J. and Herlach, D.M. (**2002**) "Direct crystallization of the peritectic $Nd_2Fe_{14}B_1$ phase by undercooling of bulk alloy melts". *Applied Physics Letters*, **80**, 1915–17.

8 Gao, J., Volkmann, T. and Herlach, D.M. (**2002**) "Undercooling-dependent solidification behavior of levitated $Nd_{14}Fe_{79}B_7$ alloy droplets". *Acta Materialia*, **50**, 3003–12.

9 Gao, J., Volkmann, T. and Herlach, D.M. (**2002**) "Critical undercoolings for different primary phase formation in $Nd_{15}Fe_{77.5}B_{7.5}$ alloy". *IEEE Transactions on Magnetics*, **38**, 2910–12.

10 Gao, J., Volkmann, T. and Herlach, D.M. (**2003**) "Solidification of levitated Nd-Fe-B alloy droplets at significant bulk undercoolings". *Journal of Alloys and Compounds*, **350**, 344–50.

11 Volkmann, T., Gao, J., Strohmenger, J. and Herlach, D.M. (**2004**) "Direct crystallization of the peritectic $Nd_2Fe_{14}B_1$ Phase by undercooling of bulk alloy melts". *Materials Science and Engineering*, **A375-377**, 1153–56.

12 Volkmann, T., Strohmenger, J., Gao, J. and Herlach, D.M. (**2004**) "Observation of a metastable phase during solidification of undercooled Nd-Fe-B alloy melts by in-situ diffraction experiments using synchrotron radiation". *Applied Physics Letters*, **85**, 2232–34.

13 Notthoff, C., Feuerbacher, B., Franz, H., Herlach, D.M. and Holland-Moritz, D. (**2001**) "Direct determination of metastable phase diagram by synchrotron radiation experiments on undercooled metallic melts". *Physical Review Letters*, **86**, 1038–41.

14 Notthoff, C., Franz, H., Hanfland, M., Herlach, D.M., Holland-Moritz, D. and Petry, W. (**2000**) "Energy-dispersive X-ray diffraction combined with electromagnetic levitation to study phase-selection in undercooled melts". *The Review of Scientific Instruments*, **71**, 3791–96.

15 Hallemans, B., Wollants, P. and Roos, J.R. (**1995**) "Thermodynamic assessment of the Fe-Nd-B phase diagram". *Journal of Phase Equilibria*, **16**, 137–49.

16 Spaepen, F. (**1975**) "A structural model for the solid-liquid interface in monatomic systems". *Acta Metallurgica*, **23**, 729–43.

17 Spaepen, F. and Meyer, R.B. (**1976**) "The surface tension in a structural

model for the solid-liquid interface". *Scripta Metallurgica*, **10**, 257–63.

18. Holland-Moritz, D. (**1998**) "Short range order and solid-liquid interfaces in undercooled melts". *International Journal of Non-Equilibrium Processing*, **11**, 169–99.

19. Holland-Moritz, D. (**2003**) *Ordnungsphänomene, fest-flüssig Grenzfläche und Phasenselektionsverhalten in unterkühlten Metallschmelzen*, Habilitation thesis, Ruhr-Universität Bochum, Bochum, Germany.

20. Lipton, J., Kurz, W. and Trivedi, R. (**1987**) "Rapid dendrite growth in undercooled melts". *Acta Metallurgica*, **35**, 957–64.

21. Bobadilla, M., Lacaze, J. and Lesoult, G. (**1988**) "Influence des conditions de solidification sur le déroulement de la solidification des acier inoxidables austénitiques". *Journal of Crystal Growth*, **89**, 531–44.

22. Coriell, S.R. and Turnbull, D. (**1982**) "Relative roles of heat transport and interface rearrangement rates in the rapid growth of crystals in undercooled melts". *Acta Metallurgica*, **30**, 2135–39.

23. Aziz, M.J. and Boettinger, W.J. (**1994**) "On the transition from short-range diffusion-limited to collision-limited growth in alloy solidification". *Acta Metallurgica Et Materialia*, **42**, 527–37.

24. Trivedi, R. and Kurz, W. (**1990**) "Modelling of solidification microstructures in concentrated solutions and intermetallic systems". *Metallurgical Transactions*, **21A**, 1311–18.

25. Umeda, T., Okane, T. and Kurz, W. (**1996**) "Phase selection during solidification of peritectic alloys". *Acta Materialia*, **44**, 4209–16.

26. Hyers, R.W. (**2005**) "Fluid flow effects in levitated droplets". *Measurement Science and Technology*, **16**, 394–401.

27. Galenko, P.K., Funke, O., Wang, J. and Herlach, D.M. (**2004**) "Kinetics of dendrite growth under the influence of convective flow in solidification of undercooled droplets". *Materials Science and Engineering*, **A375-377**, 488–92.

28. Herlach, D.M. and Galenko, P.K. (**2007**) "Rapid Solidification: In situ diagnostics and theoretical modelling". *Materials Science and Engineering*, **A449-451**, 34–41.

29. de Mooij, D.B. and Buschow, K.H.J. (**1988**) "Formation and magnetic properties of the compounds $R_2Fe_{14}C$". *Journal of the Less-Common Metals*, **142**, 349–57.

30. Herbst, J.F., Croat, J.J., Lee, R.W. and Yelon, W.B. (**1982**) "Neutron diffraction studies of $Nd_2(Co_xFe_{1-x})_{17}$ alloys: Preferential site occupation and magnetic structure". *Journal of Applied Physics*, **53**, 250–56.

31. Gabay, A.M., Popov, A.G., Gaviko, V.S., Belozerov, Ye.V., Yermolenko, A.S. and Shchegoleva, N.N. (**1996**) "Investigation of phase composition and remanence enhancement in rapidly quenched $Nd_9(Fe,Co)_{85}B_6$ alloys". *Journal of Alloys and Compounds*, **237**, 101–7.

32. Khan, Y. (**1973**) "The crystal structure of R_2Co_{17} intermetallic compounds". *Acta crystallographica. Section B: Structural Crystallography and Crystal Chemistry*, **B29**, 2502–07.

33. Strohmenger, J., Volkmann, T., Gao, J. and Herlach, D.M. (**2006**) "Metastable phase formation in undercooled Nd-Fe-B alloys investigated by in-situ diffraction using synchrotron radiation". *Material Science Forum*, **508-509**, 81–86.

34. Strohmenger, J., Volkmann, T., Gao, J. and Herlach, D.M. (**2005**) "The formation of a metastable peritectic phase in Nd-Fe-B alloys investigated by in-situ X-ray diffraction during solidification". *Materials Science and Engineering*, **A413-414**, 263–66.

35. Galenko, P. and Herlach, D.M. (**2008**) *Dendritic Growth and Grain Refinement in Undercooled Metallic Melts*, this book.

36. Barth, M., Wei, B. and Herlach, D.M. (**1995**) "Crystal growth in undercooled melts of the intermetallic compounds FeSi and CoSi". *Physical Review B*, **51**, 3422–28.

37 Eckler, K., Cochrane, R.F., Herlach, D.M., Feuerbacher, B. and Jurisch, M. (**1992**) "Evidence for a transition from diffusion- to thermally-controlled solidification in metallic alloys". *Physical Review B*, **45**, 5019–22.

38 Arnold, C.B., Aziz, M.J., Schwarz, M. and Herlach, D.M. (**1999**) "Parameter-free test of alloy dendrite-growth theory". *Physical Review B*, **59**, 334–43.

15
Effect of Varying Melt Convection on Microstructure Evolution of Nd–Fe–B and Ti–Al Peritectic Alloys

Regina Hermann, Gunter Gerbeth, Kaushik Biswas, Octavian Filip, Victor Shatrov, and Janis Priede

15.1
Introduction

Nd–Fe–B magnets are widely used for various applications in technology, biophysics, and medicine [1]. High-performance Nd–Fe–B magnets are of enormous economic interest and they will continue to play a key role in the development of electronics and motors. Rare-earth magnets are commercially produced by sintering or rapid quenching [2, 3]. The excellent ferromagnetic properties of the rare-earth magnets are due to the hard magnetic $Nd_2Fe_{14}B$ phase (ϕ-phase) with high saturation polarization and magnetocrystalline anisotropy. However, the properties of real magnets differ from the intrinsic properties of the $Nd_2Fe_{14}B$ phase. The reason for this is in the deficient microstructure. In order to obtain high $(BH)_{max}$ values, it is necessary to control the composition and the microstructure of the alloys. Mainly, the suppression of the α-Fe phase, which has a detrimental effect on the coercive field strength of the magnet, is of considerable interest for the production of near-stoichiometric compounds.

Ti–Al alloys containing γ and α_2 phases have attracted much attention owing to some properties such as low density, high specific strength, high specific stiffness, and retention of strength at high temperatures combined with adequate creep [4]. However, their limited room-temperature ductility has so far restricted their application [5]. Ti–Al alloys can be broadly classified into three categories: equiaxed single-phase γ-TiAl, fully lamellar microstructures consisting of lamellar grains composed of alternate α_2 and γ plates, and a mixture of equiaxed γ phase and lamellar $(\gamma + \alpha_2)$ phase (duplex microstructure). The two-phase $(\gamma + \alpha_2)$ alloys or the mixture of γ and lamellar $(\gamma + \alpha_2)$ alloys show better mechanical properties than single-phase γ or α_2. However, microscopic factors such as volume fraction of the lamellar phase, the distribution of orientation variants, the fineness of the lamellar structure, and their interlamellar spacing strongly affect the mechanical properties of these alloys [6].

Numerous studies have been carried out to understand the solidification behavior of the properitectic phase, peritectic phase, and the evolved microstructures, phase selection, and transformation kinetics [7]. The volume fraction of the properitectic primary phase and also the understanding of the solidification mechanisms regarding the magnetic [8] and mechanical properties are very important [9].

In metallurgical processing, it is well known that electromagnetic stirring during the solidification of metals may lead to morphology changes in the microstructure. However, systematic studies on the relation between the convection and the resulting solidified microstructure of peritectic alloys during induction melting, which is a common processing step for production of ingot alloys, are still very scarce. The change of volume fraction and morphology of the properitectic phase with varying convection state is of prime importance for the quality of the magnetic and mechanical properties of Nd–Fe–B and Ti–Al alloys.

The aim of this study is to analyze the influence of melt convection on the microstructure evolution of alloys of technological relevance using novel techniques developed on the basis of numerical simulation of fluid flow and to demonstrate the feasibility of tailoring the microstructure and the resulting alloy properties by customized melt convection using magnetic fields.

15.2
Methods Developed

15.2.1
Forced Rotation Technique

15.2.1.1 Experimental
The scheme of the experimental setup [10] is shown in Figure 15.1. An induction coil is wrapped around a quartz tube and connected to a high-frequency power supply (100 kHz, 50 kW). The quartz tube is mechanically connected to an AC motor, which can be operated at variable speeds (0–2500 rpm). The samples are sealed in a quartz ampule under a protective argon atmosphere at different

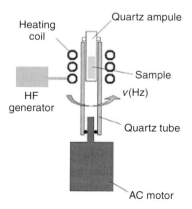

Fig. 15.1 Scheme of the experimental setup of the forced rotation technique.

pressures after the ampules were evacuated to high vacuum. The top part of the ampoule was specially designed so that it can be fixed to the quartz tube. When the sample is completely molten, the AC motor is switched on enabling the ampule to rotate at a defined rotational speed. After a fixed time, the power supply to the induction coil is switched off while the sample rotates till it completely solidifies during cooling.

15.2.1.2 Numerical Simulation

The flow in a cylindrical container has been studied theoretically by many researchers. Because the metallic melt is opaque, numerical flow simulations are necessary for understanding the flow structure in metallic liquids. Free-surface phenomena of a rotating fluid in a cylinder were investigated by Vatistas [11]. The 3D flow in a rotating lid–cylinder enclosure was analyzed by Gelfgat et al. [12]. The flow of an electrically conducting melt in a cylinder driven by a rotating magnetic field was studied by Grants and Gerbeth [13]. A completely different problem arises if the flow is driven by an alternating magnetic field in addition to container rotation. The combined action of both flow driving mechanisms leads to a complex flow pattern. The numerical simulation of this problem was carried out by Shatrov et al. [14]. The crucible rotation Ω causes an azimuthal flow, which can be characterized by the Ekman number $E = \nu/\Omega R^2$, where ν is the kinematic viscosity and R is the radius of the crucible. Induced electric currents and the magnetic field lead to a meridional Lorentz force driving a meridional flow, which can be characterized by the nondimensional field frequency $S = \mu_0 \sigma \omega R^2$ and the interaction parameter $N = \sigma \omega R^2 B^2 / \rho \nu^2$, where μ_0 is the vacuum magnetic permeability, σ is the electrical conductivity, $\omega/2\pi$ is the generator frequency, B is the magnetic field strength, and ρ is the density of the melt. S describes the skin effect of the magnetic field and the results given in the following are always for $S \gg 1$. By varying the Ekman number, the magnetic interaction parameter N, the field frequency S, as well as the contact angle between melt and crucible wall, it is found that in all considered parameter ranges the flow consists of four toroidal eddies for a melt height equal to its diameter. This is mainly caused by the nonuniformity of the magnetic field resulting in $S \gg 1$ due to the skin effect near the edges of the liquid volume. The interaction parameter N controls the intensity of the meridional flow. The spinning of the container leads to a deformation of the flow structure. At Ekman numbers $E < 10^{-2}$ the meridional flow is significantly reduced. The azimuthal flow has its maximum in the Ekman number range of $E \sim 10^{-3} - 10^{-2}$; at smaller Ekman numbers, the azimuthal flow is suppressed. The main conclusion is that a relatively *strong container spinning suppresses the flow inside the liquid* for all considered interaction parameter values. Figure 15.2 shows the resulting meridional flow Reynolds number as a function of E and N. At $E < 10^{-2}$ a faster crucible rotation Ω creates a reduced meridional flow. Figure 15.3 compares the stream lines and the angular velocity distributions for an Ekman number of slow rotation ($E = 1$) to that of fast rotation ($E = 1.5 \times 10^{-4}$) for $N = 2 \times 10^7$ and $S = 200$.

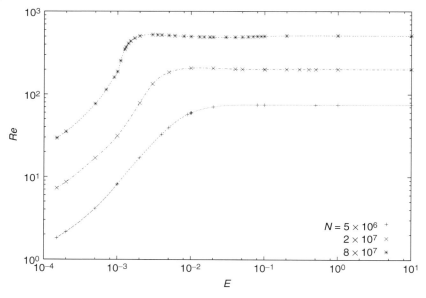

Fig. 15.2 Forced crucible rotation: the resulting meridional flow Reynolds number as function of Ekman number E and interaction parameter N.

15.2.2
Floating Zone Facility with Additional Magnetic Field

15.2.2.1 Experimental

With the aim to alter the fluid flow inside the melt over a much wider range compared to the forced rotation facility, floating zone experiments with an additional magnetic field were carried out. The floating zone facility with a double coil system (250 kHz, 30 kW) operates in two different arrangements: both coils in series connection (Figure 15.4a), or both coils separate, coupled just by induction (Figure 15.4b). The coil systems have been designed and patented [15]. The secondary coil creates an additional magnetic field, whereas its strength varies by the parameters of the coil system influencing the melt convection in a well-controlled way. The series connected coil system consists of two coils connected in series with 180° phase shift. The principal parameter to control the melt convection is the vertical gap between the coils. At small coil gaps, the flow consisting of two symmetric vortices is driven radially inwards at the mid-plane of the inductor. At large coil gaps, the flow direction changes and is now driven radially outwards at the mid-plane of the inductor. The reversal of the flow direction implies a minimum flow strength at a critical vertical coil gap with a strongly reduced melt convection in the molten zone.

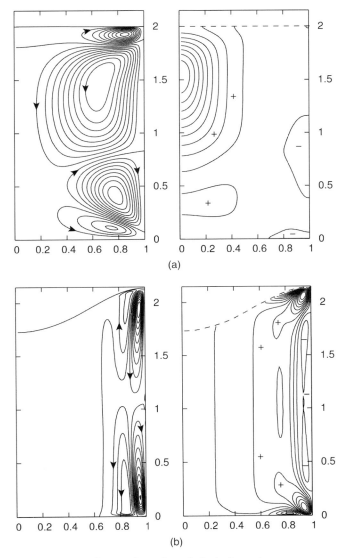

Fig. 15.3 Stream lines and angular velocity isolines at $N = 2 \times 10^7$, $S = 200$ for (a) $E = 1$ and (b) $E = 1.5 \times 10^{-4}$.

If the coils are only coupled by induction, the additional coil is part of a secondary circuit with adjustable capacitance and resistance but it is not connected to any power supply. The current in the secondary coil is induced solely by the primary coil, thereby creating the envisaged electromagnetic pump effect. Its intensity and the resulting flow direction can easily be adjusted to the process needs. The flexible

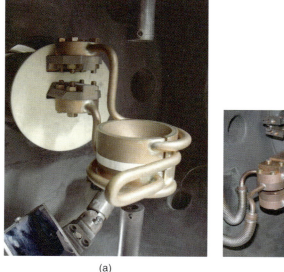
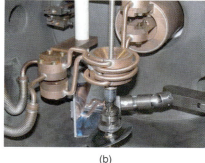

Fig. 15.4 Scheme of the experimental setup of the floating zone arrangement: (a) both coils in series connection, (b) two-phase stirrer.

system parameters are the location and distance of the secondary coil with respect to the primary one, and the capacitance and resistance of the secondary circuit. The maximum pumping action is obtained if the secondary capacitance is chosen to give resonance in the secondary circuit [16], which implies that other values of the secondary capacitance result in a lower flow driving action of this double coil system, also called a two-phase stirrer [17].

15.2.2.2 Numerical Simulation

The solution of the heat and hydrodynamic problems requires the knowledge of the electric current and the magnetic field induced by the radio frequency heating coil in the rod and in the melt. For the electromagnetic part, it was important to take into account the finite skin depth of the field both in the copper inductors as well as in the rod. This problem was solved by a control volume method on a triangular grid adapting to the shape of solid–liquid interfaces varying in the course of the solution [17]. The whole crystal was covered by a nonuniform triangular grid with up to 15 000 nodes so that there were about 13 nodes over 1 mm length. The equations were integrated in time by an implicit scheme using conjugate-gradient-type iterative solvers accelerated by incomplete LU factorization. Figure 15.5 shows the typical double vortex structure appearing for a single-coil induction heating and the change to a single convection roll by application of the magnetic two-phase stirrer. The thermophysical properties of molten Nd–Fe–B such as electrical conductivity, heat conductivity, and viscosity, necessary for numerical simulations close to real conditions, have been determined [18].

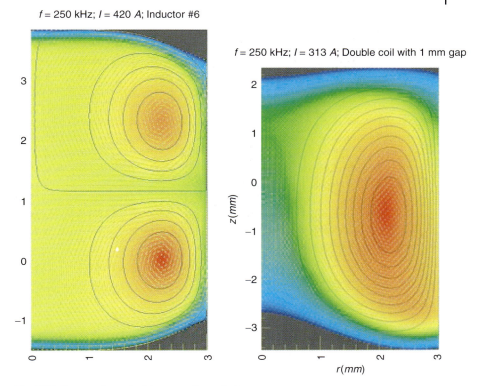

Fig. 15.5 Changes from double convection rolls to a single convection roll by application of a magnetic two-phase stirrer.

15.3
Sample Preparation

Master alloys with the nominal compositions $Nd_{10.9}Fe_{83.3}B_{5.8}$ (composition 1), $Nd_{11.8}Fe_{82.3}B_{5.9}$ (composition 2), and $Nd_{14}Fe_{80}B_{6.0}$ (composition 3) were cast to rods in a cold crucible induction furnace under argon atmosphere. For the experiments with well-defined forced melt rotation, samples with a length of 10 mm were sealed in quartz ampules under argon atmosphere at 200 and 500 mbar pressure, and fixed to a quartz tube. For the measurement of the α-Fe volume fraction, samples of about 50 mg were enclosed in a Ag foil and fixed in a boron nitride (BN) crucible. The α-Fe volume fraction was estimated by measuring the magnetic moment at 350 °C, well above the Curie temperature of the $Nd_2Fe_{14}B$ phase, using a 1.8 T vibrating sample magnetometer (VSM). The chemical analysis of the composition of the processed alloys was done by the inductively coupled plasma optical emission spectroscopy (ICP-OES) method. The ICP-OES measurements were carried out by a radial IRIS Intrepid II XUV (Thermo Corp., Franklin, MA, USA) with Echelle optics and a charge injection device (CID) detector for simultaneous recording of the selected emission lines. For the measurement of the secondary dendritic arm spacing (SDAS), an optical microscope equipped with a motorized stage was utilized. A program was created as a macro sequence (QUIPS routine) for the image analyzing system QTM 550 /QWIN (LEICA Co.), which enables inspection of a large sample area field by field.

The $Ti_{45}Al_{55}$ alloy was prepared from Ti (99.99%, Alfa Aesar) and Al (99.99%, Alfa Aesar) in a cold crucible induction furnace under argon atmosphere and cast into rods with 6 mm diameter. The average phase fraction of the properitectic phase was determined by the a4i Docu Image Analysis software from the scanning electron microscopy (SEM) micrographs and different parts of the cross section. Compression tests were performed at an initial strain rate of $1 \times 10^{-4} s^{-1}$ with an Instron 8562 electromechanical testing device under quasi-static loading at room temperature. The cylindrical sample (3 ± 0.02 mm diameter and 6 ± 0.02 mm length) was produced by electroerosive sawing from the central parts of the processed samples.

All samples were investigated by SEM. The sample compositions were measured by energy dispersive X-ray spectroscopy (EDS).

15.4
Results and Discussion

15.4.1
Nd–Fe–B Alloys

The experiments with the forced rotation technique performed at different rotation rates resulted in a drastic reduction of the α-Fe volume fraction with increasing rotation rate. The α-Fe volume fraction declined by about 39 wt% for composition 1, about 38 wt% for composition 2, and about 31 wt% for composition 3, with increasing global melt rotation rates from 0 up to 1200 rpm. Figure 15.6 shows the α-Fe volume fraction and the SDAS versus rotation rate for composition 2. The ampule diameter was chosen to be 10 mm. The argon pressure in the sealed ampules was 200 mbar. Owing to the strong gas expansion during heating and melting of the samples, the ampoules could not be filled up to normal atmospheric pressure. However, the low argon pressure and the residual oxygen in the ampule may alter the composition of the processed sample because Nd is very susceptible to oxide formation. On the other hand, the cooling rate at low gas pressure can differ from the cooling rate under normal pressure, influencing the growth of the properitectic Fe phase. Therefore, we checked the real sample composition of the processed samples. Moreover, experiments with 500 mbar argon pressure were carried out. The measurement of the real sample composition yielded a Nd loss of about 1.8 at% at 500 mbar argon pressure and about 3 at% under 200 mbar argon pressure, while the Nd loss did not change with the rotation rate. Measurements at different sample portions confirmed the homogeneity of the samples. Additionally, the steady state of the fluid flow in the melt was proved by varying the time of melt rotation prior to solidification. The time of melt rotation, which was 90 s in our

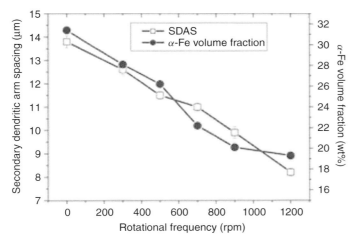

Fig. 15.6 α-Fe volume fraction and secondary dendritic arm spacing versus rotation rate for composition 2 with an ampule diameter of 10 mm.

experiments, was extended to 120 and 180 s, respectively. The α-Fe volume fraction remained constant, indicating the achievement of steady flow at 90 s forced rotation.

Since Nd–Fe–B alloys are produced under industrial conditions in large process quantities, the forced rotation technique was studied for larger crucible dimensions. Therefore, experiments with extended ampule diameters were carried out. Starting from the assumption that the internal flow field at high rotation rates is mainly determined by the Ekman number E, similar flow fields and similar microstructures in samples with different diameters can be assumed if the rotation rate is changed according to the following equation:

$$f = \frac{v}{2\pi E R^2}. \tag{15.1}$$

Samples of composition 2 with 10 and 19.4 mm diameter as well as samples of composition 3 with 6, 10, and 18 mm diameter were prepared. From Equation 15.1, compared to the processing of a sample (composition 2) with 10 mm diameter at 1200 rpm rotation rate, the resulting rotation rate for an ampule diameter of 19.4 mm amounts to 320 rpm. With the highest possible rotation rate of 2500 rpm for the 6 mm ampule (composition 3), the corresponding rates for the 10 and 18 mm ampoules amount to 900 and 278 rpm, respectively. The measured α-Fe volume fractions are summarized in Table 15.1. It can be ascertained that the compared samples exhibit similar α-Fe volume fractions, indicating the same fluid flow state prior to solidification. Equation 15.1 can therefore be applied as a simple instrument for the transmission to industrial like dimensions with prospects of improved processing routes and cost savings during final magnet production [19–21]. Figure 15.7 shows the microstructure of samples processed without any rotation and with 500 and 1200 rpm rotation. The as-solidified microstructure

Table 15.1 α-Fe volume fraction as a function of the rotation rate (rpm) for Ø 10 and 19.4 mm samples (composition 2); and for Ø 6, 10, and 18 mm samples (composition 3).

Composition 2	Ø 10 mm	Ø 19.4 mm	
f (rpm)	1200	320	
α-Fe (wt%)	18.8	19.1	
Composition 3	Ø 6 mm	Ø 10 mm	Ø 18 mm
f (rpm)	2500	900	278
α-Fe (wt%)	18.8	19.2	19.3

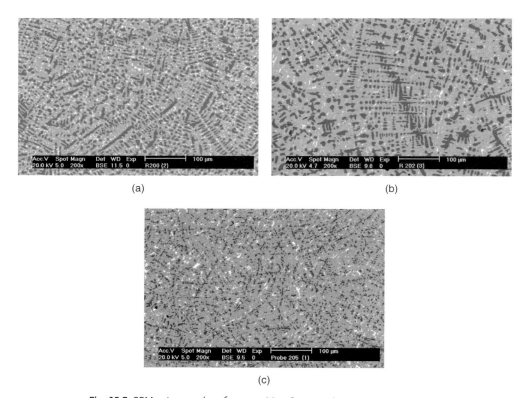

Fig. 15.7 SEM micrographs of composition 2 at rotation rates of (a) 0 rpm, (b) 500 rpm, and (c) 1200 rpm: ampule diameter 10 mm.

exhibits large α-Fe dendrites (black), simultaneously with the Nd-rich phase (white) after 0 and 500 rpm, respectively. A strong refinement of the microstructure with small-sized α-Fe dendrites is generated with increasing global melt rotation, as is clearly apparent in Figure 15.7c.

Additionally, the average SDAS was measured from SEM micrographs by the a4i Materials Image Analysis software. The SDAS ranges from 10 to 12 μm for samples without global sample rotation, whereas it reduces to 5 to 7 μm for the samples solidified at 1200 rpm. Figure 15.6 shows the variation of SDAS and the α-Fe volume fraction as a function of the rotational frequency (from 0 up to 1200 rpm). The lowering of the SDAS is clearly correlated to the drastic reduction of the α-Fe volume fraction with increasing rotational speed. No significant variation in the SDAS values was observed over the sample cross section at a particular rotational speed, indicating a homogeneous distribution of the SDAS in the microstructure and suggesting that the effective mass transfer does not vary significantly along the length or width of the investigated melt volume.

The possible reason for the reduced growth of the properitectic Fe phase under reduced internal melt motion can be attributed to the reduced convective mass transport [22, 23]. With high fluid velocity, the enhanced interdendritic flow reduces the solute boundary layer and increases the transfer of solute from the interface, removing the barrier to the growth of dendrites. Additionally, under strong convection, smaller arms dissolve into the melt, making the SDAS higher.

With the aim to alter the fluid flow inside the melt in a much wider range compared to the forced rotation facility, floating zone experiments with additional magnetic field in series coil connection, where the principal parameter to control the melt convection is the vertical gap, were carried out. Figure 15.8 shows the variation of α-Fe volume fraction with respect to the coil distance determined from VSM measurements. A pronounced minimum of the α-Fe volume fraction (14.2 wt%) was determined at 5.1 mm coil distance. Increasing or decreasing the vertical coil distance leads to a distinct increase of the α-Fe volume fraction (Figure 15.8). The minimum value of the α-Fe volume fraction with respect to the coil distance is consistent with the numerical simulation of the fluid flow, which predicts a change of the flow direction from radially outward to radially

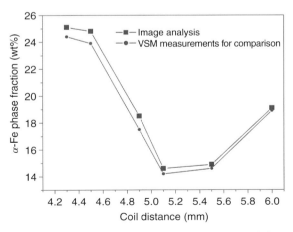

Fig. 15.8 Variation of α-Fe volume fraction versus coil distance of a series coil connection system.

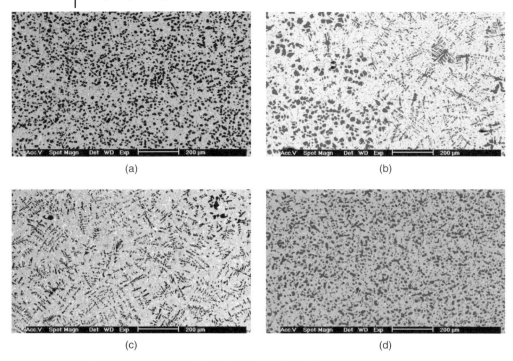

Fig. 15.9 SEM pictures of samples solidified at different coil distances: (a) 4.3 mm, (b) 4.9 mm, (c) 5.1 mm, and (d) 6.0 mm.

inward, crossing a minimum flow strength at the optimum coil gap. The value of lowest α-Fe volume fraction in the vicinity of 5.1 mm coil gap can therefore be attributed to a distinctly reduced melt convection state. Figure 15.9 shows the SEM pictures from samples solidified at different coil distances. At 4.3 mm coil distance and strong fluid flow, the morphology of α-Fe is spherical (Figure 15.9a). The morphology of the α-Fe phase becomes a mixture of dendritic and globular at 4.9 mm (Figure 15.9b) and changes further to a completely dendritic structure at 5.1 mm coil distance (Figure 15.9c). Increasing of coil gap to 6.0 mm leads again to a globular morphology (Figure 15.9d).

15.4.2
Ti–Al Alloys

$Ti_{45}Al_{55}$ alloys were subjected to solidification under two different convective conditions: common single-coil induction melting, and induction melting with additional magnetic stirring using the two-phase stirrer. The melt was maintained in the liquid state for 2 min prior to solidification to provide a steady fluid flow. The cooling rate, measured with a two-color pyrometer after switching off the induction coil system, was 40 K s^{-1}. Figure 15.10 shows the microstructures of

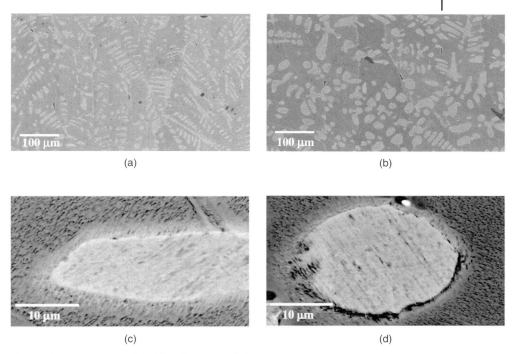

Fig. 15.10 SEM images of $Ti_{45}Al_{55}$ alloys: (a and c) common single-coil induction melting, (b and d) induction melting with the two-phase stirrer.

both sample types. The morphology change of the properitectic phase in the stirred sample is clearly evident. Dendritic morphology dominates the microstructure in Figure 15.10a, whereas a globular morphology of the properitectic phase prevails in the microstructure of the stirred alloy (Figure 15.10b). During cooling below the eutectoid temperature at 1125 °C, the properitectic α phase transforms to a lamellar structure consisting of γ and α_2 phases ($\alpha \rightarrow \alpha_2 + \gamma$). The fraction of the lamellar colonies, determined by the Image analysis software, is higher in the stirred sample and results in 26% compared to 19% with single-coil melting. The average lamellar spacing between γ and α_2 phases, measured from magnified SEM micrographs, is 0.55 µm for the stirred sample, which is twice that of the reference sample (0.27 µm). The lamellar spacing is defined as the edge-to-edge dimension perpendicular to the plate boundary of the adjacent lamella.

Additionally, EDS line-scan analysis was performed to verify the thickness of the boundary layer. Figure 15.10c and d represents the intermediate layer around the properitectic lamellar phase. The layer thickness measured from different SEM micrographs is reduced in the stirred sample under strong melt motion to 2.5 µm compared to 5 µm in the sample with single-coil melting. A broad diffusion boundary layer attenuates the solute transport [24]. In samples melted and solidified with the single induction coil, the solute distribution in the liquid is

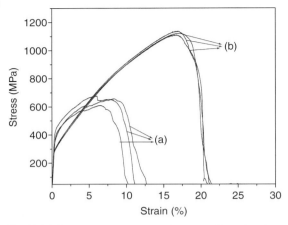

Fig. 15.11 Compression stress–strain plots of $Ti_{45}Al_{55}$ alloys: (a) common single-coil induction melting, (b) induction melting with the two-phase stirrer.

governed mainly by diffusion, and the nucleation sites grow dendritically. Under strong convection, the concentration profile is homogenized outside the diffusion boundary layer and tends to create a uniform solute concentration in the bulk liquid followed by a narrow intermediate layer. The resultant smaller diffusion boundary layer leads to an enhanced solute transport enabling globular growth. Thus, the stirring effect decreases the intermediate layer and promotes coarsening of the particles. Figure 15.11 shows the stress–strain curves obtained from the uniaxial compression tests for both types of samples. Obviously, the stirred case exhibits a much higher plastic deformation (16.9%) and compressive strength (1137 MPa) than the reference sample (7.1%, 648 MPa). But there is no considerable change in the yield strength. The improvement of the plastic deformation can be explained by the microstructural differences between the two types of samples [25]. For an improvement of the plastic deformation, a homogeneous isotropic microstructure without dendritic morphology is preferred [26]. The phase fraction of the spherical $(\alpha_2 + \gamma)$ colonies in the stirred sample is higher than the dendritic colonies in the reference sample. On the contrary, the interlamellar spacing is lower, which implies the existence of more γ/α_2 interfaces. The larger number of γ/α_2 interfaces in the stirred sample may provide an increase in the dislocation sources in order to accommodate the applied strain. Additionally, the α_2 phase in the γ/α_2 lamellae has a favorable effect on the ductility, as the semicoherent interface between γ/α_2 lamellae gives rise to an easy type of deformation.

15.5
Conclusion

The modification of melt convection and its influence on the microstructure evolution of peritectic Nd–Fe–B and Ti–Al alloys were investigated using novel techniques developed and designed on the basis of numerical simulation. Owing to the high technological relevance of these materials, the influence on the magnetic and mechanical properties is of great scientific interest.

By applying the forced rotation technique on Nd–Fe–B alloys it was found that the microstructure pattern, mainly the volume fraction of the α-Fe phase, varied strongly with the strength of the internal melt motion. From numerical simulations of the fluid flow, it could be elucidated that a global melt rotation suppresses the internal melt motion significantly. Experiments with forced rotation, corresponding to a reduced internal melt motion, resulted in a strong reduction of the soft magnetic α-Fe phase. The dendritic morphology was subjected to statistical analysis with respect to the SDAS. A strong reduction of the SDAS with reduced melt convection was found. The floating zone experiments with additional magnetic field caused by the double coil system in series connection yielded a further reduction of the α-Fe volume fraction up to 30% compared to common induction melting. The reduction of the properitectic phase fraction and the SDAS under reduced melt convection can be assigned to the reduced convective mass transfer, whereas the strong interdendritic fluid flow minimizes the thickness of the diffusion boundary layer around the properitectic phase facilitating the mass transport.

The stirring experiments on Ti–Al alloys with the double coil system as a two-phase stirrer showed a morphology change of the properitectic phase from dendritic to globular under enhanced melt convection. The volume fraction of the properitectic phase increases with increasing fluid flow. The stirring effect decreases the intermediate layer between the matrix and properitectic phase and promotes coarsening of particles and globular growth. The stirred samples show a significant improvement of the plastic deformability compared to conventionally melted samples. The reason is attributed to the isotropic spherical morphology of the lamellar α_2/γ phase, whereas the anisotropic orientation of the lamellar/dendritic phase displays undesirable plastic properties.

The study of the influence of melt convection on the microstructure formation of peritectic alloys shows the feasibility of tailoring the microstructure and the resulting alloy properties by customized melt convection using magnetic fields.

Acknowledgments

This work has been supported by Deutsche Forschungsgemeinschaft under Grant No. HE 2955/2-2,3 and in frame of the Collaborative Research Center SFB 609.

References

1 Kaneko, Y. (**2000**). *Rare-Earth Magnets with high energy products, Proceedings of the 16th International Workshop on Rare-Earth Magnets and their Applications*, Sendai, Japan, 10–14, September, pp. 83–97.

2 Sagawa, M., Fujimura, S., Togawa, N., Yamamoto, H. and Matsuura, Y. (**1984**) "New material for permanent magnets on a base of Nd and Fe". *Journal of Applied Physics*, **55**, 2083.

3 Croat, J.J., Herbst, J.F., Lee, R.W. and Pinkerton, F.E. (**1984**) "Pr-Fe and Nd-Fe based materials-a new class of high performance permanent magnets". *Journal of Applied Physics*, **55**, 2078.

4 Appel, F. and Wagner, R. (**1998**) "Microstructure and deformation of two-phase γ-titanium aluminides". *Materials Science and Engineering*, **R22**, 187–268.

5 Huang, S.C. and Hall, E. (**1991**) "Plastic deformation and fracture of binary TiAl-base alloys". *Metallurgical Transactions*, **22A**, 427–39.

6 Kim, Y.W. (**1995**) "Gamma titanium aluminides: their status and future". *Journal of Metals*, **47** (7), 39–41.

7 Boettinger, W.J., Coriell, S.R., Greer, A.L., Karma, A., Kurz, W., Rappaz, M. and Trivedi, R. (**2000**) "Solidification microstructures: recent development, future directions". *Acta Materialia*, **48**, 43–70.

8 Umeda, T., Okane, T. and Kurz, W. (**1996**) "Phase selection during solidification of peritectic alloys". *Acta Materialia*, **44**, 4209–16.

9 St John, D.H. and Hogan, L.M. (**1987**) "Simple prediction of the rate of the peritectic transformation". *Acta Metallurgica*, **35**, 171–74.

10 Priede, J., Gerbeth, G., Hermann, R. and Filip, O., (**2006**) Verfahren und Vorrichtung zur schmelzmetallurgischen Herstellung von Magnetlegierungen auf Nd-Fe-B-Basis. DE 103 31 152 B4..

11 Vatistas, G.H. (**2006**) "A note on liquid vortex sloshing and Kelvin's equilibria". *Journal of Fluid Mechanics*, **217**, 241–48.

12 Gelfgat, A.Y., Bar-Yoseph, P.Z. and Solan, A. (**2001**) "Three-dimensional instability of axisymmetric flow in a rotating lid-cylinder enclosure". *Journal of Fluid Mechanics*, **438**, 363–77.

13 Grants, I. and Gerbeth, G. (**2002**) "Linear three-dimensional instability of a magnetically driven rotating flow". *Journal of Fluid Mechanics*, **463**, 229–40.

14 Shatrov, V., Gerbeth, G. and Hermann, R. (**2008**) "Linear stability of an alternating magnetic field driven flow in a spinning cylindrical container". *Physical Review E*, 77, 046307.

15 Priede, J., Gerbeth, G., Hermann, R., Filip, O. and Schultz, L. (**2005**). Verfahren und Vorrichtung zur schmelzmetallurgischen Herstellung von Magnetlegierungen auf Nd-Fe-B-Basis. Offenlegungsschrift DE 103 28 618 A1.

16 Priede, J., Gerbeth, G., Hermann, R., Filip, O. and Behr, G. (**2003**) "Two-phase induction melting with tailored flow control". Proceedings of EPM2003, Lyon, pp. 344–9.

17 Priede, J.. Gerbeth, G., Hermann, R. and Behr, G. (**2005**) "Float-Zone Crystal Growth with a Novel Melt Flow Control". Proceedings of Joint 15th Riga and 6th PAMIR International Conference on Fundamental and Applied MHD, Jurmala, Latvia, pp. 189–92.

18 Plevachuk, Yu., Sklyarchuk, V., Hermann, R. and Gerbeth, G. (**2008**) "Thermophysical properties of Nd-, Er-, Y-Ni-alloys". *International Journal of Materials Research*, **99**, 261–64.

19 Hermann, R., Gerbeth, G., Filip, O., Priede, J. and Shatrov, V. (**2004**) Effect of hydrodynamics on microstructure evolution of Nd-Fe-B alloys, in *Solidification and Crystallization*, (ed D.M. Herlach), WILEY-VCH Verlag GmbH, pp. 185–193, ISBN 3-527-31011-8.

20 Filip, O. and Hermann, R. (**2006**) "Phase and microstructure evolution during solidification of Nd-Fe-B melts processed by novel techniques". *Journal of Optoelectronics and Advanced Materials*, **8** (2), 504–10.

21 Biswas, K., Hermann, R., Filip, O., Acker, J., Gerbeth, G. and Priede, J. (**2006**) "Influence of melt convection on microstructure evolution of Nd-Fe-B alloys using a forced crucible rotation technique". *Physical Status Solidi A*, **9**, 3277–80.

22 Hermann, R., Filip, O., Gerbeth, G., Priede, J. and Schultz, L. (**2004**) "Microstructure evolution of Nd-Fe-B

alloys in consideration of magnetohydrodynamics". *Journal of Magnetism and Magnetic Materials*, **272–276** (Suppl 1), E1855–56.

23 Filip, O., Hermann, R., Gerbeth, G., Priede, J. and Biswas, K. **(2005)** "Controlling melt convection—an innovation potential for concerted microstructure evolution of Nd-Fe-B alloys". *Materials Science and Engineering A*, **413–414**, 302–5.

24 Flemings, M.C. **(1984)** *Solidification Processing*, McGraw-Hill, New York.

25 Biswas, K., Hermann, R., Das, J., Priede, J. and Gerbeth, G. **(2006)** "Tailoring the microstructure and mechanical properties of Ti-Al alloy using a novel electromagnetic stirring method". *Scripta Materialia*, **55**, 1143–46.

26 Ramanujan, R.V. **(2000)** "Phase transformations in γ-based titanium aluminides". *International Materials Reviews*, **45**, 217–40.

16
Nanosized Magnetization Density Profiles in Hard Magnetic Nd–Fe–Co–Al Glasses

Olivier Perroud, Albrecht Wiedenmann, Mihai Stoica, and Jürgen Eckert

16.1
Introduction

The absence of defects in amorphous alloys usually gives rise to soft magnetic properties. However, many amorphous alloys composed of rare earth and transition metal do not belong to this category. In Nd–Fe alloys for example a competition between the atomic anisotropy energy and the exchange energy is expected to yield local orientations of the Nd atomic moments along the Fe moments [1, 2]. This local anisotropy induces the creation of magnetic randomly orientated domains, which can involve a significant coercivity in Nd–Fe amorphous binary alloys at room temperature [3]. It is now admitted that the hard magnetic properties of binary Nd–Fe bulk alloys result from the so-called "A1 phase". This metastable phase with a Curie temperature T_c of the order of 520 K was characterized by X-ray diffraction (XRD) as a mixture of an amorphous Fe-rich part and some crystalline parts containing Nd but with unknown microstructure. Multicomponent $Nd_{60}Fe_xCo_{30-x}Al_{10}$ alloys containing more than 5 at% Fe present hard magnetic behavior at room temperature and show high thermal stability [4]. The temperature dependence of the magnetization plotted in Figure 16.1 shows two transition temperatures: $T_{c1} \sim 30–50\,K$ and $T_{c2} \sim 450–500\,K$, whereas in NdCoAl only one transition temperature $T_{c1} \sim 40K$ occurs below which the alloy is soft magnetic. Previous XRD and High resolution transmission electron microscopy (HRTEM) studies revealed very similar microstructures for the Fe containing $Nd_{60}Fe_xCo_{30-x}Al_{10}$ alloys consisting of a Nd-rich matrix and a Fe-rich partially amorphous phase (analogous to the A1 phase reported in binary Nd–Fe alloys. The partially amorphous Fe-rich phase was found to be metastable and nonhomogeneous. The Fe content of regions that are globular and micrometer in size [5, 6] is much higher in the middle of the sphere than at the surface [7]. The presence of Nd nanoparticles in various bulk samples was discovered by Small angle neutron scattering (SANS) [8, 9] and HRTEM [10, 11]. The similarity of the Curie temperatures T_{c1} and T_{c2} in samples of different Fe content suggest that the mean composition of the different phases is almost the same for this alloys [6, 12] T_{c1} has been associated with the ferromagnetic

Phase Transformations in Multicomponent Melts. Edited by D. M. Herlach
Copyright © 2008 WILEY-VCH Verlag GmbH & Co. KGaA, Weinheim
ISBN: 978-3-527-31994-7

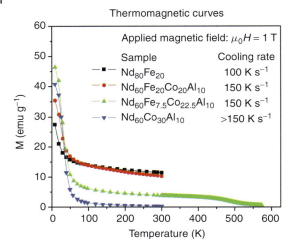

Fig. 16.1 Temperature dependence of the magnetization for different compositions obtained by mold casting (Kumar, G. private communication). The first magnetic transition ($T_{c1} \sim$ 50K) is almost the same for the different samples.

ordering of the Nd-rich phase whereas T_{c2} was assigned to the Curie temperature of the Fe-rich phase. The magnetic properties depend strongly on the preparation conditions: in fast quenched melt-spun ribbons the coercivity is much smaller than in the bulk samples obtained by copper mold casting, probably due to the completely amorphous structure of the alloys [13, 14] and/or to the absence of the metastable Fe-rich phase in the ribbons. In this context we have investigated magnetic properties and microstructures of bulk rod-like samples of $Nd_{60}Fe_{20}Co_{10}Al_{10}$. These are 3, 7, and 11 mm in diameter with the cooling rates achieved in the order of 150, 75 and 40 K s^{-1}, respectively, during mold casting. As expected, the influence of the cooling rate on the magnetization is very strong in this samples.

At room temperature and for samples with a diameter of 3 mm, the samples show a single hysteresis loop (Figure 16.2). Below T_{c1}, a second step at $\mu_0 H = 5 - 6T$ occurs, which indicates the presence of a hard magnetic phase and an additional soft magnetic phase [12]. The random anisotropic model alone cannot explain the high coercivity in these alloys where the maximum energy product $(BH)_{max}$ achieved is similar with that of the hexagonal ferrite or Alnico. The evolution of the coercivity H_c with the temperature [6, 12] suggests a strong pinning of domain walls similar to the behavior expected for random orientated objects [15]. The presence of nanoparticles combined with the metastable Fe-rich phase is supposed to be at the origin of the hard magnetic properties of such alloys. The aim of the present SANS investigation is to establish the local magnetization profiles in the bulk as a function of temperatures and magnetic fields on a nanometer length scale and to correlate them with the macroscopic magnetic properties.

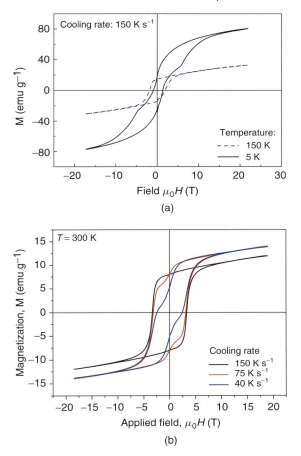

Fig. 16.2 Hysteresis loops of the sample $Nd_{60}Fe_{20}Co_{10}Al_{10}$ (Stoica, M. private communication), (a) cooling rate 150 K s^{-1}, measured at different temperatures, (b) comparison for different diameters at room temperature. The shape of the loop indicate the presence of two magnetic phases at T = 5K. A soft magnetic phase is also observed at room temperature for the samples having a lower cooling rate.

16.2
SANS with polarized neutrons in unsaturated magnetic systems

The total scattering intensity $I(Q)$ obtained by SANS is supposed to be the sum of the intensity originating from independent units. Based on SEM [7] and TEM investigations we assume here that the intensity arising from the nuclear scattering will be the sum of the scattering from nanoparticles and from a Fe-rich phase embedded in an homogeneous Nd-rich matrix. For each element j, the intensity can be written as: [16]

$$I_j(Q) = \int N_{pj} \left(\Delta \eta_j V_{pj} f_j (QR)\right)^2 N_j(R) S(Q) \, dR \tag{16.1}$$

where $N(R)$ is the size distribution of the N_{pj} particles of volume V_p and of shape defined by the factor (QR). The structure factor $S(Q)$ determines the inter-particle interaction. For the presented case of a dilute system $S(Q) = 1$ and the intensity at small angles is determined mostly by the shape of the particles themselves and by the scattering contrast $\Delta\eta$ between particles and matrix. In the case of magnetic samples, the scattering will depend on the angle ψ between the magnetic moment and the scattering vector Q. In fully random systems, the probability to find the magnetic moment in a given direction is equal for all directions. Following the definitions given in [16] the small-angle scattering with polarized neutrons (SANSPOL) intensity for polarized neutrons can be written as follows:

$$I^+(Q) = I^-(Q) = F_N^2 + \frac{2}{3} F_M^2 \tag{16.2}$$

where F_N and F_M are the nuclear and magnetic form factors, respectively. In a saturated system, when all magnetic moments are aligned along the applied magnetic field $\mu_0 H$, the intensity depends on the polarization state (+) or (−) of the incident neutrons as shown below:

$$I^+(Q) = F_N^2 + \left(F_M^2 - 2 P F_N F_M\right) \sin^2 \psi \tag{16.3}$$

$$I^-(Q) = F_N^2 + \left(F_M^2 + 2 P \varepsilon F_N F_M\right) \sin^2 \psi \tag{16.4}$$

where P is the polarization of the incident beam and ε refers to the flipper efficiency.

In the investigated alloys, the distribution of the magnetic moments is not fully random. As a consequence, we distinguish in our model between a random part and an oriented part. The orientated part of the magnetic domains will follow the Equations 16.3 and 16.4 giving rise to an anisotropic scattering contribution. The random part that leads to an isotropic scattering is purely of magnetic origin and results from the different orientations of the local magnetization. Since the orientation distribution of magnetic moments inside the domains is not known, a correlated random distribution is chosen to interpret the isotrop scattering according to $\frac{C_{V\eta} a^3}{(1+q^2 a^2)^2}$, where a corresponds to the correlation length and $C_{V\eta}$ is related to the volume and the magnetic contrast of the domains. This term has been first derived by Debye [17] and applied by Imry and Ma to micromagnetic models of domains [18]. The total scattering intensity has then been analyzed as follows:

$$I^+(Q) = F_N^2 + \frac{C_{V\eta} a^3}{(1+q^2 a^2)^2} + \left\{F_M^2 - 2 P F_N F_M\right\} \sin^2 \psi \tag{16.5}$$

$$I^-(Q) = F_N^2 + \frac{C_{V\eta}a^3}{(1+q^2a^2)^2} + \{F_M^2 + 2P\varepsilon F_N F_M\}\sin^2\psi \qquad (16.6)$$

Magnetization fluctuations extending over longer ranges similar to that predicted by [19] have not be taken into account in order to limit the number of free parameters in the model.

The presence of magnetic domains in rare earth–transition metal alloys was analyzed in terms of random local anisotropy in [20], where the competition between exchange energy and local anisotropy field leads to a local orientation of the spins along local axes, whereas the orientation of this local axes fluctuates in the sample (see also [1]). The size of magnetic domains resulting from the fits (of the order of few nanometers) is in accordance with the size, between 1 and 10 nm, expected from the calculations of [20] for 4f-based alloys.

16.3 Experimental Procedure

16.3.1 Sample Preparation

$Nd_{60}Fe_xCo_{30-x}Al_{10}$ samples were prepared under argon atmosphere to avoid oxidation by melting 99.9% pure Nd, Fe, Co, and Al in an arc-melting furnace. A copper mold was used to obtain the bulk cylinders of 3, 7, and 11 mm diameter by ejecting the melt into the mold. The cooling rates achieved in this procedure were estimated to be about 150, 75, and 40 K s^{-1}, respectively.

16.3.2 SANSPOL Measurements

SANS measurements have been performed at the Hahn-Meitner Institut Berlin using the small-angle scattering instrument V4 with polarized neutrons [16]. For the measurements in the temperature range 5 K to room temperature, cryomagnets with magnetic field in horizontal ($\mu_0 H = 1-6$ T) and vertical direction ($\mu_0 H = 10$ T) have been used. The measurements at temperature between 300 and 600 K have been performed in an electromagnet with a horizontal field applied perpendicular to the incident neutron beam. Scattering intensities have been corrected in the usual way using BerSANS software [21]. Two dimensional data have been averaged in sectors of 35° in width in the direction perpendicular to the applied magnetic field ($\psi = 90°$ in formulas 16.5 and 16.6). SANSPOL intensities $I(+)$ and $I(-)$ have been fitted simultaneously according to Equations 16.5 and 16.6. A unique set of parameters describing the chemical composition has been used at all measured temperatures and magnetic fields, that is size, size distribution, volume fractions, and nuclear scattering contrasts of the particles have been assumed to be constant for a given sample. In the following graphs, the solid lines correspond always to the fit of the curves. For reason of clarity, error bars are not always shown in the plots but in most of the cases, the error bars are not larger than the points.

16.4
Results and Discussion

Following the SEM and TEM results on similar alloys, the model considers a Nd-rich matrix, a Fe-rich region, and nanosized Nd particles. A scheme of the model is given in real space Figure 16.3. The size of the Fe-rich region is of the order of micrometers and will have some influence on the scattering only at very low Q. This contribution therefore is approximated by a simple Porod law $I = PQ^{-4}$, where P is related to the scattering contrast and to the surface. Particles of spherical shape and log-normal distribution are supposed to consist of pure Nd (radius between 1 and 10 nm, see Table 16.1). In $Nd_{60}Fe_{20}Co_{10}Al_{10}$ samples two types of nanosized particles with the same scattering contrast had to be assumed. The exact composition of these nanosized crystals is still under discussion. XRD measurements revealed peaks of pure Nd [6] or Nd_3Al [7], whereas a composition of $Nd_{69}(Fe,Co,Al)_{31}$ was found as in Xia et al. [11]. The composition of the Fe-rich metastable crystalline phase is also difficult to determine. To suit our data, we chose $Nd_{36}Fe_{32}Co_{28}Al_4$ found in [6] on the sample $Nd_{60}Fe_5Co_{25}Al_{10}$. This model is consistent with the results of previous investigations on similar samples [6, 7, 10, 11]. The fitted radii of the Nd nanoparticles, given in table 16.1 are very similar for the different samples.

16.4.1
Microstructure

Changes in the nuclear structures were supposed to be negligible for temperatures below 650 K. In order to justify this hypothesis, the thermal stability was measured at a high temperature on $Nd_{60}Fe_{20}Co_{10}Al_{10}$ (cooling rate 150 K s^{-1}). Figure 16.3b shows the volume fraction of the nanosized particles at a given temperature normalized to the value obtained at 250 K. The results were obtained from fits at different temperatures according to Equations 16.5 and 16.6, where the amount of nanoparticles was the only free fit parameter, using the same values for densities and nuclear form factors at all temperatures. Figure 16.3c shows the disappearance or growth of the nanoparticles until 670 K. Between $T = 550$ and 300 K no difference was observed (not shown here). This confirms that the microstructure remains stable until 670 K. The change in the scattering intensity confirms the DSC measurements (Figure 16.3c [7]), where the crystallization was found to start between T = 570 and 670 K.

16.4.2
Magnetic Behavior

The reconstructed sum $(I^+ + I^-)/2$ corresponding to scattering from nonpolarized neutrons is almost isotropic at all temperatures, which shows that the magnetic

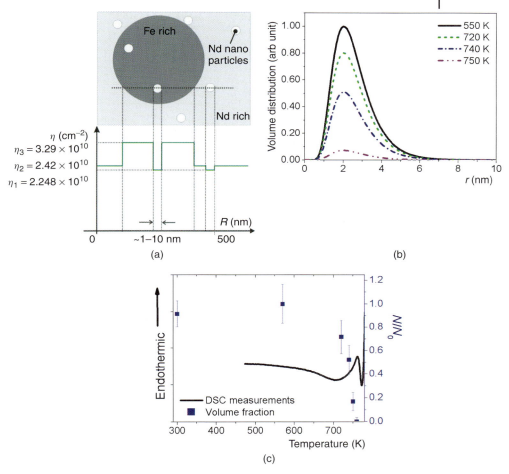

Fig. 16.3 (a) Scheme of the model in real space for all temperatures for the $Nd_{60}Fe_x Co_{30-x}Al_{10}$ alloys. The model is composed of a Fe-rich region (dark gray) and nano-sized particles (light gray) embedded in an amorphous matrix. (b) size distribution and (c) ratio of the number of particles N/N_0 (bottom), where N_0 is the number of particles at 520 K (squares) and differential scanning calorimetry (DSC) curve (continuous line). The nanoparticles disappear above 670 K, when the first exothermic reaction appears.

scattering contribution is very low compared to the nuclear one. The azimuthally averaged intensities of the sum are compared in Figure 16.4 for different temperatures and magnetic fields. At a given value of the magnetic field the intensities increase with decreasing temperatures with a pronounced step around T_{c1} that must result from an increase in the magnetic contribution. This behavior is in line with the results from the thermomagnetic curves (Figure 16.1). On the other hand, the effect of the strength of the magnetic field is even less pronounced: at a given temperature, the overall intensities decrease when the magnetic field increases

Table 16.1 Radius of the Nd nanoparticles for the different samples. The volume fraction (V) was calculated with the assumption that the scattering contrast $\Delta\eta$ is equal to: $\Delta\eta = \eta_{Fe-rich} - \eta_{Nd} = 3.89 \times 10^{10} - 2.25 \times 10^{10} = 1.64 \times 10^{10} \text{cm}^{-2}$

Sample	R_1	V_1 (%)	R_2	V_2 (%)
$Nd_{60}Fe_{20}Co_{10}Al_{10}$ at 40 K s^{-1}	6.9 nm ± 0.4	0.013	1.9 nm ± 0.2	0.039
$Nd_{60}Fe_{20}Co_{10}Al_{10}$ at 75 K s^{-1}	7.3 nm ± 0.2	0.016	2.2 nm ± 0.1	0.049
$Nd_{60}Fe_{20}Co_{10}Al_{10}$ at 150 K s^{-1}	6.7 nm ± 0.5	0.003	1.7 nm ± 0.1	0.015
$Nd_{60}Fe_{7.5}Co_{22.5}Al_{10}$ at 150 K s^{-1}	—	—	2.9 nm ± 0.1	0.013

from 1 to 6 T. This implies a decrease in the magnetic scattering contribution that is of the magnetic contrast. Such a decreasing scattering contrast could result from an alignment of magnetic domains, the occurrence of new magnetic phases, and/or to a decrease in the magnitude of the moment. In fact, this behavior suggests that the magnetic microstructure is governed by a nearly random orientation of magnetic domains. Following this concept, the curves have been fitted applying a model containing magnetic domains (corresponding to the Equations 16.5 and 16.6). The average size of the magnetic domains was found to be of the same order as the diameter of the Nd-nanoparticles (Figure 16.5). In that case, nanoparticles will interact with numerous almost randomly orientated magnetic domains, which will lead to a small anisotropy of the 2D scattering pattern (or small difference between I^+ and I^-). This behavior is effectively observed by the 2D images of the scattering intensity. We note here a similar behavior for the other samples (see the sample $Nd_{60}Fe_{20}Co_{10}Al_{10}$ with a cooling rate of 150 K s^{-1} in Figure 16.4b). The size of the magnetic domains was found to be independent of the strength of the external magnetic field (Figure 16.5) and no significant change was observed at the transition temperature T_{c1}.

Using polarized neutrons, the magnetic contribution is more reliably obtained from the difference $I^- - I^+$. The azimuthally averaged intensities (H perpendicular to Q), plotted in Figure 16.6 illustrate how the scattering behavior changes as a function of temperature and magnetic field. At 250 K, the difference $I^- - I^+$ is much larger at $\mu_0 H = 6T$ than at 1 T, indicating an alignment or rotation of the magnetic moments in the domains along the external magnetic field. Below T_{c1}, the nanoparticles become ferromagnetic and at the same time the interference term of the intensity decreases rapidly, despite an increase in the total magnetization (see the thermomagnetic curve Figure 16.1) and of the intensity of the unpolarized scattering curves. Since the nuclear scattering remains constant for all curves, the interference term $I^- - I^+$ directly yields information on the magnetic scattering contrast. The difference between I^+ and I^- in Figure 16.6 l.h.s. indicates clearly that the magnetization of the sample should be higher at $\mu_0 H = 6T$ than at $\mu_0 H = 1T$, despite decreasing the magnetic contrast by increasing the applied magnetic field (see unpolarized curves Figure 16.4).

16.4 Results and Discussion | 271

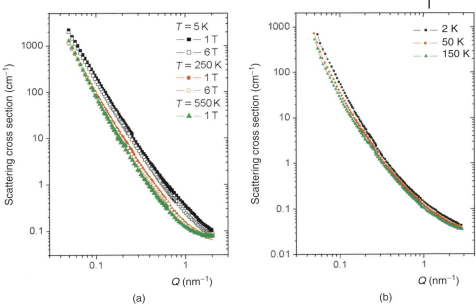

Fig. 16.4 (a) Scattering intensity $Nd_{60}Fe_{7.5}Co_{22.5}Al_{10}$, (b) $Nd_{60}Fe_{20}Co_{10}Al_{10}$ at $\mu_0 H = 6T$.

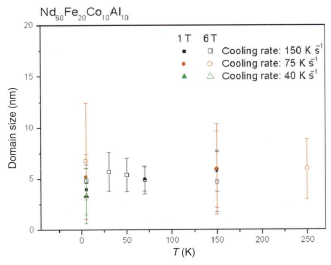

Fig. 16.5 Summary of the domain size fitted from the scattering curves of the sample with composition $Nd_{60}Fe_{20}Co_{10}Al_{10}$. The size of the magnetic domains are not affected by the temperature and the applied magnetic field till $\mu_0 H = 6T$.

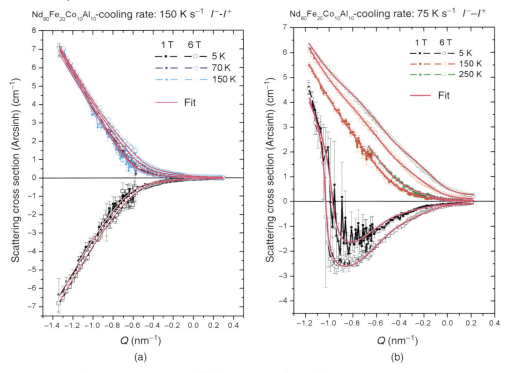

Fig. 16.6 Temperature and field ($\mu_0 H = 1$ and 6 T) behavior of the interference term of (a) $Nd_{60}Fe_{7.5}Co_{22.5}Al_{10}$ cooling rate 150 K s^{-1} and (b) $Nd_{60}Fe_{20}Co_{10}Al_{10}$ cooling rate 75 K s^{-1}.

In previous studies of the sample $Nd_{60}Fe_{20}Co_{10}Al_{10}$, we observed that the sign of the interference term below T_{c1} was negative [8, 9] whereas above T_{c1} it remained positive. The present measurements on the sample $Nd_{60}Fe_{20}Co_{10}Al_{10}$ prepared at a slower cooling rate (Figure 16.6) confirm this trend at large Q. However, even at 5 K, the interference term turns to a positive sign at Q values between 0.1 and 0.3 nm^{-1}.

According to the Equation 16.3 and with the assumption of a constant nuclear scattering, the change of the sign of $I^- - I^+$ is assigned to a change of sign of the magnetic contrast at high Q. No such change in sign occurred above T_{c1}. It results from the presence of nanoparticles in the alloy, which is paramagnetic above T_{c1} and ferromagnetic below T_{c1}. The inversion of the sign of the interference term implies that the magnetic scattering length density (SLD) of the nanoparticles ($\Delta\zeta_1$ in Figure 16.7) below T_{c1} becomes higher than the magnetic SLD of the surrounding amorphous phase and Ferric phase ($\Delta\zeta_2$ and $\Delta\zeta_3$, respectively). The resulting qualitative SLD profiles are shown in Figure 16.7. At low Q, the dominating scattering contribution results from the large Fe-rich phase embedded in the matrix. Depending on the actual values of the SLD, the SANSPOL intensity

16.4 Results and Discussion

difference $I^- - I^+$ at low Q can be positive (Figure 16.7a) or negative (Figure 16.7b) resulting from different cooling rates or compositions.

The scattering contrast between the nanoparticles and the matrix is shown in Figure 16.8. Above T_{c1} the contrast is higher at $\mu_0 H = 6\,\text{T}$ than at 1 T. This enhanced contrast is explained by an increased alignment of the magnetic domains

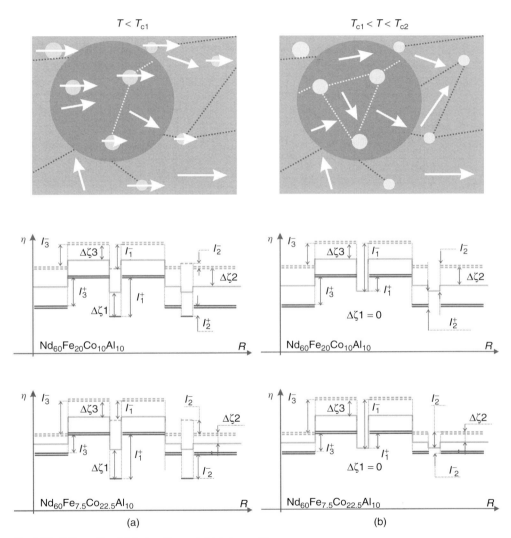

Fig. 16.7 SLD profiles (a) below T_{c1} and (b) between T_{c1} and T_{c2} depending on the compositions or cooling rate (upper and lower part). $\Delta\zeta_{1,2,3}$ are the magnetic SLD of the Nd nanoparticles, amorphous matrix and Fe-rich phase, respectively. $I^+_{1,2,3}$ and $I^-_{1,2,3}$ are the corresponding SANSPOL contributions.

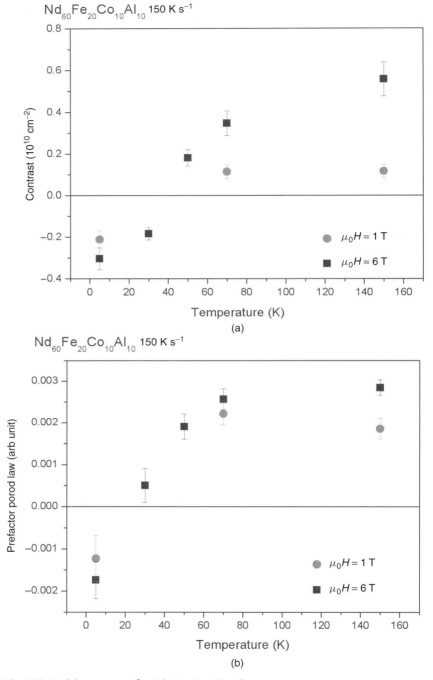

Fig. 16.8 Model parameters for $Nd_{60}Fe_{20}Co_{10}Al_{10}$ alloy prepared at a cooling rate of 150 Ks^{-1} in an external field of 1 T(cercles) and 6 T(squares).

along the applied magnetic field. Below T_{c1} the contrast is lower at $\mu_0 H = 6\,\text{T}$ since both the ferromagnetic nanoparticles and domains are more and more aligned along the magnetic field which leads to an overall smaller vector difference of the magnetization.

For all samples the same structure model has been used and all the differences observed in the scattering curves result from the different values of the magnetic parameters (Figure 16.8). The nuclear model was composed of Nd nanosized spherical particles and the Fe-rich regions on micrometer range was taken in consideration by using a background in q^{-4} shown in figure. The cooling rate does not to affect significantly the radius of the nanoparticles (Table 16.1).

Similarly, the variation of the difference of the prefactors $P^- - P^+$ corresponding to the porod law observed at low Q is represented in Figure 16.8b. Again, for $Nd_{60}Fe_{20}Co_{10}Al_{10}$ at the cooling rate of 150 K s^{-1} the sign reverses at T_{c1} corresponding to a cross over of the local magnetization in the Fe-rich phase and the amorphous matrix.

16.5
Conclusion

The magnetic behavior of the multicomponent alloys have been correlated with the microstructural model as observed by SANSPOL. The common point of the samples is the presence of the nanoparticles in the alloys, which accounts mainly for the magnetic behavior of the multicomponent alloys. Three temperature ranges are distinguished: Above T_{c2} (~500 K), the samples are paramagnetic and the observed scattering is due only to the presence of the chemical inhomogeneities in the sample alone. Between T_{c1} and T_{c2} the samples show pronounced hard magnetic properties, which is ascribed to the formation of magnetic domains. Paramagnetic nanosized particles of Nd act as pinning centers for the movement of magnetic domain walls in the sample. Below T_{c1}, Nd nanoparticles order ferromagnetically and the particle moments orient along the external magnetic field. This gives rise to the observed change in sign of the scattering contrast. The pinning effect of these ferromagnetic nanoparticles embedded in ferromagnetic environment is reduced leading to a more soft magnetic behavior.

References

1 Alben, R., Becker, J. and Chi, M. (1978) "Random anisotropy in amorphous ferromagnets". *Journal of Applied Physics*, **49**, 1653.
2 Coey, J. (1978) "Amorphous magnetic order". *Journal of Applied Physics*, **49**, 1646.
3 Nagayama, K. et al. (1990) "Magnetic properties of amorphous Fe-Nd alloys". *Journal of the Physical Society of Japan*, **59**, 2483.
4 Wei, B. et al. (2002) "Anomalous thermal stability of Nd-Fe-Co-Al bulk metallic glass". *Acta Materialia*, **50**, 4357.
5 Kumar G. et al. (2001) Formation of cluster structure and phase separation in cast Nd-Fe-Co-Al alloys. *22nd Risø International*

Symposium on Materials Science, Risoe (Denmark), 307.
6 Kumar, G., Eckert, J., Roth, S., Müller, K. and Schultz, L. (2003) "Coercivity mechanism in mold-cast $Nd_{60}Fe_xCo_{30-x}Al_{10}$ bulk amorphous alloys". *Journal of Alloys and Compounds*, **309**, 348.
7 Kumar, G., Eckert, J., Roth, S., Löser, W. and Schultz, L. (2003) "Effect of microstructure on the magnetic properties of mold-cast and melt-spun Nd-Fe-Co-Al amorphous alloys". *Acta Materialia*, **51**, 229.
8 García-Matres, E. *et al.* (2004) "Hard magnetic properties of bulk amorphous $Nd_{60}Fe_{20}Co_{10}Al_{10}$ investigated by SANSPOL". *Physica B*, **350**, e315.
9 Perroud, O. *et al.* (2006) "Small angle neutron scattering studies of hard magnetic $Nd_{60}Fe_{30-x}Co_xAl_{10}$ bulk amorphous alloys". *Materials Science Engineering A*, **449-451**, 448.
10 Schneider, S., Bracchi, A., Samwer, K., Seibt, M. and Thiyagarajan, P. (2002) "Microstructure-controlled magnetic properties of the bulk glass-forming alloy $Nd_{60}Fe_{30}Al_{10}$". *Applied Physics Letters*, **80**, 1749.
11 Xia, L. *et al.* (2003) "Primary crystallization and hard magnetic properties of $Nd_{60}Fe_{20}Co_{10}Al_{10}$ metallic glasses". *Journal of Physics D: Applied Physics*, **36**, 2954.
12 Kumar, G. (2004) "Magnetic properties of amorphous Nd-Fe-Co-Al alloys". *Materials Science Engineering A*, **375-377**, 1083.
13 Kumar, G. (2003) "Eect of Al on microstructure and magnetic properties of mould-cast $Nd_{60}Fe_{40-x}Al_x$ alloys". *Scripta Materialia*, **48**, 321.
14 Chiriac, H. and Lupu, N. (1999) *Journal of Magnetism and Magnetic Materials*, 235.
15 Gaunt, P. (1983) "Ferromagnetic domain wall pinning by a random array of inhomogeneities". *Philosophical Magazine B*, **48**, 261.
16 Wiedenmann, A. (2005) "Polarized SANS for probing magnetic nanostructures". *Physics B*, **356**, 246.
17 Debye, P., Anderson, H. and Brumberger, H. (1957) "Scattering by an inhomogeneous solid. II. The correlation function and its application". *Journal of Applied Physics*, **28**, 679.
18 Imry, Y. (1975) "Random-field instability of the ordered state of continuous symmetry". *Physical Review Letters*, **35**, 1399.
19 Weissmüller, J. *et al.* (2001) "Analysis of the small-angle neutron scattering of nanocrystalline ferromagnets using a micromagnetics model". *Physical Review B*, **63**, 214414.
20 O'Handley, R. (1987) "Physics of ferromagnetic amorphous alloys". *Journal of Applied Physics*, **62**, R15.
21 Keiderling, U. (1997) "The new 'BerSANS-PC' software for reduction and treatment of small-angle neutron scattering data". *Applied Physics A*, **74**, 1455.

17
Microstructure and Magnetic Properties of Rapidly Quenched $(Nd_{100-x}Ga_x)_{80}Fe_{20}$ ($x = 0, 5, 10,$ and 15 at%) alloys

Mihai Stoica, Golden Kumar, Mahesh Emmi, Olivier Perroud, Albrecht Wiedenmann, Annett Gebert, Shanker Ram, Ludwig Schultz and Jürgen Eckert

17.1
Introduction

Hard magnetic materials have a long history, and loadstone (Fe_3O_4) was the first permanent magnet known to mankind. The greatest strides in magnet development, however, have occurred in the last hundred years. As early as 1935, a high coercivity of 0.45 T was reported for a Nd-rich Nd–Fe alloy [1]. This remarkable observation, though playing a vital role in the discovery of the present high-energy Nd–Fe–B permanent magnets, was apparently forgotten so that it was never cited in connection with the development of Nd–Fe–B permanent magnets in the 1980s [2]. Sm–Co magnets were the focus of technological efforts before 1980s because of their superior magnetic properties and also owing to the fact that no RE–Fe (RE is rare earth) compounds suitable for permanent magnets were known up to 1978 [3].

For a long time, the only recognized stable intermetallic phase in the binary Nd–Fe system has been rhombohedral Nd_2Fe_{17}, which exhibits easy plane anisotropy and, therefore, is soft magnetic [3]. The absence of stable, hard magnetic compounds in the Nd–Fe binary alloy system opened two different areas of research in the quest for permanent magnet materials in the late 1970s. One approach was to employ nonequilibrium techniques for preparing metastable (amorphous/crystalline) phases in RE-Fe alloys. This led to the formation and identification of several previously unknown phases in the Nd–Fe alloy system such as Nd_5Fe_{17}, $NdFe_7$, $NdFe_{5+x}$, Nd_6Fe_{23}, and $A1$ [4, 5]. Another approach relied on the exploration of a variety of ternary RE–Fe–M (M = B, C, Al, Si, and Ga) alloys, which eventually led to the discovery of the highly anisotropic $Nd_2Fe_{14}B$ compound with tetragonal structure [2, 6]. The magnetic characteristics of technical $Nd_2Fe_{14}B$ magnets depend crucially on the microstructure and, hence, on the preparation method [7–9]. The microstructure involves the size, shape, and orientation of the crystallites of the compound and also the nature and

Phase Transformations in Multicomponent Melts. Edited by D. M. Herlach
Copyright © 2008 WILEY-VCH Verlag GmbH & Co. KGaA, Weinheim
ISBN: 978-3-527-31994-7

distribution of the secondary phases, which usually control domain-wall motion and reverse domain nucleation and, hence, determine the magnetization and demagnetization behavior [10, 11]. The microstructure of Nd–Fe–B magnets consists of grains of the ferromagnetic $Nd_2Fe_{14}B$ phase, surrounded by Nd-rich and B-rich intergranular phases [10]. Other phases, such as Fe-rich phases or Nd oxides and pores are found depending on the composition and processing parameters [12]. The composition and crystal structure of the Nd-rich phase observed at the grain boundaries in Nd–Fe–B magnets are not clearly established, particularly because of its high chemical reactivity [13]. Another unsolved question in the understanding of Nd–Fe–B magnets is the role of post-sintering annealing at 873 K and intergranular regions in controlling the coercivity [14]. Many investigations were carried out to clarify the microstructure of the intergranular phases in sintered Nd–Fe–B magnets [15]. These investigations revealed a number of metastable Nd–Fe binary and oxygen-stabilized Nd–Fe–O ternary phases in the Nd–Fe–B magnets [15].

The observation of metastable Nd–Fe binary phases in Nd–Fe–B magnets stimulated a careful reanalysis of Nd–Fe binary alloys in order to understand the phase formation [16]. The role of other additives such as B, C, Al, and O for stabilizing metastable Nd–Fe phases has also been intensively studied [17–19]. Despite these extensive studies of the binary Nd–Fe alloy system, the overall consistent information available is limited. Different authors have reported different results for the same composition. One of the reasons for this discrepancy is that the phase selection in the binary Nd–Fe system depends sensitively on the cooling rate, the purity of the elements, and the starting composition. In most of the studies, arc melting and copper mold casting were used as sample preparation techniques, which impose different cooling rates depending on the sample geometry and apparatus in use. Therefore, in spite of the same starting composition, different studies show different results due to a variation in the cooling conditions.

It is known that a metastable hard magnetic phase forms for compositions close to the Nd–Fe eutectic [5]. Therefore, it is essential to understand the correlation between the cooling rate and microstructure for quantification of the magnetic properties of Nd–Fe alloys. For this purpose the $Nd_{80}Fe_{20}$ alloys prepared at different cooling rates in the range of $25–150\,K\,s^{-1}$ were systematically analyzed. These alloys contained Nd and the most disputable metastable hard magnetic *A1* "phase", which justified the selection of this particular composition. Further, the addition of Ga was investigated. Ga has a positive heat of mixing with Nd and Fe [20] and helps to refine the big Nd grains. By adding Ga the formation of the hard magnetic *A1* phase may be promoted and the overall magnetic properties may become better.

The purpose of this study is to analyze the influence of the cooling rate and small compositional variation on the structure of metastable Nd–Ga–Fe alloys. Additionally, the magnetic properties and the changes induced by variation of the cooling rate and composition will be discussed in the light of the structural modification.

17.2
Sample Preparation and Experimental Investigations

The present work deals with rapidly solidified Nd–Ga–Fe alloys. High-purity elements (Nd (99.9 wt%), Fe (99.99 wt%), and Ga (99.9 wt%)) were used as starting materials. The pre-alloyed ingots were prepared from small lumps of each chemical constituent. The lumps were mechanically cleaned to remove the surface oxide layer. The elements were melted in an arc-melting furnace under an argon (99.999 wt%) pressure of 6×10^4 Pa. The ingots were crushed and remelted 3 times to ensure homogeneity. The mass of these a pre-alloyed ingots was in the range of 20–30 g. The pre-alloyed ingots were used as precursor materials for further sample preparation. In order to obtain a variety of stable and metastable phases, nonequilibrium cooling from the melt was considered essential. Techniques such as melt spinning and centrifugal copper mold casting were utilized to employ a wide range of cooling rates. High cooling rates of typically 10^6 K s^{-1} were achieved in melt spinning experiments [21]. The gradient in the cooling rate along the thickness of the ribbons was not considered in the present study. The high cooling rates employed in melt spinning limit the sample size to thin metal films or sheets of about 50 mm thickness. These specimens are difficult to be used for a large number of experiments, which require bulk samples. Bulk nanocrystalline alloys are usually prepared by copper mold casting. There, the cooling rate primarily depends on the geometry of the mold. Cooling rates can be measured both directly and indirectly [22]. Direct measurements involved temperature measurements using thermoelectric or pyrometric methods. However, uncertainties are implicit in the calculation of cooling rates because of the poor accuracy of the temperature measurement and its local fluctuations. Indirect measurements of the cooling rate are usually performed by determining its value from various microstructural features such as the dendrite arm spacing [23, 24]. In these methods, the microstructure–cooling rate correlations at low cooling rate are assumed to remain valid at higher cooling rates. Despite the accuracy criteria, these methods are widely used in measuring the local average cooling rate [22–24]. Srivastava et al. calculated maximum cooling rates of 225 and 85 K s^{-1} for $Al_{67}Cu_{33}$ mold-cast rods of 3 and 5 mm diameter, respectively [25]. However, the cooling rate is not homogeneous and varies along the length and the diameter of the specimen. According to our expertise and to keep things as simple as possible, only cylindrical rod-shaped samples were produced of various diameters but with the same length of 5 cm. Further, we assume that the rods with 3 mm diameter were cooled at 150 K s^{-1}, the 5 mm diameter rods at 100 K s^{-1}, and the 7 mm diameter rods at 50 K s^{-1}. The cooling rate of 25 K s^{-1} corresponds to the master alloy left to solidify in contact with the water-cooled bottom copper plate of the arc melter. For characterization, the specimens were cut from the bottom 5 mm region to exclude the effect of a gradient in the cooling rate along the length of the specimens. The phases were identified by X-ray diffraction (XRD) using Co Kα ($\lambda = 0.17889$ nm) radiation in the case of Nd–Fe binary alloys and Mo Kα ($\lambda = 0.07093$ nm) in the case of Nd–Ga–Fe ternary alloys. Metallographic investigations were carried out using a Zeiss polaroid optical microscope, a JEOL 6400 scanning electron microscope (SEM) equipped with an energy-dispersive X-ray (EDX) spectrometer, and a JEOL 2000FX transmission electron microscope (TEM). The magnetic properties were measured by using a vibrating sample magnetometer (VSM) with a maximum applied field of 2 T.

17.3
Binary $Nd_{80}Fe_{20}$ Rapidly Quenched Alloys

17.3.1
Structure and Cooling Rate

To study the effect of the cooling rate, $Nd_{80}Fe_{20}$ alloys were prepared at different cooling rates of ~150, 100, 50, and 25 K s^{-1} and analyzed by XRD (Figure 17.1). The samples cooled at ≥ 50 K s^{-1} show diffraction peaks of dhcp Nd- and fcc Nd-rich phases. On the other hand, the samples cooled at ~25 K s^{-1} exhibit distinct Nd_2Fe_{17} peaks in addition to the peaks of dhcp Nd and fcc Nd-rich phases. No sample shows peaks of the hexagonal Nd_5Fe_{17} phase in the as-cast state. These results define the range of cooling rates for the formation of the Nd_2Fe_{17} phase in hypereutectic $Nd_{80}Fe_{20}$ alloys. The XRD patterns for the $Nd_{80}Fe_{20}$ mold-cast alloys prepared at different cooling rates can be summarized as follows:

1. dhcp Nd, rhombohedral Nd_2Fe_{17}, and fcc Nd-rich peaks if the cooling rate is ≤ 25 K s^{-1};
2. dhcp Nd and fcc Nd-rich peaks with no indications for the formation of Nd–Fe intermetallic phases for cooling rates ≥ 50 K s^{-1}.

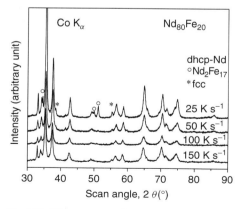

Fig. 17.1 XRD patterns of $Nd_{80}Fe_{20}$ alloys cooled at ~150, 100, 50, and 25 K s^{-1}. The samples cooled at a rate $\geq$ 50 K s^{-1} show the peaks of dhcp Nd and an fcc phase, whereas the specimens cooled at 25 K s^{-1} show additional peaks of Nd_2Fe_{17} phase.

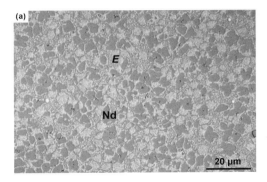

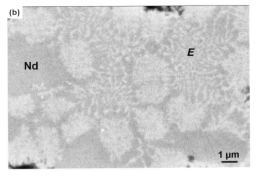

Fig. 17.2 Microstructure of $Nd_{80}Fe_{20}$ cooled at $\sim$150 K s^{-1} (mold-cast rod with 3 mm diameter) observed by (a) optical microscopy and (b) SEM in back-scattering mode. The micrographs show a fibrous eutectic (E) of Nd + Al along with primary Nd grains.

Figure 17.2a shows an optical micrograph of the cross section of a $Nd_{80}Fe_{20}$ sample cooled at $\sim$150 K s^{-1} (mold-cast rod with 3 mm diameter). Two notable features observed from the micrograph are the gray equiaxed Nd grains and the eutectic (E). The eutectic morphology can be clearly discerned in the SEM back-scattered micrograph shown in Figure 17.2b. The eutectic lamellae are irregular and discontinuous along the cross section of the specimen. The size of the eutectic lamellae and, hence, the interlamellar spacing vary from one region to another. This inhomogeneous microstructure suggests a nonuniform heat flow and different local cooling rates. According to eutectic growth theory, the interlamellar spacing is constant for a given growth rate [26]. However, if the growth rate increases, one would expect the interlamellar spacing to decrease by the formation of new lamellae. The EDX analysis identifies the primary grains as Nd crystallites. The volume fraction of primary Nd crystallites is estimated to be about 57%. This is much higher than the expected value of about 12 vol% from the Nd–Fe phase diagram [16]. This discrepancy is due to the fact that some of the eutectic Nd has submerged with the primary Nd crystallites. The size of the eutectic lamellae is smaller than 1 μm and, therefore, their composition could not be determined with high enough accuracy.

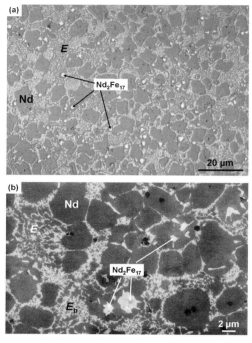

Fig. 17.3 Microstructure of $Nd_{80}Fe_{20}$ cooled at $\sim100\,K\,s^{-1}$ (mold-cast rod with 5 mm diameter) observed by (a) optical microscopy and (b) SEM (back-scattering mode). The micrographs show a fibrous eutectic of Nd + A1 (E, E_a, and E_b) and primary Nd grains. Additional Nd_2Fe_{17} grains are also observed.

A line scan along the eutectic region shows that the Fe content is higher in the bright region than in the gray one (Figure 17.2b). This type of eutectic structure observed in the Nd–Fe alloys is referred to as Nd + A1 [16].

Figure 17.3a shows an optical micrograph taken from the cross section of a $Nd_{80}Fe_{20}$ specimen cooled at $\sim100\,K\,s^{-1}$ (mold-cast rod with 5 mm diameter). The microstructure exhibits primary Nd crystallites coexisting with a eutectic (marked E). In this case the primary Nd crystallites are coarser (>5 μm) than in the case of the 3 mm diameter rod (<5 μm) but there is no remarkable change in the size of the eutectic lamellae. In addition to the primary Nd crystallites, Nd_2Fe_{17} grains with noneutectic morphology are observed, which could not be resolved in the corresponding XRD patterns due to their small volume fraction ($\sim$2 vol%). The formation of these Nd_2Fe_{17} grains is due to the partial transformation of the metastable *A1* structure into Nd_2Fe_{17} during casting. Figure 17.3b shows a high-magnification back-scattered SEM image of the eutectic region. The eutectic lamellae are inhomogeneous and their size varies from one region to the other. In the eutectic region E_a, the eutectic lamellae are coarser than in they are region E_b. The dark lamellae contain more Nd than the bright ones. This kind of diffuse

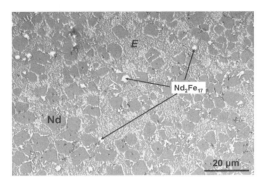

Fig. 17.4 Microstructure of $Nd_{80}Fe_{20}$ cooled at $\sim$50 K s^{-1} (mold-cast rod with 7 mm diameter). The cooled sample shows a fibrous eutectic of Nd + A1 (E) and primary Nd crystallites. Additional Nd_2Fe_{17} grains are also observed, larger than those observed in Figure 17.3a.

eutectic-like structure, known as a degenerated structure, has often been observed in Al–Cu eutectic alloys, where the eutectic is skewed owing to the limited growth of a faceted phase [26]. The EDX analysis gives compositions of $Nd_{95}Fe_5$ and $Nd_{65}Fe_{35}$ for the dark and the bright lamellae, respectively. The dark lamellae correspond to Nd-rich solid solution but the stoichiometry of the bright lamellae does not account for any known Nd–Fe phase. These Fe-containing lamellae are referred to as *A1*. They are metastable and transform into an intermediate Nd_2Fe_{17} phase after annealing at temperatures above 773 K [27] and subsequently into the stable Nd_5Fe_{17} phase after long (24 h) annealing at 873 K [27]. There is a large variation in the composition of *A1* reported by different groups, particularly because of its fine morphology, which renders an exact composition analysis difficult.

Figure 17.4 shows the microstructure of a $Nd_{80}Fe_{20}$ specimen cooled at $\sim$50 K s^{-1} (mold-cast rod with 7 mm diameter). The microstructure consists of eutectic-like Nd + A1 (*E*), primary Nd crystallites, and Nd_2Fe_{17} crystallites. It can be clearly seen in the optical micrograph that the size and the volume fraction of Nd_2Fe_{17} crystallites are larger in the specimens cooled at $\sim$50 K s^{-1} than in a sample cooled at $\sim$100 K s^{-1}. However, the composition of the eutectic ($Nd_{78}Fe_{22}$) remains fairly constant with decreasing cooling rate from 150 to 50 K s^{-1}. The formation of Nd_2Fe_{17} crystallites results from the transformation of the metastable *A1* zone due to long-time exposure to a temperature higher than 773 K. A longer exposure time (i.e. lower cooling rates) implies bigger Nd_2Fe_{17} crystallites

17.3.2
The Metastable A1 "Phase"

According to the Nd–Fe binary phase diagram shown in Figure 17.5, a hypereutectic $Nd_{80}Fe_{20}$ alloy will solidify into a eutectic of Nd + Nd_5Fe_{17} accompanied by primary Nd crystallites under equilibrium conditions. However, during copper

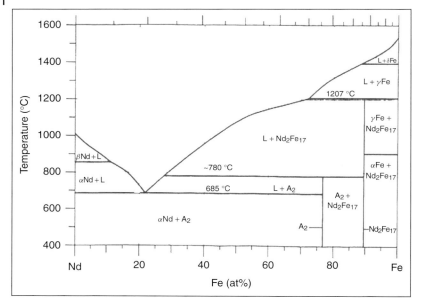

Fig. 17.5 Revised Nd–Fe phase diagram [16] that shows the existence of a hexagonal Nd_5Fe_{17} phase.

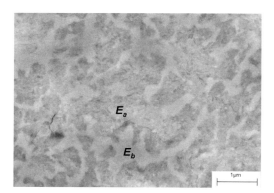

Fig. 17.6 High-magnification SEM back-scattered electron image of metastable eutectic Nd-rich + A1 observed in a mold-cast $Nd_{80}Fe_{20}$ rod of 3 mm diameter (cooled at $150\,K\,s^{-1}$). E_a is Nd-rich ($Nd_{95}Fe_5$) and E_b is an Fe-containing region termed as A1.

mold casting, which employs fast cooling ($50-150\,K\,s^{-1}$), the melt does not follow the equilibrium solidification pathway. Some regions tend to become enriched in Fe, other regions in Nd. The formation of the Fe-rich Nd_5Fe_{17} and Nd_2Fe_{17} lamellae requires long-range diffusion of Fe atoms. In contrast, the formation of the Nd lamellae requires a diffusion of atoms over a shorter distance owing to the enrichment of the eutectic liquid with Nd. Figure 17.6 shows a high-magnification

SEM image of a typical eutectic structure obtained in the $Nd_{80}Fe_{20}$ mold-cast rod (cooled at $150\,K\,s^{-1}$). Two regions, E_a ($Nd_{95}Fe_5$) and E_b ($Nd_{65}Fe_{35}$), can be clearly distinguished. The Fe-containing regions E_b are referred to as *A1*. The solidification of the Nd-rich regions (E_a) entraps Fe-richer regions (E_b) and hinders the long-range diffusion of Fe atoms required for the formation of Nd_2Fe_{17}. This promotes the formation of metastable eutectic $A1$ + Nd-rich solid solution. Since the atomic mobility of Fe and Nd is low at temperatures below 773 K [5], the alloy solidifies in a metastable state before reaching the equilibrium state during most practical casting processes. However, the hotter regions where the atomic mobility is sufficient for long-range diffusion will solidify into a structure composed of Nd_2Fe_{17} crystallites embedded in a Nd matrix. This type of morphology is called discontinuous and is usually observed when the volume fraction of one of the eutectic phases (in the present case Nd_2Fe_{17}) is very small [26]. Therefore, as-cast Nd–Fe alloys near the eutectic composition always exhibit two eutectic morphologies: that is, a discontinuous (Nd + Nd_2Fe_{17}) and a degenerated morphology (Nd or Nd-rich solid solution + $A1$) due to the limited atomic diffusivity below 873 K. With the decrease in cooling rate, the Fe-containing regions (E_b) become further enriched in iron because the diffusion is a time-dependent phenomenon. This cooling rate dependence of the diffusivity results in a wide spread in composition of the Fe-containing regions (*A1*). Furthermore TEM investigations were carried out to clarify whether *A1* is a homogeneous phase or a mixture of phases. A bright-field TEM image from the eutectic-like region in the $Nd_{80}Fe_{20}$ mold-cast rod (cooled at $150\,K\,s^{-1}$) is presented in Figure 17.7a. It shows two distinct features, that is Nd grains with an average size of 500 nm and a strong diffraction contrast, surrounded by an Fe-containing layer with an average thickness of about 200 nm

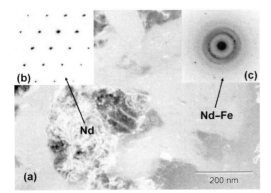

Fig. 17.7 (a) TEM bright-field image of the metastable eutectic of a $Nd_{80}Fe_{20}$ mold-cast rod (cooled at $150\,K\,s^{-1}$) showing two regions, i.e. Nd grains of about 500 nm size and Nd–Fe diffuse layers with varying thickness. The corresponding SAED patterns from the Nd and the Nd–Fe regions are shown in Figures (b) and (c), respectively. The SAED pattern from the Nd grains can be indexed as hexagonal crystals. However, the SAED pattern from the Nd–Fe regions indicates the presence of a mixture of randomly oriented nanocrystallites and an amorphous phase.

without diffraction contrast. The Nd grains show a highly deformed surface due to their oxidation during the TEM sample preparation. In contrast, the intergranular Fe-containing layers display a smooth surface. The Nd grains can be easily identified by EDX because of their large size. EDX analysis reveals that the majority of Fe is present in the intergranular layers, which are referred to as "A1". However, the EDX results from these intergranular layers vary in composition from one point to another because of the contribution of the surrounding Nd grains. Figure 17.7b and c shows selected area electron diffraction (SAED) patterns from the Nd grains and intergranular layers, respectively. The Nd grains exhibit a distinct crystalline SAED pattern, which can be indexed as dhcp Nd crystallites. On the other hand, the SAED pattern from the Fe-containing layers (Figure 17.7c) shows diffuse rings along with some superimposed diffraction spots. This suggests the coexistence of amorphous and crystalline phases in the Fe-containing regions. The spots are not periodic, indicating that the crystallites are randomly oriented with respect to the electron beam. The underlying spots may also be due to overlapping Nd crystallites encountered along the electron beam pathway due to a larger sample thickness.

Figure 17.8 presents a bright-field TEM image of the eutectic region of the $Nd_{80}Fe_{20}$ mold-cast rod (cooled at $150\,K\,s^{-1}$), showing the two phases, Nd and Nd–Fe (A1). The corresponding high-resolution transmission electron microscopy (HRTEM) images from the Nd–Fe and the Nd phases are also displayed. The HRTEM image from the Nd–Fe region clearly shows the presence of ~5 nm size crystallites (marked by arrows) embedded in an amorphous phase. In contrast, the Nd phase is crystalline and, therefore, produces clear lattice images. The boundary between the crystalline Nd and the amorphous Nd–Fe regions is clearly visible (marked with a continuous line). Thus, the eutectic-like region of the $Nd_{80}Fe_{20}$ mold-cast rod is macroscopically composed of two phases, i.e. elemental

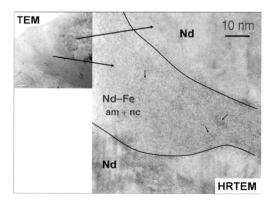

Fig. 17.8 TEM bright-field image of the eutectic-like region of the $Nd_{80}Fe_{20}$ mold-cast rod (cooled at $150\,K\,s^{-1}$) showing two regions, i.e. Nd and Nd–Fe. Corresponding HRTEM images from the Nd–Fe phase and Nd reveal that the Nd–Fe region consists of about 5 nm size crystallites (marked by arrows) and an amorphous matrix. On the other hand, the Nd phase is crystalline showing atomic columns. The interface between the crystalline Nd and an amorphous Nd–Fe region is clearly visible and marked here by continuous lines.

Nd and Nd–Fe. The Nd–Fe regions generate amorphous-like SAED patterns and no diffraction peaks in XRD. HRTEM imaging together with elemental mapping from the Nd–Fe regions reveals that these regions are composed of 5–10 nm crystallites embedded in an amorphous phase. The average composition of the Nd–Fe regions is $Nd_{56}Fe_{44}$. These regions are metastable and transform into Nd_2Fe_{17} and Nd_5Fe_{17} phases upon annealing. It is clear from the structural analysis of the $Nd_{80}Fe_{20}$ mold-cast rod that the Fe-containing region with an average composition of $Nd_{54}Fe_{46}$ is not a single phase; rather it is a mixture of amorphous and nanocrystalline phases. However, for the comparison with previous studies, the Fe-containing regions are denoted as "*A1*", as referred to in other studies [28].

17.3.3
Magnetic Properties

The main stimulus for the interest in the study of binary Nd–Fe alloys is the occurrence of room-temperature coercivity in the range from 0.2 to 0.5 T, depending on the cooling rate [28]. Since the stable intermetallic Nd_2Fe_{17} and Nd_5Fe_{17} phases exhibit an easy plane anisotropy, they are soft magnetic [29]. The high coercivity of the as-cast Nd–Fe alloys is usually attributed to the metastable Fe-containing regions referred to as *A1* [28]. As discussed in the previous section, the Fe-containing regions contain finely dispersed nanocrystallites and an amorphous phase. Therefore, they will be denoted as *A1* "regions" instead of "phase". The presence of the hard magnetic *A1* regions in Nd–Fe alloys are usually characterized by the appearance of a room-temperature coercivity of 0.5 T and a Curie temperature $T_C \sim 510$–530 K. The Curie temperature alone cannot be used to verify the formation of the *A1* regions because several other intermetallic phases such as Nd_5Fe_{17} ($T_C = 508$ K) and Nd–Fe–O ($T_C = 523$ K) have Curie temperatures in the close proximity. Owing to the difficulties in phase identification, the origin of coercivity in as-cast Nd–Fe eutectic alloys remains unclear. In the following, the correlation between the room-temperature magnetic properties and the microstructure of the Nd–Fe alloys will be shown.

Figure 17.9 presents the room-temperature demagnetization curves for the $Nd_{80}Fe_{20}$ alloys prepared at different cooling rates. The alloys cooled at rates between 75 and 150 K s^{-1} show similar features with a large coercivity of 0.44–0.48 T and a remanence of 7.4–9.7 A m^2 kg^{-1}. On the other hand, the alloys cooled at rates below 75 K s^{-1} show a two-phase magnetic behavior with a step in the demagnetization curves near zero field. This is due to the presence of the soft magnetic Nd_2Fe_{17} phase along with the hard magnetic *A1* regions. The overall coercivity remains nearly constant for alloys cooled faster than 25 K s^{-1}. As the coercivity strongly depends on the microstructure, the constant coercivity value for the samples suggests the formation of similar microstructural features resulting in the same coercivity mechanism in these alloys. These results define a limiting cooling rate (50 K s^{-1}) above which the metastable hard magnetic *A1* regions form. If we consider the entire hysteresis loops (not shown here for the $Nd_{80}Fe_{20}$ binary alloy

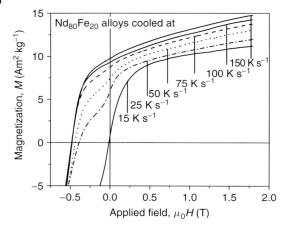

Fig. 17.9 Room-temperature demagnetization curves of $Nd_{80}Fe_{20}$ alloys cooled at different rates. The $Nd_{80}Fe_{20}$ alloys cooled faster than $50\,K\,s^{-1}$ show a smooth demagnetization curve, whereas the alloys cooled slower than $50\,K\,s^{-1}$ show a dip in the demagnetization curve due to the presence of two magnetic (soft and hard) phases.

but discussed further for the ternary alloys), they are not saturated, which indicates the presence of a strong magnetic anisotropy or paramagnetic contribution of the Nd microcrystals.

As was shown above, the hard magnetic properties are strictly related to the *A1* regions. Therefore, a higher volume fraction of *A1* will make the samples magnetically harder. We have also shown that the structure of the samples changes with the variation of the cooling rate. But it is enough to play with the cooling rate? How can the *A1* zones be extended? The first and the simplest possibility that can be considered is the addition of an element that may refine the structure. In this way, the big Nd grains may become smaller, the *A1* zone extends, and, as a consequence, the overall hard magnetic properties may improve. Such an element is Ga: it has a positive heat of mixing with both Nd and Fe [20] and, therefore, a very limited solubility.

17.4
Ternary $(Nd_{100-x}Ga_x)_{80}Fe_{20}$ ($x = 5, 10,$ and 15) Rapidly Quenched Alloys

17.4.1
XRD Studies

Figure 17.10a–d shows the XRD patterns for $(Nd_{100-x}Ga_x)_{80}Fe_{20}$ ($x = 5, 10,$ and 15) melt-spun and mold-cast alloys prepared under identical conditions (i.e., ribbons, similar rod diameter, and casting conditions). The alloys display diffraction peaks

17.4 Ternary $(Nd_{100-x}Ga_x)_{80}Fe_{20}$ (x = 5, 10, and 15) Rapidly Quenched Alloys

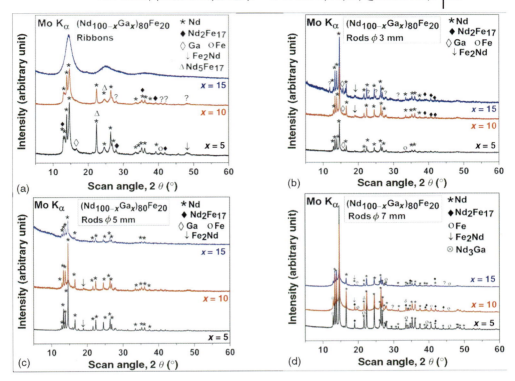

Fig. 17.10 XRD patterns of $(Nd_{100-x}Ga_x)_{80}Fe_{20}$ (x = 5, 10, and 15) alloys: (a) ribbons, (b) 3 mm diameter rods, (c) 5 mm diameter rods, and (d) 7 mm diameter rods.

corresponding to α-Nd, Nd_2Fe_{17}, Fe_2Nd, and Nd_5Fe_{17} phases. There are two observations to be noted from these XRD patterns. First, according to the revised Nd–Fe binary phase diagram (Figure 17.5), an eutectic of Nd + Nd_5Fe_{17} is expected to form along with primary crystallites of Nd. However, the observed XRD patterns give no indications for the formation of Nd_5Fe_{17} (except in case of melt-spun ribbon samples, Figure 17.10a). This implies that either the formation of Nd_5Fe_{17} is suppressed owing to nonequilibrium cooling conditions, or the fraction of the Nd_5Fe_{17} phase is below the detection limit of XRD. According to the Nd–Fe phase diagram, the expected amount of Nd_5Fe_{17} is about 26 vol% for the $Nd_{80}Fe_{20}$ alloy, which is well above the detection limit of XRD. It has been reported [27] that the Nd_5Fe_{17} phase is usually not formed in as-cast Nd–Fe alloys because of its slow formation kinetics by the peritectic reaction between the Nd-rich liquid and the Nd_2Fe_{17} phases. Therefore, the first indication from the XRD patterns is that the eutectic in mold-cast Nd–Fe–Ga alloys is not Nd + Nd_5Fe_{17}.

Figure 17.10b shows the diffraction patterns for a 3 mm diameter rod (150 K s^{-1} cooling rate). The composition with x = 5 does not show peaks corresponding to the Nd_2Fe_{17} phase, whereas the compositions with x = 10 and 15 exhibit peaks corresponding to Nd_2Fe_{17} phase. Nd_2Fe_{17} is responsible for the soft magnetic properties [28, 29].

The second feature of the XRD patterns of as-cast $(Nd_{100-x}Ga_x)_{80}Fe_{20}$ specimens is the absence of diffraction peaks corresponding to any known Nd–Fe–Ga intermetallic phase. This implies that Ga is either dissolved in α-Nd/Nd–Fe intermetallic phase(s) or present in a separate nanocrystalline or amorphous phase that cannot be resolved by XRD. Further studies (as will be seen later) have shown with no doubt that the Ga atoms are uniformly distributed in the entire volume of the sample. In fact, the structure of the samples and, consequently, the magnetic properties, can be tuned in two ways: by adjusting the cooling rate and by composition variation. In order to explain this behavior better, two extreme cases will be considered in the following: rods with 3 mm diameter (cooled at $\sim$150 K s^{-1}) and rods with 7 mm diameter ($\sim$50 K s^{-1}).

17.4.2
Tuning the Metastable Hard Magnetic A1 Zones

Figure 17.11 shows SEM micrographs of 3 mm diameter NdFe samples with different Ga contents, together with the crystalline phases as identified by XRD studies and corroborated by EDX analysis. For $x = 5$ (which corresponds to 4 at%

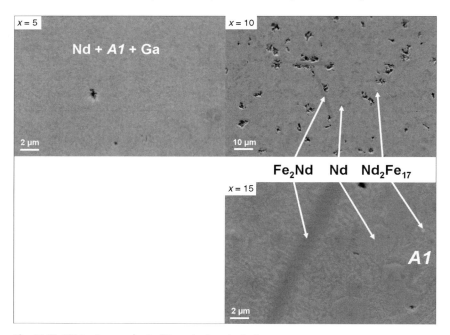

Fig. 17.11 SEM micrographs in SE mode (topological contrast) showing the microstructure of the $(Nd_{100-x}Ga_x)_{80}Fe_{20}$ ($x = 5$, 10, and 15) 3 mm diameter samples (cooled at $\sim$150 K s^{-1}).

Ga addition to binary $Nd_{80}Fe_{20}$), the structure is very refined. Small Nd and Ga grains coexist with *A1* hard magnetic zones. Once the Ga content increases, the refinement of the *A1* zones is lost and new crystalline soft magnetic phases appear. These crystalline phases were identified as Nd_2Fe_{17} and Fe_2Nd. The appearance of the Fe_2Nd phase is very interesting, because this phase does not appear in the NdFe phase diagram as an equilibrium phase, but some reports have mentioned that it may form if some conditions are fulfilled [30]. In our case it is clear that the Ga additions and, consequently, the shift of the composition from hypereutectic to hypoeutectic are the reasons. At first glance one cannot say if this phase is completely magnetically ordered; for that, additional neutron diffraction studies should be performed.

Comparing the structure as observed by SEM for the 3 mm diameter rod with $x = 15$ (which corresponds to 12 at% Ga addition) presented in Figure 17.11 with one characteristic for the starting $Nd_{80}Fe_{20}$ 3 mm diameter rod from Figure 17.2b, one can observe that the different samples are very similar (except for the finding of the Fe_2Nd phase). Apparently, for samples cast at the same cooling rate ($\sim 150\,K\,s^{-1}$ in this case), small Ga additions refine the microstructure. As a consequence, the hard magnetic properties improve. Figure 17.12 shows the hysteresis loops recorded for 3 mm diameter samples with different Ga contents. The alloy with $x = 5$ shows the highest remanence and coercivity. Its energy product is higher than for similar samples without Ga addition (see Figure 17.9). As the Ga content increases (Figure 17.12), the soft magnetic phases start to appear and the hysteresis loops show a characteristic two-phase behavior, with a reduction in coercivity and remanence.

Figure 17.13 shows the SEM micrographs of 7 mm diameter NdFe samples with different Ga contents, together with the crystalline phases as identified by XRD and corroborated by EDX analysis. The most pronounced refinement of *A1* zones is

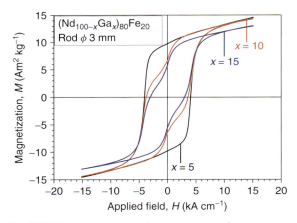

Fig. 17.12 Room-temperature hysteresis curves for $(Nd_{100-x}Ga_x)_{80}Fe_{20}$ ($x = 5$, 10, and 15) 3 mm diameter samples (cooled at $\sim 150\,K\,s^{-1}$).

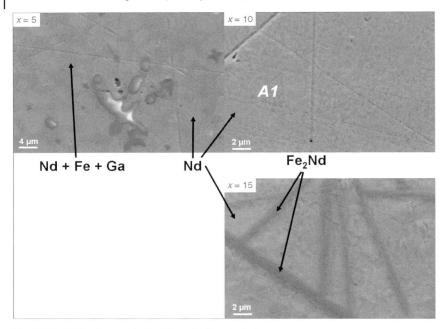

Fig. 17.13 SEM micrographs in SE mode (topological contrast) showing the microstructure of the $(Nd_{100-x}Ga_x)_{80}Fe_{20}$ ($x = 5$, 10, and 15) 7 mm diameter samples (cooled at $\sim 50\,K\,s^{-1}$).

obtained for $x = 10$ (which corresponds to 12 at% Ga addition). Figure 17.14 shows the hysteresis loops for these samples. The sample with the best hard magnetic properties is the one with $x = 10$, as expected from analyzing the SEM micrographs. The sample with $x = 5$ is also hard magnetic, but here the volume fraction of the *A1* zones is more limited. When x increases to 15, new soft magnetic phase(s) (especially Fe_2Nd) will change the magnetization/demagnetization behavior and the loop again exhibits the shape characteristic for two-phase magnets.

17.5
Conclusions

Structural and magnetic properties of $(Nd_{100-x}Ga_x)_{80}Fe_{20}$ ($x = 0$, 5, 10, and 15) alloys were investigated. According to the revised Nd–Fe binary phase diagram, the equilibrium eutectic is $Nd + Nd_5Fe_{17}$. The phase selection in Nd–Fe alloys depends sensitively on composition and cooling rate. In order to study the effect of cooling rate, $Nd_{80}Fe_{20}$ alloys cooled at different rates from 5 to $150\,K\,s^{-1}$ were investigated. The alloys cooled at $150\,K\,s^{-1}$ reveal the formation of $Nd + A1$. *A1* is a notation used for the Fe-containing regions in the metastable eutectic. Structural investigations of the *A1* regions were performed by HRTEM, showing the presence of a mixture

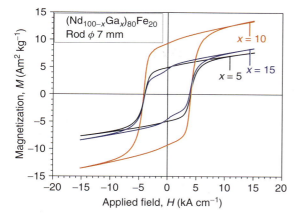

Fig. 17.14 Room-temperature hysteresis curves for $(Nd_{100-x}Ga_x)_{80}Fe_{20}$ ($x = 5$, 10, and 15) 7 mm diameter samples (cooled at $\sim 50\,K\,s^{-1}$).

of 5–10 nm sized crystallites and an amorphous phase. It was not possible to identify the crystallites due to their small size. However, EDX analysis in the STEM mode yields an average composition of almost $Nd_{50}Fe_{50}$ for the *A1* regions. The *A1* regions do not yield any diffraction peak in XRD. Therefore, the $Nd_{80}Fe_{20}$ alloy cooled at 150 K s^{-1}, which contains only Nd and *A1*, shows only hexagonal Nd peaks in XRD. Magnetic measurements show that the $Nd_{80}Fe_{20}$ samples containing the *A1* regions exhibit hard magnetic behavior at room temperature with a coercivity of 0.5 T and a remanence of 9 A m^2 kg^{-1}. The coercivity of the $Nd_{80}Fe_{20}$ alloys cooled at different rates does not change with the cooling rate as long as the *A1* regions are formed. In contrast, the remanence and the maximum magnetization decrease with decreasing cooling rate owing to a reduced volume fraction of metastable *A1* regions.

In order to increase the amount of *A1* by decreasing the size of hexagonal Nd grains, Ga additions were made to the starting binary alloy. Several samples were prepared at the same cooling rate as used for $Nd_{80}Fe_{20}$. For 4 at% Ga and a cooling rate of $\sim 150\,K\,s^{-1}$, the structure changes and the hard magnetic properties become better: a coercivity of 4.06 kA m^{-1} (which correspond to 0.51 T) and a remanence of 9.76 A m^2 kg^{-1} were measured. For this cooling rate, a further increase in the Ga content induces the appearance of soft magnetic phases and the overall magnetic behavior changes to a two-phase magnet. When the Ga content reaches 12 at% and the cooling rate is lowered to $\sim 50\,K\,s^{-1}$, the coercivity slightly increases (to 4.12 kA m^{-1} or 0.52 T), and the remanence slightly decreases (to 9.32 A m^2 kg^{-1}), but the energy product remains almost the same. However, in comparison with the starting binary alloy, the addition of a nonmagnetic element increases the hard magnetic properties just by changing the structure of the sample. Therefore, the magnetic properties determined only by Nd and Fe can be adjusted by tuning the structure, i.e. by tailoring the composition and the cooling rate.

References

1 Drozzina, V. and Janus, R. (**1935**) "A new magnetic alloy with very large coercive force". *Nature*, **135**, 36.
2 Croat, J.J., Herbst, J.F., Lee, R.W. and Pinkerton, F.E. (**1984**) "Pr-Fe and Nd-Fe-based material: a new class of high-performance permanent magnets". *Journal of Applied Physics*, **55**, 2078.
3 Wallace, W.E. (**1973**) *Rare Earth Intermetallics*, Academic Press, New York.
4 Stadelmaier, H.H., Schneider, G. and Ellner, M. (**1986**) "A CaCu$_5$-type iron-neodymium phase stabilized by rapid solidification". *Journal of the Less-Common Metals*, **115**, L11.
5 Givord, D., Nozieres, J.P., Rossignol, M.F., Taylor, D.W., Harris, I.R., Fruchart, D. and Miraglia, S. (**1991**) "Structural analysis of the hard ferromagnetic phase observed in quenched Nd-Fe alloys of hypereutectic composition". *Journal of Alloys and Compounds*, **176**, L5.
6 Gerber, J.A., Cornelison, S.G., Burmester, W.L. and Sellmyer, D.J. (**1979**) "Magnetic properties of the rare-earth glasses $(R_{65}Fe_{35})_{100-x}B_x$". *Journal of Applied Physics*, **50**, 1608.
7 Wecker, J. and Schultz, L. (**1987**) "Coercivity after heat treatment of overquenched and optimally quenched Nd-Fe-B". *Journal of Applied Physics*, **62**, 990.
8 Schultz, L., Wecker, J. and Hellstern, E. (**1987**) "Formation and properties of Nd-Fe-B prepared by mechanical alloying and solid-state reaction". *Journal of Applied Physics*, **61**, 3583.
9 Schultz, L., Schnitzke, K. and Wecker, J. (**1988**) "Mechanically alloyed isotropic and anisotropic Nd-Fe-B magnetic material". *Journal of Applied Physics*, **64**, 5302.
10 Fidler, J. and Knoch, K.G. (**1989**) "Electron microscopy of Nd-Fe-B based magnets". *Journal of Magnetism and Magnetic Materials*, **80**, 48.
11 Fidler, J. and Schrefl, T. (**1996**) "Overview of Nd-Fe-B magnets and coercivity". *Journal of Applied Physics*, **79**, 5029.
12 Altounian, Z., Ryan, D.H. and Tu, G.H. (**1988**) "A new metastable phase in the Nd-Fe-B system". *Journal of Applied Physics*, **64**, 5723.
13 Ramesh, R., Chen, J.K. and Thomas, G. (**1987**) "On the grain-boundary phase in iron-rare-earth boron magnets". *Journal of Applied Physics*, **61**, 2993.
14 Tsubokawa, Y., Shimizu, R., Hirosawa, S. and Sagawa, M. (**1988**) "Effect of heat treatment on grain-boundary microstructure in Nd-Fe-B sintered magnet". *Journal of Applied Physics*, **63**, 3319.
15 Schneider, G. (**1988**) *Konstitution und sinterverhalten von hartmagnetwerkstoffen auf Fe- Nd-B basis*, PhD thesis, Universität Stuttgart.
16 Landgraf, F.J.G. (**1990**) "Solidification and solid state transformation in Fe-Nd: a revised phase diagram". *Journal of the Less-Common Metals*, **163**, 209.
17 Amri, N., Delamare, J., Lemarchand, D. and Vigier, P. (**1991**) "Microstructural and thermomagnetic investigation of rapidly solidified Nd-Fe-Al eutectic alloys". *Journal of Magnetism and Magnetic Materials*, **101**, 352.
18 Chen, Z., Wang, N., Song, X. and Wang, X. (**1995**) "Characterization of a new Nd-Fe-O compound in $Nd_{34}Fe_{60}O_6$ alloy". *Materials Letters*, **22**, 119.
19 Lopez, G., Domingues, P.H. and Sanchez Llazamares, J.L. (**2000**) "Magnetic analysis of $Nd_{80}Fe_{15}(B_{1-x}C_x)_5$ alloys ($0.85 \leq x \leq 1$)". *Journal of Alloys and Compounds*, **302**, 209.
20 deBoer, F.R., Boom, R., Mattens, W.C.M., Miedema, A.R. and Niessen, A.K. (**1988**) *Cohesion in Metals, Transition Metal Alloys*, Cohesion and structure, Vol. **1**, North-Holland Publishing, Amsterdam, Oxford, New York, Tokyo.

21 Owen, A.E. (1985) in *The Glass Transition, Amorphous Solids and the Liquid State*, (eds N.H. March, R.A. Street and M. Tosi), Plenum Press, New York, pp. 395.

22 Hayzelden, C., Rayment, J.J. and Cantor, B. (1983) "Rapid solidification microstructures in austenitic Fe-Ni alloys". *Acta Materialia*, **31**, 379.

23 Boswell, P.G. and Chadwick, G.A. (1976) "Cellular growth in a $Pd_{781}Cu_{55}Si_{164}$ amorphous alloy". *Scripta Materialia*, **10**, 509.

24 Kurz, W. and Fisher, D.J. (1989) *Fundamentals of Solidification*, Trans Tech Publications, 63.

25 Srivastava, R., Eckert, J., Löser, W., Dhindaw, B.K. and Schultz, L. (2002) "Cooling rate evaluation for bulk amorphous alloys from eutectic microstructures in casting processes". *Materials Transactions, JIM*, **43**, 1670.

26 Chadwick, G.A. (1963) "Eutectic alloy solidification". *Progress in Materials Science*, **12**, 149.

27 Kumar, G. (2005) *Structural and magnetic characterization of Nd-based, Nd-Fe and Nd-Fe-Co-Al metastable alloys*, PhD Thesis, Technische Universität Dresden.

28 Givord, D., Nozieres, J.P., Sanchez Llazamares, J.L. and Leccabue, F. (1992) "Magnetization and anisotropy in the hard ferromagnetic phase observed in $Nd_{100-x}Fe_x$ ($x < 25$) as-cast alloys". *Journal of Magnetism and Magnetic Materials*, **111**, 164.

29 Landgraf, F.J.G., Missell, F.P., Rechenberg, H.R., Schneider, G., Villas-Boas, V., Moreau, J.M., Paccard, L. and Nozieres, J.P. (1991) "Magnetic and structural characterization of Nd_5Fe_{17}". *Journal of Applied Physics*, **70**, 6125.

30 Stadelmaier, H.H., Schneider, G., Henig, E.-Th. and Ellner, M. (1991) "Magnetic $Fe_{17}Nd_5$ in the Fe-Nd and Fe-Ti-Sm systems and other phases in Fe-Nd". *Materials Letters*, **10**, 303.

Part Four
Solidification und Simulation

18
Solidification of Binary Alloys with Compositional Stresses — A Phase-Field Approach

Bo Liu and Klaus Kassner

18.1
Introduction

It is well known that a flat solidification front moving in an imposed temperature gradient (directional solidification) turns unstable when its velocity exceeds a critical value. Owing to this morphological instability, first analyzed by Mullins and Sekerka [1], an initially planar interface will undergo bifurcation toward a cellular structure. Not too far above the threshold speed, these cells are stationary. As the pulling velocity is increased further, shallow cells turn into deep cells and then into dendritic arrays [2]. On approach of the absolute stability limit, oscillatory and chaotic cellular patterns appear, before the planar interface restabilizes [3].

The influence of elastic stresses on the morphological stability of solid–liquid interfaces has recently been addressed in two ways. One approach aims at understanding the self-stresses generated by a nonuniform composition field due to a nonzero solute expansion coefficient [4–10], whereas the other is concerned with the stability of a solid having a free surface under anisotropic stress (Grinfeld instability) [11–14], in some cases in the presence of directional solidification [15, 16].

Spencer *et al.* [5] found by linear stability analysis that composition-induced elastic stresses generate an oscillatory instability because of the phase difference between the interface shape and the profile of the solute rejected along it. This instability should result in either a traveling-wave or a standing-wave pattern for the interface. A question of this kind cannot be decided, of course, by the linear stability analysis.

At the formal level, there is a certain similarity between the mechanisms of the stress-induced oscillatory instability and that of the oscillatory instability present during rapid solidification [17, 18]. In both cases, the amount of rejected solute shows additional variations along the interface beyond those due to local thermal equilibrium as stated by the Gibbs–Thomson relation. In rapid solidification, this extra variation appears because the segregation coefficient depends on the local interface velocity, leading to a departure from equilibrium [19]. In the presence of

Phase Transformations in Multicomponent Melts. Edited by D. M. Herlach
Copyright © 2008 WILEY-VCH Verlag GmbH & Co. KGaA, Weinheim
ISBN: 978-3-527-31994-7

compositional stresses, segregation depends on the stress at the interface, which is a functional of the concentration field in the solid, hence history dependent, in particular, when diffusion in the solid is negligible. While the *local* equilibrium assumption may be upheld in this case, the equilibrium concentration depends on the additional field, again leading to a departure from the original equilibrium condition given by the Gibbs–Thomson relation. Moreover, the new condition will depend on frozen-in concentrations in the solid accumulated throughout the history of the sample, making the whole process strongly nonequilibrium and nonlocal in both space and time.

In the last decade, phase-field models have become standard tools in the simulation of microstructure evolution in materials [20, 21] involving the dynamics of one or several interfaces between different phases. They avoid explicit front tracking and are versatile enough to deal with topological changes. Information on the interface position is present implicitly. In the case of two-phase models, it is given as a level set of a particular value of the phase field, and can be recovered by computation of the appropriate level set at only those times when knowledge of the position is desired.

A phase-field model for a binary alloy can be constructed by adding to the free-energy density of a pure material the contribution of solute molecules. The simplest way to obtain this free energy is to interpolate between the known free-energy densities of the solid and liquid with a single function of ϕ, as in an early model introduced by Wheeler *et al.* [22, 23]. From a computational point of view, however, this approach places stringent constraints on the allowable interface thicknesses. One reason is that there generally is an extra contribution to the surface tension γ due to solute addition, rendering it strongly dependent on both the interface thickness and temperature. Moreover, nonequilibrium effects at the interface also have their influence on microstructural pattern formation. Numerical interfaces have, for reasons of computational efficiency, to be much thicker than physical ones, hence these nonequilibrium effects become artificially enhanced, thereby competing with, or even surpassing, capillary effects, which in turn may completely change pattern evolution. Hence, in order to meet the challenge of quantitative phase-field modeling of solidification, one needs to judiciously choose the parameters and the functional form of the model expressions in order to avoid unphysically large nonequilibrium effects at the interface.

These nonequilibrium effects were first characterized in detail by Almgren [24] using a thin interface analysis of a phase-field model for the solidification of pure melts with asymmetric diffusion. The same problems and effects emerge in alloy solidification [25], including solute diffusion along the arc length of the interface (*surface diffusion*), a modification of mass conservation associated with the local increase of arc length of a moving curved interface (*interface stretching*), and a discontinuity of the chemical potential across the interface, leading to *solute trapping* in the alloy case. We will briefly address our approach to these problems in this chapter.

So far, there have been only few studies of solids under stress using the phase-field approach. Müller and Grant [26] and Kassner and Misbah [27] independently constructed models for pure solids in contact with their melts, describing the Grinfeld instability [28].

Also coupling the Grinfeld instability with directional solidification was considered, albeit not in the context of a phase-field model [15, 16]. Nevertheless, the current work was motivated by these predecessors, in which it was found that quite moderate values of an imposed normal stress difference are sufficient to reduce the critical wavelength for the Mullins–Sekerka instability by about one order of magnitude. The question then was whether compositional or thermal stresses arising during solidification could have similarly dramatic consequences.

Here, we will deal with compositional stresses only, giving a detailed quantitative phase-field model for a dilute binary alloy in the presence of compositionally generated elastic stresses. The derivation of our model is based on the formulation by Kim et al. [29, 30], where the liquid and solid concentrations c_l and c_s are used explicitly rather than c as in the formulation by Karma et al. [25, 31], because the stresses are generated from $\Delta c_s = c_s - c_s^0$, with c_s^0 being the concentration of the unstrained reference state, so it is useful to have the solid concentration at our disposal directly.

This chapter is organized as follows. Section 18.2 describes some basic thermodynamic relations for a stressed alloy under near-equilibrium conditions and gives the equations of motion following from this. In Section 18.3, we recover a few linear stability results from the literature, in order to identify interesting regions of parameter space. In Section 18.4, we introduce our phase-field model and discuss, in view of a thin interface asymptotics not presented here, which model elements guarantee that the aforementioned spurious or rather exaggerated effects cancel. Section 18.5 gives numerical results, and finally, some conclusions are given in Section 18.6.

18.2
Equations of Motion

We consider an alloy composed of two chemical components 1 and 2, so we need only one concentration to describe its composition. Elastic constants are taken to be independent of composition and the solid–liquid interfacial energy is assumed to be isotropic. As reference state for the solid in the framework of elasticity theory, we choose the state of zero stress at uniform composition.

For such a setting, the thermodynamic equilibrium conditions at the solid–liquid interface, in the limit of zero surface stress and small-strain elasticity, have been given by Larché and Cahn [32]:

$$M_{12} = \mu_1^L - \mu_2^L \tag{18.1}$$

$$JP_l + \omega' + K'\sigma' = 0 \tag{18.2}$$

Our presentation of equilibrium results closely follows Spencer et al. [5]. M_{12} is the diffusion potential in the liquid given by the difference of the chemical potentials μ_i^l of the two components in the liquid. P_l is the pressure in the liquid, ω' the grand canonical free-energy density of the solid measured per unit volume of the reference state. K' is the sum of the principal curvatures of the interface in the reference state, σ' the interfacial free energy per area of the reference state, and $J = 1 + u_{i,i}$ is the determinant of the deformation gradient tensor, wherein u_i is the Cartesian component of the displacement vector and subscripts preceded by a comma denote derivatives with respect to the corresponding coordinate.

Assuming composition differences to induce pure dilations (i.e. an isotropic solute expansion tensor), we obtain as the stress–strain relationship for a solid with compositionally generated stresses

$$E_{ij} = S_{ijkl} T_{kl} + \eta(c_s - c_s^0)\delta_{ij} \tag{18.3}$$

E_{ij} is the strain related to the displacement gradients by $E_{ij} = \frac{1}{2}(u_{i,j} + u_{j,i})$ in the small-strain approximation. T_{ij} is the stress tensor, S_{ijkl} the compliance tensor, c_s is the concentration of component 2 in the solid, and c_s^0 is its uniform concentration in the reference state. η is the solute expansion coefficient and δ_{ij} the Kronecker symbol.

For isotropic materials, the compliance tensor is given by

$$S_{ijkl} = \frac{1+\nu}{E}\delta_{ik}\delta_{jl} - \frac{\nu}{E}\delta_{ij}\delta_{kl} \tag{18.4}$$

where E is Young's modulus and ν the Poisson ratio. Given isotropy, we may then express η for binary systems as

$$\eta = \frac{1}{3}\frac{v_2 - v_1}{v_0} \tag{18.5}$$

where v_1 and v_2 are the partial molar volumes of the two components, respectively, and v_0 is the molar volume of lattice sites in the reference state.

The diffusion potential is stress dependent via

$$M_{12}(T_{ij}, T, c_s) = \mu_1^S(0, T, c_s) - \mu_2^S(0, T, c_s) - v_0 \eta T_{kk} \tag{18.6}$$

where $\mu_i^S(0, T, c_s)$ is the chemical potential of component i in the solid at zero pressure, temperature T, and concentration c_s; T_{kk} is the trace of the stress tensor (i.e. we use Einstein's summation convention).

The stress dependence of ω' contains the elastic energy density

$$\omega'(T_{ij}, T, c_s) = \frac{\mu_2^S(0, T, c_s) - \mu_2^l(P_l, T, c_l)}{v_0} + \frac{1}{2} S_{ijkl} T_{ij} T_{kl} + c_s \eta T_{kk} \tag{18.7}$$

where $\mu_2^l(P_l, T, c_l)$ is the chemical potential of component 2 in the liquid at pressure P_l, temperature T, and concentration c_l.

Since the reference state is stress free, $P_l = 0$ from the condition of equilibrium. The equilibrium temperature is designated as T^0, and in the unstressed state, local equilibrium at the interface implies

$$c_s^0 = k c_l^0 \tag{18.8}$$

where k is the segregation coefficient.

Assuming that deviations from the local equilibrium values in the absence of stress are small and expanding all quantities in terms of these deviations, concerning the concentrations in both phases as well as the temperature, one obtains, using in addition the dilute solution approximation, the local equilibrium conditions for the interface [5]:

$$c_s = k c_l + \frac{v_0 \eta T_{kk} c_s^0}{R T_m} \tag{18.9}$$

$$T_l = T_m + m_l c_l - \frac{\gamma T_m}{L} K' \tag{18.10}$$

where R is the ideal gas constant, T_m the melting temperature of a pure substance 1, and m_l the liquidus slope.

If we adopt the frozen-temperature assumption, we need not write equations of motion for the temperature, which is instead given by the interface position via $T = G(z - z_0 - Vt) + T_0$, with G the temperature gradient and V the pulling velocity. z_0 and T_0 are position and temperature of a steady-state planar interface of the unstressed solid in the comoving frame.

The equations of motion for the concentration and strain or displacement fields are then given by

$$\frac{\partial c_l}{\partial t} = D \nabla^2 c_l \tag{18.11}$$

$$\frac{\partial c_s}{\partial t} = 0 \tag{18.12}$$

$$T_{ij,j} = 0 \tag{18.13}$$

$$(1 - 2\nu) \nabla^2 \mathbf{u} + \nabla [\nabla \cdot \mathbf{u} - 2\eta(1 + \nu) c_s] = 0 \tag{18.14}$$

The second of these four equations simply expresses the fact that there is no diffusion in the solid, that is, we consider the one-sided model. (In the comoving frame, this equation becomes $\partial c_s / \partial t = V \partial c_s / \partial z$.) The third describes mechanical equilibrium, which together with Hooke's law for an elastically isotropic body produces the fourth equation, called Lamé equation(s).

To solve these equations, they have to be supplemented with boundary conditions. Two boundary conditions at the interface are given by Equations 18.9 and 18.10. In addition, we have a Stefan condition describing solute conservation and a mechanical equilibrium condition at the interface:

$$V_n (c_l - c_s) = -D \frac{\partial c_l}{\partial n} \tag{18.15}$$

$$T_{ij} n_j = 0 \tag{18.16}$$

Finally, one must require that the concentration tends toward a fixed value c_∞ at $z = \infty$ and the displacement vector becomes zero at $z = -\infty$.

18.3
Neutral Curves

A linear stability analysis of a stress-free stationary planar front solution to the system (Equations 18.9–18.16) has been performed by Spencer *et al.* [5]. We repeat their analysis and give two results on the neutral curve in the plane pulling velocity versus wavenumber of perturbation, allowing us to discuss interesting regions of parameter space.

The pictures correspond, given the concentration at infinity c_∞, to a fixed set of liquidus slope parameter ($Mc_\infty = 10$), segregation coefficient ($k = 1$), and Poisson number ($\nu = 0.33$). The parameter that varies and takes two different values is the so-called solutal stress parameter $\varepsilon = 2E\eta^2 \nu v_0 / RT_m (1-\nu)$. A typical value of this coefficient (for gallium arsenide) would be about 0.7, corresponding to Figure 18.1a for $c_\infty = 0.17$, but there are materials, such as aluminum and calcium, where it can be an order of magnitude larger.

We note that the main effect of solutal stresses is to trigger an oscillatory instability that for small solutal stress parameters appears at large pulling velocities only. Outside of the curves delimiting the regions u and o, a planar interface is stable; inside, it becomes unstable either to a steady-state instability or to an oscillatory one. At larger stresses, a second unstable branch appears. As soon as $\varepsilon c_\infty > 1$, the planar interface becomes destabilized by stresses alone, it is unstable at *all* velocities.

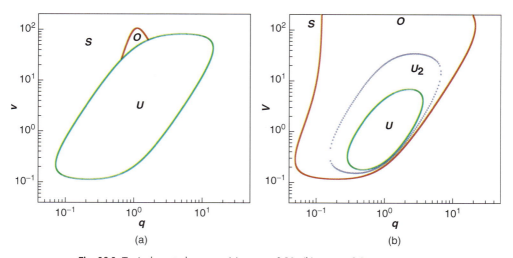

Fig. 18.1 Typical neutral curves. (a) $\varepsilon c_\infty = 0.12$, (b) $\varepsilon c_\infty = 0.9$.

Since we had difficulties simulating fast solidification with our code, we took rather large values of ε in the simulations, corresponding to a product εc_∞ close to 1.

18.4
Phase-Field Model

In order to numerically solve Equations 18.9–18.16, we have developed a phase-field description that asymptotically reduces to them. Details of the thin interface analysis of the model require more space and will be given elsewhere. The spirit of the analysis follows the work of Kim et al. [30]. Our central equations in nondimensional form are the bulk equations of motion for the phase field ϕ,

$$\tau \frac{\partial \phi}{\partial t} = W^2 \nabla^2 \phi + \phi - \phi^3 - \lambda g'(\phi)\left[c_l + \frac{z - Vt}{l_T}\right] \quad (18.17)$$

varying between $+1$ in the solid and -1 in the liquid, for the concentration field $[c = \frac{1}{2}(1 + h(\phi))c_s + \frac{1}{2}(1 - h(\phi))c_l]$

$$\frac{\partial c}{\partial t} = D\nabla \cdot \left(q(\phi)\nabla c_l + a(\phi)W[1 + c_l - c_s]\frac{\partial \phi}{\partial t}\frac{\nabla \phi}{|\nabla \phi|}\right) + \frac{1}{2}\frac{\partial h(\phi)}{\partial t} \quad (18.18)$$

and for the displacement field (F is the free-energy functional)

$$\frac{\partial}{\partial x_j}\frac{\delta F}{\delta E_{ij}} = \frac{\partial}{\partial x_j}\left\{\frac{1 + h}{2}T_{ij}\right\} = 0 \quad (18.19)$$

Here, a number of terms need to be explained: W is an effective interface thickness controlling the approach to the asymptotic limit, hence in principle a small quantity. $\phi - \phi^3$ is the derivative of the double-well potential typical in this kind of approach, and λ a coupling constant for the interaction between the phase field ϕ and the concentration field in the liquid c_l. This coupling is further described by the interpolation function

$$g(\phi) = \phi - \frac{2\phi^3}{3} + \frac{\phi^5}{5} \quad (18.20)$$

the derivative of which vanishes in the bulk phases, so the coupling is essentially different from zero in the interface region only. $l_T = |m_l|(1 - k)c_l^0/G$ is the thermal length.

To close the set of equations, we need an additional relationship between the three quantities c, c_l, and c_s, which is given by the algebraic equation

$$c_s = \left[c + \frac{1 - \phi}{2k}\frac{\varepsilon c_\infty (1 - \nu)}{\nu(1 - 2\nu)}E_{kk}\right]\left[\frac{1 + \phi}{2} + \frac{1 - \phi}{2k}(1 + \frac{2\varepsilon c_\infty (1 - \nu)}{1 - 2\nu})\right]^{-1} \quad (18.21)$$

In Equation 18.18, the first term on the right-hand side describes diffusion, $q(\phi)\nabla\phi$ being the solute current. Since we have no diffusion in the solid, $q(\phi)$ interpolates between 0 in the solid and 1 in the liquid in our nondimensional representation. The last term describes latent heat generation, $1/2(1 + h(\phi))$ is the volume fraction of the solid. The middle term is an antitrapping current and one of the features needed to avoid spurious effects of a finite interface width W. To avoid surface diffusion, surface stretching, and solute trapping effects, we must make sure that the model satisfies the following relations:

$$\int_{-\infty}^{0} dr \frac{RT_m}{v_0} \frac{\tilde{q}(\phi(r))}{f_{cc}(\phi(r))} = \int_{0}^{+\infty} dr \left[c_l - \frac{RT_m}{v_0} \frac{\tilde{q}(\phi(r))}{f_{cc}(\phi(r))} \right] \tag{18.22}$$

$$\int_{-\infty}^{0} dr[c_0(\phi(r)) - c_s] = \int_{0}^{+\infty} dr[c_l - c_0(\phi(r))] \tag{18.23}$$

$$\int_{-\infty}^{0} dr \frac{v_0}{RT_m} \frac{c_0(\phi(r)) - c_s}{\tilde{q}(\phi(r))c_0(\phi(r))} = \int_{0}^{+\infty} dr \left[1 - \frac{c_s}{c_l} - \frac{v_0}{RT_m} \frac{c_0(\phi(r)) - c_s}{\tilde{q}(\phi(r))c_0(\phi(r))} \right] \tag{18.24}$$

where r is a coordinate orthogonal to the interface appearing in the inner equations of the asymptotic analysis, $f(c)$ is the homogeneous part of the free-energy density, subscripts of f denote derivatives, and $\tilde{q}$ and q are related via

$$q(\phi) = \tilde{q}(\phi) \left[\frac{1 + h(\phi)}{2} + \frac{1 - h(\phi)}{2} \frac{f_{cc}^l(c_l)}{f_{cc}^s(c_s)} \right] \tag{18.25}$$

It turns out that the three integral conditions are satisfied by the choices

$$q(\phi) = \frac{1 - \phi}{2} \tag{18.26}$$

$$h(\phi) = \phi \tag{18.27}$$

$$a(\phi) = \frac{1}{2\sqrt{2}} \tag{18.28}$$

The interface boundary conditions will then automatically be obeyed on simulation of the bulk phase-field equations with appropriate boundary conditions at the system limits.

18.5
Simulation Results

In our simulations, we used a nine-point formula for the Laplacian to obtain better isotropy and we did the time integration via a second-order accurate midpoint scheme. The elastic equations were solved by the second-order accurate linear V type multigrid method. It uses pointwise red-black Gauss–Seidel as the smoothing operator, bilinear interpolation for the prolongation operator, and full-weighting for the restriction operator.

In Figure 18.2, we give the time evolutions of solidification cells for small and large solutal stress parameter, respectively. Spatial boundary conditions were periodic in the lateral direction.

The left panel is more or less a corroboration that our code reproduces ordinary solidification cells in the absence of stress. Moreover, simulations with stresses satisfying $\varepsilon c_\infty \leq 0.4$ essentially give the same cell shapes, the major change being a slowing down of the dynamics, that is, cells evolve more slowly with than without stress. Stress acts stabilizing in that case as is suggested by the linear stability analysis of Spencer et al. [5]. At larger solutal stress parameter, we observe oscillations. The right panel gives snapshots of an interface for $\varepsilon c_\infty = 0.95$ at various times. We observe that the system does not exhibit traveling waves, even though the boundary conditions do not suppress their appearance. In the small-velocity range that we could study, we always obtained standing waves, never traveling ones.

Figure 18.3 shows how the oscillations start to evolve at early times. It gives the concentrations in the liquid and solid phases, respectively, and the trace of the stress tensor. We notice that there is a phase shift of almost 180° between the liquid concentration and the stress and a phase shift of roughly 90° between the solid concentration and the stress.

These phase shifts drive the oscillations. A high concentration in the liquid tends to slow down interface growth, but a small stress tends to accelerate it, so stress and liquid concentration act against each other, and this conflict promotes oscillations.

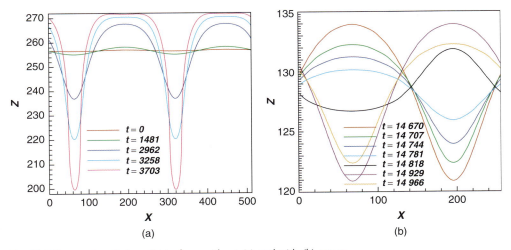

Fig. 18.2 Temporal evolution of interfaces without (a) and with (b) stress.

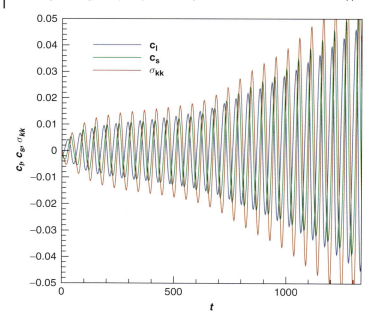

Fig. 18.3 Early dynamics of the concentrations and the trace of the stress tensor at a single interface point.

18.6
Conclusions

To summarize, we have developed a quantitative formulation of a phase-field model for directional solidification in the one-sided limit. Due to its phrasing in terms of liquid- and solid-phase concentrations, it is particularly adapted to the incorporation of stress effects, at least in the dilute-alloy limit. By introducing the coupling to mechanical stresses following compositional inhomogeneities (assuming isotropic dilation), we could perform simulations of directional solidification in the presence of concentration-induced strains. We find, in accord with the work of Spencer *et al.* [5] that small stresses stabilize a planar interface (except in rapid solidification, which we cannot yet simulate), whereas large stresses lead to oscillatory dynamics. Going beyond Spencer et al., we identify the nature of the oscillations and give the full nonlinear dynamics over a long period. At the initial stage of the dynamics, we obtain growth rates of the interface amplitude that agree with linear stability analysis. However, there remains a lot to be done. Besides an investigation into the effects of thermal stresses, it would be interesting to improve our code to be able to simulate higher growth velocities, where smaller stresses already lead to oscillatory dynamics. In order to perform realistic simulations of rapid solidification, the analysis of Section 18.2 would have to be extended to a situation where local equilibrium breaks down. Similarly, an extension of our simulations to the case of eutectic alloys, that is dropping the dilute-alloy assumption, would require a rederivation of the phase-field equations.

Acknowledgments

Support of this work by DFG Grant Nos KA 672/8-1 (within the priority program SPP 1120) and KA 672/9-1 is gratefully acknowledged.

References

1. Mullins, W.W. and Sekerka, R.F. (**1964**) *Journal of Applied Physics*, **35**, 444.
2. Esaka, H. and Kurz, W. (**1985**) *Journal of Crystal Growth*, **72**, 578.
3. Kassner, K., Misbah, C. and Müller-Krumbhaar, H. (**1991**) *Physical Review Letters*, **67**, 1551.
4. Caroli, B., Caroli, C., Roulet, B. and Voorhees, P.W. (**1989**) *Acta Metallurgica*, **37**, 257.
5. Spencer, B.J., Voorhees, P.W., Davis, S.H. and McFadden, G.B. (**1992**) *Acta Metallurgica et Materialia*, **40**, 1599.
6. Minari, F. and Billia, B. (**1994**) *Journal of Crystal Growth*, **140**, 264.
7. Spencer, B.J. and Meiron, D. (**1994**) *Acta Metallurgica et Materialia*, **42**, 3629.
8. Grange, G., Jourdan, C., Gastaldi, J. and Billia, B. (**1997**) *Acta Materialia*, **45**, 2329.
9. Spencer, B.J., Voorhees, P.W. and Tersoff, J. (**2001**) *Physical Review B*, **64**, 235318.
10. Billia, B., Bergeon, N., Nguyen Thi, H. and Jamgotchian, H. (**2004**) *Physical Review Letters*, **93**, 126105.
11. Asaro, R.J. and Tiller, W.A. (**1972**) *Metallurgical Transactions A*, **3**, 1782.
12. Grinfeld, M.A. (**1986**) *Soviet Physics Doklady*, **31**, 831.
13. Nozières, P. (**1993**) *Journal de Physique I (France)*, **3**, 681.
14. Kassner, K. and Misbah, C. (**1994**) *Europhysics Letters*, **28**, 245.
15. Durand, I., Kassner, K., Misbah, C. and Müller-Krumbhaar, H. (**1996**) *Physical Review Letters*, **76**, 3013.
16. Cantat, I., Kassner, K., Misbah, C. and Müller-Krumbhaar, H. (**1998**) *Physical Review E*, **58**, 6027.
17. Coriell, S.R. and Sekerka, R.F. (**1983**) *Journal of Crystal Growth*, **61**, 499.
18. Kelly, F.X. and Ungar, L.H. (**1986**) *Physical Review B*, **34**, 1746.
19. Aziz, M.J. (**1982**) *Journal of Applied Physics*, **53**, 1185.
20. Chen, L.-Q. (**2002**) *Annual Review of Materials Research*, **32**, 113.
21. Boettinger, W.J., Warren, J.A., Beckermann, C. and Karma, A. (**2002**) *Annual Review of Materials Research*, **32**, 163.
22. Wheeler, A.A., Boettinger, W.J. and McFadden, G.B. (**1992**) *Physical Review A*, **45**, 7424.
23. Wheeler, A.A., Boettinger, W.J. and McFadden, G.B. (**1993**) *Physical Review E*, **47**, 1893.
24. Almgren, R.F. (**1999**) *SIAM Journal on Applied Mathematics*, **59**, 2086.
25. Karma, A. (**2001**) *Physical Review Letters*, **87**, 115701.
26. Müller, J. and Grant, M. (**1999**) *Physical Review Letters*, **82**, 1736.
27. Kassner, K. and Misbah, C. (**1999**) *Europhysics Letters*, **46**, 217.
28. Kassner, K., Misbah, C., Müller, J., Kappey, J. and Kohlert, P. (**2001**) *Physical Review E*, **63**, 036117.
29. Kim, S.G., Kim, W.T. and Suzuki, T. (**1999**) *Physical Review E*, **60**, 7186.
30. Kim, S.G., Kim, W.T. and Suzuki, T. (**2004**) *Journal of Crystal Growth*, **263**, 620.
31. Echebarria, B., Folch, R., Karma, A. and Plapp, M. (**2004**) *Physical Review E*, **70**, 061604.
32. Larché, F.C. and Cahn, J.W. (**1985**) *Acta Metallurgica*, **33**, 331.

19

Elastic Effects on Phase Transitions in Multi-component Alloys

Efim A. Brener, Clemens Gugenberger, Heiner Müller-Krumbhaar, Denis Pilipenko, Robert Spatschek, and Dmitrii E. Temkin

In this chapter we focus on the influence of elastic effects on phase transitions in multi-component alloys. In many cases, these effects are small and can be ignored; however, in certain situations they are remarkably important and can significantly alter the kinetics of these processes. A particular example is the melting of a solid phase; provided that it does not occur only at the boundaries of the system, the appearance of liquid inclusions immediately leads to the emergence of elastic deformations due to the density difference between the solid and the melt phase. In this particular situation, even the geometry of the inclusions changes from a spherical to a more oblate shape, as this reduces the stored elastic energy, resulting in new coarsening laws [1, 2].

The origins of elastic effects are manifold: Apart from the mentioned density differences, also externally applied forces, concentration or thermal gradients, surface stress, as well as defects and lattice mismatches are particular reasons for the emergence of elastic effects. Usually, the elastic fields relax fast in comparison to the diffusive timescales that determine the growth kinetics, but structural transitions like martensitic transformations or crack propagation can advance with velocities of the order of the speed of sound, and then inertial effects becomes relevant.

The question of how elastic effects influence the pattern formation processes in multi-component and -phase systems is obviously intimately related to the problem of *selection*, that is, the physical mechanisms that determine the relevant time- and length-scales of the process. In fact, elastic effects can provide a strong perturbation to a system, such that conventional selection mechanisms such as the anisotropy of surface tension can take a back seat. We will therefore discuss first the problem of melting in eutectic and peritectic systems, where elastic effects turn out to be unimportant, and work out the selection mechanisms. Next, we investigate the combined motion of melting and solidification fronts in liquid film migration (LFM), where elastic effects due to coherency strain are essential, as only they provide a driving force for the process. Finally, fast crack growth is the subject of the last section; here, the full dynamical and nonlocal elastic effects are important for this moving boundary problem.

Phase Transformations in Multicomponent Melts. Edited by D. M. Herlach
Copyright © 2008 WILEY-VCH Verlag GmbH & Co. KGaA, Weinheim
ISBN: 978-3-527-31994-7

19.1
Melting of Alloys in Eutectic and Peritectic Systems

The systematic understanding of melting kinetics in alloys, particularly in eutectic and peritectic systems, is much less developed than the investigation of solidification (for a relatively recent review on solidification, see, for example, [3] and references therein). Microstructures, being at the center of materials science and engineering, are formed during the solidification process and, in this sense, the melting process is traditionally less attractive for practical applications. However, the interfacial pattern selection process during melting can be very interesting.

Here, we discuss the problem of contact melting in eutectic and peritectic systems along the boundary between two solid phases (see Figure 19.1 and Figure 19.2). The local concentrations at the L/α and L/β interfaces in such systems are different already because of the chemical difference between the α and β phases, and weak coherency strain effects are of minor relevance. If we also assume that the liquid phase extends up to the sample surfaces, the mentioned elastic deformations due the density difference are not important either, because a weak hydrodynamic flow inside the liquid phase compensates for the density difference.

We concentrate here on the velocity selection problem during melting along the solid-solid interface. The presence of the triple junction (see Figures 19.1b and Figure 19.2b) plays a crucial role in this process. In the classical problem of dendritic growth, where the triple junction is not present, the velocity selection is controlled by tiny singular effects of the anisotropy of the surface energy (for a review, see [4, 5]). The triple junction produces a very strong perturbation of the liquid-solid interfaces and controls the velocity selection. The other important difference compared to the classical dendritic problem is that the kinetics of the

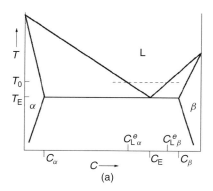
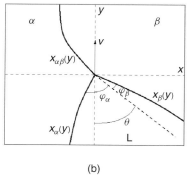
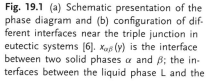

Fig. 19.1 (a) Schematic presentation of the phase diagram and (b) configuration of different interfaces near the triple junction in eutectic systems [6]. $x_{\alpha\beta}(y)$ is the interface between two solid phases α and β; the interfaces between the liquid phase L and the solid phases are denoted by $x_\alpha(y)$ and $x_\beta(y)$. In the steady-state regime this configuration moves along the y axis with a constant velocity v. The origin of the coordinate system is located at the triple junction.

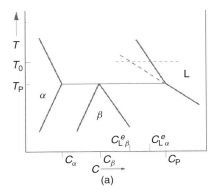 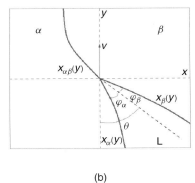

Fig. 19.2 (a) Schematic presentation of the phase diagram and (b) configuration of different interfaces near the triple junction in peritectic systems [6]. $x_{\alpha\beta}(y)$ is the interface between two solid phases α and β; the interfaces between the liquid phase L and the solid phases are denoted by $x_\alpha(y)$ and $x_\beta(y)$. In the steady-state regime this configuration moves along the y axis with a constant velocity v. The origin of the coordinate system is located at the triple junction.

contact melting in eutectic and peritectic systems is controlled by the diffusion inside the needle-like liquid phase and not in the outer phase.

All features of the contact melting process mentioned so far are common for both systems. There is, however, an important difference between eutectic and peritectic systems. In the eutectic system, (Figure 19.1), both interfaces L/α and L/β are melting fronts, while in the peritectic system, (Figure 19.2), the L/α interface is a solidification front, if the temperature T_0 is above the peritectic temperature T_P. In other words, during the melting of peritectic systems the low-temperature β-phase melts while the high-temperature α-phase solidifies. It means that, additionally to the formation of the liquid phase, the polymorphic transition $\beta \to \alpha$ occurs. In this context the contact melting in the peritectic system is similar to the process of LFM, which will be investigated in the next section, where similarly a liquid film is located between the melting and solidification fronts. The rotation of the structure in the vicinity of the triple junction has been discussed in the literature for many different processes: growth from a melt [7, 8], eutectic crystallization [9], discontinuous precipitation [10], diffusion induced grain boundary migration [11], reactive wetting [12]. In all these papers (except [12]), the surface diffusion along the free surfaces of the solid or along the inter-phase and grain boundaries has been discussed.

19.1.1
Isothermal Melting in Eutectic System

We consider now the two-dimensional problem of simultaneous melting of two eutectic phases α and β along the boundary between them. The phase diagram and the configuration of different interfaces near the triple junction are schematically presented in Figure 19.1. The concentrations of the solid phases are close to their equilibrium values, C_α and C_β, at the eutectic temperature. We assume that diffusion in the solid phases is very slow. The temperature of the sample, T_0, is slightly above the eutectic temperature T_E (for the notations see Figure 19.1). At small overheatings, the concentration field in the liquid, $C(x,y)$, obeys the Laplace equation

$$\Delta C = 0 \tag{19.1}$$

We assume local equilibrium at the liquid-solid interfaces. Then the concentrations at the L/α and L/β interfaces are

$$C_{L\alpha}(y) = C_{L\alpha}^e + (C_\beta - C_\alpha) d_\alpha x_\alpha''(y)$$
$$C_{L\beta}(y) = C_{L\beta}^e + (C_\beta - C_\alpha) d_\beta x_\beta''(y) \tag{19.2}$$

Here, the notations of the different concentrations are obvious from the phase diagram (Figure 19.1a); $d_{\alpha(\beta)}$ are the capillary lengths. The mass balance conditions at these interfaces read

$$D\partial C/\partial x = v(C_E - C_\alpha) x_\alpha'(y)$$
$$D\partial C/\partial x = -v(C_\beta - C_E) x_\beta'(y) \tag{19.3}$$

where D is the diffusion coefficient in the liquid phase. Equations 19.2 and 19.3 are written for the case of small angles θ, φ_α, and φ_β at the triple junction. This leads to small angles along the whole interface, $x_\alpha' \ll 1$ and $x_\beta' \ll 1$. In this case, we can use the so-called "lubrication" approximation for the solution of the Laplace equation. In this approximation, one neglects the derivatives with respect to the "slow" variable y in comparison to the derivatives with respect to the "fast" variable x; therefore $C(x,y) = C_0(y) + B(y)x$. The slow variable functions $C_0(y)$ and $B(y)$ can be found from the boundary conditions, Equations 19.2.

From Equations 19.3 and the expression for $C(x,y)$, we find the following relations,

$$x_\alpha' = \frac{D}{v} \frac{B(y)}{(C_E - C_\alpha)}, \quad x_\beta' = -\frac{D}{v} \frac{B(y)}{(C_\beta - C_E)} \tag{19.4}$$

which, together with the expressions for $C_0(y)$, $B(y)$, and Equations 19.2, form a closed system of two second-order differential equations for the front profiles $x_\alpha(y)$ and $x_\beta(y)$. They are subject to the boundary conditions at the triple junction

$$x_\alpha(0) = x_\beta(0) = 0 \tag{19.5}$$
$$x_\alpha'(0) = \varphi_\alpha - \theta \equiv \theta_\alpha; \quad x_\beta'(0) = -(\varphi_\beta + \theta) \equiv -\theta_\beta \tag{19.6}$$

It follows from Equations 19.4 and the boundary conditions Equations 19.5 that

$$x_\alpha(y) = -g x_\beta(y) \tag{19.7}$$

19.1 Melting of Alloys in Eutectic and Peritectic Systems

where $g = (C_\beta - C_E)/(C_E - C_\alpha)$. We also find from Equation 19.6 that

$$\theta_\alpha = \frac{g(\varphi_\alpha + \varphi_\beta)}{1+g}, \quad \theta_\beta = \frac{\varphi_\alpha + \varphi_\beta}{1+g}, \quad \theta = \frac{\varphi_\alpha - g\varphi_\beta}{1+g} \tag{19.8}$$

The relation 19.7 allows eliminating the profile x_α and to write a closed equation for the profile x_β. Moreover, one can integrate this equation once with respect to y, and finally one obtains a nonlinear first-order differential equation for the profile x_β:

$$-(d_\beta + gd_\alpha)x'_\beta = (d_\beta + gd_\alpha)\theta_\beta + \Delta\gamma + \frac{vg}{2D}x_\beta^2 \tag{19.9}$$

Here, $\Delta = (C_{L\beta}^e - C_{L\alpha}^e)/(C_\beta - C_\alpha)$ is the dimensionless overheating above the eutectic temperature. In order to find the selected velocity, one has to solve the arising nonlinear eigenvalue problem, which finally leads to the following expression:

$$v = v^* \frac{2D\Delta^2}{g(d_\beta + gd_\alpha)\theta_\beta^3}$$

where the parameter v^* equals 1.06; for further details, we refer to [6].

19.1.2
Isothermal Melting in Peritectic Systems

The phase diagram and the configuration of different interfaces near the triple junction are schematically presented in Figure 19.2. The temperature of the sample is T_0 and slightly above the peritectic temperature T_P (for the notations see Figure 19.2a). The overheating is assumed to be small and the compositions of the solid phases are close to the equilibrium values C_α and C_β at the peritectic temperature. Equations 19.1–19.3 are valid also for the peritectic system, where C_E should be replaced by C_P; in the second Equation 19.2, the sign in front of the capillary term has to be inverted,

$$C_{L\beta}(y) = C_{L\beta}^e - (C_\beta - C_\alpha)d_\beta x''_\beta(y)$$

This reflects the fact that the equilibrium between the β-phase and the liquid phase corresponds to a negative slope of the liquidus line for the peritectic phase diagram and to a positive slope for the eutectic diagram. As a result, Equations 19.7 and 19.8 remain the same and Equation 19.9 has to be replaced by

$$-(d_\beta - gd_\alpha)x'_\beta = (d_\beta - gd_\alpha)\theta_\beta - \Delta\gamma - \frac{vg}{2D}x_\beta^2 \tag{19.10}$$

We note that the parameters g and Δ for the peritectic system (Figure 19.2) are negative if we define them in the same way as for the eutectic system. Finally, for the steady-state velocity we obtain

$$v = v^* \frac{2D\Delta^2}{|g|(d_\beta + |g|d_\alpha)\theta_\beta^3}, \quad \theta_\beta = \frac{\varphi_\alpha + \varphi_\beta}{1 - |g|} \tag{19.11}$$

As we have already noted earlier, the interface $x_\alpha = |g|x_\beta$ represents now a solidification front. During melting the low-temperature β-phase melts, while the high-temperature α-phase solidifies in the peritectic system. Hence, in addition to the formation of the liquid phase, the polymorphic transition $\beta \to \alpha$ also takes place.

19.2
Combined Motion of Melting and Solidification Fronts

Our next example demonstrates that elastic effects can have significant influence on technically relevant processes. We investigate the phenomenon of LFM, which was first observed during sintering in the presence of a liquid phase [13] or during partial melting of alloys [14] (see [15] for a review). Nowadays, LFM is a well-established phenomenon of great practical importance. In LFM, one crystal is melted and another is solidified, and both solid-liquid interfaces move together with the same velocity. In the investigated alloy systems, the migration velocity is of the order of $10^{-6} - 10^{-5}$ cm s^{-1} and it is controlled by the solute diffusion through a thin liquid layer between the two interfaces [16]. The migration velocity is much smaller than the characteristic velocity of atomic kinetics at the interfaces. Therefore, at the interfaces, both solids should be locally in thermodynamic equilibrium with the liquid phase. On the other hand, these local equilibrium states should be different for the two interfaces to provide a driving force for the process. It is by now well accepted (see, for example, [15, 16]) that the difference of the equilibrium states at the melting and solidification fronts is due to the coherency strain energy. This strain energy arises only at the melting front because of the sharp concentration profile ahead this moving front (diffusion in the solid phase is very slow and the corresponding diffusion length is very small). The solute atoms diffuse ahead of the moving film and the coherency strain energy in such a frontal diffusion zone arises from the solute misfit (for the schematic diagram of LFM see Figure 11 in 17). If such a frontal diffusion zone is sufficiently small owing to the very slow diffusion in the solid phase, the coherency strain energy is not relaxed because of the possible nucleation of misfit dislocations [15] (for an alternative point of view see also [18]). Thus the equilibrium liquid composition at the melting front, which depends on the coherency strain energy and on the curvature of the front, differs from the liquid composition at the unstressed and curved solidification front. This leads to the necessary gradient of the concentration across the liquid film and the process is controlled by the diffusion in the film.

If only the melting front existed, the melting process would be controlled by the very slow diffusion in the mother solid phase and elastic effects would be irrelevant. In LFM, the system chooses a more efficient kinetic path which is controlled by the much faster diffusion in the liquid film, driven by the coherency strain. In this respect, the LFM mechanism is similar to other well-known phenomena such as diffusion-induced grain boundary migration and cellular precipitation [15]. In these processes, a relatively fast diffusion along the grain boundaries controls the kinetics, and the coherency strain energy also plays a controlling role.

Thus, a theoretical description of LFM requires the solution of a free-boundary problem for two combined moving solid-liquid interfaces with a liquid film in between. In Ref. [19] this problem was considered for simplified boundary conditions: the temperature and the chemical composition along each interface were kept constant. Their values are different for the melting and solidification fronts and differ from those far away from the migrating liquid film. This means that any capillary, kinetic, and crystallographic effects at the interfaces were neglected. It was found that under these simplified boundary conditions two co-focal parabolic fronts can move together with the same velocity. The situation is rather similar to a steady-state motion of one parabolic solidification front into a supercooled melt [20], [21] or one parabolic melting front into a superheated solid. In this approximation, the Peclet numbers were found, but the steady-state velocity remained undetermined at that stage, and the problem of velocity selection arises.

Solvability theory has been very successful in predicting certain properties of pattern selection in dendritic growth and a number of related phenomena (see, for example, [4, 5, 22]). This theory has been extended to the three-dimensional case [23], [24]. We note that capillarity is a singular perturbation and the anisotropy of the surface energy is a prerequisite for the existence of the solution.

Here, we discuss the two-dimensional problem of the steady-state motion of a thin liquid film during the process of isothermal melting of a binary alloy, see Figure 19.3. We assume that the diffusion in the solid phases is very slow and the concentration c in the liquid film obeys the Laplace equation. We introduce the normalized concentration $C = (c - c_L)/(c_L - c_S)$ with c_L and c_S being the liquidus and solidus concentrations of the equilibrium phase diagram at a given temperature. Then, the equilibrium concentration and the mass balance conditions at the solidification front read

$$C = d_2 K_2, \qquad V_n = -D \partial C/\partial \mathbf{n} \tag{19.12}$$

Here, V_n is the normal velocity; D is the diffusion coefficient in the liquid film; K is the curvature, assumed to be negative for the interfaces in Figure 19.3, and $\mathbf{n}$ is the normal vector to the interface.

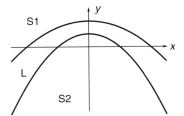

Fig. 19.3 Schematic presentation of two moving, nearly parabolic fronts; S1 and S2 are the melting and growing solids, and L is the liquid film [25].

At the melting front the equilibrium concentration is changed by the presence of the elastic coherency strain energy [1], [5], and also the diffusional flux changes, because in the solid ahead of the melting front the concentration is c_0, which is different from c_S:

$$C = -b\Delta^2 - d_1 K_1, \qquad V_n(1-\Delta) = -D\partial C/\partial \mathbf{n} \tag{19.13}$$

where $\Delta = (c_0 - c_S)/(c_L - c_S)$ is a dimensionless driving force; $b = Y\Omega(da/dc)^2/a^2 f_L''$ is the dimensionless constant which describes the coherency strain energy [16], where Ω is the atomic volume, Y is the bulk elastic modulus, a is the atomic constant, $f_L(c)$ is the free energy of the liquid phase per atom, and f_L'' is the second derivative of $f_L(c)$ at $c = c_L$; and d_i are the anisotropic chemical capillary lengths:

$$d_i(\theta) = d_0[1 - \alpha \cos 4(\theta - \theta_i)], \qquad d_0 = \gamma\Omega/f_L''(c_L - c_S)^2 \tag{19.14}$$

with the isotropic part of the surface energy γ and the anisotropy parameter $\alpha \ll 1$; θ is the angle between the normal n and the direction of motion, θ_i is the direction of the minimum of $d_i(\theta)$ for each of the interfaces.

We measure all lengths in units of the radius of curvature of the Ivantsov parabolic solidification front, R_2. Introducing a parabolic coordinate system,

$$y = (\eta^2 - \beta^2)/2, \qquad x = \eta\beta \tag{19.15}$$

we look for a solution of the Laplace equation for the concentration field $C(\eta, \beta)$. $C_0(\eta) = -b\Delta^2(\eta - 1)/(\eta_0 - 1)$ is the solution without surface tension which satisfies the boundary conditions $C_0(1) = 0$ and $C_0(\eta_0) = -b\Delta^2$. The balance equations at the interfaces, $2P = 2P(1-\Delta)\eta_0 = -\partial C_0/\partial \eta$ lead to the relations for the Peclet number and η_0:

$$P = \frac{VR_2}{2D} = \frac{b\Delta^2}{2(\eta_0 - 1)}, \qquad \eta_0 = \sqrt{\frac{R_1}{R_2}} = \frac{1}{1-\Delta} \tag{19.16}$$

These relations can also be obtained in the proper limit of Equations (27) and (28) of Ref. [19] and give the expressions for the radii of curvature for the two interfaces, R_1 and R_2, for a given velocity V, which remains undetermined at this stage.

The surface tension plays a crucial role in the velocity selection problem. The details of the solution of this selection problem are given in [25]. The result is:

$$\begin{aligned}\sigma = \sigma^* &\sim \Delta \alpha^{5/4} \quad \text{for } \Delta \ll \sqrt{\alpha} \\ \sigma = \sigma^* &\sim \alpha^{7/4} \quad \text{for } \Delta \gg \sqrt{\alpha}\end{aligned} \tag{19.17}$$

where $\sigma = d_0/PR_2$ is the usual stability parameter.

In conclusion, we developed a selection theory for the process of LFM where the strong diffusive interaction between melting and solidification fronts plays a crucial role. This process is very important in practical applications, in particular during sintering in the presence of a liquid phase [13, 15].

19.3
Continuum Theory of Fast Crack Propagation

Fracture and material failure are encountered in various scientific and industrial fields. Despite their importance for many real-life applications, their description still remains a major challenge for solid state physics and materials science. The reason is that the extremely long-ranging elastic forces responsible for fracture are as important as microscopic effects such as atomic bond breaking. Atomistic approaches require more detailed information, either theoretical or empirical, about the atomic properties and interactions in the given systems than we can usually provide reliably. This in turn makes any quantitative description depend heavily on the chosen model and its parameters.

Experiments show that due to dissipative effects, real crack tips are often not sharp, but rounded with a finite tip radius r_0. Other experimental results also indicate that certain features of crack growth are generic [26]. For example, it has been observed that the steady-state velocity of the crack tip saturates well below the Rayleigh speed v_R, which is the theoretical upper limit [27]. When high tensions are applied, tip blunting sets in until finally, tip splitting occurs.

In the picture of Griffith [28], crack growth is interpreted as a competition between two different effects: the release of the elastic energy stored in the material on one hand and the increase of surface or fracture energy due to crack growth on the other hand.

Following this idea, we present here a generic theory of crack propagation that exhibits many important features of cracks. The only ingredients of the model are the linear theory of dynamical elasticity and a melting-solidification transition at the crack front. It turns out that this model has strong links to elastic instabilities at surfaces and interfaces: If a solid is uniaxially stressed, it can release elastic energy by developing morphological perturbations of the interface. These perturbations lead to fast propagating notches that look and behave very much like cracks. This behavior, historically known as the Asaro-Tiller-Grinfeld (ATG) instability, and the idea of how to regularize it in order to describe a steady-state crack growth regime provide the starting point for developing the model.

We restrict ourselves to the two-dimensional case and plane strain. The system under investigation is a solid matrix that contains one crack. Since diffusive bulk transport is not important here, we can consider a pure material. To simplify the theory, we imagine that the crack is filled with a soft condensed phase instead of vacuum, and the growth is then interpreted as a first-order phase transformation of the hard solid matrix to this soft phase [29, 32]. While we assume the mass density ρ of both phases to be equal, we can, however, model the soft phase to have vanishing elastic constants. This way, it does not support any elastic stresses, mimicking an "elastic vacuum". In the bulk, the elastic displacements u_i have to fulfill Newton's equation of motion in its fully dynamical version,

$$\frac{\partial \sigma_{ij}}{\partial x_j} = \rho \ddot{u}_i$$

with a nonvanishing right-hand side. These dynamical effects provide the mechanism to avoid the finite-time cusp singularity one encounters in the framework of elastostatics. Since we work in the limit of linear elasticity, the stress tensor σ_{jk} is connected to the strain tensor ϵ_{ik} via Hooke's law for isotropic elasticity, $\sigma_{kj} = 2\mu\epsilon_{kj} + \lambda\epsilon_{ll}\delta_{kj}$, with the Lamé coefficient λ and the shear modulus μ.

Using only well-accepted thermodynamical principles, it is our goal to construct a minimal model to describe fast crack propagation fully self-consistently in the spirit of an interfacial pattern process. To these means, we assign each phase a chemical potential μ, and the difference

$$\delta\mu = \Omega\left(\frac{1}{2}\sigma_{jk}\epsilon_{jk} - \gamma\kappa\right) \tag{19.18}$$

between two phases at the interface, assuming that the soft phase is stress free because of negligible elastic moduli, provides the driving force for the phase transformation process [33]. Here, γ denotes the interfacial energy per unit area, the interface curvature κ is positive if the crack shape is convex, and Ω denotes the atomic volume. Mechanical equilibrium implies that the solid matrix is free of normal and shear stresses at the crack contour, that is $\sigma_{nn} = \sigma_{n\tau} = 0$, which serve as boundary conditions. We note that these boundary conditions are not entirely trivial and depend in general on surface tension, as well as inertial and coherency effects, as discussed in [34].

Using a pattern formation description, the interface motion has to be expressed through the local driving forces for all surface points of the extended crack. In our model of phase transition kinetics, the motion of the interface is locally expressed by the normal velocity

$$v_n = \frac{D}{\gamma\Omega}\delta\mu \tag{19.19}$$

with a kinetic coefficient D with dimension $[D] = \mathrm{m}^2\mathrm{s}^{-1}$.

Our aim is the determination of not only the crack speed, but also the entire crack shape, including the size of the tip region. Of course, the difficulty arises that the elastic fields have to be determined in domains that continuously change their shape in the course of time and are not known in advance. Those free moving boundary problem are notoriously hard to solve, and we will use two different approaches to tackle this problem numerically.

The first method is based on a sharp-interface approach and is designed in particular for steady-state growth. In contrast to other methods, the limit of fully separated length scales is performed analytically, leading to a very efficient numerical scheme. The macroscopic length of the crack is not considered here, and instead the stress intensity factor K is given [27]. The external elastic driving force is given by the dimensionless parameter $\Delta = K^2(1 - \nu^2)/(2\gamma E)$, with $\Delta = 1$ being the Griffith point.

In order to solve the elastodynamic problem of a crack with finite tip radius r_0, we use a series expansion[35]

$$\sigma_{ij} = \frac{K}{(2\pi r)^{1/2}} \left(f_{ij}^{(0)} + \sum_{n=1}^{N=\infty} \frac{A_n f_{ij,d}^{(n)} + B_n f_{ij,s}^{(n)}}{r^n} \right)$$

The functions $f_{ij,d}^{(n)}(\theta_d, v)$ and $f_{ij,s}^{(n)}(\theta_s, v)$ are the known universal angular distributions for the dilatational and shear contributions which also depend on the propagation velocity.

For the unknown coefficients of expansion A_n and B_n, one has to solve the linear problem of fulfilling the boundary conditions $\sigma_{nn} = \sigma_{n\tau} = 0$ on the crack contour. The tangential stress $\sigma_{\tau\tau}$ is determined only through the solution of the elastic problem. As it enters into the equations of motion (19.18) and (19.19), it leads to a complicated coupled and nonlocal problem.

The idea for solving this problems has been developed originally in the context of dendritic growth [36, 37]. The strategy is as follows: first, for a given guessed initial crack shape and velocity, we determine the unknown coefficients A_n and B_n from the boundary conditions. In a second step, the chemical potential and the normal velocity at each point of the interface are calculated. Afterwards, we obtain the new shape by advancing the crack according to the local interface velocities. This three-step procedure is repeated until the shape of the crack remains unchanged in the co-moving frame of reference, providing a natural way to solve the problem, as it follows the physical configurations to reach the steady state.

For low driving forces, steady-state solutions with a finite tip radius r and a propagation velocity below the Rayleigh speed exist, see Figure 19.4. This demonstrates that inertial effects suppress the finite-time cusp singularity of the ATG instability. By increasing the driving force Δ, we see that the tip curvature becomes negative, hinting at tip splitting.

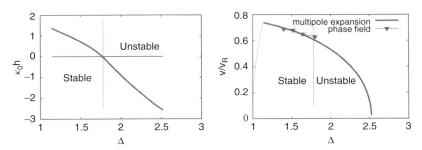

Fig. 19.4 (a) Dimensionless tip curvature as a function of the driving force Δ (h is the asymptotic crack opening in the Lagrangian reference frame). (b) Steady-state velocity as a function of the dimensionless driving force Δ. The results obtained with the phase-field method are included for comparison.

To get further insights into dynamical aspects of crack growth, including the branching regime, we also developed a second method, based on a phase-field description to solve the equations of motion. However, it requires much larger computational efforts, and therefore it is much more difficult to obtain quantitative results of the same quality as the series expansion technique.

The idea of the phase-field method is to introduce a new scalar field ϕ with values $\phi = 0$ for the soft and $\phi = 1$ for the hard phase and a smooth transition between the two phases over an interface width ξ instead of a sharp jump. The elastodynamic equations are then derived from the proper potential energy U by variation with respect to the displacements u_i,

$$\rho \ddot{u}_i = -\frac{\delta U}{\delta u_i} \tag{19.20}$$

and the dissipative phase-field dynamics follows from

$$\frac{\partial \phi}{\partial t} = -\frac{D}{3\gamma \xi} \frac{\delta U}{\delta \phi} \tag{19.21}$$

The potential energy is given by

$$U = \int dV \left(f_s + f_{dw} + f_{el} \right) \tag{19.22}$$

The first two energy density contributions are well known also from other phase-field approaches: $f_{dw} = 6\gamma \phi^2 (1-\phi)^2/\xi$ is the double well potential to ensure that the system stays in one of the two possible bulk phases, and the surface energy density $f_s(\phi) = 3\gamma \xi (\nabla \phi)^2/2$ suppresses unwanted interface fluctuations. The coupling of the phase-field and the elastic terms is given by

$$f_{el} = \mu(\phi) \epsilon_{ij}^2 + \frac{\lambda(\phi)(\epsilon_{ii})^2}{2}$$

with $\mu(\phi) = h(\phi)\mu^{(1)} + (1 - h(\phi))\mu^{(2)}$ and $\lambda(\phi) = h(\phi)\lambda^{(1)} + (1 - h(\phi))\lambda^{(2)}$, where $h(\phi) = \phi^2(3 - 2\phi)$ interpolates between the phases; the superscripts denote the bulk values in the different phases.

In the limit $\xi \to 0$, these equations lead to the correct sharp interface limit above, as described by Equations 19.18 and 19.19. This was carefully proven in [29] for the case of static elasticity. Note that after rescaling, the equations of motion depend only on the driving force Δ and the kinetic coefficient D; in the numerical realization, also the phase-field width ξ and the strip width L appear. If the driving force Δ is very small, the tip radius is determined by a microscopic length scale that can be mimicked by the phase-field interface width ξ [31]; in general, ξ cuts off the tip radius only for very low kinetic coefficients. Otherwise it is fairly big and selected by the parameter D/v_R; for high kinetic coefficients or crack growth in narrow sheets, the saturation is induced by the system size.

In order to get quantitative results that can be compared to the ones obtained by the multipole expansion method, one has to make sure that all length scales are well separated [32].

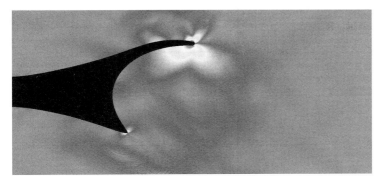

Fig. 19.5 Snapshot of a crack shape in the branching regime. The grey scale visualizes the elastic stress in the solid part of the system.

Figure 19.5 shows a snapshot of a typical branching scenario obtained from phase-field simulation. Here, irregular splitting of the crack tip occurs, followed by a symmetrical growth of the side branches that ends when one finger wins the competition and becomes itself the main branch of the crack.

19.4 Summary

Elastic effects can play a crucial role in phase transitions of multi-component alloys. We have illustrated their influence on various pattern-formation processes and demonstrated that selection of the spatiotemporal structures can be significantly changed accordingly. In cases where coherency effects and elastic stresses are small, especially since long-range liquid transport allows relaxing them, or other chemical driving forces are present, the elastic effects typically give only minor corrections, as we have shown for melting of eutectic and peritectic systems.

The coherency strain due to sharp concentration gradients in solid phases, where bulk diffusion is slow and full equilibrium cannot be established, can, however, act as efficient driving force for example in LFM. Instead of reaching the thermodynamical equilibrium by slow diffusion in a solid matrix, the propagation of a thin melt layer between the solid phases turns out to be kinetically more efficient, and elastic effects propel this process.

Finally, in diffusionless processes like fast crack propagation, elastic effects inevitably play a dominant role, and their accurate treatment poses a major challenge to their full numerical treatment. Despite the long-range interactions and the free-boundary nature of these problems, the emergence of several, very different length scales is a key ingredient. Phase-field methods are a feasible way to tackle crack propagation models in the fully dynamical regime, and eigenmode expansion techniques, which separate the scales already on an analytical level, provide powerful tools for a precise numerical treatment.

References

1 Brener, E.A. and Marchenko, V.I. (**1992**) *JETP Letters*, **56**, 368.
2 Brener, E.A., Iordanskii, S.V. and Marchenko, V.I. (**1998**) *Physical Review Letters*, **82**, 1506.
3 Müller-Krumbhaar, H., Kurz, W. and Brener, E.A. (**2001**) Solidification, in *Phase Transformations in Materials*, (ed G. Kostorz), Willey-VCH Verlag GmbH, Weinheim, p. 81
4 Kessler, D., Koplik, J. and Levine, H. (**1988**) *Advances in Physics*, **37**, 255.
5 Brener, E.A. and Mel'nikov, V.I. (**1991**) *Advances in Physics*, **40**, 53.
6 Brener, E.A. and Temkin, D.M. (**2007**) *Acta Materialia*, **55**, 2785.
7 Voronkov, V.V. (**1978**) *Crystallographica*, **23**, 249.
8 Brener, E.A., Tatarchenko, V.A. and Fradkov, V.E. (**1982**) *Crystallographica*, **27**, 205.
9 Temkin, D.E. (**1986**) *Crystallographica*, **31**, 12.
10 Brener, E.A. and Temkin, D.E. (**2003**) *Acta Materialia*, **51**, 797.
11 Brener, E.A. and Temkin, D.E. (**2002**) *Acta Materialia*, **50**, 1707.
12 Warren, J.A., Boettinger, W.J. and Roosen, A.R. (**1998**) *Acta Materialia*, **46**, 3247.
13 Yoon, D.N. and Hupmann, W.J. (**1979**) *Acta Metallurgica*, **27**, 973.
14 Muschik, T., Kaysser, W.A. and Hehenkamp, T. (**1989**) *Acta Metallurgica*, **37**, 603.
15 Yoon, D.N. (**1995**) *International Materials Reviews*, **40**, 1149.
16 Yoon, D.N., Cahn, J.W., Handwerker, C.A., Blendell, J.E. and Baik, Y.J. (**1986**) *Interface Migration and Control of Microstrucutres*, American Society for Metals, Ohio, p. 19.
17 McPhee, W.A.G., Schaffer, G.B. and Drennan, J. (**2003**) *Acta Metallurgica*, **51**, 3701.
18 Kirkaldy, J.S. (**1998**) *Acta Materialia*, **46**, 5127.
19 Temkin, D.E. (**2005**) *Acta Materialia*, **53**, 2733.
20 Ivantsov, G.P. (**1947**) *Doklady Akademii Nauk SSSR*, **58**, 567.
21 Ivantsov, G.P. (**1952**) *Doklady Akademii Nauk SSSR*, **83**, 573.p.
22 Saito, Y. (**1996**) *Statistical Physics of Crystal Growth*, World Scientific Publishing, Singapore.
23 Ben Amar, M. and Brener, E.A. (**1993**) *Physical Review Letters*, **71**, 589.
24 Brener, E.A. (**1993**) *Physical Review Letters*, **71**, 3653.
25 Brener, E.A. and Temkin, D.M. (**2005**) *Physical Review Letters*, **94**, 184501.
26 Fineberg, J. and Marder, M. (**1999**) *Physics Reports*, **313**, 1.
27 Freund, L.B. (**1998**) *Dynamic Fracture Mechanics*, Cambridge University Press.
28 Griffith, A.A. (**1921**) *Philosophical Transactions of the Royal Society of London, Series A*, **21**, 163.
29 Kassner, K., Misbah, C., Müller, J., Kappey, J. and Kohlert, P. (**2001**) *Physical Review E*, **63**, 036117.
30 Brener, E.A. and Spatschek, R. (**2003**) *Physical Review E*, **67**, 016112.
31 Spatschek, R., Hartmann, M., Brener, E.A., Müller-Krumbhaar, H. and Kassner, K. (**2006**) *Physical Review Letters*, **96**, 015502.
32 Spatschek, R., Müller-Gugenberger, C., Brener, E.A. and Nestler, B. (**2007**) *Physical Review E*, **75**, 066111.
33 Noziéres, P. (**1993**) *Journal de Physique I*, **3**, 681.
34 Spatschek, R. and Fleck, M. (**2007**) *Philosophical Magazine Letters*, **87**, 909.
35 Pilipenko, D., Spatschek, R. and Brener, E.A. (**2007**) *Physical Review Letters*, **98**, 015503.
36 Saito, Y., Goldbeck-Wood, G. and Müller-Krumbhaar, H. (**1987**) *Physical Review Letters*, **58**, 1541.
37 Saito, Y., Goldbeck-Wood, G. and Müller-Krumbhaar, H. (**1988**) *Physical Review A*, **38**, 2148.

20
Modeling of Nonisothermal Multi-component, Multi-phase Systems with Convection

Harald Garcke and Robert Haas

20.1
Introduction

Phase-field models have been successfully used to describe solidification phenomena. Most models studied so far have been restricted to single-component systems or isothermal multicomponent systems. In this chapter, we develop in a systematic way phase-field models for nonisothermal, multicomponent, multiphase systems which also allow for convection. In particular, we want to allow for arbitrary phase diagrams.

In phase-field models, interfaces have a positive interfacial thickness, and quantities such as surface tension or surface energy density do not enter the models in a direct way. This is in particular true in cases where more than two phases appear. Therefore, a calibration of parameters in the model can be difficult. So far, only ad hoc approaches have been used, see, for example, [1, 2]. In this paper we will present models in which the calibration of parameters can be achieved directly. In particular, we can avoid a third phase field attaining nonzero values in an interface between two phases.

The outline of the chapter is as follows. We first review some basic facts of phase-field models for multicomponent systems. In Section 20.3 we describe how Ginzburg–Landau energies for multiphase systems can be constructed in a way guaranteeing that in an interfacial layer between two phases only two phase-field functions appear. This is an important issue not shared by energies that have been used so far. In Section 20.4, thermodynamically consistent phase-field models for convective multicomponent systems are developed. Related sharp-interface models have been discussed in [3]. We will finish with some comments concerning the well-posedness of multiphase-field systems.

Phase Transformations in Multicomponent Melts. Edited by D. M. Herlach
Copyright © 2008 WILEY-VCH Verlag GmbH & Co. KGaA, Weinheim
ISBN: 978-3-527-31994-7

20.2
Phase-field Models for Multicomponent, Multiphase Systems

We are going to formulate a model that allows for N components whose concentrations will be described by a vector $\mathbf{c} = (c_i)_{i=1}^N$ and M phases, where the local phase concentration is given by a vector $\boldsymbol{\phi} = (\phi_\alpha)_{\alpha=1}^M$. Both vectors have to fulfill the constraints

$$\sum_{i=1}^N c_i = 1 \quad \text{and} \quad \sum_{\alpha=1}^M \phi_\alpha = 1$$

For the free energy of the system, we make the ansatz

$$\mathcal{F}(T, \mathbf{c}, \boldsymbol{\phi}) = \int_\Omega (f(T, \mathbf{c}, \boldsymbol{\phi}) + T(\varepsilon a(\boldsymbol{\phi}, \nabla \boldsymbol{\phi}) + \tfrac{1}{\varepsilon} w(\boldsymbol{\phi})))$$

where f is the bulk free-energy density, T is the absolute temperature, and the a- and w-terms describe the interfacial free energy of the phase boundaries. Defining the internal energy density $e = f + Ts$ with the entropy density $s = -f_{,T}$, the energy flux $\mathbf{J}_0$, and the mass fluxes $\mathbf{J}_1, \ldots, \mathbf{J}_k$ we obtain the conservation laws

$$\frac{\partial e}{\partial t} = -\nabla \cdot \mathbf{J}_0 \tag{20.1}$$

$$\frac{\partial c_i}{\partial t} = -\nabla \cdot \mathbf{J}_i, \quad i = 1, \ldots, N \tag{20.2}$$

The fluxes are assumed to be linear in the thermodynamic driving forces and we hence postulate

$$\mathbf{J}_0 = L_{00}(T, \mathbf{c}, \boldsymbol{\phi}) \nabla (\tfrac{1}{T}) + \sum_{j=1}^N L_{0j}(T, \mathbf{c}, \boldsymbol{\phi}) \nabla \left(\frac{-\mu_j}{T}\right)$$

$$\mathbf{J}_i = L_{i0}(T, \mathbf{c}, \boldsymbol{\phi}) \nabla (\tfrac{1}{T}) + \sum_{j=1}^N L_{ij}(T, \mathbf{c}, \boldsymbol{\phi}) \nabla \left(-\frac{\mu_j}{T}\right)$$

Here, $(L_{ij}(T, \mathbf{c}, \boldsymbol{\phi}))_{i,j=0}^N$ is a symmetric and positive semidefinite matrix with $\sum_{i=1}^N L_{ij} = 0$. The conservation laws have to be coupled to equations for the phase field as follows, see [4, 5]:

$$\varepsilon \omega(\boldsymbol{\phi}, \nabla \boldsymbol{\phi}) \frac{\partial \phi_\alpha}{\partial t} = \varepsilon (\nabla \cdot a_{,\nabla \phi_\alpha}(\boldsymbol{\phi}, \nabla \boldsymbol{\phi}) $$
$$- a_{,\phi_\alpha}(\boldsymbol{\phi}, \nabla \boldsymbol{\phi})) - \frac{1}{\varepsilon} w_{,\phi_\alpha}(\boldsymbol{\phi}) - \frac{f_{,\phi_\alpha}(T, \mathbf{c}, \boldsymbol{\phi})}{T} - \lambda \tag{20.3}$$

where ω is possibly an anisotropic kinetic term and λ is a Lagrange multiplier, see also Section 20.3. To be able to separate the kinetics of the different interfaces, we developed a kinetic term of the form

$$\omega(\boldsymbol{\phi}, \nabla \boldsymbol{\phi}) = \omega_0 + \sum_{\alpha < \beta} \omega_{\alpha\beta}(\phi_\alpha \nabla \phi_\beta - \phi_\beta \nabla \phi_\alpha) \tag{20.4}$$

with $\omega_{\alpha\beta}(\mathbf{q}) = 0$ if $\mathbf{q} = 0$. The $\omega_{\alpha\beta}$ can be direction dependent, and with the help of Equation 20.4 we are able to have a specific constitutive form for each possible interface. Examples are

$$\omega_{\alpha\beta}(\mathbf{q}) = \tau_{\alpha\beta}^0 \left(1 + \xi_{\alpha\beta}\left(3 \pm 4\frac{|\mathbf{q}|_4^4}{|\mathbf{q}|^4}\right)\right) - \tau_0$$

or

$$\omega_{\alpha\beta}(\mathbf{q}) = \tau_{\alpha\beta}^0 \left(\max_{1 \leq k \leq r_{\alpha\beta}} \left(\frac{\mathbf{q}}{|\mathbf{q}|} \cdot \xi_{\alpha\beta}^k\right)\right) - \tau_0$$

with $|\mathbf{q}|_4^4 = \sum_{i=1}^d q_i^4$ and $|\mathbf{q}|^4 = \left(\sum_{i=1}^d |q_i^2|\right)^2$. The first example corresponds to cubic anisotropy, whereas the second one leads to a facetted anisotropy with $r_{\alpha\beta}$ corners $\xi_{\alpha\beta}^1, \ldots, \xi_{\alpha\beta}^N$, see [4, 5]. For possible ways to incorporate anisotropy in the interfacial energy, we refer to Section 20.3.

In [4, 5] it was shown that the above system (Equations 20.2, 20.2, 20.3) is thermodynamically consistent. In fact, it was shown that an entropy inequality in the local form

$$\frac{\partial}{\partial t}(s - \varepsilon a(\boldsymbol{\phi}, \nabla \boldsymbol{\phi}) - \frac{1}{\varepsilon}w(\boldsymbol{\phi})) \geq -\nabla \cdot \left(\frac{1}{T}J_0 + \sum_{i=1}^N \left(-\frac{\mu_i}{T}\right) J_i + \varepsilon \sum_{\alpha=1}^N a_{,\nabla\phi_\alpha} \frac{\partial \phi_\alpha}{\partial t}\right)$$

holds. One goal of this work is to develop Ginzburg–Landau energy densities, that is, the functions a and w, such that realistic interfacial energy densities can be incorporated, see Section 20.3, and to generalize the model (Equations 20.2–20.3) to incorporate convection (Section 20.4).

20.3
Multiphase Ginzburg–Landau Energies

We use Ginzburg–Landau type energies of the form

$$\mathcal{F} = \int_\Omega (\varepsilon a(\boldsymbol{\phi}, \nabla \boldsymbol{\phi}) + \tfrac{1}{\varepsilon} w(\boldsymbol{\phi})) \, dx$$

in order to model interfacial energies. In the sharp-interface limit, this energy can be related to classical energies on interfaces. An ad hoc generalization of the double-well potential for two-phase systems to the multiphase case leads to

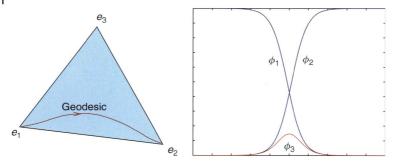

Fig. 20.1 Behavior of the phase-field vector across an interface. In general, third-phase contributions occur.

interfaces in which more than two phase fields are nonzero. Of course, such third-phase contributions, in general, are physically unrealistic.

The behavior of the phase-field vector across an interface is shown in Figure 20.1 where $e_1, \ldots, e_M$ are the standard basic vectors. The behavior of $\boldsymbol{\phi}$ across the interface when considered in the phase space

$$\mathcal{G} := \left\{ \boldsymbol{\phi} \in \mathbb{R}^M \mid \sum_{\alpha=1}^{M} \phi_\alpha = 1, \ \phi_\beta \geq 0, \ \beta = 1, \ldots, M \right\}$$

can be interpreted as a geodesic with respect to a weighted distance in $\mathbb{R}^M$, see [6].

We wish to construct Ginzburg–Landau energies with the following properties:
1. Avoidance of third-phase contributions: In an interface from phase α to phase β only the phase fields α to β are different from zero.
2. Simple calibration of surface energies and model parameters: in particular, we want explicit relations between surface energy densities and parameters in the Ginzburg–Landau energy.

The interfacial energy of an interface can be computed as

$$\gamma_{\alpha\beta} = 2 \inf_{\mathbf{p}} \left\{ \int_{-1}^{1} \sqrt{w(\mathbf{p}) a(\mathbf{p}, \mathbf{p}' \otimes v)} \, dy \right\}$$

where we take the infimum over all vector-valued functions $\mathbf{p}$ in the interval $[-1, 1]$ with $\mathbf{p}(-1) = e_\alpha$ and $\mathbf{p}(1) = e_\beta$. From the corresponding Euler–Lagrange equations, we compute after a suitable reparametrization

$$w_{,\boldsymbol{\phi}}(\boldsymbol{\phi}) + a_{,\boldsymbol{\phi}}(\boldsymbol{\phi}, \partial_z \boldsymbol{\phi} \otimes v) - \frac{d}{dz}(a_{,\nabla\boldsymbol{\phi}}(\boldsymbol{\phi}, \partial_z \boldsymbol{\phi} \otimes v)v) = \boldsymbol{\lambda} \tag{20.5}$$

with a vector-valued Lagrange parameter $\boldsymbol{\lambda} = \lambda(1, \ldots, 1)$ and a critical point $\boldsymbol{\phi}(z)$. This is the equation which one obtains to leading order in an inner expansion

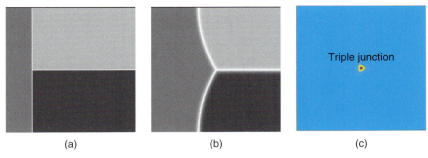

Fig. 20.2 Phase profiles of a phase-field computation with $N = 3$ at time $t = 0.0$ (a) and $t = 0.2$ (b) and the product of all three phase fields at time $t = 0.2$ (c). Away from the triple junction, the product of the phase fields is zero, that is, only up to two phase fields are different from zero.

close to an interface, see [6]. In the case of an obstacle potential, that is, in the case where w is set to be infinity outside of the set $\mathcal{G}$, one has to replace Equation 20.5 by a variational inequality ([1, 2]). We now look for solutions of Equation 20.5 that fulfill (1) and (2), and in [3, 7] we derived specific conditions on the functions a and w. Here we just state a few examples for possible a and w and give the result of a numerical computation for the phase-field equation, see Figure 20.2 and [3] for more details.

20.3.1
Some Examples of Ginzburg–Landau Energies

For a multiphase system with $a = a(\nabla \boldsymbol{\phi})$ as the gradient energy density and w as potential energy, we obtain for the case of equal surface energy densities $\gamma_{\alpha\beta} = \gamma$

$$a(\nabla \boldsymbol{\phi}) = \frac{\gamma}{2} \sum_{\alpha=1}^{M} |\nabla \phi_\alpha|^2 \quad \text{and} \quad w(\boldsymbol{\phi}) = 9\gamma \sum_{\alpha=1}^{M} \phi_\alpha^2 \left(\sum_{\substack{\beta \leq \delta \\ \beta,\delta \neq \alpha}} \phi_\beta \phi_\delta \right)$$

whereas in the case of unequal surface energy densities the following functions fulfill all the desired properties:

$$a(\nabla \boldsymbol{\phi}) = -\frac{1}{2} \sum_{\alpha < \beta} \gamma_{\alpha\beta} \nabla \phi_\alpha \cdot \nabla \phi_\beta \quad \text{and}$$

$$w(\boldsymbol{\phi}) = 9 \sum_{\alpha < \beta} \gamma_{\alpha\beta} \phi_\alpha^2 \phi_\beta^2 + 9 \sum_{\alpha < \beta, \delta} \gamma_{\alpha\beta\delta} \phi_\alpha \phi_\beta \phi_\delta^2$$

where $\gamma_{\alpha\beta\delta} = \gamma_{\alpha\delta} + \gamma_{\beta\delta} - \gamma_{\alpha\beta}$. If we want to consider anisotropy, we choose gradient energies a that depend more generally on $\nabla \boldsymbol{\phi}$ but also depend on $\boldsymbol{\phi}$, see [6].

Typical choices are

$$a(\boldsymbol{\phi}, \nabla\boldsymbol{\phi}) = \sum_{\alpha<\beta} a_{\alpha\beta}\sigma_{\alpha\beta}(\phi_\alpha\nabla\phi_\beta - \phi_\beta\nabla\phi_\alpha) \quad \text{and}$$

$$w(\boldsymbol{\phi}) = 9\sum_{\alpha<\beta} \phi_\alpha^2\phi_\beta^2 \left(1 + 8\sum_{\delta\notin\{\alpha,\beta\}} \phi_\delta\right)$$

We suppose here that the functions $\sigma_{\alpha\beta}$ are two-homogeneous.

An example for an obstacle potential is

$$a(\boldsymbol{\phi}, \nabla\boldsymbol{\phi}) = \sum_{\alpha,\beta} \gamma_{\alpha\beta} A_{\alpha\beta}(\phi_\alpha\nabla\phi_\beta - \phi_\beta\nabla\phi_\alpha) \quad \text{and}$$

$$w(\boldsymbol{\phi}) = \frac{16}{\pi^2}\sum_{\alpha<\beta} \gamma_{\alpha\beta}\phi_\alpha\phi_\beta + \sum_{\alpha<\beta<\delta} \gamma_{\alpha\beta\delta}\phi_\alpha\phi_\beta\phi_\delta$$

where w is defined to be infinity if $\boldsymbol{\phi} \notin \mathcal{G}$. Here, the coefficients $\gamma_{\alpha\beta\delta}$ have to be chosen appropriately large. We refer to Refs [3, 7] for more information on the results in this section.

20.4
Convective Phase-Field Models

In many applications, phase transitions occur in interaction with fluid flow. Therefore one is interested in a mathematical description of phase transitions in convective systems. For isothermal single- and multicomponent systems as well as for nonisothermal one-component systems much work has already been done [8–11]. Finally, a wide selection of topics and references is presented in [12].

In this section, we derive a phase-field model and a sharp-interface model for multicomponent systems with convection in a nonisothermal regime.

20.4.1
Conservation Laws and Entropy Inequality

To postulate the classical balance laws, we assume that $R = R(t)$ is an arbitrary material volume. Thus *mass conservation* is given by

$$\frac{\mathrm{d}}{\mathrm{d}t}\int_{R(t)} \varrho\,\mathrm{d}\mathbf{x} = 0 \tag{20.6}$$

Using Reynold's transport theorem we obtain

$$\int_{R(t)} \left(\frac{\partial\varrho}{\partial t} + \nabla\cdot(\varrho\mathbf{v})\right)\mathrm{d}\mathbf{x} = 0$$

Since $R(t)$ is an arbitrary material volume, we obtain the local version

$$\frac{\partial \varrho}{\partial t} + \nabla \cdot (\varrho \mathbf{v}) = 0 \qquad (20.7)$$

We proceed by postulating that changes in the total momentum of $R(t)$ are due to forces $\mathbf{k}$ acting in the volume and acting on the surface. The latter forces are caused by *mechanical interactions* along the boundary $\partial R(t)$, such as frictional forces or shear forces. Thus the *momentum balance* is given by

$$\frac{d}{dt} \int_{R(t)} \varrho \mathbf{v} \, d\mathbf{x} = \int_{\partial R(t)} \mathbf{T} \mathbf{v}_R d\mathcal{H}^{d-1} + \int_{R(t)} \varrho \mathbf{k} \, d\mathbf{x} \qquad (20.8)$$

where $\mathbf{v}_R$ is the outer unit normal to $\partial R(t)$ and $d\mathcal{H}^{d-1}$ denotes integration with respect to the $(d-1)$-dimensional surface measure. Now we can derive in a standard manner (see e.g. [13]) the momentum balance, in the local form, using Reynold's transport theorem and the mass balance (Equation 20.7), that is:

$$\varrho \left(\frac{\partial \mathbf{v}}{\partial t} + (D \mathbf{v}) \mathbf{v} \right) = \nabla \cdot \mathbf{T} + \varrho \mathbf{k} \qquad (20.9)$$

where $\nabla \cdot \mathbf{T}$ is the *divergence* of the stress tensor $\mathbf{T}$ (for a definition see e.g. [13]). We now postulate that the total energy of $R(t)$ consists of internal energy (with density E) and kinetic energy depending on the material velocity $\mathbf{v}$. Furthermore, we assume that changes of this total energy are due to work by the volume force density $\mathbf{k}$ and stress forces $\mathbf{T}\mathbf{v}_R$ as well as the energy flux density J_E. Finally, we will neglect external heat sources. Thus the *energy balance* is given by

$$\frac{d}{dt} \int_{R(t)} \varrho \left(E + \frac{1}{2} |\mathbf{v}|^2 \right) d\mathbf{x} \qquad (20.10)$$
$$= -\int_{\partial R(t)} J_E \cdot \mathbf{v}_R d\mathcal{H}^{d-1} + \int_{\partial R(t)} (\mathbf{T} \mathbf{v}_R) \cdot \mathbf{v} d\mathcal{H}^{d-1} + \int_{R(t)} \varrho (\mathbf{k} \cdot \mathbf{v}) \, d\mathbf{x}$$

The first term on the right hand-side describes *energy outflow* with energy flux density J_E, the second term accounts for the work by the *surface stress*, and the third term accounts for the work by the *body forces*. The energy identity in its local form is given by

$$\frac{\partial}{\partial t} \left(\varrho \left(E + \frac{|\mathbf{v}|^2}{2} \right) \right) + \nabla \cdot \left(\varrho \mathbf{v} \left(E + \frac{|\mathbf{v}|^2}{2} \right) + J_E - \mathbf{T}^\top \mathbf{v} \right) = \varrho \mathbf{k} \cdot \mathbf{v}$$

and using Equations 20.7 and 20.9 we have

$$\varrho \left(\partial_t E + \mathbf{v} \cdot \nabla E \right) = -\nabla \cdot J_E + \mathbf{T}^\top : \nabla \mathbf{v} \qquad (20.11)$$

We complete our balance laws by the *conservation of species* assuming that concentration changes are due to the concentration fluxes J_i. We note that no chemical reactions take place. Then the *conservation of species* is given by

$$\frac{d}{dt} \int_{R(t)} \varrho c_i \, d\mathbf{x} = -\int_{\partial R(t)} J_i \cdot \mathbf{v}_R d\mathcal{H}^{d-1} \tag{20.12}$$

where J_i, $i = 1, \ldots, N$, denotes the *mass flux* of component i. Again, using Reynold's transport theorem we obtain

$$\varrho \left(\frac{\partial c_i}{\partial t} + \mathbf{v} \cdot \nabla c_i \right) + \nabla \cdot J_i = 0 \tag{20.13}$$

For the fluxes J_i we require $\sum_{i=1}^{N} J_i = 0$ in order to guarantee the constraint $\sum_{i=1}^{N} c_i = 1$ *during the evolution*. Besides, an important requirement of *irreversible thermodynamics* is that the *second law of thermodynamics* holds. This fundamental law follows from the following *entropy inequality*:

$$\frac{d}{dt} \int_{R(t)} \varrho S \, d\mathbf{x} \geq -\int_{\partial R(t)} J_S \cdot \mathbf{v}_R d\mathcal{H}^{d-1} \tag{20.14}$$

which has the local form

$$\varrho \left(\frac{\partial S}{\partial t} + \mathbf{v} \cdot \nabla S \right) + \nabla \cdot J_S \geq 0 \tag{20.15}$$

Here J_S denotes the entropy flux. Using the notion of material derivatives, we obtain our system of balance laws:

$$D_t \varrho = -\varrho \nabla \cdot \mathbf{v} \tag{20.16}$$

$$\varrho D_t c_i = -\nabla \cdot J_i \tag{20.17}$$

$$\varrho D_t \mathbf{v} = \nabla \cdot T + \varrho \mathbf{k} \tag{20.18}$$

$$\varrho D_t E = -\nabla \cdot J_E + T^\top : \nabla \mathbf{v} \tag{20.19}$$

supplemented by the entropy inequality

$$\varrho D_t S \geq -\nabla \cdot J_S \tag{20.20}$$

20.4.2
Exploitation of the Entropy Principle

In order to obtain phase-field-type equations that are derived from *free energies including gradients of the phase fields* we include $\nabla \phi$ in the list of variables which we base our *constitutive theory* on. Since in classical phase-field theories time derivatives of the phase field enter the *entropy inequality* (see [4, 9]) or the *energy balance* (see

[14]) we also include the time derivative $D_t \phi$ into the list of variables. Precisely, we assume that $S, E, T, J_E, J_1, \ldots, J_N$ depend on the variables

$$Y = (\varrho, \mathbf{c}, \nabla \mathbf{c}, \nabla \mathbf{v}, T, \nabla T, \phi, \nabla \phi, D_t \phi), \quad (20.21)$$

where T is the absolute temperature. Analogous to the ideas of Liu and Müller (see [15, 16]), we now use the method of *Lagrange multipliers* to derive restrictions on the constitutive relations that are enforced by the entropy inequality. Under suitable conditions, the existence of Lagrange multipliers can be guaranteed such that

$$\varrho D_t S + \nabla \cdot J_S - \lambda_E(\varrho D_t E + \nabla \cdot J_E - T : \nabla \mathbf{v}) - \lambda_\varrho(D_t \varrho + \varrho \nabla \cdot \mathbf{v}) \quad (20.22)$$
$$- \sum_{i=1}^{N} \lambda_{c_i}(\varrho D_t c_i + \nabla \cdot J_i) - \lambda_\mathbf{v} \cdot (\varrho D_t \mathbf{v} - \nabla \cdot T - \mathbf{k}) \geq 0$$

holds for all fields $(\varrho, \mathbf{c}, \mathbf{v}, T, \phi)$, see [3]. In the following we will assume that $\lambda_E = \frac{1}{T}$, which can be obtained by an appropriate normalization of the temperature (see Alt and Pawlow [17] or arguments according to Müller, see [16], pp. 16, 184). Defining the *free energy* $F = E - TS$, we obtain from Equation 20.22 after *multiplying by* $-T$:

$$\varrho D_t F + \nabla \cdot (J_E - TJ_S) + \varrho S D_t T + J_S \cdot \nabla T - T : \nabla \mathbf{v} + T\lambda_\varrho(D_t \varrho + \varrho \nabla \cdot \mathbf{v})$$
$$+ T \sum_{i=1}^{N} \lambda_{c_i}(\varrho D_t c_i + \nabla \cdot J_i) + T\lambda_\mathbf{v} \cdot (\varrho D_t \mathbf{v} - \nabla \cdot T - \mathbf{k}) \leq 0 \quad (20.23)$$

Using the chain rule for material derivatives, we derive

$$\varrho \left(F_{,\varrho} + \frac{\lambda_\varrho}{\varrho} T \right) D_t \varrho + \varrho(F_{,T} + S) D_t T + \varrho F_{,\nabla T} \cdot D_t \nabla T + \varrho(F_{,\mathbf{c}} + T\lambda_\mathbf{c}) \cdot D_t \mathbf{c}$$
$$+ \varrho(F_{,\nabla \mathbf{c}} : D_t \nabla \mathbf{c} + F_{,\nabla \mathbf{v}} : D_t \nabla \mathbf{v} + F_{,\phi} \cdot D_t \phi + F_{,\nabla \phi} : D_t \nabla \phi + F_{,D_t \phi} \cdot D_t^2 \phi)$$
$$+ \nabla \cdot (J_E - TJ_S) + J_S \cdot \nabla T - T : \nabla \mathbf{v} + T\lambda_\varrho \varrho \nabla \cdot \mathbf{v}$$
$$+ T \sum_{i=1}^{N} \lambda_{c_i} \nabla \cdot J_i + T\lambda_\mathbf{v} \cdot (\varrho D_t \mathbf{v} - \nabla \cdot T - \mathbf{k}) \leq 0$$

where $F_{,T}$, $F_{,\nabla T}$, and so on. denote the derivatives with respect to variables corresponding to T, ∇T, and so on. Since this inequality has to hold *for all fields* with $\mathbf{c} \in \Sigma^N$ and $\phi \in \Sigma^M$, we obtain that the terms appearing linear will vanish. Hence we obtain

$$\lambda_\varrho = -\frac{\varrho F_{,\varrho}}{T}, \quad S = -F_{,T}, \quad F_{,\nabla T} = 0$$
$$\Pi^N \lambda_\mathbf{c} = -\frac{1}{T} \Pi^N F_{,\mathbf{c}}$$
$$\Pi^N F_{,\nabla \mathbf{c}} = 0, \quad F_{,\nabla \mathbf{v}} = 0, \quad \Pi^M F_{,D_t \phi} = 0, \quad \lambda_\mathbf{v} = 0$$

where $\mathbf{\Pi}^K$ is the projection on $T\Sigma^K$, and for a matrix $A = (A_{ij})_{i=1,\ldots,K, j=1,\ldots,d}$ we define

$$\left(\mathbf{\Pi}^K A\right)_{ij} = a_{ij} - \frac{1}{K}\sum_{k=1}^{K} a_{kj}$$

for $K \in \{M, N\}$.

Using the commutator rule and defining the *chemical potentials* $\boldsymbol{\mu}' = \mathbf{\Pi}^N F_{,c}$ as well as $\boldsymbol{\mu} = (-1, \boldsymbol{\mu}')$ we obtain the inequality

$$\varrho(F_{,\phi}\cdot D_t\phi + F_{,\nabla\phi} : \nabla D_t\phi) - (T + F_{,\varrho}\varrho^2 I + F_{,\nabla\phi}\otimes\nabla\phi) : \nabla\mathbf{v}$$
$$+ \nabla\cdot\left(J_E - TJ_S - \sum_{i=1}^{N}\mu_i J_i\right) + J_S\cdot\nabla T + \sum_{i=1}^{N} J_i\cdot\nabla\mu_i \leq 0$$

where we have set $F_{,\nabla\phi}\otimes\nabla\phi = \sum_{i=1}^{M} F_{,\nabla\phi_i}\otimes\nabla\phi_i$. For simplicity we set $S = T + F_{,\varrho}\varrho^2 I + F_{,\nabla\phi}\otimes\nabla\phi$, and after elementary calculations we obtain

$$(\varrho F_{,\phi} - \nabla\cdot(\varrho F_{,\nabla\phi}))\cdot D_t\phi - S : \nabla\mathbf{v} + J_S\cdot\nabla T + \sum_{i=1}^{N} J_i\cdot\nabla\mu_i \quad (20.24)$$
$$+ \nabla\cdot\left(J_E - TJ_S - \sum_{i=1}^{N}\mu_i J_i + \varrho F_{,\nabla\phi} D_t\phi\right) \leq 0$$

Since $\mathbf{\Pi}^M F_{,D_t\phi} = 0$ we have

$$0 = \left(\mathbf{\Pi}^M F_{,D_t\phi}\right)_{,c_i} = \mathbf{\Pi}^M F_{,c_i, D_t\phi}$$

Hence $\mathbf{\Pi}^M \mu_{i,D_t\phi} = 0$ and $\mu_{i,D_t\phi}\cdot\partial_{x_k} D_t\phi = 0$, and then $\nabla\mu$ does not depend on $\nabla D_t\phi$. In order to obtain a model with $F_{,\nabla\phi} \neq 0$, we need that

$$J = J_E - TJ_S - \sum_{i=1}^{N}\mu_i J_i$$

depends on $D_t\phi$. We do not aim to derive the most general models and hence we assume that J is *affine linear* in $D_t\phi$ (see also [18]), that is

$$J = J^1 + J^2 D_t\phi$$

where J^1, J^2 do not depend on $D_t\phi$. Then we obtain

$$J^2 = -\varrho F_{,\nabla\phi}$$

We assume the representation

$$J_E = J_E^1 + J_E^2 \, D_t \phi$$
$$J_S = J_S^1 + J_S^2 \, D_t \phi$$
$$J_i = J_i^1 + J_i^2 \, D_t \phi, \; i = 1, \ldots, N$$

where $J_E^1, J_E^2, J_S^1, J_S^2, J_i^1, J_i^2$, $(i = 1, \ldots, N)$ do not depend on $D_t \phi$. Hence we obtain

$$\left(\varrho F_{,\phi} - \nabla \cdot (\varrho F_{,\nabla \phi}) + \sum_{i=1}^{N} \nabla \mu_i J_i^2 + \nabla T J_S^2 \right) \cdot D_t \phi$$
$$- S : \nabla v + \nabla \cdot J^1 + \sum_{i=1}^{N} \nabla \mu_i \cdot J_i^1 + \nabla T \cdot J_S^1 \leq 0$$

Suppose that the fluxes for $D_t \phi$ have the standard form, that is, $J^1 = 0$, that is,

$$J_S^1 = \frac{1}{T} J_E^1 - \sum_{i=1}^{N} \frac{\mu_i}{T} \cdot J_i^1$$

Then we obtain

$$\left(\varrho F_{,\phi} - \nabla \cdot (\varrho F_{,\nabla \phi}) + \sum_{i=1}^{N} \nabla \mu_i J_i^2 + \nabla T J_S^2 \right) \cdot D_t \phi \qquad (20.25)$$
$$- S : \nabla v + T \sum_{i=1}^{N} \nabla \frac{\mu_i}{T} \cdot J_i^1 - T \nabla \frac{1}{T} \cdot J_E^1 \leq 0$$

and this inequality must be fulfilled for all values of $D_t \phi$, ∇v, $\nabla \frac{\mu_0}{T}, \ldots, \nabla \frac{\mu_N}{T}$ where $\mu_0 = -1$. If we define $\frac{\mu}{T} = (\frac{\mu_0}{T}, \ldots, \frac{\mu_N}{T})$ and $X = (D_t \phi, \nabla v, \nabla \frac{\mu}{T})$ inequality (Equation 20.25) admits the abstract form $-A(X) \cdot X \leq 0$, which has to hold for all tuples $X \in \Sigma^M \times \mathbb{R}^{(d \times d) \times ((N+1) \times d)}$, where M denotes the number of phases. Then Equation 20.25 yields that the existence of functions such that, see [3],

$$A_1(Y') : \Sigma^M \to \mathbb{R}^M, \qquad B_1(Y') : \Sigma^M \to \mathbb{R}^{d \times d},$$
$$A_2(Y') : \mathbb{R}^{d \times d} \to \mathbb{R}^M, \qquad B_2(Y') : \mathbb{R}^{d \times d} \to \mathbb{R}^{d \times d},$$
$$A_3(Y') : \mathbb{R}^{(N+1) \times d} \to \mathbb{R}^M \qquad B_3(Y') : \mathbb{R}^{(N+1) \times d} \to \mathbb{R}^{d \times d}$$
$$C_1(Y') : \Sigma^M \to \mathbb{R}^{(N+1) \times d},$$
$$C_2(Y') : \mathbb{R}^{d \times d} \to \mathbb{R}^{(N+1) \times d},$$
$$C_3(Y') : \mathbb{R}^{(N+1) \times d} \to \mathbb{R}^{(N+1) \times d}$$

which are linear for all tuples $Y' = (\varrho, c, \nabla c, \nabla v, T, \nabla T, \phi, \nabla \phi, D_t \phi, \nabla \frac{\mu}{T})$ and these functions fulfill

$$\nabla \cdot (\varrho F_{,\nabla \phi}) - \varrho F_{,\phi} - J_S^2 \cdot \nabla T - \sum_{i=1}^{N} J_i^2 \cdot \nabla \mu_i$$
$$= A_1 \, D_t \phi + A_2 \nabla v + A_3 \nabla \frac{\mu}{T} \qquad (20.26)$$

$$S = B_1 D_t \varphi + B_2 \nabla \mathbf{v} + B_3 \nabla \frac{\mu}{T} \qquad (20.27)$$

$$(J_E^1, J_1^1, \ldots, J_N^1) = C_1 D_t \varphi + C_2 \nabla \mathbf{v} + C_3 \nabla \frac{\mu}{T} \qquad (20.28)$$

for all tuples $(D_t \varphi, \nabla \mathbf{v}, \nabla \frac{\mu}{T})$. In order to fulfill the second law, the matrix

$$\begin{pmatrix} A_1 & A_2 & A_3 \\ B_1 & B_2 & B_3 \\ C_1 & C_2 & C_3 \end{pmatrix}$$

is positive semidefinite on an appropriate subspace, see [3]. Equation 20.26 gives a relation between thermodynamical driving forces on the left-hand side and the derivatives of φ, $\mathbf{v}$, $\frac{\mu}{T}$ on the right-hand side. It turns out that (Equation 20.26) is a generalized phase-field equation. Equations 20.27 and 20.28 give very general representations of S and the fluxes $J_E, J_1, \ldots, J_N$ in terms of the quantities $D_t \varphi$, $\nabla \mathbf{v}$ and $\nabla \frac{\mu}{T}$ where the functions B_i and C_i may depend on Y' for $i \in \{1, 2, 3\}$. The following example shows that the usual choices for the phase-field equations, the tensors, and fluxes are special cases of Equations 20.26–20.28.

20.4.2.1 Example

We set A_2, A_3, B_1, B_3, C_1, C_2 equal to zero. Furthermore, let

$$A_1(X) = \beta(X) T I$$
$$B_2(X) \nabla \mathbf{v} = 2\nu E + \lambda \operatorname{tr}(E) I$$
$$\mathbf{e}_l C_3(X) \mathbf{e}_m = -(L_{ij})_{ij=0}^N, \text{ for all } l, m \in \{1, \ldots, d\}$$

with positive $\beta \in \mathbb{R}$ as well as $2E = \nabla \mathbf{v} + (\nabla \mathbf{v})^T$ and appropriately chosen $\nu = \nu(T, \varphi)$, $\lambda = \lambda(T, \varphi) \in \mathbb{R}$ such that S becomes a positive definite tensor. Finally, let $(L_{ij}(T, \mathbf{c}, \varphi))_{i,j=0}^N$ be a symmetric and positive semidefinite matrix with $\sum_{i=1}^N L_{ij} = 0$ as well as $J_S^2 = J^2$ and $J_i^2 = 0$ for $i = 1, \ldots, N$. Then we extract the following equations:

$$\beta D_t \varphi = \nabla \cdot \left(\frac{\varrho}{T} F_{,\nabla \varphi} \right) - \frac{\varrho}{T} F_{,\varphi} \qquad (20.29)$$

$$J_i = \sum_{j=0}^N L_{ij} \nabla \frac{-\mu_j}{T}$$

$$J_S = \frac{1}{T} J_E - \sum_{i,j=1}^N \frac{\mu_i}{T} L_{ij} \nabla \frac{-\mu_j}{T} + \frac{\varrho F_{,\nabla \varphi}}{T} D_t \varphi \qquad (20.30)$$

Finally, we derive from Equation 20.25 that

$$-\beta T (D_t \varphi)^2 - S : \nabla \mathbf{v} - T \sum_{i,j=0}^N \nabla \frac{\mu_i}{T} L_{ij} \nabla \frac{\mu_j}{T} \leq 0$$

holds. Thus the entropy inequality (Equation 20.15) is fulfilled.

20.5
Mathematical Analysis

In Ref. 3, Haas considered a system of phase-field equations for which coupling to other fields was neglected and proved an existence result. The resulting system is nonlinear parabolic and the main mathematical difficulty lies in the fact that quadratic terms in $\nabla \phi$ appear in the equation. Such equations are difficult to handle mathematically and Haas [3] was only able to show existence of solutions in one space dimension.

References

1 Garcke, H., Nestler, B. and Stoth, B. (**1999**) "A multi phase field concept: numerical simulations of moving phase boundaries and multiple junctions". *SIAM Journal on Applied Mathematics*, **60** (1), 295–315.

2 Garcke, H., Nestler, B. and Stoth, B. (**1994**) "Anisotropy in multi-phase systems: a phase field approach". *Interfaces and Free Boundaries*, **1**, 175–98.

3 Haas, R. (**2007**) Modeling and analysis for general non-isothermal convective phase field systems, Ph.D. thesis, University Regensburg, *http://www.opus-bayern.de/uni-regensburg/volltexte/2007/783*.

4 Garcke, H., Nestler, B. and Stinner, B. (**2004**) "A diffuse interface model for alloys with multiple components and phases". *SIAM Journal on Applied Mathematics*, **64** (3), 775–99.

5 Garcke, H., Nestler, B. and Stinner, B. (**2005**) "Multicomponent alloy solidification: phase-field modelling and simulations". *Physical Review E*, **71** (4), 041609-1–041609-6.

6 Garcke, H., Nestler, B. and Stoth, B. (**1998**) "On anisotropic order parameter models for multi phase systems and their sharp interface limits". *Physica D*, **115**, 87–108.

7 Garcke, H., Haas, R. and Stinner, B. "On Ginzburg-Landau free energies for multi-phase systems" (in preparation).

8 Anderson, D.M., McFadden, G.B. and Wheeler, A.A. (**2000**) "A phase-field model of solidification with convection". *Physica D*, **135**, 175–94.

9 Blesgen, T. (**1999**) "A generalization of the Navier-Stokes equations to two-phase flows". *Journal of Physics D: Applied Physics*, **32**, 1119–23.

10 Beckermann, C. et al. (**1999**) "Modelling melt convection in phase-field simulations of solidification". *Journal of Computational Physics*, **154**, 468–96.

11 Gurtin, M.E., Polignone, D. and Viñals, J. (**1996**) "Two-phase binary fluids and immiscible fluids described by an order parameter". *Mathematical Models and Methods in Applied Sciences*, **6**, 815–31.

12 Chen, L.Q. (**2002**) "Phase-field models for microstructural evolution". *Annual Review of Materials Research*, **32**, 113–40.

13 Gurtin, M.E. (**1981**) *An Introduction to Continuum Mechanics*, Academic Press, New York.

14 Fried, E. and Gurtin, M.E. (**1993**) "Continuum theory of thermally induced phase transitions based on an order parameter". *Physica D*, **68**, 326–43.

15 Liu, I.S. (**1972**) "Method of Lagrange multipliers for exploitation of the entropy principle". *Archive for Rational Mechanics and Analysis*, **46** (2), 131–48.

16 Müller, I. (**1985**) *Thermodynamics, Interactions of Mechanics and Mathematics Series*, Pitman, Boston.

17 Alt, H.W. and Pawlow, I. (**1994**) "On the entropy principle of phase transition models with a conserved order parameter". *Advances in Mathematical Sciences and Applications*, **59**, 389–416.

18 Alt, H.W. and Pawlow, I. (**1992**) "A mathematical model of dynamics of non-isothermal phase separation". *Physica D*, **59**, 389–416.

21
Phase-field Modeling of Solidification in Multi-component and Multi-phase Alloys

Denis Danilov and Britta Nestler

21.1
Introduction

Alloy systems with multiple components are an important class of materials, in particular for technological applications and processes. The microstructure formation of a material plays a central role for a broad range of mechanical properties and, hence, for the quality and the durability of the material. From the point of view of continuous optimization of material properties, the study of pattern formation during alloy solidification has been a focus of many experimental and, recently, computational work. Since the microstructure characteristics are a result of the process conditions used during production, the analysis of the fundamental correlation between the processing pathway and the microstructure is of fundamental importance. Multiple components in alloys are combined with the appearance of multiple phases leading to complex phase diagrams, various phase transformations, and different types of solidification. Modeling and numerical simulations aim at predicting microstructure evolution in multi-component alloys in order to virtually design materials. However, the great number of material parameters and of physical variables involved in systems yields a complexity that remains a big challenge for future work. In particular, gaining statistically meaningful data from computations requires simulations in sufficiently large domains with a tremendous need of memory and computing time resources. To treat complex systems, high-performance computing, parallelization, and optimized algorithms including adaptive mesh generators are mandatory.

21.2
Phase-field Model for Multicomponent and Multiphase Systems

In this section, we briefly introduce a phase-field model for a general class of multi-component and multi-phase alloy systems consisting of K components and N different phases in a domain Ω. Details of the model construction are described in [1, 2].

It is assumed that the system is in mechanical equilibrium and, for simplicity, that pressure and mass density are constant. The concentrations of the components are represented by a vector $c(\mathbf{x}, t) = (c_1(\mathbf{x}, t), \ldots, c_K(\mathbf{x}, t))$. Similarly, the phase fractions are described by a vector-valued order parameter $\boldsymbol{\phi}(\mathbf{x}, t) = (\phi_1(\mathbf{x}, t), \ldots, \phi_N(\mathbf{x}, t))$. The variable $\phi_\alpha(\mathbf{x}, t)$ denotes the local fraction of phase α. It is required that the concentrations of the components and the phase-field variables fulfill the constraints $\sum_{i=1}^{K} c_i = 1$ and $\sum_{\alpha=1}^{N} \phi_\alpha = 1$. The physical effects occurring during solidification such as heat and mass transfer, the release of latent heat, the Gibbs–Thomson relation, and interface kinetics are obtained on the basis of an entropy functional $S(e, c, \boldsymbol{\phi})$ of the form

$$S(e, c, \boldsymbol{\phi}) = \int_\Omega \left(s(e, c, \boldsymbol{\phi}) - \left(\varepsilon a(\boldsymbol{\phi}, \nabla \boldsymbol{\phi}) + \frac{1}{\varepsilon} w(\boldsymbol{\phi}) \right) \right) dx \qquad (21.1)$$

depending on the internal energy e, the concentrations $c_i, i = 1, \ldots, K$, and the phase fields $\phi_\alpha, \alpha = 1, \ldots, N$. The first term in the entropy functional $s(e, c, \boldsymbol{\phi})$ is a bulk entropy density. The second and third summands $a(\boldsymbol{\phi}, \nabla \boldsymbol{\phi})$ and $w(\boldsymbol{\phi})$ model surface entropy densities taking into account the thermodynamics of the interfaces. As typical in diffuse-interface models, ε is a small length scale parameter related to the thickness of the diffuse interface.

Knowing the free-energy densities of the pure phases $f_\alpha(T, c)$, the total free energy $f(T, c, \boldsymbol{\phi})$ is obtained as a suitable interpolation of the f_α. The Gibbs relation reads (observe that owing to the assumptions the Gibbs free energy and the Helmholtz free energy coincide up to a constant)

$$df = f_{,T} \, dT + \sum_i f_{,c_i} \, dc_i + \sum_\alpha f_{,\phi_\alpha} \, d\phi_\alpha = -s \, dT + \sum_i \mu_i \, dc_i + \sum_\alpha r_\alpha \, d\phi_\alpha$$
(21.2)

where T is the temperature, $s = -f_{,T}$ is the entropy density, $\mu_i = f_{,c_i}$ are the chemical potentials, and $r_\alpha = f_{,\phi_\alpha}$ are the potentials due to the appearance of different phases. Here, $f_{,X}$ denotes the partial derivative of the free energy f with respect to X. The internal energy density is given by $e = f + sT$. From this relation it can be derived that $s_{,e} = \frac{1}{T}$ and $s_{,c_i} = \frac{-\mu_i}{T}$. Through the free energies f_α, a general class of phase diagrams for multi-phase multi-component alloy systems can be incorporated into the phase-field model. The model allows for systems with general free energies $f_\alpha(T, c)$, being convex in c and concave in T.

The gradient entropies $a(\boldsymbol{\phi}, \nabla \boldsymbol{\phi})$ in Equation 21.1 can be expressed in terms of a generalized antisymmetric gradient vector $q_{\alpha\beta} = \phi_\alpha \nabla \phi_\beta - \phi_\beta \nabla \phi_\alpha$ by

$$a(\boldsymbol{\phi}, \nabla \boldsymbol{\phi}) = \sum_{\alpha < \beta} A_{\alpha\beta}(q_{\alpha\beta}) = \sum_{\alpha < \beta} \gamma_{\alpha\beta} (a_{\alpha\beta}(q_{\alpha\beta}))^2 |q_{\alpha\beta}|^2 \qquad (21.3)$$

where $\gamma_{\alpha\beta}$ represent surface entropy densities. The formulation using the generalized gradient vector $q_{\alpha\beta}$ allows distinguishing between the physics of each phase (or grain) boundary by providing enough degrees of freedom. Anisotropy of the surface entropy density is modeled by the factor $(a_{\alpha\beta}(q_{\alpha\beta}))^2$ depending on the orientation of the interface. Phase boundaries with isotropic surface entropies are realized by $a_{\alpha\beta}(q_{\alpha\beta}) = 1$. The function $w(\phi)$ can be chosen in the form of a multiwell or multiobstacle potential. For details, we refer to [1, 2].

The governing set of equations follows from conservation laws for the internal energy e and the concentrations $c_i, i = 1, \ldots, K$ coupled to a gradient flow for the nonconserved phase-field variables $\phi_\alpha, \alpha = 1, \ldots, N$. The equations are derived by variational differentiation of the entropy functional $S(e, c, \phi)$ ensuring energy and mass conservation and the increase of total entropy. They read

$$\frac{\partial e}{\partial t} = -\nabla \cdot \left(L_{00} \nabla \frac{1}{T} + \sum_{j=1}^{K} L_{0j} \nabla \left(\frac{-\mu_j}{T} \right) \right) \quad (21.4)$$

$$\frac{\partial c_i}{\partial t} = -\nabla \cdot \left(L_{i0} \nabla \frac{1}{T} + \sum_{j=1}^{K} L_{ij} \nabla \left(\frac{-\mu_j}{T} \right) \right) \quad (21.5)$$

$$\tau \varepsilon \frac{\partial \phi_\alpha}{\partial t} = \varepsilon \left(\nabla \cdot a_{,\nabla\phi_\alpha}(\phi, \nabla\phi) - a_{,\phi_\alpha}(\phi, \nabla\phi) \right) - \frac{w_{,\phi_\alpha}(\phi)}{\varepsilon} - \frac{f_{,\phi_\alpha}(T, c, \phi)}{T} - \lambda \quad (21.6)$$

where $\nabla \cdot (\ldots)$ denotes the divergence of the term in the brackets. $a_{,\phi_\alpha}$, $w_{,\phi_\alpha}$, $f_{,\phi_\alpha}$, and $a_{,\nabla\phi_\alpha}$ are the derivatives of the energy contributions with respect to ϕ_α and $\nabla\phi_\alpha$, respectively. The parameter λ in Equation 21.6 is a Lagrange multiplier guaranteeing that the constraint $\sum_{\alpha=1}^{N} \phi_\alpha = 1$ is preserved, that is

$$\lambda = \frac{1}{N} \sum_{\alpha=1}^{N} \left[\varepsilon \left(\nabla \cdot a_{,\nabla\phi_\alpha}(\phi, \nabla\phi) - a_{,\phi_\alpha}(\phi, \nabla\phi) \right) - \frac{w_{,\phi_\alpha}(\phi)}{\varepsilon} - \frac{f_{,\phi_\alpha}(T, c, \phi)}{T} \right] \quad (21.7)$$

In referring to nonequilibrium thermodynamics, we postulate the fluxes for the conserved quantities to be linear combinations of the thermodynamical driving forces $\nabla \frac{\delta S}{\delta e} = \nabla \frac{1}{T}$ and $\nabla \frac{\delta S}{\delta c_i} = \nabla \frac{-\mu_i}{T}$.

For computations based on diffuse-interface models, it is required that the spatial resolution of the numerical method must be greater than the thickness of the diffusive phase boundary layer. The interfacial thickness itself must be less than the characteristic scale of the growing microstructure. In this case, a nonuniform grid with adaptive refinement can dramatically reduce the use of computational resources against a uniform grid with the same spatial resolution, Figure 21.1.

21.3
Modeling of Dendritic Growth

As first results, we describe the application of the phase-field model to simulate dendritic growth structures in pure Ni at low, moderate, and high undercoolings [3]. The results of the numerical simulations have been compared with the available experimental data [4] and with the analytical Brener theory [5], see Figure 21.2. We find

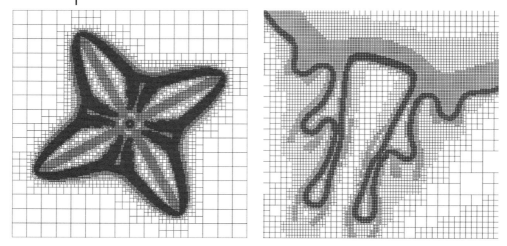

Fig. 21.1 Adaptive mesh used for the simulation of a dendritic structure: complete dendrite (left image) and magnification of the sidebranch region (right image).

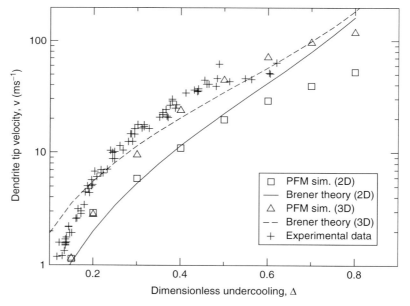

Fig. 21.2 Tip velocity of nickel dendrites plotted against the dimensionless undercooling Δ. The data of phase-field simulations (squares and triangles) are shown in comparison with the theoretical predictions by Brener [5]. The recently measured experimental data of pure nickel solidification (crosses) [4] confirm the simulation results in the undercooling range $0.4 \leqslant \Delta \leqslant 0.6$.

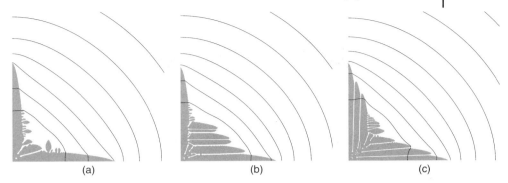

Fig. 21.3 Nonisothermal phase-field simulations of dendritic solidification in $Ni_{99}Zr_1$ for three different undercoolings: (a) $\Delta T = 60$ K, (b) $\Delta T = 75$ K, and (c) $\Delta T = 90$ K. The isolines refer to the temperature field. Only one-quarter of the dendrite is shown in a domain of size $(4.8 \times 10^{-6}) \times (4.8 \times 10^{-6})$ m.

a good agreement between the simulated dendrite velocities and the appropriate 2D and 3D Brener theory [5] for dimensionless undercoolings $\Delta = 0.3, \ldots, 0.6$. Furthermore, our simulation results are consistent with recent experimental measurements [4] within the same range of undercoolings.

The results have been extended further to modeling of dendritic solidification in dilute binary Ni–Zr alloy [6]. Predictions of a sharp-interface model and of the diffuse-interface model describing the phase transition under the consideration of both thermal and solutal diffusion are compared with the experimental results evaluating the dendritic tip velocity in electromagnetically levitated Ni–Zr samples. For the nonisothermal simulations, there is a disagreement between the phase-field and the sharp-interface predictions. In the sharp-interface model, the temperature field is assumed to be in the form of an Ivantsov solution with isolines having a parabolic form. In the simulation corresponding to Figure 21.3, the temperature field is not compatible with this idealization of the sharp-interface model, leading to the disagreement in the "velocity undercooling". Further, in the sharp-interface model, the solute and the temperature fields are coupled only through a balance of undercoolings, which is assumed only at the dendritic tip, whereas in the phase-field model, these fields are coupled along the whole solid–liquid interface. The growth of the dendritic tip is influenced by the other regions of the solid–liquid interface. It should also be noted that the characteristic length scale of the diffusion field is small and comparable to the radius of the dendritic tip. Thus in the isothermal case, only a region in the vicinity of the tip contributes to the tip velocity. In the nonisothermal case, the coupling length scale, that is, the characteristic length scale of the temperature field, is much larger than in the isothermal approximation. This leads to the coupling between the tip velocity and the development of the dendritic structure. In Figure 21.3, the thermal boundary layer in the liquid is much larger than the solutal boundary layer.

Multicomponent alloys form the most important class of metallic materials for technological and industrial processes. Combined with the number of components is a wealth of different phases, phase transformations, complex thermodynamic interactions and pattern formations. In Refs. [7, 8], the influence of changing the alloy composition on the interface stability, on the characteristic morphology, and on the growth velocity is investigated by performing numerical simulations of a general multicomponent phase-field model with interacting and coupled diffusion fields. We carried out a series of numerical computations for different alloy compositions varying from $Ni_{60}Cu_{36}Cr_4$ to $Ni_{60}Cu_4Cr_{36}$. The concentration of Ni was kept at 60 at% and the initial undercooling was fixed at 20 K measured from the equilibrium liquidus line in the phase diagram at a given composition of the melt. A morphological transition from dendritic to globular growth occurs at a melt composition of about $Ni_{60}Cu_{20}Cr_{20}$. Figure 21.4 shows the dendritic morphologies observed for Cr concentrations less than 20 at% and the globular morphologies for Cr concentration crossing this threshold. The velocity of the dendritic/globular tip increases linearly from 1.19 to 3.24 cm/s with increasing concentration of Cr. The morphological transition can be quantitatively measured using the deviation $\delta = (S - S_0)/S_0$ of area S covered by the solid phase from the triangular area S_0 between the dendritic tips and the center of the initial seed, Figure 21.4. With the change of melt composition, the deviation δ changes the sign from a negative value for dendritic morphologies to a positive value for globular shapes. The transitional value $\delta = 0$ corresponds to the composition $Ni_{60}Cu_{20}Cr_{20}$ and is hence in good agreement with the transition point given by the vanishing diagonal part of the isoline. The transition takes place in a smooth manner.

The growth process of a crystalline dendritic solid phase from an undercooled melt in the metallic glass composite Zr–Ti–Nb–Ni–Cu–Be has been analyzed

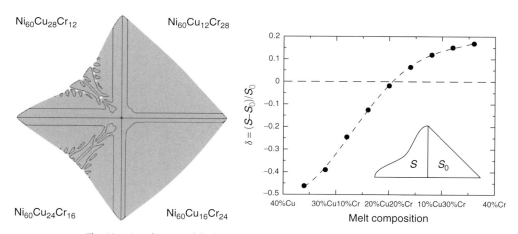

Fig. 21.4 Dendritic to globular transition for different ternary alloy compositions of a Ni–Cu–Cr system. The atomic percentages of Cu and Cr are exchanged keeping Ni fixed at 60 at%.

by experimental investigations and by adaptive finite element simulations in [9]. The phase transition: L(liquid) $\to$ S(dendrite) $+$ M(matrix) can be described by considering a pseudoternary system of late transition metals $A =$ (Zr, Ti, Nb), of early transition metals $B =$ (Ni, Cu), and of the component $C =$ Be. Scanning and transmission electron microscopies are employed to experimentally determine structural properties of the two-phase system and to provide concentration distributions of the components across the dendrite/matrix interface. The computed and experimentally observed microstructures, length scales, and chemical compositions are in good agreement. We have demonstrated that the complex multicomponent system is well described by a pseudo-ternary phase diagram and by the solidification path $L \to S + M$ which could be well understood introducing the pseudo components $A =$ (Zr,Ti,Nb), $B =$ (Ni,Cu), $C =$ (Be) and in terms of the transformation $A_{75}B_{13}C_{12} \to A_{94}B_6 + A_{65}B_{17}C_{18}$.

21.4
Solute Trapping During Rapid Solidification

In Ref. [10], we examined the concentration profiles and corresponding solute trapping effects at a moving planar interface using the thermodynamically consistent phase-field model. We compared the results with available experimental data [11] for nonequilibrium partition coefficients in Si–As alloys. We focused on the high-growth-velocity regime, $V > 1\,\mathrm{m\,s^{-1}}$, because for low and medium velocities different models had already described the experimental data with equal success. A definition of the partition coefficient at the diffuse interface was suggested, predicting a steeper profile for high growth velocities, $V > 1\,\mathrm{m\,s^{-1}}$, in accordance with the experimentally measured data of the nonequilibrium partition coefficient.

Figure 21.5 displays the different behaviors of the partition coefficients given by various models as a function of the interface velocity V. The definition of k suggested in [10] has a steeper profile for high velocities $V > 1\,\mathrm{m\,s^{-1}}$ compared to both the behavior of k_m in the phase-field model for rapid solidification of Wheeler *et al.* [12] and the predictions for k_A of the continuous growth model in [13, 14]. The experimental data for rapid solidification can be well described by the sharp-interface approach of Sobolev and by the diffuse-interface approach suggested in our work [10]. However, these models predict a qualitatively different character of the transition to establish complete solute trapping: a sharp transition as predicted by the local-non-equilibrium sharp-interface model or a smooth continuous transition as predicted by the phase-field model.

21.5
Comparison of Molecular Dynamics and Phase-field Simulations

In a recent paper [15], results were presented from phase-field modeling and molecular dynamics simulations concerning the relaxation dynamics in a finite-

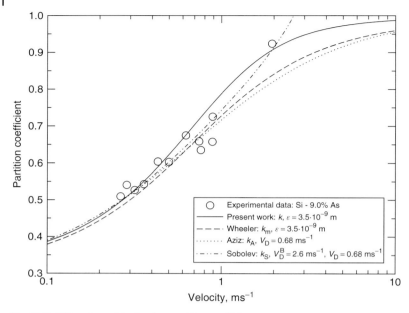

Fig. 21.5 Different models for the partition coefficient as a function of the interface velocity applied to the alloy system Si-9 at.% As in comparison with experimental data taken from [11].

temperature two-phase crystal–liquid sample subjected to an abrupt temperature drop. Relaxation takes place by propagation of the solidification front under the formation of a spatially varying concentration profile in the melt. The molecular dynamics simulations are carried out with an interatomic model appropriate for the NiZr alloy system and provide the thermophysical data required for setting up the phase-field simulations. Regarding the concentration profile and velocity of the solidification front, best agreement between the phase-field model and the molecular dynamics simulations is obtained when increasing the apparent diffusion coefficients in the phase-field treatment by a factor of four against their molecular dynamics estimates, Figure 21.6.

21.6
Modeling of Eutectic Growth

In Refs. [8, 2], microstructure simulations of binary and ternary eutectic phase transformation processes were shown to illustrate the wide variety of realistic growth structures and morphologies that can be described and investigated by the phase-field model.

Figure 21.7 shows the dependence of the diffuse-interface thickness on the phase-field parameter ε. For pure substances, this parameter controls the thickness

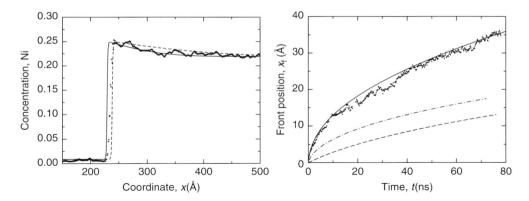

Fig. 21.6 (left) Concentration field of Ni across the solid–liquid interface: molecular dynamics data ("+"), phase-field model with $D_L = 1.64 \times 10^{-9}$ m^2 s^{-1} (solid line) and with $D_L = 6.47 \times 10^{-9}$ m^2 s^{-1} (dashed line). (right) Computed interface position as a function of time, dashed line corresponding to input data: $D_L = 1.64 \times 10^{-9}$ m^2 s^{-1}, $D_S = 1.33 \times 10^{-8}$ m^2 s^{-1}, and $\nu = 2.5 \times 10^{-10}$ m^2 s^{-1}, the dotted-dashed line shows the result for a larger kinetic coefficient of $\nu = 10^{-7}$ m^2 s^{-1} and the solid line represents the simulated curve for $D_L = 6.47 \times 10^{-9}$ m^2 s^{-1}, $D_S = 5.25 \times 10^{-8}$ m^2 s^{-1}, and $\nu = 10^{-7}$ m^2 s^{-1}.

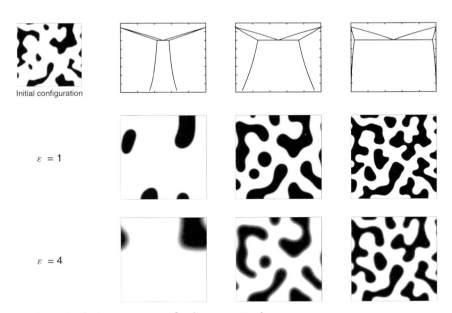

Fig. 21.7 The final microstructure for three eutectic phase diagrams and two values of the phase-field parameter ε. The initial configuration is shown in the top left corner.

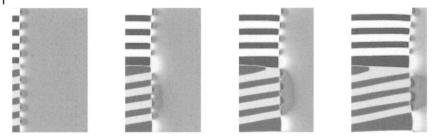

Fig. 21.8 Growth of two eutectic grains (white/black and light/dark-gray) of a binary A–B alloy with different crystal orientations into an isothermally undercooled melt (continuous gray scale) at four time steps.

of diffuse interface in a linear manner. In the case of binary (and in general multicomponent) alloys, the thickness of the phase boundaries depends on the level of segregation of alloy components at the interface and, hence, on the specific phase diagram of the alloy. If the phase diagram leads to a strong concentration difference between the phases, then the interface thickness exhibits only a weak dependence on the control parameter ε [16].

In Figure 21.8, the formation of two eutectic grains in the binary A–B "edge" system of initial composition $(c_A, c_B, c_C) = (0.5, 0.5, 0.0)$ has been simulated in 2D. The simulation involves pattern formation on different length scales. On a larger scale, grains with different orientations due to anisotropy of the surface entropy densities $\gamma_{\alpha\beta}$ grow and form a eutectic grain boundary. To include anisotropic effects, we used the facetted formulation of Equation 21.3 for a cubic crystal symmetry and defined two sets of four corners for the upper and lower grain, whereas the corners of the lower grain are rotated by 10° with respect to the growth direction. On a smaller scale, a lamellar eutectic substructure solidifies. Below a critical eutectic temperature T_e (here $T_e = 1.0$), a parent liquid phase L transforms into two solid phases α and β in a binary eutectic reaction: $L \rightarrow \alpha + \beta$. The white and light gray colored regions as well as the black and dark-gray colored regions represent the same solid phases, namely α and β, with just different orientations. The results illustrate the capability of the model to distinguish between several phases and grains at the same time. The images visualize the phase evolution and the concentration profile of the alloy component B in the liquid ahead of the growing solid phases at different time steps. Concentration-depleted zones occur in dark gray and concentration-enriched zones appear in light gray.

Figure 21.9 shows a time sequence of a 3D simulation of ternary eutectic solidification. The computation was initialized with cubic crystal shapes. During the evolution, a regular hexagonal structure of the three isotropic solid phases with 120° angles between the solid phases is established as the steady-growth configuration in 3D in analogy to the lamellar structure in 2D. This symmetry breaks if anisotropy is included.

The simulation in Figure 21.10 was conducted with an initial composition vector of $(c_A, c_B, c_C) = (0.47, 0.47, 0.06)$ so that the concentration component c_C acts as a

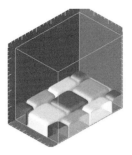  

Fig. 21.9 Formation of a 3D hexagonal rod-like structure in a ternary eutectic system with isotropic surface energies and three different solid phases α, β, and γ.

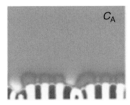 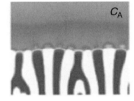 

Fig. 21.10 Simulation of lamellar eutectic growth in a ternary system with an impurity component c_C: The concentration profile of the main component c_A in melt is shown at the left and center images for two time steps. The ternary impurity c_C is pushed ahead of the growing eutectic front so that concentration enriched zones of component c_C can be observed at the solid–liquid interface in the right image.

ternary impurity of minor amount. As can be seen in the first two images, the solid phase α in white color is formed by using the concentration c_A, whereas the solid phase β rejects A atoms. If a γ solid phase containing c_C as its major composition is introduced, it is unstable and immediately dissolves for these concentration proportions. Neither the α phase nor the β phase engulfs the concentration c_C so that it increases all along the solid–liquid interface. The simulated evolution process recovers the experimentally observed effect that the impurity becomes enriched ahead of the solidifying lamellae and builds up. At larger computational domains, we expect the effect of cell/colony formation to occur.

Figure 21.11 illustrates the pattern formations obtained from finite-element simulations of eutectic solidification in the Al–Cu system for different initial melt composition and different initial lamellar spacings of the two solid phases α and θ. The concentration field in the melt shows enriched zones in light color ahead of the α solid phase (black) and depleted zones in dark color ahead of the θ solid phase (white). The lamellar spacings obtained lie in the range of 30–140 nm. This length scale of the solidification microstructure is comparable with experimental results of oscillatory eutectic Al–Cu patterns, reported in [17], for the same initial melt composition. For large lamellar spacings, that is $\lambda = 140$ nm, the lamellar growth morphology becomes unstable and an irregular oscillatory growth mode is established (Figure 21.11c).

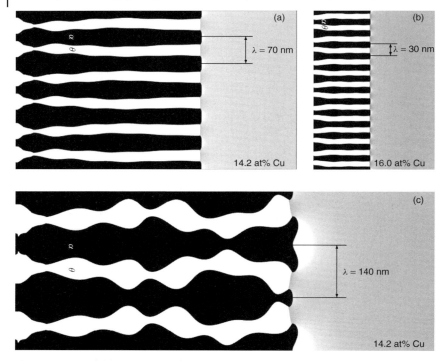

Fig. 21.11 Phase-field simulations of lamellar eutectic structures in Al–Cu for different initial melt compositions and different spacings between the lamellae of the two solid phases.

Acknowledgment

The authors thank the German Research Foundation (DFG) for funding the project Ne 822/2 within the priority program SPP 1120.

References

1 Garcke, H., Nestler, B. and Stinner, B. (**2004**) "A diffuse interface model for alloys with multiple components and phases". *SIAM Journal on Applied Mathematics*, **64** (3), 775–99.
2 Nestler, B., Garcke, H. and Stinner, B. (**2005**) "Multicomponent alloy solidification: phase-field modeling and simulations". *Physical Review E*, **71** (4), 041609.
3 Nestler, B., Danilov, D. and Galenko, P.K. (**2005**) "Crystal growth of pure substances: phase-field simulations in comparison with analytical and experimental results". *Journal Of Computational Physics*, **207**, 221–39.
4 Funke, O., Phanikumar, G., Galenko, P.K., Chernova, L., Reutzel, S., Kolbe, M. and Herlach, D.M. (**2006**) "Dendrite growth velocity in levitated undercooled nickel melts". *Journal of Crystal Growth*, **297** (1), 211–22.
5 Brener, E. (**1990**) "Effect of surface energy and kinetics on the growth of needle like dendrites". *Journal of Crystal Growth*, **99**, 165–70.

6 Galenko, P.K., Reutzel, S., Herlach, D.M., Danilov, D. and Nestler, B. (2007) "Modelling of dendritic solidification in undercooled dilute Ni-Zr melts". *Acta Materialia*, **55** (20), 6834–42.

7 Danilov, D. and Nestler, B. (2004) "Dendritic to globular morphology transition in ternary alloy solidification". *Physical Review Letters*, **93** (21), 215501.

8 Danilov, D. and Nestler, B. (2004) "Phase-field simulations of solidification in binary and ternary systems using a finite element method". *Journal of Crystal Growth*, **275**, e177–82.

9 Huang, Y.-L., Bracchi, A., Niermann, T., Seibt, M., Danilov, D., Nestler, B. and Schneider, S. (2005) "Dendritic microstructure in the metallic glass matrix composite $Zr_{56}Ti_{14}Nb_5Cu_7Ni_6Be_{12}$". *Scripta Materialia*, **53** (1), 93–97.

10 Danilov, D. and Nestler, B. (2006) "Phase-field modelling of solute trapping during rapid solidification of a Si-As alloy". *Acta Materialia*, **54**, 4659–64.

11 Kittl, J.A., Aziz, M.J., Brunco, D.P. and Thompson, M.O. (1995) "Nonequilibrium partitioning during rapid solidification of Si-As alloys". *Journal of Crystal Growth*, **148**, 172–82.

12 Wheeler, A.A., Boettinger, W.J. and McFadden, G.B. (1993) "Phase-field model of solute trapping during solidification". *Physical Review E*, **47** (3), 1893–909.

13 Aziz, M.J. (1982) "Model for solute redistribution during rapid solidification". *Journal of Applied Physics*, **53** (2), 1158–68.

14 Aziz, M.J. and Kaplan, T. (1988) "Continuous growth model for interface motion during alloy solidification". *Acta Metallurgica*, **36** (8), 2335–47.

15 Danilov, D., Nestler, B., Guerdane, M. and Teichler, H. "Bridging the gap between molecular dynamics simulations and phase-field modelling: Dynamics of a $[Ni_xZr_{1-x}]_{liquid} - Zr_{crystal}$ solidification front". *Physica D*, submitted.

16 Danilov, D., Selzer, M. and Nestler B. (2008) *Influence of parameters of phase diagram on diffuse interface thickness and microstructure formation in phase-field model for binary alloy.* Unpublished.

17 Zimmermann, M., Karma, A. and Carrard, M. (1990) "Oscillatory lamellar microstructure in off-eutectic Al-Cu alloys". *Physical Review B*, **42** (1), 833–37.

22
Dendrite Growth and Grain Refinement in Undercooled Melts
Peter K. Galenko and Dieter M. Herlach

22.1
Introduction

By far most materials of daily human life are solidified from the liquid state. Solidification is therefore one of the most important phase transformations in industrial production routes. It plays an important role in the casting and foundry industry. In these cases, solidification takes place under near-equilibrium conditions. Solidification under near equilibrium means that there is one solidification pathway from the liquid to the stable solid. It is well known that the spectrum of materials with different properties is greatly enhanced by undercooling the melt prior to solidification [1]. An undercooled melt is a metastable system of enhanced free energy. This makes it possible that the system can select various solidification pathways in solid states such as supersaturated solid solutions, disordered superlattice structures, grain-refined alloys, and so on, which can differ essentially in their physical and chemical properties. If solidification from undercooled melts is considered, the most important growth mechanism is dendritic growth. It is generated by a negative temperature gradient and, in case of alloys a concentration gradient ahead the solid–liquid interface. Dendrite growth controls the microstructure evolution during solidification of undercooled melts [2].

Present attempts are directed toward quantitative observation of the dynamics and morphology of dendrite growth in order to develop experimentally verified theoretical models. In the present work, modern concepts of experimental diagnostics during rapid solidification on undercooled drops without a container are presented. The experimental results will serve to verify theoretical concepts to describe rapid dendrite growth taking into account heat and mass transport by forced convection in electromagnetically processed melts. Phase-field modeling as well as sharp-interface models are applied and critically assessed to describe the experimental results of measurements of dendrite growth velocities as a function of undercooling on levitation-processed samples. Dendrite growth of nominally pure Ni is studied, which should be exclusively governed by heat transport due to heat conduction and possibly convection. During rapid solidification of deeply

Phase Transformations in Multicomponent Melts. Edited by D. M. Herlach
Copyright © 2008 WILEY-VCH Verlag GmbH & Co. KGaA, Weinheim
ISBN: 978-3-527-31994-7

undercooled melts, deviations from local equilibrium at the solid/liquid interface occur, which are considered by a kinetic undercooling of the solidification front. In case of congruently melting alloys forming superlattice structures as intermetallic alloys such as $Ni_{50}Al_{50}$, the kinetic interface undercooling is essentially enhanced because crystallization is limited by atomic diffusion. In case of noncongruently melting alloys, in particular in alloys containing a strongly partitioning element, segregation occurs because of different solubilities of the solvent in the solid and liquid state. As a consequence, it gives rise to a concentration gradient that controls the growth dynamics of "chemical" dendrites until nonequilibrium effects at high growth velocities, such as solute trapping, set in. Solute trapping causes the concentration gradient to disappear, and growth dynamics of the alloy is controlled by the thermal gradient eventually at large undercoolings. As a suitable system to study such nonequilibrium solidification of alloys, we investigate Ni–Zr alloys where Zr has a very small equilibrium partition coefficient. In the present work we extend our experimental and theoretical investigations to a ternary alloy system, in which Al is added to Ni–Zr. The sharp-interface model is extended to be applicable to ternary alloys. Results of modeling are compared to experimentally determined growth velocities measured as a function of undercooling. The impact of undercooling and growth of dendrites on microstructure evolution will be demonstrated for Ni–Zr alloys.

22.2
Solidification of Pure (One-Component) System

22.2.1
Diffuse Interface Model

The phase-field model via "thin-interface" analysis [3] has been used for the present investigations. In this model, the thickness W_0 of a diffuse interface between solid and liquid phases is assumed to be small compared to the scale of the crystal but not smaller than the microscopic capillary length d_0. The phase-field and energy equations were taken from Ref. 4 with the momentum and continuity equations for the liquid taken from Ref. 5. Furthermore, in the momentum equation, the Lorentz force caused by the alternating electromagnetic field has been introduced for an undercooled levitated droplet. A system of governing equations is described by

- energy conservation

$$\frac{\partial T}{\partial t} + (1 - \phi)(\vec{v} \cdot \nabla) T = a \nabla^2 T + \frac{T_Q}{2} \cdot \frac{\partial \Phi}{\partial t} \qquad (22.1)$$

- continuity of the liquid phase

$$\nabla \cdot [(1 - \phi) \vec{v}] = 0 \qquad (22.2)$$

- momentum transport

$$(1-\phi)(\vec{v}\cdot\nabla)\vec{v} = -\frac{1-\phi}{\rho}\nabla p + \frac{(1-\phi)}{\rho}\vec{F}_{LZ}$$
$$+\nabla\cdot\left[\frac{\eta}{\rho}\nabla(1-\phi)\vec{v}\right] + \vec{F}_D \quad (22.3)$$

- phase-field evolution

$$\tau(\vec{n})\frac{\partial \Phi}{\partial t} = \nabla\cdot\left(W^2(\vec{n})\nabla\Phi\right) + \sum_{w=x,y,z}\frac{\partial}{\partial w}$$
$$\left(|\nabla\Phi|^2\, W(\vec{n})\frac{\partial W(\vec{n})}{\partial(\partial_w\Phi)}\right) - \frac{\partial F}{\partial\Phi} \quad (22.4)$$

The symbols in Equations 22.1–22.4 denote: T the temperature; T_Q the adiabatic temperature of solidification defined by $T_Q = Q/c_p$; Q the latent heat of solidification; c_p the specific heat; a the thermal diffusivity; Φ the phase-field variable ($\Phi = -1$ is the liquid phase and $\Phi = 1$ is for the solid phase); $\phi = (1+\Phi)/2$ the fraction of the solid phase ($\phi = 0$ is for the liquid and $\phi = 1$ is for the solid); $\vec{v}$ the fluid flow velocity in the liquid; x, y, z the Cartesian coordinates; t the time; ρ the density; η the dynamic viscosity; and p is the pressure. The dissipative force F_D in the Navier–Stokes Equation 22.3 is taken from [5]. Furthermore, in solving Equation 22.3, the Lorentz force has been averaged in time: $F_{LZ} \approx |B|^2/(4\pi\delta)$, where $|B| = B_0\exp[(r - R_0)/\delta]$ is the modulus of the magnetic induction vector; B_0 is the time averaged value of the magnetic induction; r is the radial distance of a droplet of radius R_0; $\delta = [2/(\omega\sigma_R\mu_0)]^{1/2}$ is considered as a skin depth for the alternating magnetic field in the droplet, which decreases for a short distance at which the modulus of magnetic induction $|B|$ decays exponentially (where ω is a frequency of the applied current, σ_R is the electrical conductivity, and μ_0 is the magnetic permeability). The free energy F is defined by $F(T, \Phi) = f(\Phi) + \lambda(T - T_M)g(\Phi)/T_Q$, where T_M is the equilibrium melting temperature. With the inclusion of the double-well function $f(\Phi) = -\Phi^2/2 + \Phi^4/4$ and the odd function $g(\Phi) = \Phi - 2\Phi^3/3 + \Phi^4/5$ itself, the free energy F is constructed in such a way that a tilt λ of an energy well controls the coupling between T and Φ.

The time $\tau(\vec{n})$ of the phase-field kinetics and the thickness $W(\vec{n})$ of the anisotropic interface are given by

$$\tau(\vec{n}) = \tau_0\, a_c(\vec{n})\, a_k(\vec{n}) \left[1 + a_2\frac{\lambda\, d_0}{a\,\beta_0}\frac{a_c(\vec{n})}{a_k(\vec{n})}\right]; \quad W(\vec{n}) = W_0\, a_c(\vec{n}) \quad (22.5)$$

where τ_0 is the time scale for the phase-field kinetics, W_0 is the parameter of the interface thickness with $W_0 = \lambda d_0/a_1$, and $a_1 = (5/8)2^{1/2}$. The second term in brackets of Equation 22.5 for $\tau(\vec{n})$ defines a correction $a_2 = 0.6267$ for the "thin-interface" asymptotic [6]. The anisotropy of interfacial energy is given by

$$a_c(\vec{n}) = \frac{\sigma(\vec{n})}{\sigma_0} = (1 - 3\varepsilon_c)\left[1 + \frac{4\varepsilon_c}{1 - 3\varepsilon_c}\left(n_x^4 + n_y^4 + n_z^4\right)\right] \quad (22.6)$$

where $\sigma(\vec{n})$ is the interface energy depending on the normal vector $\vec{n}$ to the interface, σ_0 is the mean value of the interfacial energy along the interface, and ε_c is the anisotropy parameter. The anisotropy of kinetics of atomic attachment to the interface is given by

$$a_k(\vec{n}) = \frac{\beta(\vec{n})}{\beta_0} = (1 - 3\varepsilon_k)\left[1 - \frac{4\varepsilon_k}{1 - 3\varepsilon_k}\left(n_x^4 + n_y^4 + n_z^4\right)\right] \tag{22.7}$$

where $\beta(\vec{n}) = \mu^{-1}$ is the reciprocal kinetic coefficient dependent on the normal vector $\vec{n}$ to the interface, β_0 is the averaged kinetic coefficient along the interface which is defined by $\beta_0 = (1/\mu_{100} - 1/\mu_{110})/(2T_Q)$, and $\varepsilon_k = (\mu_{100} - \mu_{110})/(1/\mu_{100} + 1/\mu_{110})$ is the kinetic anisotropy parameter in which μ_{100} and μ_{110} are the kinetic coefficients in the $\langle 100 \rangle$ and $\langle 110 \rangle$ direction, respectively. In Equations 22.5–22.7, the normal vector has the components (n_x, n_y, n_z) defined by the gradients of the phase-field as follows:

$$n_x^4 + n_y^4 + n_z^4 = \left[(\partial\Phi/\partial x)^4 + (\partial\Phi/\partial y)^4 + (\partial\Phi/\partial z)^4\right]/|\Phi|^4$$

22.2.2
Sharp-Interface Model

We extend the Lipton-Glicksman-Trivedi/ Lipton-Kurz-Trivedi (LGT/LKT) model [7, 8] by including changes of heat transport by convection in the solution of the heat transport equation [9]. Accordingly, the thermal undercooling $\Delta T_T = T_i - T_\infty$, with T_i the temperature at the tip of the dendrite and T_∞ the temperature of the undercooled melt far from the interface is expressed by:

$$\Delta T_T = T_Q Pe_g \cdot \exp(Pe_g + Pe_f) \cdot \int_1^\infty q^{-1} \cdot \exp\left[-qPe_g + (\ln q - q)Pe_f\right] dq \tag{22.8}$$

where R is the tip radius of the parabolic dendrite, $Pe_g = VR/(2a)$ is the thermal Péclet number, $Pe_f = U_0 R/(2a)$ is the flow Péclet number, V is the velocity and R the radius of curvature of the tip of the dendrite, U_0 is the velocity of the uniformly forced flow far from the dendrite tip, and a is the thermal diffusivity. The flow velocity U_0 can be estimated from the energy balance for the energetics of the electromagnetic field, the gravitational field, and the viscous dissipation. This yields

$$U_0 = \left[\frac{2}{\rho}\left(\rho g R_0 + \frac{B_0^2(1 - \exp(-2R_0/\delta))}{8\pi} + \frac{\rho \eta^2}{2\delta^2}\right)\right]^{1/2} \tag{22.9}$$

where g is the modulus of the vector of the acceleration due to gravity, ρ is the mass density, η is the dynamic viscosity of the liquid phase, δ is the skin depth, R_0 is the radius of the sample, and B_0 is the time-averaged value of the magnetic field inside the levitation coil. The curvature undercooling $\Delta T_R = T_L - T_i$ (with T_L the liquidus temperature) due to the Gibbs–Thomson effect is described by

$$\Delta T_R = 2\Gamma(1 - 15\varepsilon \cos 4\theta)/R \tag{22.10}$$

where $\Gamma = \sigma/\Delta S_f$ (σ the interface energy and ΔS_f the entropy of fusion) is the capillary constant (Gibbs–Thomson parameter), ε_c is the parameter of anisotropy of the surface energy, and θ is the angle between the normal to the interface and the direction of growth along the growth axis. The kinetic undercooling ΔT_K at large velocities is given by

$$\Delta T_k = V/\mu_k, \quad \mu_k = \mu_{ko}(1 - \varepsilon_k \cos 4\theta) \tag{22.11}$$

where μ_k is the kinetic coefficient for growth of the dendrite tip, and ε_k is the parameter of anisotropy for the growth kinetics. The total undercooling ΔT measured in the experiment is the sum of the individual contributions:

$$\Delta T = \Delta T_T + \Delta T_R + \Delta T_k \tag{22.12}$$

According to Equation 22.12, ΔT is expressed in terms of the Péclet numbers: that is, the product of growth velocity V and radius of curvature of the dendrite tip R. An independent expression for the dendrite tip radius R is obtained from solvability theory as

$$R = \frac{\Gamma}{\sigma^* \, T_Q \, Pe_g} \tag{22.13}$$

with the stability parameter σ^* given by

$$\sigma^* = \sigma_0 \cdot \varepsilon_c^{7/4} \left[1 + \chi(Re)\frac{U_0 \Gamma}{a \, T_Q}\right]^{-1} \tag{22.14}$$

where σ_0 is a constant; $Re = U_0 R\rho/\eta$ is the Reynolds number with ρ the mass density. For computation of the stability parameter σ^* we choose the results of phase-field modeling [10] with $\sigma_0 \varepsilon_c^{7/4}/\sigma^* = 1.675$ for the 3D upstream fluid flow imposed on the scale of a freely growing dendrite. Thus, from the two main Equations 22.12 and 22.13 the velocity V and the tip radius R of the dendrite can be calculated as a function of the initial undercooling ΔT.

22.2.3
Results of Nominally Pure Ni

We obtained the morphological spectrum of interfacial crystal structures for a wide range of undercooling. According to the results of modeling, the spectrum of the microstructures obtained exhibits a change from grained crystals at very small undercoolings ($\Delta T < 0.15 T_Q$) to dendritic patterns at intermediate under-

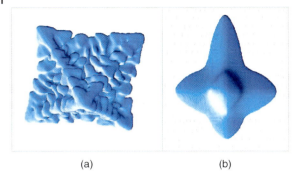

(a) (b)

Fig. 22.1 (a) Dendrite growing into pure nickel undercooled at $\Delta T = 0.55 T_Q$ K. Pattern has been simulated on a grid of size 650^3 nodes. Details of modeling are given in Ref. 11. (b) Growth of nickel dendrite under convective flow at $\Delta T = 0.30 T_Q$ K and $U_0 = 0.7\,\mathrm{m\,s^{-1}}$. Growth velocity of the upstream branch is pronounced in comparison with the downstream branch owing to forced convection. Dashed lines around the dendrite indicate the flow velocity vectors in the vertical cross-section. The pattern has been simulated on a grid of size $230 \times 230 \times 330$ nodes.

coolings ($0.1 T_Q < \Delta T < 1.0 T_Q$) to grained crystals again at high undercoolings ($\Delta T \rangle 1.0 T_Q$). Within the range of intermediate undercoolings, the shape of dendrites is dictated by the preferable crystallographic direction (which is $\langle 100 \rangle$ for the case of Ni).

Stochastic noise plays a crucial role in the formation of branched crystal patterns of dendritic type. Figure 22.1a shows the dendritic crystal with secondary branches with the application of thermal noise. Another effect has to be outlined as well: solidification under the influence of forced convective flow in a droplet produces dendritic growth pronounced in the direction opposite to that of the far-field flow velocity U_0 (see [9]). The present results of modeling also confirm this outcome: with the imposition of the fluid flow, the growth becomes pronounced in the direction opposite to the flow as shown in Figure 22.1b. For these structures, that is, with the thermal noise in a stagnant melt, and also with the melt flow (Figure 22.1) we compared the results for dendrite growth velocity V in pure Ni versus undercooling ΔT quantitatively.

As shown in Figure 22.1b, the results of phase-field modeling exhibit an increase of the velocity of the upstream dendrite branch. As soon as the thermal boundary layer shrinks ahead of the upstream branch because of the flow, the heat of solidification is removed better, and the growth velocity enhances. In Figure 22.2 we have compared the predictions of the present phase-field modeling with new experimental data [12] for growth kinetics of nickel dendrites. The enhanced dendrite velocity due to the melt flow decreases the discrepancy between theory and experimental data at small undercoolings. However, Figure 22.2 also clearly shows that comparison of experimental findings (open squares) with the results of the phase-field modeling (22.1)–(22.7) (open circles) and the predictions of the sharp-interface model (22.8)–(22.14) confirms that convection alone cannot

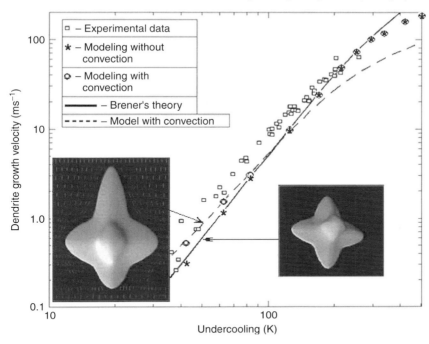

Fig. 22.2 Dendritic solidification of undercooled nickel melt. Experimental data (squares) [12] are compared with phase-field modeling without including liquid convection (circles), phase-field modeling with inclusion of liquid convection (stars), sharp-interface model of Brener (solid line) [14], and sharp-interface model with convection (dashed line) [9].

describe the experimental results satisfactorily (dashed line). The reason for the discrepancy that still exists might be due to the presence of small amounts of impurity (see Ref. 13 and references therein). It is shown in the next section that small amounts of impurities indeed may enhance the growth velocity leading to additional contribution in the growth kinetics together with the convective flow. This contribution principally improves the quantitative model's predictions in comparison with experimental data obtained on undercooled droplets processed in electromagnetic levitation (EML) techniques.

22.2.4
Results on Congruently Melting Intermetallic Alloy Ni$_{50}$Al$_{50}$

Ni$_{50}$Al$_{50}$ alloy forms a stoichometric NiAl intermetallic phase that shows congruent melting and solidification behavior. This means that similar as in a pure metal constitutional effects do not play a role either in the growth kinetics or in the stability of a growing dendrite. But different to dendrite growth velocities of pure metals, which are as large as 100 m s^{-1} at the highest undercoolings [15], growth dynamics of intermetallics is expected to be more sluggish since the formation of a superlattice structure of an intermetallic phase requires short-range diffusion of atoms to sort them out on the correct lattice site. As a consequence, the kinetic undercooling controlled by atomic attachment kinetics at the interface will be larger than in the case of a pure metal and may become comparable to the thermal undercooling at large undercoolings. Owing to the larger kinetic interface undercooling, the growth velocity of intermetallic alloy should be less than the atomic diffusive speed, which is in the order of $1–10\,\mathrm{m\,s^{-1}}$ [16]. Estimations by hydrodynamic computations result in fluid flow velocities induced by strong alternating electromagnetic fields in levitation-processed melts in the order of $0.1–1\,\mathrm{m\,s^{-1}}$ [17]. Therefore, effects by forced convection on the growth dynamics should be more pronounced in levitation-processed melts of NiAl since the fluid flow velocities will be comparable to or even larger than the dendrite growth velocity over a larger undercooling range if compared with pure Ni.

The dendrite growth velocity was measured by using a high-speed video camera system, observing the advancement of the growing solid–liquid interface of the undercooled droplet. Details of the experiments to measure the dendrite growth velocity on levitation-processed molten drops are given elsewhere [12]. In terrestrial EML, fluid flow velocities are high because of the high levitation fields needed to compensate the gravitational force. They are reduced if EML is applied in reduced gravity since the positioning fields to compensate disturbing accelerations are orders of magnitude smaller than levitation forces. Analogous experimental investigations were performed during parabolic flights under the conditions of reduced gravity by using a facility for containerless processing of metals in space (TEMPUS: Tiegelfreies Elektro-Magnetisches Prozessieren Unter Schwerelosigkeit) on board the Airbus A300 Zero-G, which is operated by DLR and ESA (European Space Agency) [18]. The results of comparative measurements of the dendrite growth velocity as a function of undercooling of NiAl intermetallic compound are depicted in Figure 22.3. The growth velocities measured in reduced gravity are much smaller than those determined under terrestrial conditions in an undercooling range $\Delta T < 100\,\mathrm{K}$. In this undercooling range, the growth velocities are comparable with the fluid flow velocity U_0 induced by strong levitation forces in EML experiments on Earth. This indicates that convection leads to an enhancement of the heat and mass transport ahead the solid–liquid interface, and hence to an increase of the dendrite growth velocity as discussed in the previous section.

Insofar as the sharp-interface modeling neglects the influence of fluid flow, the experimental results obtained under reduced gravity are in line with modeling (solid line in Figure 22.3) at undercoolings $\Delta T < 100\,\mathrm{K}$. At an undercooling of $\Delta T \approx 100\,\mathrm{K}$, the dendrite growth velocity becomes comparable to the fluid flow velocity U_0. In this undercooling range, the curve calculated by the sharp-interface model without taking into account convection meets the experimental data obtained under conditions of reduced gravity. This curve joins the measurements of growth velocities at 1g at an undercooling of $\Delta T \approx 100\,\mathrm{K}$: that is, at a dendrite growth velocity of about $0.8\,\mathrm{m\,s^{-1}}$, which becomes comparable with the fluid flow velocity. At this undercooling and higher undercoolings, heat transport by convection loses its influence on growth kinetics, and modeling

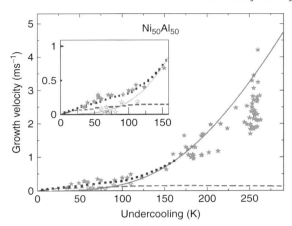

Fig. 22.3 Growth velocity as a function of undercooling for intermetallic compound NiAl measured in reduced gravity (open stars) and on Earth (closed stars), compared with results of modeling within sharp-interface theory without convection (solid line) [18] and with convection (squares). For comparison, also results of interface velocity computed by molecular dynamic simulations are given (dashed line) [19]. The inset shows an enlargement of a section of the figure.

without taking into account convection describes the data up to undercoolings $\Delta T < 175$ K.

If convection effects are taken into consideration in sharp-interface modeling with assuming a fluid flow velocity $U_0 = 0.8 \, \text{m s}^{-1}$, also the results of growth velocity measurements on Earth are described within the uncertainty of measurements in the undercooling range $\Delta T < 100$ K (squares in Figure 22.3). For comparison, results of interface velocity computed by molecular dynamic simulations for the same alloy are shown (dashed line in Figure 22.3) [19]. At small undercoolings, molecular dynamic simulations reproduce the experimental results obtained under reduced gravity within the scatter of the measurements. The molecular dynamic simulations lead to an underestimation of the measured growth velocities at undercoolings $\Delta T > 100$ K and show a maximum in the $V(\Delta T)$ relation at $\Delta T \approx 150$ K opposite to the experimental findings and results of the sharp-interface modeling.

At larger undercoolings, the experimentally determined growth velocities continuously increase with undercooling and the experiments reveal an abrupt increase of the growth velocity at a critical undercooling $\Delta T^* \approx 250$ K. Such a drastic change of growth kinetics may indicate disorder trapping in the growth of the intermetallic compound leading to solidification of a disordered superlattice structure as described by Aziz and Boettinger [20]. A similar growth behavior was found in other intermetallic compounds [21, 22]. So far, the sharp-interface model is not able to reproduce the sharp rise of the measured growth velocity at $\Delta T^* \approx 250$ K. For future work it may be important to integrate a velocity-dependent kinetic growth coefficient and/or order parameter into models of dendrite growth.

22.3
Solidification of Binary Alloys with Constitutional Effects

22.3.1
Diffuse Interface Model

Using the diffuse interface model explained in Ref. 23, the equations for diffusion and phase field were solved using a finite-element method with Lagrange elements and with linear test functions. A nonuniform adaptive mesh has been used to describe the high-order spatial resolution in the vicinity of the solid–liquid interface. The qualitative features of dendritic growth and quantitative comparison with data for solidifying Ni–Zr dendrites from undercooled melts were made in comparison with the results of the phase-field modeling [24].

22.3.2
Sharp-Interface Model and Growth Velocities of Binary Ni–Zr Alloys

For the growth of an axisymmetric parabolic dendrite of an alloy into the forced flow we refer to a method of solution by Ben Amar et al. [25]. Following this method, analytical solutions can be found for the temperature T_i and the concentration C_i at the tip of a chemical dendrite as:

$$T_i = T_0 + T_Q Pe_t \cdot \exp(Pe_t + Pe_{ft}) \cdot \int_1^\infty \eta^{-1}$$
$$\cdot \exp\left[-\eta Pe_t + (\ln \eta - \eta) Pe_{ft}\right] d\eta \quad (22.15)$$

$$C_i = C_0 + (1 - k_e)C_0 Pe_c \cdot \exp(Pe_c + Pe_{fc}) \cdot \int_1^\infty \eta^{-1}$$
$$\cdot \exp\left[-\eta Pe_c + (\ln \eta - \eta) Pe_{fc}\right] d\eta \quad (22.16)$$

with $Pe_c = VR/2D$ and $Pe_{fc} = U_0 R/2D$ the solution growth and flow solution Péclet number, respectively. D denotes the chemical diffusion coefficient. Stability analysis of the chemical dendrite delivers the relation:

$$\frac{2d_0 a}{V R^2} = \sigma_0 \cdot \varepsilon_c^{7/4} \left[\frac{1}{2} + \frac{a}{D} \cdot \frac{\Delta_e k_e}{[1 - (1 - k_e)Iv(Pe_c)] \cdot T_Q}\right] \quad (22.17)$$

with d_0 the capillary constant, and $\Delta_e = m_e C_0 (k_e - 1)/k_e$ is the equilibrium interval of solidification with k_e the equilibrium partition coefficient. Since we are mainly interested in the low undercooling range, the kinetic undercooling can be neglected. The total undercooling then consists of thermal undercooling ΔT_t, curvature undercooling ΔT_r, and constitutional undercooling ΔT_C:

$$\Delta T = \Delta T_T + \Delta T_R + \Delta T_C \quad (22.18)$$

The thermal undercooling $\Delta T_T = T_i - T_\infty$ with T_i the temperature at the tip of the dendrite and T_∞ the temperature of the undercooled melt far from the interface is expressed by:

$$\Delta T_T = T_Q Iv(Pe_t) \qquad (22.19)$$

with Iv the Ivantsov function depending on the thermal Péclet number $Pe_t = (VR)/2a$. V is the velocity of the tip of the dendrite, R is the radius of curvature at the tip of the dendrite, and a is the thermal diffusivity. Owing to the strong curvature of the dendrite tip, a depression of the melting temperature due to the Gibbs–Thomson effect has to be taken into account by the curvature undercooling: $\Delta T_r = T_L - T_i$, with T_L the liquidus temperature and T_i the temperature at the tip. It is described by Equation 22.10. Finally, the constitutional undercooling is described by

$$\Delta T_C = m_e(C_i - C_0) \qquad (22.20)$$

The dendrite growth velocities of dilute $Ni_{100-x}Zr_x$ ($x = 0, 0.1, 0.5, 1$) alloys were calculated as a function of undercooling and compared with the experimental results obtained from measurements on drops processed by the electromagnetic levitation technique [13]. In particular, it is apparent that growth velocity of the very dilute alloy (Ni–0.1 at% Zr) is enhanced compared to pure Ni, in particular at small undercooling, while the growth velocity of the other alloys (Ni–0.5 at% Zr and Ni–1 at% Zr) is reduced in relation to the growth kinetics of the pure metal.

The effect of tiny amounts of strongly partitioning impurity present in nominally pure nickel on the solidification kinetics, in particular at small undercoolings, may explain the qualitative disagreement of modeling predictions with existing experimental data shown in Figure 22.2. For the whole range of the undercooling investigated, it is seen from Figure 22.4 that the experimental results of dendrite growth measurements on dilute Ni–1 at% Zr alloy samples are well described as a function of undercooling. As can be seen from Figure 22.4, the experimental results of dendrite growth measurements on dilute Ni–Zr alloys are well described as a function of undercooling by the sharp-interface theory of dendrite growth. In particular, it is apparent from the calculated and measured growth dynamics that the dendrite growth velocity of the very dilute $Ni_{99.0}Zr_{0.1}$ alloy is enhanced compared to pure Ni, in particular at small undercoolings, whereas the growth velocity of the higher concentrated alloys is reduced compared with pure Ni. At small undercoolings the growth of a dendrite is controlled by chemical diffusion of the solute element, and at high undercoolings by the thermal field. In the range in between, solute trapping decreases the concentration gradient in front of the dendrite and the gradient eventually disappears if the growth velocity reaches the atomic diffusive speed: that is, complete partitionless solidification takes place. Since the chemical diffusion coefficient is about 3 to 4 orders of magnitude smaller than the thermal diffusivity, the chemical dendrites become narrow because of the strong concentration gradient and broaden when reaching the limit of absolute chemical stability. This effect is responsible for the enhancement of the dendrite growth velocity of very dilute alloys at small undercooling, since a thin dendrite

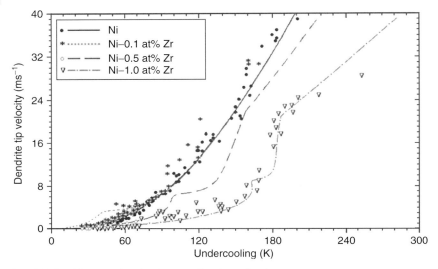

Fig. 22.4 Dendrite growth velocities as a function of undercooling measured for pure Ni (closed circles), $Ni_{99.9}Zr_{0.1}$ (stars), $Ni_{99.5}Zr_{0.5}$ (open circles), and $Ni_{99}Zr_1$ (triangles) compared with predictions of dendrite growth theory (solid lines), as calculated within the sharp-interface model.

propagates faster than a thick one. In addition to this constitutional effect in the stability of a growing dendrite, the solute causes a constitutional undercooling that requires solute redistribution, which in turn requires energy, and therefore makes the dendrite growth more sluggish. The effect of velocity enhancement dominates at small concentrations, whereas the effect of reduction of growth velocity due to constitutional undercooling predominates at larger concentrations.

22.3.3
Grain Refinement Through Undercooling

Undercooling has a strong impact on grain refinement in microstructure evolution [26]. In particular, it is well known that several microstructural transitions from coarse-grained to equiaxed-grain-refined microstructure occur if the undercooling is changed prior to solidification. Such microstructural transitions have been described by Karma within a model of dendrite fragmentation due to a Rayleigh-like instability of a growing dendrite [27]. According to this model, fragmentation of primarily formed dendrites needs a characteristic break-up time Δt_{bu}. The fragmentation process itself requires atomic diffusion in the interdendritic liquid regime to take place within reasonable experimental times.

Considering the solidification of an undercooled melt, there are two different solidification intervals. First, nucleation initiates crystallization. During the growth of nuclei, dendrites are formed that propagate very rapidly under nonequilibrium conditions into the undercooled melt. Owing to the rapid heat release during this crystallization period and small heat transfer of the levitated drop to the environment during the short-time period of recalescence (quasiadiabatic conditions), the undercooled melt heats up very rapidly close to the liquidus temperature (recalescence). In a second step, the interdendritic liquid solidifies under near-equilibrium conditions during the postrecalescence period Δt_{pl} before the completely solidified sample cools down to ambient temperature.[1] Fragmentation of dendrites takes place if the break-up time Δt_{bu} is smaller than the postrecalescence period Δt_{pl} during which solid dendrites and liquid interdendritic phase still coexist. This means if $\Delta t_{bu} > \Delta t_{pl}$, dendrites do not have enough time to fragment and a coarse-grained dendritic microstructure solidifies, whereas in the case $\Delta t_{bu} < \Delta t_{pl}$ dendrites break up with resulting in crystallization of a grain-refined equiaxed microstructure. Since the postrecalescence time Δt_{pl} depends on heat transfer of the sample to the environment, fragmentation is more likely at small cooling rates (larger Δt_{pl}) than at rapid cooling (smaller Δt_{pl}) [28].

Within the frame of the model by Karma [27], the critical undercoolings for microstructural transitions are determined by calculating the break-up time and determine the undercooling at which Δt_{bu} equals the postrecalescence time $\Delta t_{pl} = \Delta t_{bu}$. Assuming that side branches of dendrites perturb the trunk with an initial amplitude that scales with the dendrite tip radius R, and the main driving force for the perturbation is the concentration gradient in the dendrite trunk that counteracts the stabilizing effect of minimum interface area (energy) between solid and liquid phase, the break-up time is calculated as:

$$\Delta t_{bu}(\Delta T) \approx \frac{3}{2} \frac{R_T^3(\Delta T)}{d_0} \frac{c_p k_e \Delta_e}{DQ} \tag{22.21}$$

with R_T the dendrite trunk radius, which depends strongly on undercooling and heat and mass transport in the liquid, and $\Delta_e = m_e C_0 (k_e - 1)/k_e$. R_T is related to the dendrite tip radius $R(\Delta T)$ via an empirical relation taken from investigations of dendrite growth in transparent noble gases as $z = R_T/R \approx 20$ [29].

From dendrite growth model, Equations (22.16)–(22.20), the dependence of R from forced convection is estimated so that the change of the critical undercoolings for microstructural transitions by fluid flow is determined. Postrecalescence times Δt_{pl} are inferred from temperature–time profiles measured during cooling and solidification of levitated drops. The break-up time Δt_{bu} is calculated within the Karma model described by Equation 22.21. Taking R from solution of Equations (22.14)–(22.20), one may predict Δt_{bu} without and with convection (with the free parameter of a fluid flow velocity of $0 \leq U_0 \text{ m s}^{-1} \leq 1.4$ [9]). Applying the condition of $\Delta t_{bu} = \Delta t_{pl}$, the critical undercoolings for microstructural transitions from coarse-grained dendritic to grain-refined-equiaxed microstructures (and vice versa) are determined with and without convection.

Microstructures of the $Ni_{99}Zr_1$ alloy were investigated on samples undercooled by levitation and solidified after preselected undercoolings [30]. As is shown in

---

[1] If undercooling reaches the hypercooling limit or the unit undercooling, the entire melt solidifies under non-equilibrium conditions and $\Delta T_{pl} = 0$.

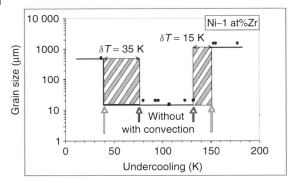

Fig. 22.5 Grain size as a function of undercooling for the $Ni_{99}Zr_1$ alloy. The microstructure changes from coarse-grained dendritic (left) at a lower critical undercooling to grain-refined equiaxed (middle) and at an upper critical undercooling to a coarse-grained dendritic microstructure (right). The dots represent results of microstructure investigations, which show a unique microstructure either coarse-grained or grain-refined. The shadowed regions represent the span of critical undercoolings for dendrite fragmentation calculated within the Karma model with and without convection.

Figure 22.5, two transitions are found: one at smaller undercoolings and another at higher undercoolings. The results of microstructure investigations are given by the closed circles, which represent only samples showing a unique microstructure either grain-refined or coarse-grained. The shadowed regions correspond to the range of the respective critical undercoolings if forced convection is taken into account or neglected. Samples solidified after undercoolings in the range of the shadowed regions reveal mixed microstructures consisting of both coarse-grained and grain-refined microstructures. The different microstructures are distributed inhomogeneously across the levitation-processed samples. This is due to the fact that in levitated drops convection roles are present, with fluid flow velocities changing across the sample. The span of the lower critical undercooling range with $\delta T = 35$ K is broader than that of the upper critical undercooling with $\delta T = 15$ K. This is easily understood by the fact that at smaller undercoolings the growth velocity is smaller and fluid flow has a stronger effect than in the upper undercooling range where the dendrite growth velocity exceeds the fluid flow velocity by about 1 order of magnitude.

22.4
Solidification of a Ternary Alloy

22.4.1
Diffuse Interface Model

A multicomponent, multiphase field model [31] for the direct simulation of dendritic growth structures has been applied. The model and its implementation allows an online coupling to thermodynamic databases and therefore offers the possibility to consider temperature- and composition-dependent solute partitioning at the interface. A detailed description of the model can be found in Ref. 32. In applying this model to analyze the dendritic growth of the ternary $Ni_{98}Zr_1Al_1$ alloy, the phase-field equations in the antisymmetric approximation and their simulation details are given in Ref. 33.

22.4.2
Sharp-Interface Model and Dendrite Growth Velocities of Ni–Zr–Al

An extended version of the sharp-interface model to multicomponent alloy systems was suggested in Refs. 34, 35. Using these works the model has been further advanced to the case of rapid solidification using analysis and solutions of Refs. 36, 37. Accordingly, the total undercooling ΔT measured in the experiment is expressed as the sum of various contributions:

$$\Delta T = \Delta T_t + \Delta T_R + \Delta T_k + \Delta T_c + \Delta T_n \tag{22.22}$$

with the thermal undercooling ΔT_t, the curvature undercooling ΔT_R, the kinetic undercooling, ΔT_k, the constitutional undercooling ΔT_c, and the undercooling due to the shift of the liquidus line slope from its equilibrium position at large growth velocities ΔT_n, respectively. The kinetic undercooling ΔT_k at large velocities is given by

$$\Delta T_k = V/\mu; \quad \mu = \mu_0(1 - \varepsilon_k \cos 4\theta) \tag{22.23}$$

where μ is the kinetic coefficient for growth of the dendrite tip, and ε_k is the parameter of anisotropy for the growth kinetics. In alloys also mass transport has to be considered. The constitutional undercooling in alloys is given by

$$\Delta T_{ci} = \sum_i \frac{k_{Vi}\Delta_{Vi} Iv(Pe_{ci})}{((1-(1-k_{Vi}))Iv(Pe_{ci}))} \tag{22.24}$$

with the Péclet number of chemical diffusion $Pe_{ci} = (VR)/2D_i$, D_i being the diffusion coefficient, $\Delta_{vi} = m_{Vi}C_{oi}(k_{Vi}-1)/k_{Vi}$ the nonequilibrium solidification interval, m is the tangent of the liquidus line slope in the kinetic phase diagram of the alloy's solidification, and k_{Vi} the velocity-dependent partition coefficient. Under the conditions of rapid solidification, for the range of growth velocity $V < V_D$ (where V_D is the atomic diffusive speed in bulk liquid), the liquidus slope is described by [38]

$$m_{vi} = \frac{m_{Ei}}{1-k_{Ei}}\left\{1 - k_{Vi} + \ln\left(\frac{k_{vi}}{k_{Ei}}\right) + (1-k_{vi})^2\frac{V}{V_{Di}}\right\} \tag{22.25}$$

where m_{Ei} and k_{Ei} are the slopes of the liquidus line and the solute partitioning coefficient, respectively, in the phase diagram of the alloy's equilibrium state. The solute partitioning is described by the nonequilibrium partition coefficient, which becomes dependent on the growth velocity for the case of rapid solidification. This yields [39]

$$k_{vi} = \frac{(1-V^2/V_{Di}^2)k_{Ei}+V/V_{Dli}}{(1-V^2/V_{Di}^2)\left[1-(1-k_{Ei})C_i^*\right]+V/V_{Dli}}, \quad V < V_{Di}$$
$$k_{vi} = 1; \quad V \geq V_{Di}$$
(22.26)

with the liquid concentration C^* at the tip of a paraboloid of revolution

$$C_i^* = \frac{C_{0i}}{1-(1-k_{vi})Iv(P_{Ci})}, \quad V < V_{Di}$$
$$C_{Vi}^* = C_{0i}^*; \quad V \geq V_{Di}$$
(22.27)

and V_{Di} the interface diffusion coefficient.

Equation 22.22 describes the relation of undercooling in terms of the Péclet numbers, that is as a function of the product $V \times R$. For unique determination of the growth velocity and tip radius as a function of undercooling, one needs a second equation for the tip radius, which comes from stability analysis and is given for multicomponent alloys:

$$\frac{2d_0 a}{VR^2} = \sigma_0 \varepsilon_c^{7/4} \left[\frac{1}{2}\xi_T(P_T) + \frac{c_p}{Q}\sum_{i=Zr,Al}(a/D_i)f_i(C_i)\xi_{Ci}(P_{Ci})\right]$$
(22.28)

where ξ_T and ξ_{Ci} are the stability functions depending on thermal and chemical Péclet numbers (see [33] and references therein).

Figure 22.6 shows the experimental data of the dendrite growth velocity as a function of undercooling as obtained from levitation-undercooled ternary alloy

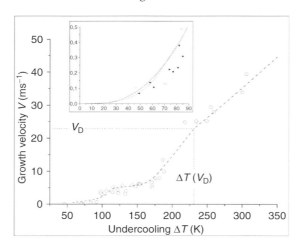

Fig. 22.6 Predictions of the sharp-interface model (dashed line) for the dendrite growth velocity versus the undercooling in comparison with data (open circles) measured on ternary $Ni_{98}Zr_1Al_1$ alloy [33]. Inset shows the "velocity–undercooling" relationship for the two dimensional modeling using the sharp-interface model (dashed line) and diffuse interface model (solid line) in the range of small undercooling. Solid circles are related to experimental measurements on $Ni_{99}Zr_1$ samples.

$Ni_{98}Zr_1Al_1$ applying the high-speed camera technique. The dashed curves (including inset) give predictions of the sharp-interface model (22.22)–(22.28). It is seen that the model is able to satisfactorily predict the solute diffusion-limited growth (for $\Delta T < 115$ K), both for solution and thermal dendrite growth (for 115 K $< \Delta T < 229$ K) and purely thermally controlled regime of dendritic solidification (for $\Delta T > 229$ K) in comparison with experimental findings. In the latter regime, tips of dendrites grow by the diffusionless mechanism, which occurs owing to complete solute trapping for the velocity equals to or greater than the solute diffusion speed in bulk liquid: that is, $V \geq V_D$. In such a case, a core of the main stems of dendrites should form with the initial chemical composition of the liquid melt.

Note that the parameters of the sharp-interface model (22.22)–(22.28) were compared and adjusted to the ones existing in the diffuse-interface model [32]. It offered a special ability to get a good agreement between these two models in two dimensions and for the diffusion-limited solidification (see inset in Figure 22.6).

22.5
Summary and Conclusions

We have presented experimental results of measurements of dendrite growth velocities in undercooled melts of nickel, NiAl compound, as well as dilute binary Ni–Zr and ternary Ni–Zr–Al alloys processed in a containerless scheme by EML. High-speed camera technique has been applied for these studies to measure the growth dynamics with high reliability in the entire undercooling range 30 K $< \Delta T <$ 310 K accessible by EML and the capacitance proximity sensor.

The sharp-interface model of free dendrite growth has been extended to include effects of fluid flow on mass and heat transport in drops of melts, which are processed without a container in strong alternating electromagnetic fields. The experimental results were analyzed within this model leading to the conclusions that fluid flow in the melt by forced convection due to electromagnetic stirring enhances the growth velocity in the small undercooling range at which the dendrite growth velocity is comparable to the speed of fluid flow. In addition, it is demonstrated that even small amounts of impurities lead to an increase of the growth velocity in the range of small undercoolings at which the growth is then controlled by diffusion of the impurities. The consequences of forced convection on critical undercoolings for microstructural transitions were demonstrated as estimated within a model of dendrite fragmentation. The critical undercoolings for microstructural transitions in undercooled melts are essentially controlled by forced convection.

The in situ diagnostics of rapid solidification of containerless undercooled melts may be combined with advanced computer modeling of solidification processes on the basis of experimentally proved physical models to develop the perspective of virtual materials design of high functional materials. In such a way, production routes of metallic materials may be improved and even new materials may be designed using computer experiments without the need of time- and energy-consuming trial and error studies in the experiment.

Acknowledgments

We thank Markus Apel (ACCESS, Aachen), Denis Danilov (FH Karlsruhe), Britta Nestler (FH Karlsruhe), and Ingo Steinbach (ACCESS, Aachen) for very intense and highly productive collaboration. We acknowledge the experimental findings on kinetics of rapid solidification and investigations of microstructure of as-solidified samples provided by Oliver Funke, Phanikumar Gandham, Matthias Kolbe, and Sven Reutzel (DLR Köln). We are grateful to the Deutsche Forschungsgemeinschaft for financial support within the priority programme SPP 1120 under grant No. HE 1601/13-1,2,3.

References

1 Herlach, D.M., Galenko, P. and Holland-Moritz, D. (2007) *Metastable Solids from Undercooled Melts*, Pergamon Materials Series, 1st edn (ed. R. Cahn) Oxford, Elsevier.
2 Kurz, W. and Fisher, D.J. (1992) *Fundamentals of Solidification*, 3rd edn, Trans Tech Publications, Aedermannsdorf.
3 Karma, A. and Rappel, J.-W. (1998) "Quantitative phase-field modeling of dendritic growth in two and three dimensions". *Physical Review E*, 57, 4323–49.
4 Bragard, J., Karma, A., Lee, Y.H. and Plapp, M. (2002) "Linking phase-field and atomistic simulations to model dendritic solidification in highly undercooled melts". *Interface Science*, 10, 121–36.
5 Beckermann, C., Diepers, H.-J., Steinbach, I., Karma, A. and Tong, X. (1999) "Modeling melt convection in phase-field simulations of solidification". *Journal of Computational Physics*, 154, 468–96.
6 Karma, A. and Rappel, J.-W. (1999) "Phase-field model of dendritic sidebranching with thermal noise". *Physical Review E*, 60, 3614–25.
7 Lipton, J., Glicksman, M.E. and Kurz, W. (1987) "Equiaxed dendrite growth in alloys at small supercooling". *Metallurgical Transactions A*, 18, 341–45.
8 Lipton, J., Kurz, W. and Trivedi, R. (1987) "Rapid dendrite growth in undercooled alloys". *Acta Metallurgica*, 35, 957–64.
9 Galenko, P.K., Funke, O., Wang, J. and Herlach, D.M. (2004) "Kinetics of dendritic growth under the influence of convective flow in solidification of undercooled droplets". *Materials Science and Engineering A*, 375-377, 488–92.
10 Jeong, J.-H., Goldenfeld, N. and Danzig, J.A. (2001) "Phase field model for three-dimensional dendritic growth with fluid flow". *Physical Review E*, 64, 041602-1–14.
11 Nestler, B., Danilov, D. and Galenko, P.K. (2005) "Crystal growth of pure substances: phase-field simulations in comparison with analytical and experimental results". *Journal of Computational Physics*, 207, 221–39.
12 Funke, O., Phanikumar, G., Galenko, P.K., Chernova, L., Reutzel, S., Kolbe, M. and Herlach, D.M. (2006) "Dendrite growth velocity in levitated undercooled nickel melts". *Journal of Crystal Growth*, 297, 211–22.
13 Herlach, D.M. and Galenko, P.K. (2007) "Rapid solidification: in situ diagnostics and theoretical

modelling". *Materials Science and Engineering A*, **449-451**, 34–41.
14. Brener, E. (**1990**) "Effects of surface energy and kinetics on the growth of needle-like dendrites". *Journal of Crystal Growth*, **99**, 165–70.
15. Herlach, D.M. (**1991**) "Containerless undercooling and solidification of pure metals". *Annual Review of Materials Science*, **21**, 23–44.
16. Aziz, M.J. and Kaplan, T. (**1988**) "Continuous growth model for interface motion during alloy solidification". *Acta Metallurgica*, **36**, 2335–47.
17. Hyers, R.W., Matson, D.M., Kelton, K.F. and Rogers, J. (**2004**) "Convection in containerless processing". *Annals of the New York Academy of Sciences*, **1027**, 474–94.
18. Reutzel, S., Hartmann, H., Galenko, P.K., Schneider, S. and Herlach, D.M. (**2007**) "Change of the kinetics of solidification and microstructure formation induced by convection in the Ni–Al system". *Applied Physics Letters*, **91**, 041913-1–3.
19. Kerrache, A., Horbach, J. and Binder, K. (**2008**) "Molecular-dynamics computer simulation of crystal growth and melting in $Al_{50}Ni_{50}$", *Europhysics Letters*, **81**, 58001–6.
20. Aziz, M.J. and Boettinger, W.J. (**1994**) "On the transition from short-range diffusion-limited to collision-limited growth in alloy solidification". *Acta Metallurgica and Materialia*, **42**, 527–37.
21. Barth, M., Wei, B. and Herlach, D.M. (**1995**) "Crystal growth in undercooled melts of the intermetallic compounds FeSi and CoSi". *Physical Review B*, **51**, 3422–28.
22. Gandham, P., Biswas, K., Funke, O., Holland-Moritz, D., Herlach, D.M. and Chattopadhyay, K. (**2005**) "Solidification of undercooled peritectic Fe–Ge alloy". *Acta Materialia*, **53**, 3591–600.
23. Garcke, H., Nestler, B. and Stinner, B. (**2004**) "A diffuse interface model for alloys with multiple components and phases". *SIAM Journal on Applied Mathematics*, **64**, 775–99.
24. Galenko, P.K., Reutzel, S., Herlach, D.M., Danilov, D. and Nestler, B. (**2007**) "Modelling of dendritic solidification in undercooled dilute Ni–Zr melts". *Acta Materialia*, **55**, 6834–42.
25. Ben-Amar, M., Bouisou, Ph. and Pelce, P. (**1988**) "An exact solution for the shape of a crystal growing in a forced flow". *Journal of Crystal Growth*, **92**, 97–100.
26. Herlach, D.H., Eckler, K., Karma, A. and Schwarz, M. (**2001**) "Grain refinement through fragmentation of dendrites in undercooled melts". *Materials Science and Engineering A*, **304-306**, 20–25.
27. Karma, A. (**1998**) "Model of grain refinement in solidification of undercooled melts". *International Journal of Non-Equilibrium Processing*, **11**, 201–33.
28. Schwarz, M., Karma, A., Eckler, K. and Herlach, D.M. (**1994**) "Physical mechanism of grain refinement in solidification of undercooled melts". *Physical Review Letters*, **73**, 1380–3.
29. Kaufmann, E. (**2000**) Sidebranch development of xenon dendrites, Ph.D. Thesis, Technical University Zürich.
30. Galenko, P.K., Phanikumar, G., Funke, O., Chernova, L., Reutzel, S., Kolbe, M. and Herlach, D.M. (**2007**) "Dendritic solidification and fragmentation in undercooled Ni–Zr alloys". *Materials Science and Engineering A*, **449-451**, 649–53.
31. Steinbach, I., Pezzolla, F., Nestler, B., Seeβelberg, M., Prieler, R., Schmitz, G.J. and Rezende, J.L.L. (**1996**) "A phase field concept for multiphase systems". *Physica D*, **94**, 135–47.
32. Eiken, J., Böttger, B. and Steinbach, I. (**2006**) "Multiphase-field approach for multicomponent alloys with extrapolation scheme for numerical application". *Physical Review E*, **73**, 066122-01–9.
33. Galenko, P.K., Reutzel, S., Herlach, D.M., Fries, S., Steinbach, I. and Apel, M. (**2008**) "Dendritic solidification in undercooled Ni–Zr–Al melts: experiments and modeling". *Acta Materialia*, in preparation.

34 Bobadilla, M., Lacaze, J. and Lesoult, G. (**1988**) "Influence des conditions de solidification sur le déroulement de la solidification des aciers inoxydables austénitiques". *Journal of Crystal Growth*, **531**, 992–1000.

35 Rappaz, M. and Boettinger, W.J. (**1999**) "On dendritic solidification of multicomponent alloys with unequal liquid diffusion coefficients". *Acta Materialia*, **47**, 3205–19.

36 Galenko, P. and Sobolev, S. (**1997**) "Local nonequilibrium effect on undercooling in rapid solidification of alloys". *Physical Review E*, **55**, 343–52.

37 Galenko, P.K. and Danilov, D.A. (**1997**) "Local nonequilibrium effect on rapid dendritic growth in binary alloy melt". *Physics Letters A*, **235**, 271–80.

38 Galenko, P. (**2002**) "Extended thermodynamical analysis of a motion of the solid–liquid interface in a rapidly solidifying alloy". *Physical Review B*, **65**, 144103-1–13

39 Galenko, P. (**2007**) "Solute trapping and a transition to diffusionless solidification in binary systems". *Physical Review E*, **76**, 031606-01–9.

23
Dendritic Solidification in the Diffuse Regime and Under the Influence of Buoyancy-Driven Melt Convection

Markus Apel and Ingo Steinbach

23.1
Introduction

The fundamentals of dendritic pattern formation have been theoretically well understood since the development of the microscopic solvability theory [1, 2]. Quantitative phase-field simulations of dendrites in 3D are possible today utilizing the schemes introduced by Karma [3, 4]. However, some issues for a complete understanding of dendritic solidification structures in technical alloy solidification are still open. These include the extension to multicomponent alloys and the impact of fluid flow on dendritic growth structures.

From the open issues, this chapter addresses the following items: First, we will outline an extension to the Karma corrections to multicomponent alloys and its application to equiaxed dendritic growth in $Ni_{98}Al_1Zr_1$ up to 80 K undercooling. Second, we will discuss the impact of fluid flow on the primary dendrite spacing selection. Furthermore, we will discuss the extension of the phase-field method coupled to buoyancy-driven fluid flow [5] to multicomponent melts. As applications we will show examples for buoyancy-driven flow pattern in the interdendritic region under directional growth conditions. Simulation results for 2D as well as for 3D arrangements are presented.

23.2
The Multiphase-field Model

We start with a general multicomponent multiphase-field model [6–9] developed during the last decade. This model and its numerical implementation [10] allow phase-field calculations for technical alloys via a coupling of the Gibbs energy calculations to thermodynamic databases [11]. The basic phase-field equation for a dual phase change problem reads

$$\dot{\phi} = \mu^*(\mathbf{n}) \left[\sigma^*(\mathbf{n}) \left(\nabla^2 \phi + \frac{\pi^2}{\eta^2} \left(\phi - \frac{1}{2} \right) \right) + \frac{\pi}{\eta} \sqrt{\phi(1-\phi)} \Delta G \right] \quad (23.1)$$

In this equation ϕ denotes the phase-field order parameter where $\phi = 1$ in solid and $\phi = 0$ in liquid, $\mu^*(\mathbf{n})$ the effective interfacial mobility (see next section) dependent on the interface normal in the coordinate system of the crystal, σ^* the interfacial stiffness, and ΔG the thermodynamic driving force acting on the interface. The parameter η denotes the interfacial thickness, which can be large compared to the physical interface thickness, but must be selected small compared to the length scale to be resolved. For dendritic structures the relevant length scale is the principal radius of the dendrite tip.

23.2.1
Extension of the Karma corrections to Multicomponent Alloys

For simulation of solidification microstructures it is usually desirable to work in the so-called "Gibbs–Thomson" limit. In this limit the phase-field method with its diffuse interfaces yields the sharp-interface solution of the Gibbs–Thomson equation. In general, the diffuse interface may introduce deviations from a sharp-interface solution due to different artifacts, namely (i) interface-thickness-dependent driving force, (ii) artificial solute trapping, and (iii) interface stretching. These unwanted effects result in a rather slow convergence of the diffuse-interface solution when the numerical discretization and the interface width are decreased, which makes quantitative simulations difficult. Karma introduced correction schemes for spurious effects in the thin-interface limit of the phase-field method [4]. However, the corrections given are limited to a symmetric or a one-side diffusion model and are also restricted to binary alloys. In the following, we will propose a generalized formulation for the correction terms.

The main problem in extending the kinetic correction to multicomponent alloys, and to the case where we have concentration-dependent diffusivities in both phases, is that an analytical integration is no longer possible. Instead, we use an averaging procedure at the interface that defines the required model parameters. The effective mobility μ^* in Equation 23.1 reads

$$\mu^*(\mathbf{n}) = \frac{\mu(\mathbf{n})}{1 - A\mu(\mathbf{n})} \quad (23.2)$$

$$A = D^{ij} \left\langle \left(c^j_{\text{liquid}}(x) - c^j_{\text{solid}}(x) \right) m^j(x) \right\rangle \Delta S \frac{a}{\eta} \quad (23.3)$$

$$D^{ij} = \left\langle D^{ij}_{\text{liquid}}(x) \left(1 - \phi(x)\right) + D^{ij}_{\text{solid}}(x) \phi(x) \right\rangle \quad (23.4)$$

where $\mu(\mathbf{n})$ is the physical interface mobility, $D^{ij}_{\text{liquid}}(x)$ and $D^{ij}_{\text{solid}}(x)$ are the diffusion matrices in liquid and solid, $c^j_{\text{liquid}(x)}$ and $c^j_{\text{solid}}(x)$ the j component of the local concentration in liquid and solid, $m^j(x)$ the liquidus slope, ΔS the entropy jump between solid and liquid, and a an integration constant. The average $\langle \rangle$ is taken over interface cells that are arranged in the normal direction in the interface as described in [12]. This ensures that the coefficients of the correction scheme are constant in the direction of the dominating solutal fluxes.

The multicomponent diffusion equation with a generalized *antitrapping* flux is written

$$\dot{c}^i + \mathbf{u}\nabla c_{\text{liquid}}^i = \nabla \cdot \left(\phi D_{\text{solid}}^{ij} \nabla c_{\text{solid}}^j + (1-\phi) D_{\text{liquid}}^{ij} \nabla c_{\text{liquid}}^j + j_{\text{antitrap}}^i \right) \quad (23.5)$$

and

$$j_{\text{antitrap}}^i = b\eta \left((c_l^j(x) - c_s^j(x)) \frac{D_{\text{liquid}}^{ij} - D_{\text{solid}}^{ij}}{D_{\text{liquid}}^{ij} + D_{\text{solid}}^{ij}} \right) \dot{\phi} \cdot \frac{\nabla \phi}{|\nabla \phi|} \quad (23.6)$$

The additional advective transport term $\mathbf{u}\nabla c_{\text{liquid}}^i$ in Equation 23.5 considers the fluid flow and we restrict ourselves to advection in the liquid melt only with liquid velocity $\mathbf{u}$. The antitrapping flux reduces in the binary case and with vanishing solid diffusivity to the formulation proposed in [4]. The integration constants a and b are selected such that in a set of benchmark simulations for equiaxed dendritic growth in a highly undercooled melt, the tip radius and velocity converge when the numerical discretization and, therefore, the interface width η are reduced. (see below).

23.2.2
Fluid Flow Coupling for Multicomponent Alloys

Fluid flow modeling is treated on the basis of the diffuse interface concept introduced in [5]. It relys on a modification of the Navier–Stokes equation in relation to diffuse boundaries, namely, an additional interfacial friction term. The specific formulation of the interface friction satisfies the no-slip condition at a solid–liquid interface regardless of interface width. In addition, buoyancy forces in the Buissinesque approximation are added. Because the coupling between fluid flow and the phase-field equation is rather weak (negligible dependence of the driving force on the melt pressure for standard conditions), the interaction between fluid flow and phase transformation is only mediated through an advective term in the solute transport equation. The equations for fluid flow and mass conservation read

$$\dot{\mathbf{u}} + \mathbf{u}\nabla \cdot \left[\frac{\mathbf{u}}{1-\phi} \right] = \nabla[\nu \nabla \mathbf{u}] \quad (23.7)$$

$$-\frac{1-\phi}{\rho_0}\left[\nabla p + \sum_i \frac{\partial \rho}{\partial c^i}(c_{\text{liquid}}^i - \langle c_{\text{liquid}}^i \rangle_g)\mathbf{g} \right] - h^* \frac{\mu \phi^2 (1-\phi)\mathbf{u}}{\eta^2}$$

$$\nabla \cdot [(1-\phi)\mathbf{u}] = 0 \quad (23.8)$$

where ν is the liquid viscosity, ρ_0 the average melt density, p the melt pressure, $\frac{\partial \rho}{\partial c^i}$ the linear coefficient of density change with the i component, $\langle c_{\text{liquid}}^i \rangle_g$ the global average melt concentration, $\mathbf{g}$ the gravity vector, and h^* an integration constant to ensure the no-slip condition at in the diffuse interface (see [5]). The only change compared to the binary case is that one has to sum up the buoyancy forces for all components i.

For the calculations presented in this chapter a dedicated version of the software MICRESS was used. This version covered the model extensions introduced above.

23.3
Rapid Solidification in $Ni_{98}Al_1Zr_1$

As a first example we will discuss equiaxed dendritic growth in $Ni_{98}Al_1Zr_1$. Experimental studies concerning dendritic growth velocities in binary Ni–Zr [13] and ternary Ni–Al–Zr are also reported in this book [14]. In order to make a quantitative comparison of model predictions to experimental results, phase-field simulations were performed for a melt undercooling between 20 and 80 K. Within this undercooling regime the growth kinetics should be determined solely by solute diffusion and interfacial curvature, and other contributions to the overall undercooling, such as thermal or kinetic contributions, should be negligibly small. This assumption can be verified, for example, by an estimation of the undercooling contributions according to the LKT model [15]. A comparative discussion between phase-field and sharp-interface modeling results can be found in [14] and [16].

The following material parameters enter the phase-field model. First, the free energies of the liquid and the solid phase as a function of temperature and composition are necessary to calculate the driving force. This data is derived by Dupin using the CALPHAD method and provided as a database (Dupin, N., private communication, 2005). The interfacial properties and diffusion coefficients are adapted from [17] and summarized in Table 23.1. No diffusion in the solid phase was considered.

As mentioned before, for quantitative phase-field simulations it is necessary to ensure convergence of the results. We demonstrate this by the calculation results plotted in Figure 23.1.

A series of 2D simulations of the dendritic growth at a fixed melt undercooling of 20 K were performed using varying grid spacings Δx. It can be seen that dendrite tip radius and growth velocity start to converge to a constant value at approximately $\Delta x = 15$ nm, corresponding to an interface width to tip radius ratio η/r of approximately 1/3. For the following calculations with varying melt undercooling we adapt our numerical discretization such that the ratio η/r is always approximately 0.2. To select a correct discretization for given conditions a

Table 23.1 Material data used for phase field simulations of dendritic growth in $Ni_{98}Al_1Zr_1$

	Material data		
Interfacial energy and anisotropy	2.7×10^{-5} J cm^{-2}	1.8%	
Diffusion constant in liquid Ni for Al and Zr	3×10^{-5} cm^2 s^{-1}	2.1×10^{-5} cm^2 s^{-1}	

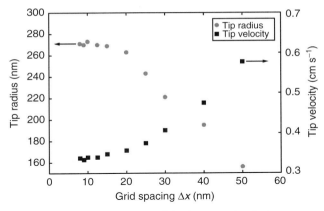

Fig. 23.1 Convergence behavior of dendrite growth velocity and tip radius with grid spacing. Interface width is always six grid spacings.

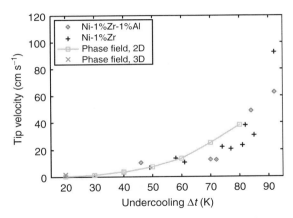

Fig. 23.2 Dendrite growth velocity in $Ni_{98}Al_1Zr_1$ as a function of undercooling.

test calculation with an approximate estimate for Δx was performed. The tip radius from this calculation was then used to determine the discretization required for the final simulation.

The focus of interest is now how the dendrite growth velocity increases with increasing melt undercooling. Figure 23.2 shows the calculated growth velocities as a function of the melt undercooling together with the experimental data. It can be seen that the experimental data could be reproduced reasonably well by the phase-field simulations without any fitting parameter.

The corresponding growth structures are shown in Figure 23.3. The upper half of the figure shows the Al composition, whereas the lower half shows Zr. The segregation behavior of the two elements is different. Al slightly enriches in the solid with a segregation coefficient $k \approx 1.2$ close to 1. In contrast, Zr shows a pronounced segregation into the melt with $k \approx 0.011$. Owing to its very slight segregation, Al

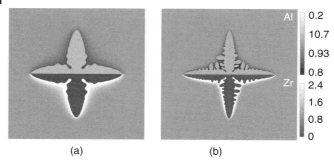

Fig. 23.3 Dendrite growth structures in $Ni_{98}Al_1Zr_1$ calculated for a melt undercooling of 40 and 60 K.

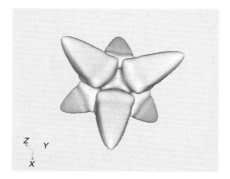

Fig. 23.4 3D simulation of a Ni dendrite in $Ni_{98}Al_1Zr_1$ grown at 20 K melt undercooling.

practically does not affect the growth kinetics in the solution-controlled regime, which agrees well with the experimental results. As undercooling increases, the tip radius decreases from $r = 270$ nm at $\Delta T = 20$ K to $r = 20$ nm at $\Delta T = 80$ K.

For intermediate undercoolings, 3D calculations are possible on a single CPU. A calculated 3D dendrite is shown in Figure 23.4, were the calculation time until growth velocity reaches steady state was approximately 48 h.

23.4
Directional Solidification with Buoyancy-driven Interdendritic Flow

In this section the models developed in the previous section are applied to the study of dendritic growth in a directional solidification environment with interdendritic flow. In this configuration, a 2D approximation is very appropriate, since flow is mostly confined within the interdendritic region, or the region in front of the dendritic array, and flow around the dendrites in lateral direction can be neglected. First, the influence of interdendritic flow on the selection of mean spacing will be studied for a binary alloy. Flow pattern formation in a ternary alloy is the final application in Section 23.4.2.

23.4.1
Spacing Selection in Binary Alloy

Buoyancy-driven convection in dendritic alloy solidification in the gravity field is inevitable because of the density differences in the liquid melt. The density of the metallic material depends strongly on the solute content, and owing to the solute redistribution at the growing solidification front, the concentration gradient around the dendritic tips is steep. Although the magnitude of this effect varies according to the alloy and the growth conditions, there will be radial gradients with respect to the direction of gravity, and there is no stable regime against the onset of flow. Also, in most alloys and under technically feasible temperature gradients, the solutal density change is 2 orders of magnitude higher than the thermal density change, and therefore a stabilization of flow due to a stable temperature configuration is not effective. We use here as an example an Al–Cu alloy with a Cu concentration $c_0 = 4$ at%. The material data are summarized in Table 23.2.

The temperature is set by a gradient in z-direction $G_z = 100$ K cm^{-1} and a cooling rate $\dot{T} = 0.4$ K s^{-1}, resulting in a steady-state growth speed of $v_g = 40\,\mu$m s^{-1}. For all calculations the rectangular calculation box has a height of 600 μm and variable dimensions in lateral directions. The calculations are started from spherical seeds at the bottom of the calculation domain which defines the initial spacing. The dendrite spacing is then free to adjust according to the growth conditions. Boundary conditions are periodic in the x direction. For the concentration field, the top boundary is of fixed concentration and the bottom boundary is adiabatic. Flow is constrained to be parallel to top and bottom boundaries. The calculations are performed in a moving frame, where the foremost dendrite tip is kept in fixed position relative to the top boundary [18].

Figure 23.5 shows a snapshot of typical simulation results. In Figure 23.5a the gravity vector is pointing in a positive z-direction with a magnitude of $g = 3$ in units $g_t = 9.81$ m s^{-2} of the terrestrial gravity constant. Because of segregation of the heavy copper into the interdendritic melt, the density increases close to the dendrite and the melt is upwardly buoyant. In the following, this will be termed upward flow. Unstable plumes form and the copper-enriched melt is washed out

Table 23.2 Material data used for phase field simulations of dendritic growth in Al$_{96}$Cu$_4$

		Material data
Melting point Al	T_m	933.6 K
Liquidus slope	m_l	1.6 K/%
Partition coefficient	k	0.14
Diffusivity (in liquid)	D_l	3×10^{-9} m^2 s^{-1}
Interfacial energy	σ_0	2.4×10^{-9} J m^{-2}
Anisotropy in interfacial stiffness	ϵ	0.3
Density variation	$\frac{1}{\rho}\frac{\partial \rho}{\partial c}$	0.01 1/%
Kinematic viscosity	ν	0.57 10^{-6} m^2 s^{-1}

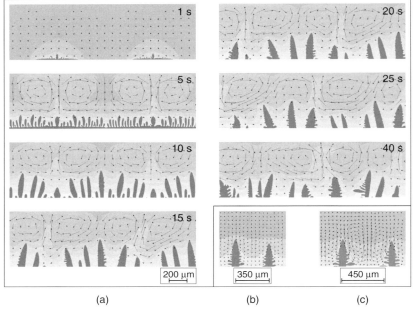

Fig. 23.5 Concentration and flow profile in directional dendritic growth. The solid dendrites are of lower Cu concentration and appear dark gray. (a) Upward buoyancy +3g. The time sequence starts after seeding of two crystals at the bottom of the domain. The solid spreads and forms side branches that evolve into a dendritic array. Between 5 and 10s the moving frame sets in to keep the leading tip at fixed position, withdrawing the whole domain one grid spacing to the bottom and adding a new layer with initial concentration c_0 at the top. The dendrite spacing adjusts to approximately 200 μm. Maximum flow speed 800 μms^{-1}. (b) Downward buoyancy −1g. 350 μm spacing is metastable. Maximum flow speed 33 μms^{-1}. (c) Downward buoyancy −1g. 450 μm spacing is stable. Maximum flow speed 50 μms^{-1}. The calculations were performed with a dedicated version of the software MICRESS [10].

into the bulk liquid region. In Figure 23.5a we can follow the transient from initial seeding of two solids at the bottom of the calculation domain into a fully developed dendritic array with mean spacing around 200 μm. Obviously neither the solid structure nor the convective pattern reaches a steady state. This is because the temporarily leading dendrites trigger the melt flow by stopping transversal flow and supporting new upward flow in a low friction area. On the other hand, the upward flow transports segregated copper along the dendrite into the tip region, which in turn slows down growth. The dendrites that have fallen back now face downward flow, created by neighboring dendrites, which are more advanced. This downward flow transports melt that is relatively low in copper and enhances growth. In this manner, convection and growth of individual tips are connected by an oscillating interaction. This picture explains the experimental findings by Mathissen and Arnberg [19], performing *in situ* synchrotron radiation imaging of the dendritic solidification structures in a thin sample.

Reverting the vector of gravity in the downward direction leads to a completely different picture (see Figure 23.5b and c). Downwardly buoyanced melt leads to an enrichment of copper in the interdendritic region. In contrast to the case with upward buoyancy, the mean spacing is significantly increased (>400 µm for −1g), which is clearly due to the enrichment of copper at the base of the dendrites. The convecting rolls are now confined to the interdendritic region and a stable flow pattern is established. To characterize the dependence of the spacing on the magnitude of flow more closely, a number of simulations were performed with a fixed domain size and two initial seeds set in regular spacing seeking for the minimum stable spacing. Stable growth in downward flow is characterized by symmetrical convection rolls and the two tips at equal positions in the moving frame of the calculation. Close to the minimum spacing, metastable states were also observed with a broken symmetry of the convection rolls and one tip behind the front tip. Such a state could persist for times that were comparable to and longer than the time needed to reach the steady state for a stable spacing. However, because of computational limitations, a closer investigation was not possible.

Figure 23.6 provides a stability map of all calculations performed, classified into unstable, metastable, and stable states. Taking the minimum stable spacing for all gravity levels under consideration, and noting that the average spacing is a multiple close to 1.5 of the minimum spacing, we can plot the average spacing, normalized by the average spacing at 0g versus the gravity level. Figure 23.7 gives the calculated spacings together with experimental results from solidification experiments in a centrifuge by Battaile et al. [20] and a scaling relation recently derived by the authors [21]. The purely diffusive case ($g = 0$) is discussed in [22].

23.4.2
Buoyancy-driven Fluid Flow in a Ternary Alloy

In the case of a binary alloy we were able to see that the stability of the melt convection pattern depends on the orientation of the gravity vector with respect to

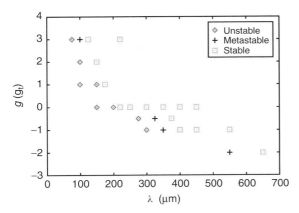

Fig. 23.6 Stability diagram of spacings for different levels of gravity.

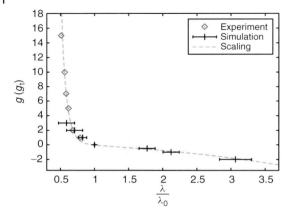

Fig. 23.7 Comparison between experiment, simulation, and scaling of the spacing λ, normalized by the spacing λ_0 at gravity $g = 0$.

growth direction. In mould-casting, where growth direction varies over the casting, regions of stable as well as unstable convection pattern may appear in one casting. This variation may result in a complex flow pattern on the macroscopic scale, but would also lead to variations of the dendritic microstructures depending on growth orientation.

Concerning the stability of the convection pattern, the situation is different for alloys with three or more components. In a binary alloy, the composition of the alloying element will impact the strength of the buoyancy-driven flow, but not the qualitative characteristics of the flow pattern. In multicomponent alloys the situation may arise in which some alloying elements decrease, while others increase, the density of the melt. In such a situation only a change in composition can lead to a transition from a stable to an unstable flow pattern. We will demonstrate this effect by a series of simulations for a ternary alloy. The thermodynamic data are those of the Al–Mg–Cu alloy system. We take Al as the main component and add a few percent of Mg and Cu. In order to demonstrate the effect of composition change on the convection qualitatively, we assume that Cu increases melt density by 1% per at% and Mg decreases melt density by 1% per at%. In the calculation domain (300 × 300 cells, $\Delta x = 1\,\mu m$), the gravity vector is pointed downwards with 1g; the thermal conditions and the initial situation is similar to the one in the previous section. The simulation starts with two seeds at the bottom of the domain. A temperature gradient in z-direction was imposed. During growth, Cu and Mg are segregated in the melt and lead to melt density variations around the dendrites.

In Figure 23.8 for two different alloy compositions, the calculated interdendritic flow patterns are shown for a steady-state situation. Figure 23.8a the nominal alloy composition is Al2%Cu4%Mg, and Figure 23.8b Al0.5%Cu8%Mg. As one can see in the figure, the direction of rotation for the convection rolls changes. For the first alloy composition the "heavy" Cu dominates the buoyancy force and

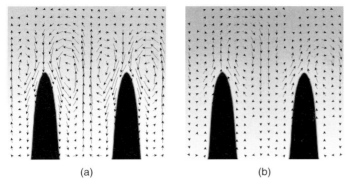

Fig. 23.8 (a) Steady-state convection rolls in Al-2%Cu-4%Mg.
(b) Convection roll in Al-0.5%Cu-8%Mg.

the melt-flow around the dendrite tip is pointed downwards in the direction of **g**. Reducing Cu and increasing Mg finally lead to a situation where the melt density is smaller around the dendrites and the flow direction points upward with an opposite rotation of the convection roll. Because the segregation coefficient for the two elements are different, $k_{Cu} \approx 11$, $k_{Mg} \approx 3$, higher quantities of Mg are required to overcompensate the effect of Cu on the melt density.

One can conclude that a composition somewhere in between will show a minimum density variation in the melt, and therefore only weak convection. We may note that, in general, a perfect compensation of the lighter and heavier components is not possible because different diffusivities in the melt can also result in density variations.

23.5
Summary and Conclusion

The present work represents a further step towards quantitative phase-field simulation of dendritic microstructures in technical alloys and for conditions encountered in technical casting processes. In this context, the notation *quantitative* means that the phase-field simulation depends only on physical parameters, and a sound numerical solution without any free parameters is achievable. Addressing this matter, a general formulation for interface mobility correction and antitrapping current was proposed for multicomponent alloys. Calculated growth velocities for dendrites in a $Ni_{98}Al_1Zr_1$ melt, growing at undercoolings up to 80 K agree well with experimental data. Calculations for directional growth conditions led to quantitative predictions for the primary dendrite spacing in binary alloys.

In a technical environment, melt flow plays a significant role in microstructure formation and therefore cannot be neglected. The effect of interdendritic fluid flow on primary arm spacing is discussed quantitatively for a binary alloy. Depending on the orientation and strength of the interdendritic convection rolls, spacing can increase or decrease by a factor of 3. Furthermore, we have extended the recent fluid flow model to buoyancy forces in multicomponent alloys. First calculations demonstrate that the transition of the rotation direction of interdendritic convection rolls is dependent on the alloy composition.

Acknowledgment

This work was funded by the Deutsche Forschungsgemeinschaft within the scope of Priority Program 1120.

References

1. Langer, J.S. (**1987**) in *Chance and Matter, Lectures on the theory of pattern formation, Les Houches, session XLVI*, (eds J. Souletie, J. Vannimenus and R. Stora), North Holland Amsterdam, pp. 692–711.
2. Ben-Amar, M. and Brenner, E. (**1993**) *Physical Review Letters*, **71**, 589.
3. Karma, A. and Rappel, W.-J. (**1996**) "Phase-fiels method for computational efficient modeling of solidification with arbitrary interface kinetics". *Physical Review A*, **53**, 3017–20.
4. Karma, A. (**2001**) "Phase-field formulation for quantitative modeling of alloy solidification". *Physical Review Letters*, **87** (11), 115701–4.
5. Beckermann, C., Diepers, H.-J., Steinbach, I., Karma, A. and Tong, X. (**1999**) "Modeling melt convection in phase-field simulations of solidification". *Journal of Computational Physics*, **154**, 468–96.
6. Steinbach, I., Pezzolla, F., Nestler, B., Seeßelberg, M., Prieler, R., Schmitz, G.J. and Rezende, J.L.L. (**1996**) "A phase field concept for multiphase systems". *Physica D*, **94**, 135–47.
7. Tiaden, J., Nestler, B., Diepers, H.J. and Steinbach, I. (**1998**) "The multiphase-field model with an integrated concept for modeling solute diffusion". *Physica D*, **115**, 73–86.
8. Steinbach, I. and Pezzolla, F. (**1999**) "A generalized field method for multiphase transformations using interface fields". *Physica D*, **134**, 385–93.
9. Eiken, J., Böttger, B. and Steinbach, I. (**2006**) "Multi phase field approach for alloy solidification". *Physical Review E*, **73**, 066122-1–9.
10. www.micress.de.
11. www.thermocalc.se.
12. Steinbach, I., Diepers, H.-J. and Beckermann, C. (**2005**) "Transient growth and interaction of equiaxed dendrites". *Journal of Crystal Growth*, **275**, 624–38.
13. Schwarz M. (**1998**) Ph.D. Thesis, "Kornfeinung durch Fragmentierung von Dendriten". Ruhr-Universitaet Bochum.
14. Galenko, P., Reutzel, S. and Herlach, D. (**2008**) *Phase Transformations in Multicomponent Melts*, Wiley-VCH Verlag GmbH.
15. Lipton, J., Kurz, W. and Trivedi, R. (**1987**) "Rapid dendrite growth in undercooled alloys". *Acta Metallurgica*, **35** (4), 957–64.
16. Galenko, P., Reutzel, S., Herlach, D., Gommez-Fries, S., Steinbach, I. and Apel, M. (**2008**) In preparation.
17. Galenko, P., Reutzel, S., Herlach, D., Danilov, D. and Nestler, B. (**2007**) *Acta Materialia*, **55**, 6834–42.
18. Diepers, H.-J., Ma, D. and Steinbach, I. (**2002**) "History effects during the selection of primary dendrite spacing. Comparison of phase-field simulations with experimental observations". *Journal of Crystal Growth*, 149, 237–239.
19. Arnberg, L. and Mathisen, R.H. (**2005**) "X-ray radiography observation of columnar dendritic

growth and constitutional undercooling in an Al-30 wt% Cu alloy". *Acta Materialia*, **53**, 947–56.
20 Battaile, C.C., Grugel, R.N., Hmelo, A.B. and Wang, T.G. **(1994)** "The effect of enhanced gravity levels on microstructural development in Pb-50 wt% Sn alloys during controlled solidification". *Metallurgical Transactions A*, **25**, 865–70.
21 Steinbach, I. and Apel, M. **(2008)** "Pattern formation in constraint dendritic growth with solutal buoyancy". in preparation
22 Steinbach, I. **(2008)** "Effect of interface anisotropy on spacing selection in constrained dendrite growth". *Acta Materialia*, in press.

24
Stationary and Instationary Morphology Formation During Directional Solidification of Ternary Eutectic Alloys

Bernd Böttger, Victor T. Witusiewicz, Markus Apel, Anne Drevermann, Ulrike Hecht, and Stephan Rex

24.1
Introduction

The investigation of ternary alloys is an important link between binary systems on one hand, which are quite well understood today and technical multicomponent alloys on the other hand, which may consist of up to 20 elements. In principle, the full complexity of multicomponent and multiphase alloys is already present in ternary systems and understanding ternary systems having more than two phases is the most important step for a successful description of technical alloys.

This concept is also implemented in the CALPHAD technique [1], which allows the construction of multicomponent Gibbs energy databases by extrapolation of ternary descriptions as most important basic elements. Only for a small number of phases quaternary interactions have to be included. The same holds for diffusion coefficient data. Despite their importance, for example for microstructure simulation, only few mobility data are available today, especially for the liquid phase. Again, the ternary system is the basic element for construction of multicomponent mobility databases.

24.1.1
About the Project

This chapter basically summarizes the main findings of a basic research project funded by the Deutsche Forschungsgemeinschaft (DFG) in the framework of the Priority Program 1120. The aim of this project is to provide description of coupled three-phase growth in ternary eutectic alloys by combination of assessment and optimization of the thermodynamic description, directional solidification experiments, and numerical phase-field simulation of morphology formation. Especially by using direct coupling of a phase-field model to thermodynamic CALPHAD databases [2–5] a well-fitted quantitative comparison of the microstructure between experiments and simulation was achieved.

Phase Transformations in Multicomponent Melts. Edited by D. M. Herlach
Copyright © 2008 WILEY-VCH Verlag GmbH & Co. KGaA, Weinheim
ISBN: 978-3-527-31994-7

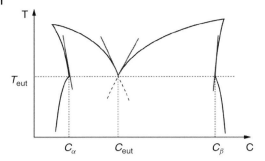

Fig. 24.1 Schematic binary eutectic phase diagram. Invariant growth is possible only at the eutectic temperature T_{eut} and at the eutectic composition c_{eut}. The four tangent lines correspond to the linearized phase diagram description.

24.1.2
General Aspects of Ternary Eutectic Systems

Binary eutectic systems have been studied intensively in the past [6]. Figure 24.1 shows a schematic phase diagram for a binary eutectic system. With only one compositional degree of freedom, the eutectic transformation occurs at the fixed temperature T_{eut}. This fact makes the incorporation of thermodynamics into modeling software quite straightforward, as the solidus and liquidus lines at the eutectic temperature can be approximated by straight lines (see Figure 24.1). Invariant eutectic growth at T_{eut} in directional solidification typically leads to lamellar structures (Figure 24.2). Only if the phase fractions of the two solid phases involved in the eutectic transformation are highly different, fibrous structures can be more favorable owing to the smaller interfacial energy contribution. Then, 3D simulation has to be used for microstructure modeling (Figure 24.3) [7].

In contrast, in ternary eutectics there is one extra degree of freedom in composition. The usual graphical representation of the phase diagram is the liquidus projection onto the Gibbs triangle. In the case of the Ag–Cu–Zn system given in Figure 24.8 three univariant lines ("grooves") can be seen which disembogue into an invariant point, labeled E_1. Thus, two types of eutectic reactions are possible in ternary eutectic systems: two-phase univariant reactions along the three univariant grooves and three-phase invariant growth at the ternary eutectic point.

Two-phase univariant growth typically shows lamellar microstructures, similar to binary eutectic reaction. However, owing to additional degree of freedom in composition, there is a long-range pile-up in the concentration field, which can lead to morphological instabilities resulting in typical eutectic cells or dendrites (Figure 24.4). Owing to varying solidification temperatures along the groove, phase-field simulation cannot use a linear phase diagram approximation, and hence, online coupling to thermodynamic databases becomes essential [2].

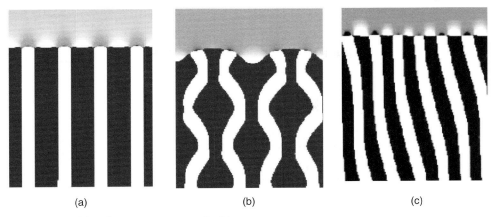

Fig. 24.2 Typical lamellar microstructures of a binary eutectic alloy using the phase-field code MICRESS and the linearized phase diagram description in Figure 24.1. Stable (a), oscillating (b), and tilted (c) modes can be observed.

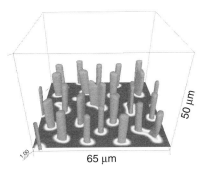

Fig. 24.3 Three-dimensional simulation of fibrous binary eutectic growth. In the representation, one solid phase has been removed.

Fig. 24.4 Simulation of cellular structures in univariant eutectic growth of a ternary alloy.

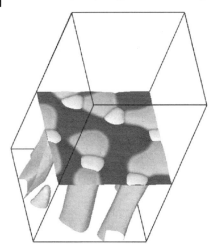

Fig. 24.5 3D simulation of invariant ternary eutectic growth. The interfaces between the solid phases and the melt as well as between the solid phases are shown.

At the invariant ternary eutectic point, three phases are growing simultaneously at a constant temperature forming a fibrous 3D microstructure with a planar front (Figure 24.5). So, linearized phase diagram data can be used for simulation.

24.2
Investigations on the Eutectic System Ag–Cu–Zn

The ternary Ag–Cu–Zn system was selected as a first model system to observe the intended coupled growth of three phases. The alloy system shows some technological interests, because different alloy compositions are used as high-temperature solders. A precise thermodynamic description of the system provides quantitative information about phase equilibria, the driving forces for liquid–solid and solid–solid phase transformations, heat capacity, latent heat evolution, and the thermodynamic factor for constructing the diffusivity matrices for diffusion.

24.2.1
Thermodynamic Properties and Thermodynamic Assessment

Knowledge of the thermodynamic properties of liquid, β, and γ phases in the ternary Ag–Cu–Zn alloys, is critical to obtain an accurate thermodynamic description of the system due to their large composition extensions. Since such data were not available in literature, the enthalpy of formation of liquid β, ζ, and γ phases as well as their transformation enthalpies and temperatures were measured by direct isoperibolic solution calorimetry and differential scanning calorimetry (DSC), respectively.

Details on the preparation of the samples, calorimetric measurements, as well as main results are published in [8, 9]. The thermodynamic description of the entire ternary Ag–Cu–Zn system was obtained by modeling the Gibbs energy of all individual phases in the system using the CALPHAD approach. The model parameters have been evaluated by means of a computer optimization technique based on own measured thermodynamic properties and phase transformation data, as well as previous literature data on phase equilibria for the ternary system [10, 11].

Figure 24.6a–c illustrates the main results obtained calorimetrically for the integral enthalpy of formation of the liquid phase. Figure 24.7 compares the melting, order/disorder, and other solid-state transformations of β, (β°), and ζ in Ag–Cu–Zn samples obtained by DSC with results of calculations using the performed CALPHAD description of the Ag–Cu–Zn system. A reasonable correspondence was achieved between the measured and optimized thermodynamic properties of the alloys. Figure 24.8 shows the projection of the liquidus surface onto the Gibbs triangle where existence of one E-type and one U-type reaction is clearly distinguished.

24.2.2
Bridgman Experiments

The sample alloy of the eutectic composition were prepared using an induction furnace for melting and consecutively casting under high-purity 3 bar Ar overpressure to cope with the high vapor pressure of Zn. Preweighted amounts of Ag 99.97 wt% and Cu, respectively Zn, both of 99.99 wt% purity, have been used for the alloys. The castings were made at about 900°C melt temperature in a steel mold with finger-shaped cavities of diameter 12 mm and length 220 mm, respectively. For sample preparation, these ingots were machined into cylindrical samples of 7.5 mm in diameter and 160 mm in length.

Directional solidification of the cylindrically shaped samples was performed in alumina-tube crucibles (Ø 10/8 mm) using a vertical Bridgman-type furnace. This Bridgman furnace is equipped with a heater zone and a highly effective liquid metal cooling zone, separated by an adiabatic zone 40 mm in length. The furnace assembly is vertically moved by a stepper motor along the axis of the fixed sample. Quenching of the sample at the end of the pulling length is achieved by a rapid raising of the furnace in such a way that the sample/crucible is suddenly immersed into the liquid metal bath. During the experiment, the vacuum-tight furnace casing is set under 1 bar pure Ar atmosphere. To cope with evaporation of Zn in the course of the experiments, the heater temperatures were carefully adjusted under the boundary condition of keeping the solid/liquid interface within the region of the adiabatic zone, and the sample inside the crucible was covered by a floss of low melting glass.

As a result of directional solidification, however, the processed samples did not show the expected coupled growth of three phases, but regular binary lamellae of

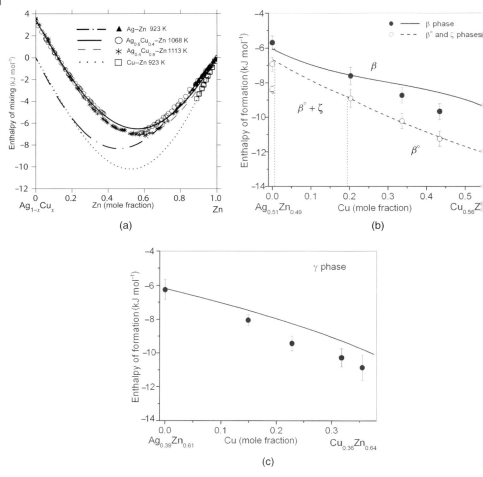

Fig. 24.6 Integral enthalpy of formation of liquid phase referred to liquid Ag, Cu, and Zn (a) and standard enthalpy of formation at 298 K of β, β° and ξ samples (b) as well as γ samples (c) referred to fcc Ag, fcc Cu, and hcp Zn. Points are experimental data of the present work and lines result from proposed CALPHAD description of the ternary system.

α_2 or (Ag) and β phases with only very tiny amount of α_1 or (Cu) incorporated (Figure 24.9). This was confirmed by the thermodynamic assessment. The isothermal cut of the Gibbs triangle at 666°C, just 1° above the eutectic temperature, given in Figure 24.9a, shows that the eutectic point E_1 is located very close to the tie-line (Ag)-β, and the calculated fraction of phases in the eutectic alloy as function of temperature (Figure 24.9c) reveals that the amount of the α-Cu phase at the

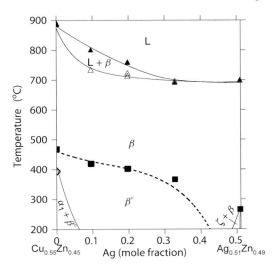

Fig. 24.7 Melting, order/disorder, and other solid-state transformations of the β Ag–Cu–Zn samples: points are DSC data; lines result from the present CALPHAD description of the system; the dashed line denotes the order/disorder transformation.

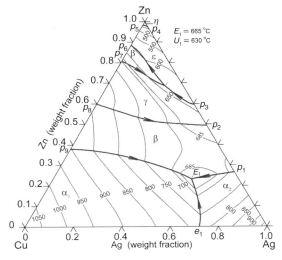

Fig. 24.8 Calculated liquidus surface of the Ag–Cu–Zn system according to the proposed CALPHAD description; isotherms are given in °C.

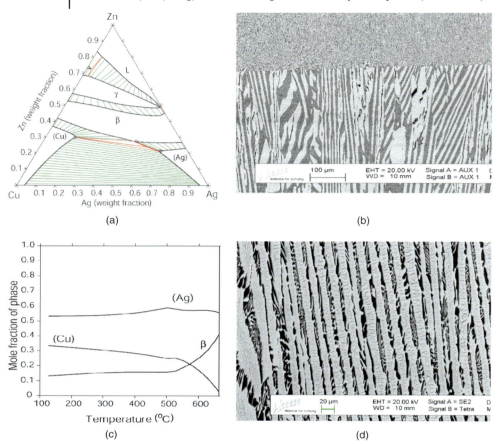

Fig. 24.9 Isothermal section at 666 °C (a), microstructure in vicinity of solid–liquid interface of the quenched unidirectional solidified (UDS) sample processed with a pulling rate $V = 2.03\,\mu m\,s^{-1}$ and temperature gradient $G = 12 \pm 3\,°C\,mm^{-1}$ (b), calculated variation of phase fraction with temperature (c), and microstructure of the same UDS sample but pulled through the temperature field down to $\sim 200\,°C$ and quenched (d): white phase is (Ag) solid solution, gray phase is β, and black phase is (Cu) solid solution.

eutectic temperature is in the range of only 2–4%. However, when cooled down to 200 °C, the amount increases up to 30%. Both were found also in the quenched unidirectionally solidified sample (Figure 24.9b and d) and thus confirmed the thermodynamic assessment.

However, for the intended investigation of eutectic coupled three-phase growth, the alloys system Ag–Cu–Zn turned out not to be best suited.

24.3
Investigation on the Ternary Alloy System In–Bi–Sn

The In–Bi–Sn system was selected as alternative eutectic sample alloy to the Ag–Cu–Zn system for further investigations for reasons mentioned above. The system is of some importance for technical application. Since lead is a cause for great environmental and health concerns, it is safer and more economical to replace lead in the solder materials than cleaning up the waste. Substantial efforts are being made to develop new lead-free solder materials. Currently lead-free solders based on the In–Bi–Sn alloy system are combined with a number of other elements like zinc, copper, silver, or gold to get materials which have a low melting point, good electrical and thermal conductivity, mechanical strength, good wettability, and corrosion resistance. Thus the In–Bi–Sn ternary system is the core system of new generation solders, and its precise thermodynamic description is of great technological importance.

24.3.1
Measurement of Thermodynamic Properties and Thermodynamic Assessment

The work was focused on a more precise determination of the solubility of the components in the coexisting phases at 77, 59, and 25 °C and the measurement of the enthalpies of melting of selected ternary alloys by energy-dispersive X-ray spectrometry (EDS) and DSC methods, respectively, as to provide a more reliable description of the phase equilibria for the entire Bi–In–Sn system. The experimental procedures in detail and main results are published in [12, 13].

The new thermodynamic description of the entire Bi–In–Sn system was obtained by fitting experimental information obtained in the present work and published in Refs. 14–18.

Examples of some calculated vertical sections in comparison with experimental data are presented in Figure 24.10. The calculated phase transformation temperatures fit reasonably well with the experimental data. Moreover, the present experiments and calculations suggest that both invariant equilibria L-γ-β-InBi$_2$ and L–(Bi)–(Sn)–InBi correspond to eutectic reactions.

This is in contradiction to the models proposed by Yoon *et al.* [19] and Moelans *et al.* [20], which propose U-type reactions, but agrees well with the model of Ohnuma *et al.* [21] and also with most of the experimental data on phase equilibria [14–18]. Direct observations of growth using unidirectional solidification techniques [12, 22–24] also confirm the eutectic nature of these reactions.

Calculated isothermal phase equilibria at 59.3 °C and the liquidus surface in the Bi–In–Sn system are presented in Figure 24.11a and b, respectively. The relative extension of the different phase fields are reproduced with the present thermodynamic description. The extended phase fields of γ, β, InBi$_2$, and bct-A5 (Sn) found by us experimentally are confirmed by the present calculations. The calculated temperatures and compositions of the invariant reactions approach the experimental values reported in literature.

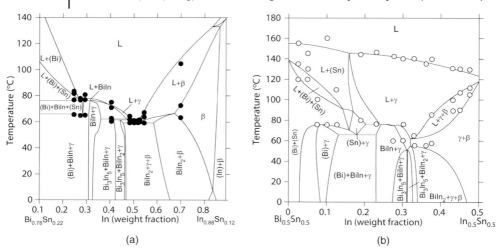

Fig. 24.10 Vertical sections in the Bi–In–Sn system through alloy compositions Bi-22 wt%Sn, and In-12 wt%Sn (a), and isopleth for 50 wt%Sn (b). Black points represent experimental values obtained in the present work (DSC) and lines are calculated with the present thermodynamic description; open points are experimental data of [19].

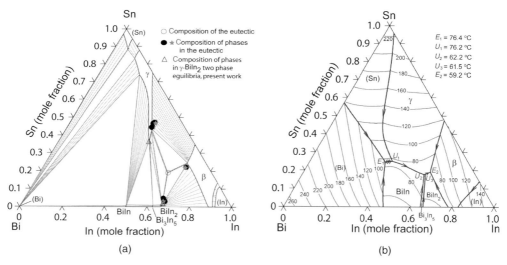

Fig. 24.11 Isothermal section at 59.3 °C (a) and liquidus surface projection in the Bi–In–Sn system (b): isotherms are given in °C.

24.3.2
Micro-Bridgman Assembly

Since the Bi–In–Sn system shows a rather low temperature at the invariant eutectic E_2 (cf. Figure 24.11b) we could use a micro-Bridgman assembly built for transparent organic alloys. To enable *in situ* observation of the microstructure evolution by light microscopy, the sample alloy had to be processed in a flat transparent containment showing a sample thickness or "depth" smaller than the spacing of the growing microstructure. Then the sample alloy was solidified in a 2D pattern, which could be observed representatively on the surface of the opaque sample.

The sample alloys used in the study were prepared from pure substances purchased from commercial suppliers and used in the as-received state with at least 99.999% purity. The preparation and alloying of the different samples were carried out by weighting the required quantities on an analytic balance with an accuracy of ±0.1 mg and melting together under vacuum in an alumina crucible. A sample of the resulting solid ingot was studied using DSC and the alloy composition was checked by an integral measurement with scanning electron microscopy/EDS (Gemini 1550).

The glass cells used for the investigations (Figure 24.12a) are composed of two oversized standard polished microscopy glass plates (70 and 150 × 15 × 1 mm). Then, two plates were glued together along the long sides by filling the gap in between the beveled edges with silicone resin. As a spacer a calibrated stainless steel strip of thickness 5 or 10 μm was placed between the glasses. When the silicone resin became solid enough, the spacer was removed from the cell. The prepared alloy was melted in an alumina crucible under Ar protection gas and sucked into the preheated sample cell. After filling, the two short edges of the cell are closed by silicon resin as well.

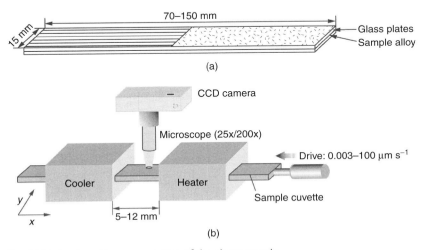

Fig. 24.12 A schematic representation of the glassy sample cuvette (a) and the micro-Bridgman assembly (b).

The cells were introduced in a horizontal micro-Bridgman furnace configuration mounted on the movable stage of a metallographic microscope. Details of the Bridgman furnace and the means of observation (Figure 24.12b) are given in [12]. Cooler temperatures in the range of 20–30 °C were used, while heater temperatures were adjusted between 75 and 85 °C, so that the solidification interface was located in the gap between the copper blocks. The temperature gradients at the solid–liquid interface were directly determined by means of calibration using 20-µm type-K thermocouples mounted in empty glass cells made as mentioned before. In general, temperature gradients in the range of 4.9 and 8 K mm^{-1} were applied. The growth rate was controlled by a PC via a PI Mercury M-227.50 stepper drive allowing a velocity range between 0.003 and 100 µm s^{-1}.

The microstructure evolution in the gap between the two copper blocks was observed with an automatic CCD photo camera (1280 × 960 pixels) mounted on top of a standard metallographic microscope, which provides also the illumination of the sample. The used magnifications of the microscope were 25×, 50× and 200×. The video images were afterwards analyzed with standard image processing software.

24.3.3
Stationary Coupled Growth

Three phase eutectic growth in 2D is very different compared to the "normal" three-dimensional modes, where typically rod-like structures are formed which allow the formation of quadruple junctions. In the thin glass capillary experiments, the formation of such a configuration would imply extremely high curvatures, so lamellar structures with alternating phases are preferred (Figure 24.13). Figure 24.14 shows a phase configuration which was achieved after a long time of continuous growth in the micro-Bridgman assembly. The stable phase sequence ABACp ("p" denotes periodic lamellar sequence) has been selected, where A corresponds to BiIn$_2$, B to γ-Sn, and C to β-In.

Like in the case of lamellar binary eutectics, the most stable steady state microstructure can be regarded as the result of a competence between diffusion and curvature. But the presence of three phases gives rise to additional complexity in the form of different possible stacking sequences. To compare the stability of these stacking sequences, several phase-field simulations have been performed for different stacking and periodic length λ. Figure 24.15 shows the resulting v-shaped curves for several stacking sequences, which were obtained by plotting the average front undercooling evaluated from the simulation runs against the applied periodic length λ. As is generally assumed [6], the growth mode with the lowest undercooling corresponds to the most stable mode. While for each stacking sequence a preferred spacing λ can be assigned, there is also a difference in the value between these minima of the undercooling curves. From all the tested sequences ABACp shows clearly the lowest front undercooling at its preferred spacing. So, from simulation, it can be confirmed that this stacking order should imply the most stable growth conditions for the 2D ternary eutectic front.

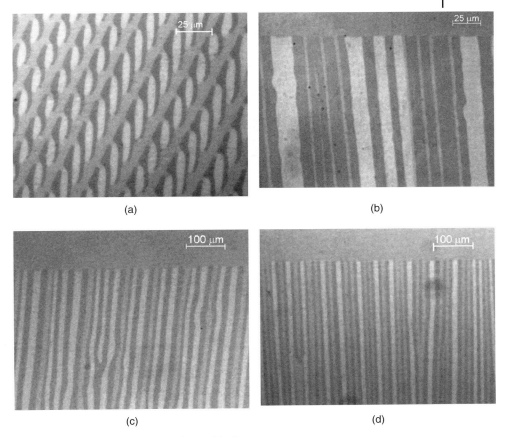

Fig. 24.13 Various microstructures observed in the micro-Bridgman apparatus as intermediate stage of solidification for compositions near to the ternary eutectic point (a–d): black phase is BiIn$_2$, gray phase is γ-Sn, and white phase is β-In.

24.4
Transient Growth

The micro-Bridgman assembly allows direct *in situ* observation of the microstructure formation process. This permits for the examination of transient growth. Starting from an equilibrated sample with composition inside the coupled ternary eutectic zone, one can observe the transition from primary planar growth to ternary coupled growth via the corresponding univariant grooves. The average front temperature can be obtained by measuring the front position in the constant temperature gradient overtime. The dynamics of this transient process should depend mainly on the segregation behavior (thermodynamic data) and the diffusion

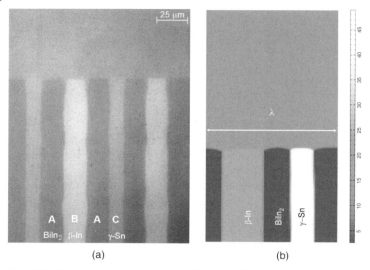

Fig. 24.14 Stationary three-phase coupled growth of ternary eutectic In–Bi–Sn observed after a long-term run in the micro-Bridgman (a) and simulated using the new thermodynamic description (b).

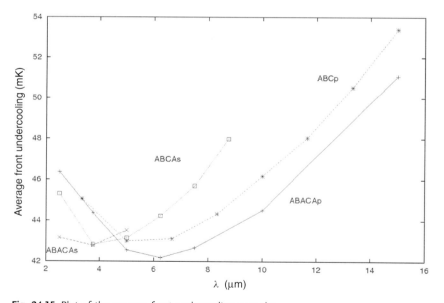

Fig. 24.15 Plot of the average front undercooling over the periodic length λ for various stacking sequences ("p" and "s" denote periodic and symmetric boundary conditions) for fixed values of G and V. ABACp is clearly the stacking sequence with lowest front undercooling at stable growth.

coefficients in the melt, while curvature undercooling should be about two orders of magnitude smaller compared to the solutal effects, and thus can be neglected.

24.4.1
Solidification Path During Transient Solidification

Figure 24.16 shows the measured development of the front undercooling over the solidified length, which is proportional to the elapsed time. In contrary to what has been expected, for the given initial composition of the sample In-21-5 at % Bi-17.8 at % Sn, (Figure 24.17a) the solidification path passes through different univariant grooves before reaching ternary coupled growth. This obviously is due to a high nucleation barrier for the β-In phase on BiIn$_2$: Instead of formation of γ-Sn as second phase the front is undercooled even far below the ternary eutectic temperature and γ-Sn is nucleating. Now, more than a length of about 1 mm, univariant growth of BiIn$_2$ and γ-Sn is observed. Finally, β-In is formed with high undercooling, resulting in a steeply increasing front temperature due to extremely rapid growth. But still ternary eutectic growth is not established, because during the explosive growth after β-In nucleation γ-Sn has been overgrown. So, a second univariant period in the BiIn$_2$/β-In groove is passed until a thorough renucleation of γ-Sn ternary eutectic growth is established.

Interestingly, exactly the same behavior can be observed in the phase-field simulation (Figure 24.18), if a high nucleation undercooling for β-In on BiIn$_2$ is assumed and the diffusion coefficients are calibrated correctly (see below). By evaluation of the melt composition at the solidification front, the solidification path can be plotted. In Figure 24.17b the path from the initial composition (square) to

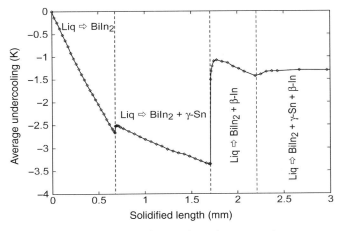

Fig. 24.16 Plot of the average front undercooling versus the solidified length for a transient experiment, showing different reactions along the solidification path.

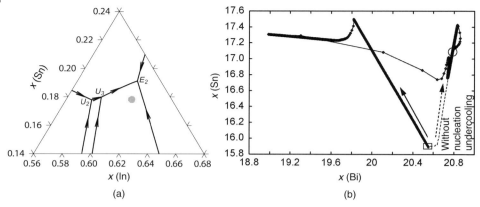

Fig. 24.17 Initial composition marked as dashed circle (a) for the transient experiment of Figure 24.16 and solidification path according to simulation (b).

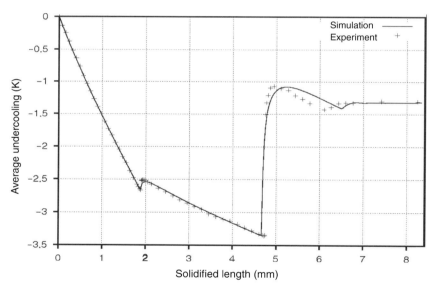

Fig. 24.18 Comparison of the average front undercooling between experiment (plus symbols) and simulation (line) after calibration.

the ternary eutectic (circle) is shown. The dark linear parts correspond to the three solidification regimes (primary $InBi_2$, univariant $BiIn_2/\gamma$-Sn, and univariant $BiIn_2/\beta$-In); the curved regions show the composition changes immediately connected to nucleation events. For comparison, the dashed line shows the solidification path, which would be expected if no nucleation barriers were in effect.

24.4.2
Quantitative Comparison to Simulation: Calibration of Diffusion Data

On the other hand, quantitative comparison of the observed growth kinetics in the transient experiment to simulation allows calibrating diffusion data, which cannot be easily measured experimentally and were unknown at this stage [24, 25]. This calibration has been done exemplarily for the experiment in Figure 24.16. Due to the fact that three nearly linear growth regions are formed, in principle, three independent diffusion parameters can be fitted by comparison of the corresponding slopes between experiment and simulation. These three parameters are the mobility values of the elements in the melt, which are assumed to be constant for the compositions near to the ternary eutectic point and which, via the phenomenological coefficients and together with the thermodynamic factors, allow for the evaluation of the reduced diffusion matrix in the melt. The resulting diffusion parameters are shown in Table 24.1.

At the same time, quantitative matching of the measured undercooling curve in Figure 24.18 with simulation results gives information about the nucleation undercooling which was effective in the experiment. The results are given in Table 24.2. The high value of nearly 10 K for the nucleation undercooling of β-In on the eutectic $BiIn_2/\gamma$-Sn front explains the strange solidification path described in Section 24.4.1.

24.4.3
Stationary Univariant Growth: Calibration of Interfacial Energies

Furthermore, choosing suitable starting compositions, stationary univariant growth could be observed for all three univariant eutectic grooves. The lamellar spacing

Table 24.1 Diffusion coefficients of the melt obtained from calibration of phase-field simulation with transient experiment in Figure 24.16.

Diffusion matrix coefficient in 10^{-5} cm^2 s^{-1}	In	Sn
In	9.09	2.95
Sn	−0.27	3.00

Table 24.2 Nucleation undercooling obtained from calibration of phase-field simulation with transient experiment in figure 24.16.

Nucleation undercooling in K	
γ-Sn on $BiIn_2$	2.0
β-In on $BiIn_2/\gamma$-Sn	9.9

Table 24.3 Interfacial energy between solid phases and melt obtained from comparison of measured and simulated lamellar spacing in the monovariant regions.

Interfacial energy in 10^{-6} J cm^{-2}	
BiIn$_2$/melt	0.9
β-In/melt	1.3
γ-Sn/melt	5.1

over a wide velocity range λ was obtained to be proportional to $v^{-1/2}$, as predicted in 1966 by Jackson and Hunt [6]:

$$\lambda_{\alpha\beta}^2 v = \text{const.} = k_\alpha^{\alpha\beta}\sigma_\alpha + k_\beta^{\alpha\beta}\sigma_\beta \tag{24.1}$$

According to Equation 24.1, for each univariant reaction, the parameters k_α, and k_β can be obtained by variation of the surface energy values σ_α, and σ_β, respectively, if selection of the lamellar spacing λ is simulated using the phase-field technique and correct thermodynamic and diffusion data. As mentioned in Section 24.1.2, long-range diffusion fields that are typical for univariant growth have to be taken into account using a 1D far-field approximation. This procedure leads to three equations for three univariant reactions, which resulted in three unknown surface energies for the three involved solid phases against the melt (Table 24.3).

24.5
Summary and Conclusion

In this work, two ternary eutectic alloy systems, Ag–Cu–Zn and In–Bi–Sn, have been investigated. The thermodynamic description was elaborated for Ag–Cu–Zn combining experimental data from isoperibolic calorimeter and DSC experiments and literature, which predicts an invariant ternary eutectic reaction in the silver-rich corner involving the phases α-Cu, α-Ag, and β-Zn. For In–Bi–Sn system, already existing descriptions were optimized using the results of experimental equilibrium phase compositions obtained in the ternary eutectic region, phase transformation temperatures, and enthalpy of melting different alloys.

For both systems Bridgman unidirectional solidification experiments were carried out. For Ag–Cu–Zn system, no coupled ternary eutectic growth could be obtained. The reason for this unexpected finding was identified to be the very low phase fraction of α-Cu formed during solidification at the ternary eutectic composition, confirmed by the new thermodynamic description.

Instead, Bi–In–Sn system showed a nicely coupled ternary eutectic growth in directional solidification experiments. Due to the very low melting temperature of 59 °C at the ternary eutectic composition, experiments could be performed where microstructure formation could be observed *in situ* using a conventional metallographic light microscope. Stationary and transient experiments show a wide variety of coupled and uncoupled growth modes.

Phase-field simulations with direct coupling to the assessed thermodynamic data were used for comparison to the 2D experiments. Through parameter calibration, diffusion coefficients of the melt, nucleation parameter, and interfacial energies could be obtained, which otherwise cannot be measured directly.

It can be concluded that the combination of computational thermodynamics (CALPHAD) and phase-field simulations forms a powerful tool for understanding microstructure formation processes in multicomponent and multiphase systems. Even quantitative comparisons or predictions are possible if an extensive calibration with well-defined experimental data is performed.

References

1 Saunders, N. and Miodownik, A. (1998) *CALPHAD Calculation of Phase Diagrams: A Comprehensive Guide*, Elsevier.
2 Eiken, J., Böttger, B. and Steinbach, I. (2006) *Physical Review E*, **73**, 066122–1–066122–9.
3 Böttger, B., Eiken, J. and Steinbach, I. (2006) *Acta Materialia*, **54**, 2697–2704.
4 Steinbach, I., Böttger, B., Eiken, J. and Fries, S.G. (2007) *Journal of Phase Equilibrium and Diffusion*, **28**, 101–6.
5 www.micress.de.
6 Jackson, K.A. and Hunt, J.D. (1966) *Transactions of the Metallurgical Society of AIME*, **236**, 1129–42.
7 Apel, M., Böttger, B. and Steinbach, I. (2003) in *MCWASP-X*, (eds D. Stefanescu, J.A. Warren, M.R. Jolly and M.J.M. Krane), TMS, Destin, FL, pp. 37–44.
8 Witusiewicz, V.T., Hecht, U., Rex, S. and Sommer, F. (2002) *Journal of Alloys and Compounds*, **337**, 189–201.
9 Witusiewicz, V.T., Fries, S.G., Hecht, U., Drevermann, A. and Rex, S. (2006) *International Journal of Materials Research*, **97**, 1–13.
10 Gebhard, E., Petzow, G. and Krauβ, W. (1962) *Zeitschrift fur Metallkunde*, **53**, 372–79.
11 Rönkä, K.J., van Loo, F.J.J. and Kivilahti, J.K. (1997) *Zeitschrift fur Metallkunde*, **88**, 9–13.
12 Witusiewicz, V.T., Hecht, U., Rex, S. and Apel, M. (2005) *Acta Materialia*, **53**, 3663–69.
13 Witusiewicz, V.T., Hecht, U., Böttger, B. and Rex, S. (2007) *Journal of Alloys and Compounds*, **428**, 115–24.
14 Dooley, G.J. and Peretti, E.A. (1964) *Journal of Chemical and Engineering Data*, **9**, 90–91.
15 Scherpereel, L.R. and Peretti, E.A. (1967) *Journal of Materials Science*, **2**, 256–59.
16 Stel'makh, S.I., Tsimmergakl, V.A. and Sheka, I.A. (1972) *Ukrainskii Khimicheskii Zhurnal*, **38**, 631–33 (in Russian).
17 Stel'makh, S.I., Tsimmergakl, V.A. and Sheka, I.A. (1972) *Ukrainskii Khimicheskii Zhurnal*, **38**, 855–59 (in Russian).
18 Kabassis, H., Rutter, J.W. and Winegard, W.C. (1986) *Materials Science and Technology*, **2**, 985–88.
19 Yoon, S.W., Rho, B.-S., Lee, F.M. and Lee, B.-J. (1999) *Metallurgical and Materials Transactions A*, **30**, 1503–15.
20 Moelans, N., Hari Kumar, K.C. and Wollants, P. (2003) *Journal of Alloys and Compounds*, **360**, 98–106.
21 Ohnuma, I., Cui, Y., Liu, X.J., Inohana, Y., Ishihara, S., Ohtani, H., Kainuma, R. and Ishida, K. (2000) *Journal of Electronic Materials*, **29**, 1113–21.

22 Ruggiero, M.A. and Rutter, J.W. (1995) *Materials Science and Technology*, **11**, 136–42.
23 Ruggiero, M.A. and Rutter, J.W. (1997) *Materials Science and Technology*, **13**, 5–11.
24 Rex, S., Böttger, B., Witusiewicz, V. and Hecht, U. (2005) *Materials Science and Engineering A*, **413-414**, 249–54.
25 Böttger, B., Witusiewicz, V. and Rex, S. (2006) in *MCWASP-XI* (eds C.-A. Gandin and M. Bellet), TMS, Sophia Antipolis, pp. 425–32.

25
Dendritic Microstructure, Decomposition, and Metastable Phase Formation in the Bulk Metallic Glass Matrix Composite $Zr_{56}Ti_{14}Nb_5Cu_7Ni_6Be_{12}$

Susanne Schneider, Alberto Bracchi, Yue-Lin Huang, Michael Seibt, and Pappannan Thiyagarajan

25.1
Introduction

Since their discovery in the early 1990s [1], a large number of bulk metallic glass (BMG) alloys have been progressively found. These materials allow slow cooling of the melt to form completely amorphous materials. They have evoked considerable scientific and industrial interest as promising candidates for technological applications. A common feature of these alloys is their excellent glass-forming ability (GFA), which has been shown to be related to their high reduced glass transition temperatures [2].

On the basis of the alloys with high GFA, several BMG matrix composites, consisting of a glassy matrix that embeds *in situ* formed second-phase crystals, have recently been found to exhibit interesting physical properties [3–5] and to combine valuable plastic properties with the well-known high strength of single-phase BMGs [6]. The existence of dendritic second phase precipitates embedded in the amorphous matrix has been shown to be responsible for the plastic behavior, and the deformation mechanisms have been indicated to depend on the composite microstructure [3, 7]. This suggested the possibility of controlling the mechanical properties of a material through tailoring its microstructural features (structural length scales, crystalline morphology, matrix–crystal compositions, etc.) and, furthermore, the chance of understanding the former by characterizing the latter.

The excellent thermal stability of BMGs has allowed extensive investigations that have been carried out on Zr-based alloys to learn about the mechanisms involved in the transformation of fully amorphous samples to their thermodynamically more stable crystalline states [8–13]. Various mechanisms such as primary crystallization at heterogeneous nucleation sites, quasi-crystallization accompanied by polymorphous transformation, quasi- and nano-crystallization preceded by (spinodal) decomposition (phase separation) have been proposed for the devitrification

Phase Transformations in Multicomponent Melts. Edited by D. M. Herlach
Copyright © 2008 WILEY-VCH Verlag GmbH & Co. KGaA, Weinheim
ISBN: 978-3-527-31994-7

observed in Zr-based BMGs [14, 15]. These mechanisms depend strongly on the chemical composition of the alloys and on the cooling/reheating parameters used during the experiments. Thus the investigation of the devitrification processes in these alloys has been a great challenge. Recently it has been shown that some of the conclusions derived for the crystallization behavior of fully homogeneous BMGs may be applied to other metastable materials. For instance, studies on Zr-based metallic glass matrix composites (intrinsic composites) have shown a devitrification behavior upon annealing similar to that of fully amorphous systems. It has been shown that partially crystalline samples of $Zr_{66.4}Nb_{6.4}Cu_{10.5}Ni_{8.7}Al_8$ containing dendritic body-centered cubic (bcc) phase form a quasi-crystalline phase in the amorphous matrix after thermal treatment [7]. Since Zr–Nb–Cu–Ni–Al and Zr–Ti–Nb–Cu–Ni–Be constitute a promising class of materials with interesting mechanical properties mediated by a strong interplay between their deformation behavior and their microstructure [3, 7, 16], it will be of great interest to study the crystallization mechanisms in these composites. We believe that with the knowledge gained by this study on the relationship between their crystallization behavior and mechanical properties it might be possible to improve the properties of these advanced materials [17].

Among the devitrification mechanisms cited above, phase separation is of particular interest since it has been claimed to play a primary role in the microstructure formation of other alloys [5, 18]. The term *phase separation* implies the formation of two phases with different chemical compositions in the undercooled liquid and, hence, different glass transition and melting temperatures. Since the GFA (i.e. a measure of the thermal stability of an undercooled liquid) is generally expressed by the ratio T_g/T_m (i.e. the so-called reduced glass transition temperature), the crystallization of the decomposed phase with lower T_g/T_m becomes more favorable, while the phase with higher T_g/T_m ratio solidifies as an amorphous matrix that is embedded with the crystalline precipitates. Decomposition could occur during cooling from the melt at moderate cooling rates as well as during reheating at temperatures in the undercooled liquid region. In both cases, phase separation controls the microstructural properties of the alloys through the crystallization pathway that is determined by the chemical composition and thermal stability of the decomposed phases.

In this chapter, we first present experimental observations on the dendritic phases and the chemical composition profiles in as-prepared $Zr_{56}Ti_{14}Nb_5Cu_7Ni_6Be_{12}$ intrinsic composites by high-resolution and analytical transmission electron microscopy (TEM). In the second part of the chapter, results on the crystallization process of $Zr_{56}Ti_{14}Nb_5Cu_7Ni_6Be_{12}$ intrinsic composites upon thermal treatment by using a range of complementary techniques, such as differential scanning calorimetry (DSC), X-ray diffraction (XRD), scanning electron microscopy (SEM), TEM with local energy dispersive X-ray (EDX) analysis, and small-angle neutron scattering (SANS) are discussed. Particular attention has been paid to the identification of phase separation involved in the devitrification pathway of this alloy.

25.2
Experimental Procedures

Zr–Ti–Nb–Cu–Ni–Be intrinsic composites were produced by cooling a homogeneous melt of the excellent glass-forming alloy $Zr_{56}Ti_{14}Nb_5Cu_7Ni_6Be_{12}$ (at%) with a rate slightly above its critical cooling rate [3]. The samples consist of an amorphous matrix embedding dendrites that are formed during cooling of the melt by nucleation and dendritic growth of a crystalline bcc phase and followed by solidification of the remaining liquid alloy [19].

The structural properties of the samples and the chemical compositions of both the dendrites and the matrix were investigated by SEM, XRD, high-resolution transmission electron microscopy (HTEM), and EDX. The SEM images were obtained on mechanically polished specimens with the use of a Philips SEM515. XRD was performed with Cu Kα radiation using a Siemens D5000 diffractometer, while HTEM employed a Philips CM200-UT-FEG microscope at 200 kV. Samples for HTEM were mechanically thinned and, afterwards, prepared by ion milling with a liquid nitrogen cooling stage. The thermodynamic properties of the specimens were determined by DSC using a Perkin Elmer DSC7 at a heating rate of $20\,K\,min^{-1}$. The DSC7 was further employed for isothermal annealing under protective atmosphere (Ar gas flux) at temperatures determined by calorimetric studies (see below). The microstructure of the samples was investigated by SANS using the time-of-flight small-angle neutron diffractometer (SAND) at the Intense Pulsed Neutron Source (IPNS) at Argonne National Laboratory.

25.3
Results and Discussion

Solidification of the $Zr_{56}Ti_{14}Nb_5Cu_7Ni_6Be_{12}$ alloy has been recently shown to proceed through primary crystallization of a bcc crystalline phase followed by the glass transition of the remnant melt [3, 19].

The composite nature of the samples can be clearly seen from the SEM image in Figure 25.1, which shows the presence of dendrites (bright regions) embedded in an inter-dendritic matrix (dark regions) in the as-cast specimens. The intensity contrast in Figure 25.1 is due to the different chemical compositions of the two regions. The bcc structure identified in the dendrite by selected area electron diffraction (SAD) is consistent with XRD patterns (Figure 25.2) which show Bragg peaks of a bcc phase with a lattice constant of about 0.351 nm.

In order to qualitatively determine the elemental distribution in the composite, including the lightest species Be, TEM measurements were performed in combination with electron energy loss spectroscopy (EELS). Energy-filtered transmission electron microscopy (EFTEM) images were taken using electrons with energy loss within a window enclosing the absorption edges of each atom species. Figure 25.3

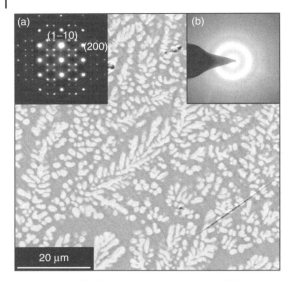

Fig. 25.1 SEM backscattered electron image of the as-quenched composite $Zr_{56}Ti_{14}Nb_5Cu_7Ni_6Be_{12}$. TEM-SAD patterns for a dendrite and the matrix are shown in the insets (a) and (b), respectively.

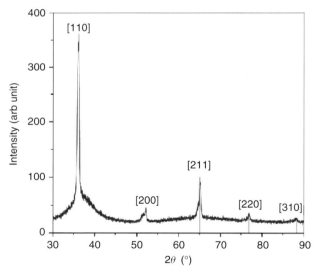

Fig. 25.2 Cu Kα X-ray diffraction pattern of a $Zr_{56}Ti_{14}Nb_5Cu_7Ni_6Be_{12}$ composite. The vertical lines indicate the Bragg peaks relative to a bcc structure of lattice constant $a = 0.351$ nm.

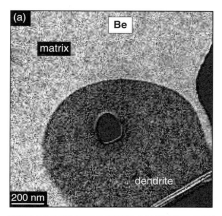
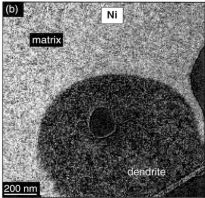
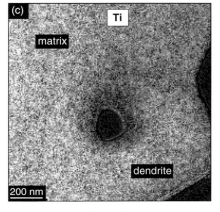

Fig. 25.3 Energy-filtered TEM images of a $Zr_{56}Ti_{14}Nb_5Cu_7Ni_6Be_{12}$ composite showing the contrast at the interface dendrite/matrix due to local variations of Be, Ni, and Ti concentration, respectively.

shows the images relative to the element Be, Ni, and Ti, respectively. Since the recorded intensity is proportional to the local concentration of the element in the composite, Figure 25.3 suggests that the matrix is enriched in Be and Ni, while the dendrites are Be- and Ni-depleted with an electron signal that has been found to be below the detection limit of the technique. Ti and Zr are distributed almost homogeneously in the composite since only a light contrast could be resolved in the corresponding energy-filtered image (see, for example, Figure 25.3c). The energy-filtered images recorded for the other elements (not reported here) indicate that the concentration of Cu is also higher in the matrix, whereas Nb is enriched in the dendrite.

Quantitatively, the concentration distributions of Zr, Ti, Nb, Cu, and Ni were determined by EDX combined with a TEM nanoprobe. Figure 25.4 shows, for instance, the local concentration of Cu, Ni, and Ti measured along a line across the interface between the dendrite and the matrix. In accordance with the EELS–TEM results, the matrix is Cu- and Ni-rich, while the Ti concentration is slightly higher in the dendrites. Similarly, Zr and Nb are depleted in the matrix within the experimental error. The interface dendrite/matrix (marked by a gray area in Figure 25.4) evidently defines a composition transition with steep Ni and Cu gradients in a region whose thickness is approximately 40 nm.

Taking the dendrite volume fraction measured by SEM (Figure 25.1), and assuming that all Be atoms are contained in the matrix, as concluded from the EELS–TEM analysis described above, the composition of the dendrites can be estimated to be $Zr_{70}Ti_{15}Nb_9Cu_4Ni_2$, while that of the matrix is likely $Zr_{49}Ti_{13}Nb_3Cu_9Ni_8Be_{18}$, indicating the tendency of the $Zr_{56}Ti_{14}Nb_5Cu_7Ni_6Be_{12}$ alloy to form a two-phase composite during cooling from the melt. The crystallization of the dendritic phase

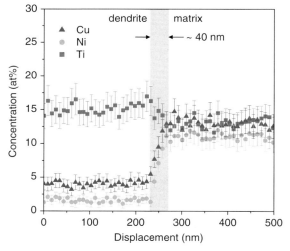

Fig. 25.4 Concentration profiles for Cu, Ni, and Ti in the $Zr_{56}Ti_{14}Nb_5Cu_7Ni_6Be_{12}$ sample determined by TEM-EDX along a line across the dendrite/matrix interface.

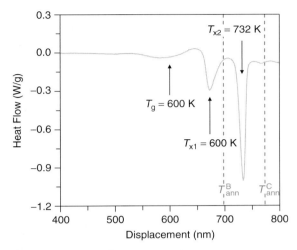

Fig. 25.5 DSC scan of an as-prepared $Zr_{56}Ti_{14}Nb_5Cu_7Ni_6Be_{12}$ metallic glass matrix composite recorded with a heating rate of 20 K min^{-1}. The arrows indicate the onset of the glass transition T_g and two different exothermic events T_{x1} and T_{x2}, respectively.

in the bcc structure, through nucleation and growth mechanisms, then stabilizes the system prior to the solidification of the embedding undercooled liquid matrix. Since the liquid phase should be homogeneous on the length scale of the observed dendrites, the microstructure is the result of atomic diffusion across distances of the dendrite arm diameter during cooling from the melt. This process has been successfully simulated using phase-field simulation techniques [20].

Figure 25.5 shows a DSC scan of an as-cast sample recorded at constant heating rate (20 K min^{-1}). A glass transition with the onset at $T_g = 600$ K and two exothermic events at $T_{x1} = 673$ K and $T_{x2} = 732$ K have been observed. The two exothermic peaks indicate clearly a two-step crystallization process related likely to a precursor phase formation at T_{x1} and to the complete crystallization of the remnant glassy matrix at T_{x2}. Similar scans have been already measured in other Zr-based BMGs, which showed a complex devitrification pathway upon annealing [7]. In order to elucidate the crystallization steps involved in this alloy and to identify the formed phases, several samples were heated to different temperatures below and above T_{x1} during DSC measurements at 20 K min^{-1}. Following that, XRD measurements were made on the as-cast and annealed samples to identify the formed phases through the analysis of the resulting diffraction peaks.

Figure 25.6 shows the XRD patterns recorded on an as-cast sample (sample A) and on two representative annealed specimens (sample B and C), which were reheated isothermally for 30 min at $T_{ann}^B = 698$ K (T_{x1} ¡ T_{ann}^B ¡ T_{x2}) and at $T_{ann}^C = 773$ K ($T_{x2} > T_{ann}^C$), respectively.

The bcc structure identified in the as-cast sample by SAD and associated to the dendritic phase by HTEM is consistent with the XRD pattern in Figure 25.6a,

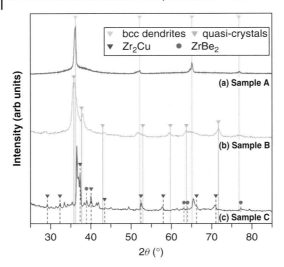

Fig. 25.6 X-ray diffraction patterns (Cu Kα radiation) of $Zr_{56}Ti_{14}Nb_5Cu_7Ni_6Be_{12}$ as-cast sample (a), and of two specimens annealed for 30 min at 698 K (b) and at 773 K (c), respectively. The symbols and the relative vertical lines indicate the peaks associated with different ordered phases.

which shows Bragg reflections ascribable to a bcc unit cell with a lattice constant $a = 3.507$ Å and overlapping broad maxima arising from the amorphous matrix. Neither XRD nor HTEM revealed the presence of an additional crystalline phase in the as-cast specimens. Instead, additional Bragg peaks emerge in the diffraction pattern of sample B as shown in Figure 25.6b, revealing the formation of an ordered phase in the specimen that was annealed at T_{ann}^B (slightly above the first exothermic DSC event).

In accordance with other works on the devitrification process in Zr-based BMGs [7, 21, 22], the observed reflections can be attributed to an icosahedral phase with a quasi-lattice constant $a_{\text{qc}} = 4.775$ Å. This value is in good agreement with that observed by Wanderka et al. [21] and recently by Van de Moortèle et al. (i.e. $a_{\text{qc}} = 4.779$ Å) in $Zr_{46.75}Ti_{8.25}Cu_{7.5}Ni_{10}Be_{27.5}$ (Vit4) [22], suggesting, furthermore, that the chemical composition of the quasi-crystals measured by Van de Moortèle (i.e, $Zr_{63.0}Ti_{14.4}Cu_{10.2}Ni_{12.4}$) may be close to that of the icosahedral phase found here. As the XRD pattern in Figure 25.6b clearly indicates, the bcc dendritic phase present in the as-cast sample is found to be stable upon thermal treatments, reflecting the fact that the observed first exothermic peak in the DSC scan is, indeed, due to the formation of quasi-crystals in the amorphous phase. It appears that the quasi-crystalline precipitates are metastable as can be deduced from Figure 25.6c, where the Bragg peaks present in Figure 25.6b disappear after the isothermal treatment at 773 K, that is, above T_{x2}. An almost identical XRD pattern was observed for a sample heated up to 873 K (not shown) indicating that the exothermic DSC event at T_{x2} can be associated with both the transformation of the quasi-crystals and the full crystallization of the glassy matrix. From the XRD pattern in Figure 25.6c two phases can be identified: the tetragonal Zr_2Cu (space group $I4/mmm$; lattice constants $a = 3.18$ Å, $c = 11.08$ Å) and the hexagonal Be_2Zr (space group $P6/mmm$; lattice constants $a = 3.82$ Å, $c = 3.24$ Å). Additional peaks could not be well identified but they presumably originate from Ti_2Ni, Zr_2Ni, or a Zr_2Ni- type ternary phase. These results are in good agreement with other works showing that Zr_2Cu, Be_2Zr, and Zr_2Ni are the main products of the devitrification process in Zr-based amorphous alloys (e.g. Vit1 [23], Vit4 [21, 22], and $Zr_{58.5}Ti_{8.2}Cu_{14.2}Ni_{11.4}Al_{7.7}$ [24]).

We would like to point out that quasi-crystals as a precursor phase, and intermetallic phases with almost identical structure as those mentioned above have been found to form mostly in fully amorphous alloys. Therefore, our results seem to indicate that the transformation pathway, (amorphous matrix) → (quasi-crystals + remnant glass) → (intermetallic phases), does not depend on the microstructure of the as-cast specimens (e.g. in our case a complex intrinsic composite with dendritic precipitates) but rather on the composition of the amorphous matrix. No quasi-crystals have been observed yet for $Zr_{56}Ti_{14}Nb_5Cu_7Ni_6Be_{12}$ alloys, but since the primary crystallization of the dendrites shifts the composition of the remnant amorphous matrix toward the compositions of Vit1 ($Zr_{41.2}Ti_{13.8}Cu_{12.5}Ni_{10}Be_{22.5}$) and Vit4 ($Zr_{46.75}Ti_{8.25}Cu_{7.5}Ni_{10}Be_{27.5}$) [19], we conclude that dendritic precipitates and, hence, the formation of the $Zr_{49}Ti_{13}Nb_3Cu_9Ni_8Be_{18}$ matrix are likely the prerequisites for the formation of the quasi-crystals in alloys with the given chemical composition.

In order to further investigate the formation of quasi-crystals in $Zr_{56}Ti_{14}Nb_5Cu_7Ni_6Be_{12}$ glass matrix composite, HTEM investigations were performed on a sample annealed at 678 K, just above the exothermic DSC peak at T_{x1}. Figure 25.7a shows a bright-field TEM image of an interdendritic region indicating the presence of small crystalline clusters in the amorphous matrix. The clusters have a size between 5 and 20 nm. These values are consistent with the broad XRD reflections shown in Figure 25.6b, which have been associated with the quasi-crystalline phases. Furthermore, it is interesting to note that no crystalline cluster is present in a region within about 100 nm from the dendrite/matrix interface into the matrix. Since the matrix of the as-cast sample has been found to be homogeneous by EDX and energy-filtered TEM [19], the higher stability of the interface region against crystallization, compared to that of the off-dendrite regions, may be related to a reduced atomic transport kinetics. This is likely dependent on the presence of the dendrites that might hamper the cluster formation when the matrix is annealed above its glass transition temperature into the undercooled liquid state.

Figure 25.7b shows a direct image of an interdendritic cluster embedded in the amorphous matrix. The fast Fourier transform (FFT) of the HTEM picture is presented in Figure 25.7c. It clearly shows a typical fivefold rotational symmetry of an icosahedral phase. According to the indexing scheme proposed by Bancel et al. [25], and by using six independent vectors as expressed by $q_1 = (1, +\delta, 0)$, $q_2 = (1, -\delta, 0)$, $q_3 = (0, 1, +\delta)$, $q_4 = (0, 1, -\delta)$, $q_5 = (+\delta, 0, 1)$, and $q_6 = (-\delta, 0, 1)$, where $\delta = (1 + \sqrt{5})/2$ is the golden mean, the reflections can be identified as

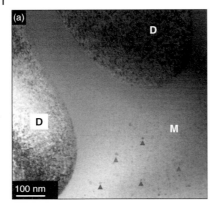
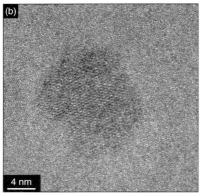
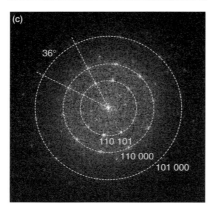

Fig. 25.7 TEM investigation of the devitrification process in off-dendrite regions in a $Zr_{56}Ti_{14}Nb_5Cu_7Ni_6Be_{12}$ sample annealed at 678 K: (a) bright-field TEM image (D: dendrite, M: matrix, solid triangles: precipitates embedded in the glassy matrix), (b) HTEM image, and (c) the relative fast Fourier transform (FFT) indicating the fivefold symmetry typical of a quasi-crystal.

(110 $\bar{1}$01), (110 000), and (101 000), respectively. The quasi-lattice constant is found to be 4.78 Å, which is in good agreement with the value from the X-ray diffraction data.

In addition to XRD and HTEM, SANS was employed to characterize the microstructure of the as-cast and the annealed specimens. Figure 25.8a shows the scattering data for an as-cast sample and three specimens annealed at temperatures 633, 663, and 703 K that lie between T_g and T_{x2}. For the as-cast sample, the high intensity at low scattering vector ($Q = 4\pi \cdot \sin\theta/\lambda$, where 2θ is the scattering angle and λ is the wavelength of neutrons) can be attributed to the presence of large inhomogeneities in the form of dendrites embedded in a matrix with different chemical compositions.

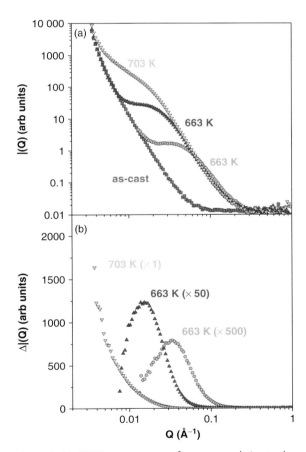

Fig. 25.8 (a) SANS measurements for a $Zr_{56}Ti_{14}Nb_5Cu_7Ni_6Be_{12}$ as-cast sample and three other specimens annealed at 633, 663, and 703 K, respectively. (b) Interference peaks obtained by subtracting the curve relative to the as-cast sample from the measurements of the annealed specimens. The peaks indicate the presence of scattering inhomogeneities that are distributed in a quasi-periodic arrangement.

Assuming that the dendrites remain unchanged during annealing, as evidenced by the XRD results described above, the scattering signals for the formed inhomogeneities can be obtained by subtracting the SANS signal of the as-cast sample from the data for the annealed specimens. As shown in Figure 25.8b, the resulting curves exhibit broad peaks, suggesting that the scattering inhomogeneities are distributed in a quasi-periodic arrangement. Interestingly, the sample annealed at 633 K exhibits an interference peak at $Q \approx 0.05$ Å^{-1}, which shifts to lower Q with increasing annealing temperature. We believe that this arises owing to the growth of inhomogeneities in the amorphous matrix. Similar interference peaks have been previously observed by SANS in isothermally annealed Vit1 [8] and Vit105 [14] samples, which were interpreted as due to a modulated chemical decomposition process preceding the formation of spatially correlated nanocrystals.

The SANS results and the formation of quasi-crystals observed by XRD and HTEM suggest that the amorphous matrix may decompose upon annealing near the glass transition region into two phases with different Be contents. The modulated chemical decomposition may be strongly influenced by the high diffusivity of Be and anticorrelation fluctuations of Be and Ti [26]. The lower thermal stability of the Be-depleted phase [27, 28] may drive the polymorphous reaction and the formation of Be-free quasi-crystals embedded in a Be-rich matrix. These conclusions are supported by the structural and chemical analysis of the quasi-crystalline phase presented above. Similar transformation pathways have recently been proposed for the devitrification processes in $Zr_{65}Al_{7.5}Cu_{7.5}Ni_{10}Ag_{10}$ and in Vit4 with the experimental support of compositional analysis and thermodynamic data [15, 22].

Annealing at higher temperatures (above T_{x2}) induces the transformation of the icosahedral phase into Zr_2TM (with TM = Nb, Cu, Ni, and Ti) [29] and of the Be-rich matrix into $ZrBe_2$ and other unknown ternary phases. Assuming that the composition of the quasi-crystals is $Zr_{63.0}Ti_{14.4}Cu_{10.2}Ni_{12.4}$ (i.e., $Zr_{63}TM_{37}$ or Zr_2TM) as discussed above, the observed transformation of the quasi-crystals into a crystalline phase at high temperature is likely the result of a polymorphous reaction (Zr_2TM quasi-crystals) → (Zr_2TM crystals), which further stabilizes the alloy. On the contrary, the formation of quasi-crystals in the amorphous matrix upon low-temperature annealing is thermodynamically more favorable than that of the crystalline Zr_2TM-type phases since, as demonstrated for $Zr_{65}Al_{7.5}Cu_{12.5}Ni_{10}Ag_5$ [30], the activation energy calculated with the Kissinger method for the (glass) → (icosahedral phase) transformation is much lower than that necessary for the reaction (residual glass) → (crystalline phases). Thus, the formation of the quasi-crystalline phase at $T = T_{x1}$, as a result of a chemical decomposition upon reheating in the glass transition region, must precede the devitrification of the amorphous matrix.

25.4
Summary

$Zr_{56}Ti_{14}Nb_5Cu_7Ni_6Be_{12}$ alloy shows phase separation into a complex two-phase microstructure composed of a bcc dendritic phase embedded in a remnant amorphous matrix which forms during the cooling process of the melt. The experimental results obtained by high-resolution and analytical TEM permitted the determination of the compositions of the two phases and the concentration gradient at the dendrite/matrix interface. The complex devitrification behavior observed in $Zr_{56}Ti_{14}Nb_5Cu_7Ni_6Be_{12}$ is strongly influenced by the precipitation of the dendritic phase during cooling from the melt since this shifts the chemical composition of the glassy matrix towards that for Vit1 and Vit4, according to the scheme proposed in our previous work [19]. In terms of calorimetric properties, the amorphous matrix in $Zr_{56}Ti_{14}Nb_5Cu_7Ni_6Be_{12}$ intrinsic composites behaves very similar to Vit1: their glass transition temperatures at a heating rate of $20\,\text{K}\,\text{min}^{-1}$ are found to be the same ($T_g = 623$ K), while the crystallization behavior is similar showing two exothermic events in the DSC scans at almost the same onset temperatures (i.e. $T_{x1} = 723$ K, $T_{x2} \approx 740$ K) [8]. Hence, it is likely that the precursor quasi-crystalline phase is present in both $Zr_{56}Ti_{14}Nb_5Cu_7Ni_6Be_{12}$ alloy and Vit1.

Our SANS study indicates that the formation of the quasi-crystals is likely to be preceded by a phase separation upon reheating in the undercooled liquid state. Furthermore, the decomposed phases of the glassy matrix with different Be contents are distributed quasi-periodically. The phase with lower Be concentration has a lower thermal stability and reacts polymorphously at about 673 K to form Be-free Zr_2TM-type quasi-crystals consistent with that reported in [23]. Upon annealing at temperatures above 732 K, the Be-rich matrix crystallizes into a mixture of $ZrBe_2$ and other ternary phases, while the metastable quasi-crystals form a Zr_2TM-type crystalline phase.

Acknowledgments

The authors are grateful to C.C. Hays for providing the samples. The use of the IPNS was made available by the DOE under contract W-31-109-ENG-38 to the University of Chicago. We gratefully acknowledge D. Wozniak (IPNS, Argonne National Lab) in the SANS measurements. This work was financially supported by Deutsche Forschungsgemeinschaft (SP1120 Schn-466/3).

References

1 Peker, A. and Johnson, W.L. (**1993**) *Applied Physics Letters*, **63**, 2342.

2 Inoue, A. (**2000**) *Acta Materialia*, **48**, 279.

3 Hays, C.C., Kim, C.P. and Johnson, W.L. (**2000**) *Physical Review Letters*, **84**, 2900.

4 Schneider, S., Bracchi, A., Samwer, K., Seibt, M. and Thiyagarajan, P. (**2002**) *Applied Physics Letters*, **80**, 1749.

5 Bracchi, A., Samwer, K., Almer, J., Aravinda Narayanan, R., Thiyagarajan, P. and Schneider, S. (**2004**) *Physical Review B*, **70**, 172105.

6 Das, J., Löser, W., Kühn, U., Eckert, J., Roy, S.K. and Schultz, L. (**2003**) *Applied Physics Letters*, **82**, 4690.

7 Kühn, U., Eckert, J., Mattern, N. and Schultz, L. (**2002**) *Applied Physics Letters*, **80**, 2478.

8 Schneider, S., Thiyagarajan, P. and Johnson, W.L. (**1996**) *Applied Physics Letters*, **68**, 493.
9 Busch, R., Schneider, S., Peker, A. and Johnson, W.L. (**1995**) *Applied Physics Letters*, **67**, 1544.
10 Löffler, J.F. and Johnson, W.L. (**2000**) *Applied Physics Letters*, **76**, 3394.
11 Löffler, J.F., Bossuyt, S., Glade, S.C., Johnson, W.L., Wagner, W. and Thiyagarajan, P. (**2000**) "Crystallization of bulk amorphous Zr-Ti(Nb)-Cu-Ni-Al". *Applied Physics Letters*, **77**, 525.
12 Kündig, A.A., Löffler, J.F., Johnson, W.L., Uggowitzer, P. and Thiyagarajan, P. (**2001**) *Scripta Materialia*, **44**, 1269.
13 Löffler, J.F., Thiyagarajan, P. and Johnson, W.L. (**2000**) *Journal of Applied Crystallography*, **33**, 500.
14 Pekarskaya, E., Löffler, J.F. and Johnson, W.L. (**2003**) *Acta Materialia*, **51**, 4045.
15 Chen, M.W., Inoue, A., Sakurai, T., Menon, E.S.K., Nagarajan, R. and Dutta, I. (**2005**) *Physical Review B*, **71**, 092202.
16 Hays, C.C., Kim, C.P. and Johnson, W.L. (**2000**) *Physical Review Letters*, **84**, 2901.
17 Oh, J.L., Ohkubo, T., Kim, Y.C., Fleury, E. and Hono, K. (**2005**) *Scripta Materialia*, **165**, 93.
18 Bracchi, A., Samwer, K., Niermann, T., Seibt, M. and Schneider, S. (**2004**) *Applied Physics Letters*, **85**, 2565.
19 Huang, Y.-L., Bracchi, A., Niermann, T., Seibt, M., Danilov, D., Nestler, B. and Schneider, S. (**2005**) *Scripta Materialia*, **53**, 93.
20 Nestler, B., Danilov, D., Bracchi, A., Huang, Y.-L., Niermann, T., Seibt, M. and Schneider, S. (**2007**) *Materials Science and Engineering A*, **452-453**, 8–14.
21 Wanderka, N., Macht, M.-P., Seidel, M., Mechler, S., Ståhl, K. and Jiang, J.Z. (**2000**) *Applied Physics Letters*, **77**, 3935.
22 Van de Moortèle, B., Epicier, T., Soubeyroux, J.L. and Pelletier, J.M. (**2004**) *Philosophical Magazine Letters*, **84**, 245.
23 Tang, X.-P., Löffler, J.F., Johnson, W.L. and Wu, Y. (**2003**) *Journal of Non-Crystalline Solids*, **317**, 118.
24 Scudino, S., Eckert, J., Mickel, C. and Schultz, L. (**2005**) *Journal of Non-Crystalline Solids*, **351**, 856.
25 Bancel, P.A., Heiney, P.A., Stephens, P.W., Goldman, A.I. and Horn, P.M. (**1985**) *Physical Review Letters*, **54**, 2422.
26 Macht, M.-P., Wanderka, N., Wiedenmann, A., Wollenberger, H., Wei, Q., Fecht, H.J. and Klose, S.G. (**1996**) *Materials Science Forum*, **225-227**, 65.
27 Johnson, W.L. (**1996**) *Materials Science Forum*, **225-227**, 35.
28 Wang, W.H. and Bai, H.Y. (**1998**) *Journal of Applied Physics*, **84**, 5961.
29 Audier, M., Brechet, Y., De Boissieu, M. and Guyot, P. (**1991**) *Philosophical Magazine B*, **63**, 1375. The transformation of a quasi-crystalline phase into a crystalline one has been experimentally observed in good accordance with theoretical works, for example, in the Al-Fe-Cu system.
30 Chen, M.W., Dutta, I., Zhang, T., Inoue, A. and Sakurai, T. (**2001**) *Applied Physics Letters*, **79**, 42.

Further Reading

Sun, Y.F., Wei, B.C., Wang, Y.R., Li, W.H., Cheung, T.L. and Shek, C.H. (**2005**) *Applied Physics Letters*, **87**, 051905.

Index

a

A1, metastable "phase" 283–287
AAS, *see* atomic absorption spectroscopy
ab initio molecular dynamics (MD) simulation 172
abrasion 76
absorption spectroscopy, atomic 133
Ackerson's approximation 201
activation
— energy 120
— thermal 203
adaptive mesh generators 339, 342–343
adsorption 32–33
aerogel furnace device 41
Ag
— Al-based alloys 81–83
— Cu-based alloys 55, 61–69
Ag–Cu–Zn 391–395
— liquidus surface 393
a4i Materials Image Analysis 255
Al
— Al–Ni alloys, *see* Al–Ni alloys
— Cu-based alloys 55, 60–69
— interfacial tension 24
— Nd–Fe–Co–Al 263–276
— Ti–Al 245–261
Al-based alloys
— binary 66
— growth morphology 40–41
— immiscible, *see* immiscible Al-based alloys
— monotectic 3–17
— soft metals 19
— ternary 6–10
Al–Bi–Cd 104
Al–Bi–Cu 9–10
Al–Bi–Cu–Sn 11–15
Al–Bi–Sn 6–8
Al–Bi–Zn 6

Al–Cu
— eutectic composition 80–81
— spacing selection 379
Al–Cu–Ag, invariant eutectic composition 81–83
Al–Fe 66–68
Al–Mg–Cu 382
Al–Ni alloys 66–68, 148–154, 367–369, 376–378
— entropy and free enthalpy 165–168
— mixing tendency 136–137
Al–Pb 89–90
Al–Sn–Cu 8–9
algorithms and software
— adaptive mesh generators 339, 342–343
— a4i Materials Image Analysis 255
— computer-aided design (CAD) 76
— edge detection 59
— FIT2D 192
— FLUENT 77
— force biased 172–173
— Metropolis method 145
— MICRESS 376, 389
— Monte Carlo simulation, *see* Monte Carlo simulation
— Nelder–Mead 173, 177
alloy melts
— crystal growth 157–170
— glass-forming 118–119
— *see also* melts
alloys
— Al-based, *see* Al-based alloys
— amorphous 177–178, 263
— binary, *see* binary alloys
— Cu-based, *see* Cu-based alloys
— hypermonotectic 51
— interstitial 88
— melting 312–316

alloys (contd.)
— Nd–Fe-based, see Nd–Fe-based alloys
— phase equilibria 87–107
— ternary, see ternary alloys
amorphous alloys 263
— Cu–Zr 177–178
anisotropic chemical capillary lengths 318
anisotropy
— magnetocrystalline 87, 245
— multicomponent multiphase systems 327
— PF simulations 217, 221–223
annealing 91
— LC method 134
— modeling 164–165
— Zr–Ti–Nb–Cu–Ni–Be 418
antisymmetric approximation 367
antitrapping flux 375
approximation
— Ackerson's 201
— antisymmetric 367
— dilute solution 303
— "lubrication" 314
— Percus–Yevick 121–123
— Würth's 201
arbitrary phase diagrams 325
arc melting 278
arm spacing, secondary dendritic 252–255
— secondary dendritic 259
Arrhenius fit 117
Arrhenius plot 137, 204–205
ARTEMIS facility 41–42
as-cast sample 417
as-quenched composite 410
Asaro–Tiller–Grinfeld (ATG) instability 319, 321
asymmetry energy 31
atmosphere, protecting 58
atomic absorption spectroscopy (AAS) 133
atomic attachment 356
— kinetics 111
atomic diffusion 111–129, 413–414
atomic dynamics
— liquid Ni 115–118
— Ni–P-based alloy melts 118–119
— Zr-based alloys 119–120
atomic radius, metals 178
atomic size distribution 179
atomic volume 149–150, 153
atomistic first-principles calculations 157
Avrami model, Kolmogorov–Johnson–Mehl– 203
azeotrope phase diagram 196
azimuthal flow 247

b

B (boron), see Nd–Fe–B
back-scattered electron (BSE) contrast 6–9, 14
ball-milled material 90–92
barrier-crossing 204
basic phase-field equation 373
bcc dendrites 414
Be
— Zr–Ti–Nb–Cu–Ni–Be 407–420
— Zr–Ti–Ni–Cu–Be 119–120
bearings, self-lubricating 19
Bernal's liquids, multicomponent 171–183
Bhatia–Thornton formalism 120–122
Bi
— Al-based alloys 104
— Al–Bi 40–41
— Al–Bi–Cu–Sn 11–15
— In–Bi–Sn 395–400
— interfacial tension 24
— ternary alloys 6–11
Bi–Cd 101–103
Bi–Cu–Sn 11
bifurcation 299
bimodal curve 40
binary Al-based alloys 66
binary alloys 40–41
— compositional stresses 299–309
— constitutional effects 362–366
— diffuse-interface model 362
— eutectic composition 80–81
— rapid quenching 280–288
— sharp-interface model 362–364
— simulation 141–156
— solidification 362–366
— solute trapping 363
— spacing selection 379–381
binary colloidal crystals, growth 199–202
binary colloidal melts, solidification 185–211
binary data sets, thermodynamic extrapolation 11
binary eutectic growth 390
binary eutectic phase diagram 388
binary interaction parameters 62
binary melts 112
binary nanoalloys 97–105
Blackburn, sum rule of Cummings and 59
BMGs, see bulk metallic glasses
boron, see Nd–Fe–B
boundary groove, see "grain boundary groove in an applied temperature gradient" method
boundary layer, solutal 343
Bragg peaks 410
Bragg reflection 91, 199

Index

branching regime 323
Brener theory 341, 359
Bridgman experiments 391
—micro-Bridgman assembly 396–398, 401
Bridgman furnace 82
BSE, see back-scattered electron...
bulk metallic glasses (BMGs) 171
—Zr–Ti–Nb–Cu–Ni–Be 407–420
buoyancy convection 373–385
—gravity-driven 112, 117
—PF modeling 373–376
buoyancy-driven interdendritic flow, directional solidification 378–383
Burgers circuit 93–95
Butler equation 57, 65

c

Cahn and Hilliard model 31
calibration
—diffusion data 402–403
—interfacial energies 403–404
calibration of parameters 325
calorimetric melting 102–103
calorimetry, differential scanning, see differential scanning calorimetry
Calphad method 391–393, 405
—PF modeling 157, 376
—thermodynamic modeling 3, 16
capillarity, Young–Laplace equation 21–23
capillary, graphite 132
capillary constant 22–23
capillary lengths 316–318
capillary method, long-, see long-capillary method
capillary supercooling 48
capillary term 99
casting
—as-cast sample 417
—Cu mold, see Cu mold casting
casting industry 353
Cd
—Al-based alloys 104
—Al–Cd 40–41
—Bi–Cd 101–103
cell, Voronoi-like 43
cellular structure 299
centrifugal Cu mold casting 279
charge, effective 189
charge density, surface 195
charge variable systems 193–196
charged colloidal suspensions, tunable interactions 186–190
charged sphere mixtures 196–199

chemical capillary lengths, anisotropic 318
"chemical" dendrites 354
chemical diffusion, see interdiffusion
—Peclet number 367
chemical reactions, invariant 13
chromium, see Cr
circuit, Burgers 93–95
classical nucleation theory (CNT) 19–20
—binary colloidal crystals 200, 203–207
—undercooled melts 240–241
clusters
—crystal-like 207
—interdendritic 415
—Mackay 176
Co
—binary alloys 66–68
—Cu-based alloys 55, 61–69
—Nd–Fe–Co–Al 263–276
co-focal parabolic fronts 317
coarse-grained microstructure 364–366
coercive field strength 245
coexisting phases 100
coherency strain 316
—elastic energy 318
colloidal crystals, binary 199–202
colloidal melts, solidification 185–211
computer-aided design (CAD) 76
concentration dependence, solid-liquid interface energy 83–84
concentration profiles 412
—planar crystallization front 163–165
concentric heater 24
conductivity, thermal 74
—thermal 81–82
configuration entropy 167–168
congruently melting Ni–Al 359–361
conservation laws 330–332, 354–355
constitutional effects, binary alloys 362–366
container scattering 114
continuum theory, fast crack propagation 319–323
contrast
—BSE 6–9, 14
—magnetic 272
—topological 290
convection
—buoyancy-driven 373–385
—electromagnetically induced 231
—gravity-driven buoyancy 112, 117
—hydrodynamic 217–220
—nonisothermal systems 325–338
—rolls 251, 383
—varying 245–261

convective flow, dendrite growth 358
convective PF modeling 330–336
convergence behavior, dendrite growth velocity 377
cooling conditions, nonequilibrium 289
cooling rate, rapidly quenched Nd–Fe 280–283
coordination number 121, 178
copper, see Cu
correlation functions, partial pair 148–149
correlation length, ferromagnetic 87
Coulomb potential 185
counterions 186
counting experiments, microscopic 203
coupled growth 53, 398–400
crucible rotation 247–248
crystal growth 73
— binary colloidal crystals 199–202
— crystal–liquid interfaces 154
— MD simulation 157–170
crystal-like clusters 207
crystal–liquid interfaces 152
— mass diffusion controlled crystal growth 154
crystallization
— Al–Ni alloys 148–154
— nano- 407
— planar front 162–165
— single step 235
— thermodynamics 158–162
crystals
— binary colloidal 199–202
— quasi- 414–418
Cu 119–120
— Ag–Cu–Zn 391–395
— Al-based alloys 11–15, 80–83, 379, 382
— interfacial tension 24
— Ni–Cu–Cr 344
— Pd–Cu–Ni–P 118, 137–139
— ternary alloys 8–11
— Zr–Nb–Cu–Ni–Al 408
— Zr–Ti–Nb–Cu–Ni–Be 407–420
Cu-based alloys 55
Cu–Co 66–68
Cu mold casting 264, 278–279
Cu–Ni 66–68
Cu–Zr amorphous alloy 177–178
"cube-on-cube" orientation 95
cubic anisotropy 327
Cummings and Blackburn, sum rule of 59
Curie temperature
— hard magnetic glasses 263
— Nd–Fe–B 251

curvature
— interface 320
— local 75–79

d
Darken equation 142
Darken–Manning equation 131, 135–136
Debye–Hückel potential 186
Debye–Waller factor 122
decomposition, Zr–Ti–Nb–Cu–Ni–Be 407–420
deformation, plastic 258
deionized conditions 194
deionized mixture 206
demagnetization curves, room-temperature 287–288
demixing
— liquid–liquid 141
— multiple 3
— Pd–Cu–Ni–P 137–139
dendrite growth 236–240
— convective flow 358
— directional 380
— Nd–Fe–B 252–255
— Ni–Cu–Cr 344
— Ni–Zr 343
— PF modeling 341–345
— Rayleigh-like instability 364
— undercooled melts 353–372
— velocity 231, 241, 377
dendrite–matrix interface 411–412
dendrites
— bcc 414
— "chemical" 354
dendritic arm spacing, secondary 252–255
— secondary 259
dendritic microstructure 382, 407–420
dendritic precipitates 415
dendritic solidification
— diffuse regime 373–385
— PF modeling 373–376
dense random packing 175–176
density 56
— correlation functions 116
— elastic energy 302
— measurement 59
— scattering length 272–273
— surface charge 195
— surface energy 329
— surface entropy 340–341
devitrification process 416
differential scanning calorimetry (DSC)
— interfacial tension 24–26
— morphology formation 396

—solid-liquid interface energy   81
—ternary alloys   4, 6–7
—Zr–Ti–Nb–Cu–Ni–Be   408–411, 413–414
differential sedimentation   198
differential thermal analysis (DTA), Al-based alloys   8
—Al-based alloys   12–14
diffraction
—selected area electron   285–287
—X-ray, see X-ray diffraction
diffraction patterns   191–192, 288–289
diffuse-interface model   341
—binary alloys   362
—one-component systems   354–356
—ternary alloys   366–367
diffusion
—atomic   111–129, 413–414
—chemical, see interdiffusion
—data calibration   402–403
—experimental techniques   132–135
—influence of thermodynamic forces   135–139
—inter-, see interdiffusion
—multicomponent metallic melts   131–140
—solid phase   316
—surface   300
diffusion capillary, graphite   132
diffusion coefficients
—PF simulations   403
—self-, see self-diffusion coefficients
diffusion equation, multicomponent   375
diffusion-limited solidification   369
diffusion potential   302
diffusion velocity, lateral   52
diffusionless processes   323
diffusivity, self-   111
dilute solution approximation   303
directional dendritic growth   380
directional solidification   51, 299
—buoyancy-driven interdendritic flow   378–383
—ternary eutectic alloys   387–406
discontinuous morphology   285
dislocations, edge   95–96
displacement, mean-squared   143
dissipation
—PF dynamics   322
—viscous   356
domains, magnetic   267–273
double convection rolls   251
double-tangent construction   159–160

droplets
—critical radius   20
—irregularly spaced   45
drops, irregular   50–53
DSC, see differential scanning calorimetry
DTA, see differential thermal analysis
Du Noüy ring method   21
duplex microstructure   245
dynamics
—dissipative phase-field   322
—macroscopic/microscopic   111–214
—Ni–Zr crystallization front   162–165

e
edge detection algorithm   59
edge dislocations   95–96
EDX, see energy-dispersive X-ray spectrometry
EELS, see electron energy loss spectroscopy
effective charge   189
effective interface thickness   305
effective mobility   374
Einstein relation, Stokes–   126
Ekman number   247–248, 253
elastic coherency strain energy   318
elastic constants   301–304
elastic effects, phase transitions   311–324
elastic energy, coherency strain   318
elastic energy density   302
elastic moduli   96
elastic neutron scattering   115
"elastic vacuum"   319
electromagnetic levitation   58, 114–117, 353, 359–360
—undercooled melts   228
electromagnetic pump effect   249
electromagnetic stirring   246
electromagnetically induced convection   231
electromagnetically processed melts   353
electron diffraction, selected area   285–287
electron energy loss spectroscopy (EELS)   89
elemental solids   97
energy
—activation   120
—asymmetry   31
—elastic   318
—free-energy functional   217, 305
—Gibbs, see Gibbs energy
—Gibbs free   99–100
—Ginzburg–Landau   327–330
—interfacial   31, 328
—magnetic   263–264
—pair interaction   186
—solid-liquid interfaces, see solid-liquid interface

energy
energy density
— elastic 302
— surface 329
energy-dispersive X-ray spectrometry (EDX) 133, 236, 252
— rapidly quenched alloys 280–283, 286, 290–293
ensemble, semi-grandcanonical 142
— semi-grandcanonical 145
enthalpy
— free, see free enthalpy
— of formation 392
— temperature dependence 159
entropy
— Al–Ni–Zr 165–168
— configuration 167–168
— multicomponent multiphase systems 327, 340–341
— Ni–Zr 158–162
— structural 167–168
— surface density 340–341
— vibration 159
entropy inequality 329–332
entropy of fusion 73, 79
equations of motion
— compositional stress 301–304
— Newton's 141, 319
equiaxed-grain-refined microstructure 364
equilibration, samples 75
equilibrium, local 314
equilibrium partition coefficient 162
Euler–Lagrange equations 328
eutectic alloys, ternary 387–406
eutectic composition
— Al–Cu system 80–81
— invariant 81–83
eutectic grooves, univariant 403
eutectic growth 346–350, 390
eutectic Nd grains, fibrous 281–283
eutectic phase diagram 347, 388
eutectic point, invariant ternary 389
eutectic systems
— Jackson–Hunt model 39, 46–48
— melting in 312–316
— phase diagrams 312
— ternary 349
"eutectic" temperature 104
expansion, thermal, see thermal expansion
extended sharp interface theory 240
external interfaces, single-phase material 88–97
extrapolation

— Redlich–Kister/Muggianu-type 12
— thermodynamic 11

f

Faber–Ziman formalism 120–122
facetted anisotropy 327
facetted nucleus 222
fast crack propagation, continuum theory 319–323
fcc short-range order 175–176
Fe
— Cu-based alloys 55, 61–69
— Nd–Fe 280–288
— Nd–Fe-based alloys, see Nd–Fe-based alloys
— Nd–Fe–Co–Al 263–276
— Nd–Ga–Fe 288–295
γ-Fe 227–240
Fe–Ni 66–68
ferromagnetic correlation length 87
fibrous binary eutectic growth 390
fibrous eutectic Nd grains 281–283
fibrous monotectic growth 47–48
Fick's laws 131
Fick's second equation 133
field strength, coercive 245
filtering, Fourier 91
— Fourier 93
finite system size 101
first-order phase transitions 88
first-principles calculations 157
FIT2D 192
fivefold symmetry 172
fixed-charge single-component systems 186
floating zone facility 248–251
flow, heat, see heat flow
fluctuations, magnetization 267
FLUENT 77
fluid flow, buoyancy-driven 381–383
fluid flow coupling 375–376
force biased algorithm 172–173
forced rotation technique 246–248
forces
— Lorentz 247
— thermodynamic 135–139
form factors 266
foundry industry 353
Fourier filtering 91, 93
Fourier's law 74
fourth-order cumulant 145–146
free energy
— Gibbs 99–100
— multicomponent multiphase systems 326

—nucleation   220–221
free-energy functional   217, 305
free enthalpy
    —Al–Ni–Zr   165–168
    —Ni–Zr   158–162
free-surface phenomena   247
Frenkel law, Wilson–   200–201
Frenkel model, Wilson–   142
    —Wilson–   153
Freundlich-type equations, Gibbs–Thompson–   98
fringes, Moiré   89
front
    —co-focal parabolic   317–318
    —planar   162–165
    —solidification   44–45
full phase diagram, charge variable systems   193–196
furnace
    —aerogel   41
    —Bridgman   82

g

Ga, Nd–Ga–Fe   288–295
gap, ternary miscibility   5–7
Gauss–Seidel smoothing operator   306–307
geometrical correction, groove coordinates   76–77
getter, niobium   24
Gibbs adsorption   32
Gibbs energy
    —surface tension   57
    —ternary alloys   6, 11–12
Gibbs free energy   99–100
Gibbs' phase rule   98
Gibbs–Thompson–Freundlich-type equations   98
Gibbs–Thomson coefficient   79–83
Gibbs–Thomson equation   78–79, 299–300, 340
Gibbs triangle   388, 395
Ginzburg–Landau energy, multiphase   327–330
glass-forming ability   407
glass-forming alloy melts   118–119
glass matrix composite   407–420
glasses
    —bulk metallic, see bulk metallic glasses
    —hard magnetic   263–276
global packing fraction   177
goniometer   190
"grain boundary groove in an applied temperature gradient" method   73

grain refinement, undercooled melts   353–372
grains, fibrous eutectic Nd   281–283
graphite diffusion capillary   132
gravity, reduced   360
gravity-driven buoyancy convection   112, 117
Green–Kubo relations   144
Griffith point   320
grinding   76
Grinfeld instability   301
    —Tiller–Asaro–   319, 321
grooves
    —coordinates   76–77
    —eutectic   403
    —grain boundary   75–79
growth
    —binary colloidal crystals   199–202
    —coupled   53
    —crystal, see crystal growth
    —dendritic, see dendrite growth
    —eutectic, see eutectic growth
    —peritectic   215–225
    —stationary   398–400, 403–404
    —three-phase coupled   400
    —transient   400–404
growth kinetics   215–225
growth morphology   39–54, 216
growth velocity   199–200
    —Ni–Zr   362–364
    —Ni–Zr–Al   367–369

h

hard magnetic glasses   263–276
    —microstructure   268
hard magnetic phase
    —A1 zones   290–292
    —microstructure evolution   245
    —Nd–Fe–B   227–240
hard sphere model   123
    —Bernal's   175–176
head crystal   198
heat flow, radial   74
heat of mixing, negative   171
heat transport equation   356
heater, concentric   24
heterogeneous nucleation, PF modeling   220–223
heterophase interfaces, internal   97–105
hexagonal lattice   43
hexagonal phase, Ni–Fe   284
hexagonal rod-like structure   349
high-pressure torsion straining   92
Hilliard, model of Cahn and   31
holes, icosahedral 12-neighbor   166

homogeneous nucleation   203
Hooke's law   303, 320
hopping processes   119
Hückel potential, Debye–   186
Hunt..., see Jackson–Hunt...
hydrodynamic convection, PF modeling   217–220
hyperbolic tessellation   174
hypermonotectic alloys   51
hysteresis   265, 287–288, 291

i
icosahedral 12-neighbor holes   166
icosahedral short-range order   123, 175–176
ICP-MS, see inductively coupled plasma MS
ideal solid solution   197
ideal solutions   61–62
immiscible Al-based alloys, interfacial tension   19–38
In, Al–In   40–41
In–Bi–Sn   395–400
in situ X-ray radiography   112
inclusions, liquid   311
incoherent scattering   113
induction melting   257
inductively coupled plasma MS (ICP-MS)   133, 138
inherent structure approach   169
instability
— Asaro–Tiller–Grinfeld   319, 321
— Grinfeld   301
— Mullins–Sekerka   301
— Rayleigh   39, 364
instationary morphology formation   387–406
insulator, thermal   41
integral enthalpy of formation   392
interactions, tunable   186–190
interdendritic cluster   415
interdendritic convection rolls   384
interdendritic flow, buoyancy-driven   378–383
interdendritic liquid   365
interdendritic region   6, 381
interdiffusion   131, 367
interdiffusion coefficients   124
— Arrhenius plot   137
— vanishing   142
interdiffusion constant   143
interface curvature   320
interface mobility   162, 230
interface thickness, effective   305
interface velocity   153
interfaces
— crystal–liquid   154
— crystal–melt   152
— dendrite–matrix   411–412
— diffuse   341
— extended sharp interface theory   240
— external   88–97
— internal heterophase   97–105
— "ordered"   89
— particle–matrix   98
— sharp-interface model, see sharp-interface model
— solid-liquid, see solid-liquid interfaces
— static   163
— stretching   300
interfacial energy   31, 328
— calibration   403–404
— γ phase nucleation   238
interfacial segregation   98
interfacial tension
— and miscibility gap   31–33
— composition dependence   30–32
— liquid-liquid   19–38
intergrowth   15
intermetallic compounds, formation   69
intermetallic phases, superlattice structure   240
internal heterophase interfaces   97–105
interstitial alloys   88
invariant eutectic composition, Al–Cu–Ag   81–83
invariant reactions, Al–Bi–Cu–Sn system   13
invariant ternary eutectic point   389
inverse relaxation time   116
iron, see Fe
irregular drops   50–53
irregularly spaced droplets   45
irreversible thermodynamics   332
Ising behavior   146–147
isothermal melting   313–316
isothermal section   394, 397
isothermal surface tension curves   66
isothermally undercooled melt   348
isothermals   80
isotropic scattering   266

j
jacket, water cooled   74
Jackson–Hunt constant   49–50
Jackson–Hunt model   39, 46–48, 404
Jackson–Hunt plot   45–46
Johnson–Mehl–Avrami model, Kolmogorov–   203
Jones, see Lennard–Jones...

k

Karma corrections 374–375
kinetic coefficient, see interface mobility
kinetic undercooling 367
kinetics
   —atomic attachment 111
   —growth 215–225
   —nucleation 203–206, 215–225
   —phase-field (PF) 355
Kissinger method 418
Kister form, Redlich– 58
Kister/Muggianu, see Redlich–Kister/Muggianu-type extrapolation
Kolmogorov–Johnson–Mehl–Avrami model 203

l

Lagrange equations, Euler– 328
Lagrange multiplier 326, 333
lamellar eutectic growth 349–350
lamellar microstructures 389
Landau energy, Ginzburg– 327–330
Laplace equation 314, 317
   —Young–, see Young–Laplace equation
lateral diffusion velocity 52
lattice, hexagonal 43
laws and equations
   —basic phase-field equation 373
   —Butler equation 57, 65
   —conservation laws 330–332, 354–355
   —Darken equation 142
   —Darken–Manning equation 131, 135–136
   —entropy inequality 329–332
   —equations of motion 301–304
   —Euler–Lagrange equations 328
   —Fick's laws 131
   —Fick's second equation 133
   —Fourier's law 74
   —Gibbs' phase rule 98
   —Gibbs–Thompson–Freundlich-type equations 98
   —Gibbs–Thomson equation 78–79, 299–300, 340
   —Green–Kubo relations 144
   —heat transport equation 356
   —Hooke's law 303, 320
   —interface mobility 230
   —interfacial energy 328
   —Laplace equation 314, 317
   —modified Navier–Stokes equation 218
   —multicomponent diffusion equation 375
   —Navier–Stokes equation 355
   —Newton's equation of motion 319
   —Newton's equations of motion 141
   —Ornstein–Zernicke form 147
   —Porod law 268, 274–275
   —Reynold's transport theorem 330–331
   —scattering law 114
   —second law of thermodynamics 332–336
   —Stokes–Einstein relation 126
   —stress–strain relationship 302
   —sum rule of Cummings and Blackburn 59
   —Vegard's law 56
   —Wilson–Frenkel 153, 200–201
   —Young–Laplace equation of capillarity 21–23
   —see also model, theory
layer structure, two-phase 161
lead, see Pb
Legendre polynomials 59
lengths, capillary 316–318
Lennard–Jones (LJ) mixture, symmetric 144–148
levitation, electromagnetic 58, 114–117, 228, 353, 359–360
levitation coil 114
light scattering, static 190–193
liquid film migration (LFM) 311–313, 316–318
liquid inclusions 311
liquid–liquid demixing 141
liquid–liquid demixing transition 144–148
liquid–liquid interfaces, adsorption 32–33
liquid–liquid interfacial tension 19–38
   —composition dependence 30–32
   —temperature dependence 33
liquid Ni, atomic dynamics 115–118
liquid–solid interfaces, see solid–liquid interfaces
liquid Zr–Ni, short-range order 120–124
liquids
   —Bernal's 171–183
   —interdendritic 365
   —mass transport 111
   —metastable 117
   —monoatomic 175–177
   —solid-liquid interfaces, see solid-liquid interfaces
liquidus curve 52, 160, 162
liquidus slope 303–304

liquidus surface 6–8, 10–11, 393
liquidus temperature 56
LJ, see Lennard–Jones . . .
local curvature, grain boundary grooves 75–79
local equilibrium, liquid–solid interfaces 314
local packing fraction 174
local undercooling 77–78
long-capillary (LC) method 132–134
   —simulation 150
   —X-ray radiography 134–135
Lorentz force 247
lubrication, self- 19
"lubrication" approximation 314

m

Mackay cluster 176
macroscale modeling 162
macroscopic dynamics 111–214
magnetic contrast 272
magnetic domains 267–273
   —size 271
magnetic energy 263–264
magnetic field, floating zone facility 248–251
magnetic form factors 266
magnetic glasses 263–276
magnetic phase
   —hard, see hard magnetic phase
   —soft, see soft magnetic phase
magnetic SLD 272–273
magnetic systems, unsaturated 265–268
magnetization
   —density profiles 263–276
   —fluctuations 267
   —temperature dependence 264
magnetocrystalline anisotropy 87, 245
magnetometer, vibrating sample 251, 255, 280
magnets
   —Nd–Fe–B 245
   —two-phase 292
Manning equation, Darken– 131
   —Darken– 135–136
Manning factor 136, 144
   —simulation 151
Marangoni motions 19
mass conservation 330
mass diffusion, controlling crystal growth 154
mass spectrometry (MS), inductively coupled plasma 133
   —inductively coupled plasma 138
mass transport, liquids 111
matrix interfaces 98, 411–412

matrix strains 97
MD, see molecular dynamics
Mead algorithm, Nelder– 173
   —Nelder– 177
mean-squared displacement 143
Mehl–Avrami model, Kolmogorov–Johnson– 203
melt spinning 279
melt structure 111–129
melting 88
   —arc 278
   —calorimetric 102–103
   —combined motion of fronts 316–318
   —eutectic and peritectic systems 312–316
   —induction 257
   —isothermal 313–316
melting point 56
melting temperature 131–140
melts
   —alloy, see alloy melts
   —buoyancy-driven convection 373–385
   —colloidal, see colloidal melts
   —congruent 359–361
   —electromagnetically processed 353
   —glass-forming 118–119
   —isothermally undercooled 348
   —multiphase 3–108
   —Ni–Zr 158–162
   —undercooled 206–207
   —varying convection 245–261
meniscus volume 22
meridional flow 247
mesh generators, adaptive 339
   —adaptive 342–343
mesoscale modeling 162
metallic glasses 407–420
   —bulk, see bulk metallic glasses
metallic melts, multicomponent, see multicomponent metallic melts
metals
   —atomic radius 178
   —nanoscale 87–107
   —soft 19
   —transition 68
metastable A1 "phase" 283–287
metastable hard magnetic A1 zones, tuning 290–292
metastable liquids 117
metastable χ phase 227–240
metastable phase formation 407–420
Metropolis method 145
Mg, Al-based alloys 382

MICRESS   376, 389
micro-Bridgman assembly   396–398
    —transient growth   401
microexamination   75
microgravity   112, 117
micromagnetic model of domains   266
microscopic counting experiments   203
microscopic dynamics   111–214
microscopic temperature field, simulation   77
microscopic transport processes, simulation   141–156
microstructure
    —coarse-grained   364–366
    —dendritic   382, 407–420
    —duplex   245
    —equiaxed-grain-refined   364
    —hard magnetic glasses   268
    —monotectic growth   42–45
    —Nd–Fe–B   278
    —peritectic alloys   245–261
    —rapidly quenched Nd–Fe   280–283
migration, liquid film   311–313
    —liquid film   316–318
Miller indices   193
milli-Q water   190
miscibility gap   5–7, 31–33
mixing
    —Al–Ni   136–137
    —negative heat of   171
mixtures
    —charged spheres   196–199
    —deionized   206
    —organic   34–35
mobility
    —effective   374
    —interface   162, 230
mode coupling theory   119, 124–125
    —critical temperature   152
    —self-diffusion coefficients   124–125
model
    —Bhatia–Thornton formalism   120–122
    —Cahn and Hilliard   31
    —diffuse-interface, see diffuse-interface model
    —Faber–Ziman formalism   120–122
    —hard sphere   123
    —Jackson–Hunt   39, 46–48, 404
    —Kolmogorov–Johnson–Mehl–Avrami   203
    —micromagnetic   266
    —multilayer   35
    —negentropic   229
    —nucleation and dendrite growth   236–240
    —sharp-interface, see sharp-interface model
    —spherical cap   216
    —two-phases   169
    —Wilson–Frenkel   142
    —see also laws and equations, theory
modeling
    —annealing   164–165
    —Calphad method, see Calphad method
    —macro-/mesoscale   162
    —nonisothermal systems with convection   325–338
    —PF, see phase field modeling
modified Navier–Stokes equation   218
Moiré effect   103
Moiré fringes   89
molar Gibbs free energy   99
mold casting, Cu, see Cu mold casting
molecular dynamics (MD) simulation   141–142, 147
    —ab initio   172
    —and phase field modeling   162–165, 345–346
    —crystal growth   157–170
    —numerical reliability   166–167
molecular layers, solid–liquid interface   215
monoatomic liquids, computational optimization   175–177
monotectic alloys, Al-based   3–17
monotectic growth   39–54
    —fibrous   48–50
monotectic phase diagrams   4–5
Monte Carlo simulations   142, 145, 147
    —nucleation kinetics   221
morphological stability, solid–liquid interfaces   299
morphology formation, (in)stationary   387–406
motion
    —combined, see combined motion
    —equations of, see equations of motion
    —Stokes/Maragoni   19
    —tagged particles   143
MS, see mass spectrometry
Muggianu, see Redlich–Kister/Muggianu-type extrapolation
Mullins–Sekerka instability   301
multicomponent alloys
    —elastic effects   311–324
    —monotectic Al-based   3–17
    —surface tension   55–71
    —thermal expansion   55–71

multicomponent Bernal's liquids 171–183
multicomponent diffusion equation 375
multicomponent immiscible Al-based alloys
  19–38
multicomponent metallic melts 111–129
  —diffusion 131–140
multicomponent multiphase systems
  325–338
    —anisotropy 327
    —free energy 326
    —PF modeling 326–327
    —solidification 339–351
multilayer model 35
multiphase Ginzburg–Landau energies
  327–330
multiphase melts, thermodynamics 3–108
multiple demixing 3
multiple scattering 115
multiplier, Lagrange 326
    —Lagrange 333

n

nanoalloys, binary 97–105
nanocrystalline phase 285–287, 290
nanocrystallization 407
nanoparticles 91–95, 268–273
nanoscale metals, phase equilibria 87–107
nanosized magnetization density profiles
  263–276
Navier–Stokes equation 218, 355
Nb
    —Zr–Nb–Cu–Ni–Al 408
    —Zr–Ti–Nb–Cu–Ni–Be 407–420
Nb getter 24
Nd, oxide formation 252
Nd–Fe 280–288, 287–288
Nd–Fe–B
    —Curie temperature 251
    —dendrite growth 252–255
    —forced rotation experiments 252–256
    —hard magnetic phase 227
    —magnets 245
    —microstructure 278
    —microstructure evolution 245–261
    —nonperitectic 236
    —PF simulations 215–225
    —undercooled melts 227–244
Nd–Fe-based alloys 215–295
Nd–Fe–Co–Al 263–276
Nd–Ga–Fe 288–295
    —soft magnetic phase 291–292
Nd grains, fibrous eutectic 281–283
Nd nanoparticles 268–273
    —radius 270

nearest neighbor coordination number 121
negative heat of mixing 171
negentropic model 229
12-neighbor holes, icosahedral 167
Nelder–Mead algorithm 173, 177
Nelder–Mead simplex 173, 180
neutral curves 304–305
neutron scattering 113–115
    —simulation 150
    —small-angle 265–268, 409
    —structure factors 207
neutrons, polarized 265–268
    —polarized 270
Newton's equations of motion 141, 319
Ni
    —Al–Ni alloys 136–137, 148–154
    —Al–Ni–Zr 165–168
    —binary alloys 66–68
    —dendrite growth and grain refinement
      356–359
    —liquid 115–118
    —Pd–Cu–Ni–P 118, 137–139
    —Zr-based alloys 119–120
    —Zr–Nb–Cu–Ni–Al 408
    —Zr–Ti–Nb–Cu–Ni–Be 407–420
Ni–Al, congruently melting 359–361
Ni–Al–Zr, rapid solidification 376–378
Ni–Cu–Cr, dendrite growth 344
Ni–Fe 277, 284
Ni–P-based glass-forming alloy melts
  118–119
Ni–Zr
    —crystallization front dynamics
      162–165
    —dendrite growth 343
    —entropy and free enthalpy 158–162
    —growth velocity 362–364
    —phase diagram 158, 161–162
Ni–Zr–Al, growth velocity 367–369
niobium, see Nb
nonequilibrium cooling conditions 289
nonisothermal PF simulations 343
nonisothermal systems 325–338
nonperitectic Nd–Fe–B 236
nuclear form factors 266
nucleation
    —classical theory, see classical nucleation
      theory
    —free energy 220–221
    —heterogeneous 220–223
    —kinetics 203–206, 215–225
    —phase 73
    —γ phase 238
nucleation curve 52

nucleation undercooling 404
nucleus, (un)faceted 222–223
number density profiles 152, 154
numerical reliability, MD simulation 166–167
numerical simulation, *see* simulation

o

obstacle potential 330
off-dendrite regions 416
one-component systems, solidification 354–361
Onsager coefficient 125–126, 143–144, 147
optimization, computational, *see* computational optimization
order, short-range, *see* short-range order
order parameter, phase-field 374
"ordered" interfaces 89
organic mixtures 34–35
Ornstein–Zernicke form 147
oscillating drop technique 59
overheating 315
oxide formation, Nd 252

p

P (phosphor)
 —Ni–P 118–119
 —Pd–Cu–Ni–P 118, 137–139
packing, dense random 175–176
packing fraction 174, 177
pair interaction energy 186
parabolic fronts, co-focal 317
partial coordination numbers 178
partial pair correlation functions 148–149
partial structure factors 120–121, 143
partial wetting 34
particle motion, tagged 143
partition coefficient, equilibrium 162
pattern formation, dendritic 373
Pb, Al–Pb 40–41
 —Al–Pb 89–90
Pb nanoparticles 91–95
Pd–Cu–Ni–P 118
 —demixing tendency 137–139
pearls, string of, *see* string of pearls
Peclet number 49, 317–318, 357
 —chemical diffusion 367
Percus–Yevick approximation 121–123
peritectic growth 215–225
peritectic systems
 —melting in 312–316
 —microstructure evolution 245–261
 —PF modeling 220–223
 —phase diagrams 313

phase
 —coexisting 100
 —hard magnetic, *see* hard magnetic phase
 —metastable A1 "phase" 283–287
 —soft magnetic, *see* soft magnetic phase
Φ phase, *see* hard magnetic phase
χ phase, metastable 227–240
phase behavior, simulation 141–156
phase diagram
 —arbitrary 325
 —azeotrope 196
 —binary eutectic 388
 —charge variable systems 193–196
 —charged sphere mixtures 196–199
 —eutectic 347
 —Ni–Fe 284
 —Ni–Zr 158, 161
 —Ni–Zr melts 161–162
 —peritectic materials 215
 —quasi-binary 228, 237
 —ternary monotectic 4–5
 —triple junction 312–315
phase equilibria, nanoscale metals and alloys 87–107
phase-field (PF) equation, basic 373
phase-field (PF) kinetics 355
phase field (PF) modeling
 —and MD simulation 162–165, 345–346
 —compositional stress 299–309
 —convective 330–336
 —dendrite growth 341–345
 —dissipative dynamics 322
 —eutectic growth 346–350
 —heterogeneous nucleation 220–223
 —hydrodynamic convection 217–220
 —input parameter 162–163
 —Karma corrections 374–375
 —multicomponent multiphase systems 326–327, 373–376
 —peritectic alloys 220–223
 —solidification 339–351
phase-field (PF) order parameter 374
phase-field (PF) simulations
 —diffusion coefficients 403
 —Nd–Fe–B 215–225
 —nonisothermal 343
phase-field (PF) vector 328
phase formation, multicomponent monotectic Al-based alloys 3–17
phase nucleation 73
γ phase nucleation, interfacial energy 238
phase rule, Gibbs' 98

phase selection in undercooled melts 227–244
phase separation 408
—temperature 26
phase stability 88–105
phase transitions 88–105
—elastic effects 311–324
—first-order 88
planar crystallization front, Ni–Zr 162–165
planar solidification front 44
plastic deformation 258
Poisson ratio 302
polarization (magnetic), saturation 245
polarized neutrons 265–268, 270
polydispersity 185, 189
polymorphic transition 313
polystyrene 188
polytetrahedral structure 166
Porod law 268, 274–275
potential
—Coulomb 185
—Debye–Hückel 186
—diffusion 302
precipitates, dendritic 415
propagation
—fast crack 319–323
—planar crystallization front 163–165
properitectic phase 219, 234
—microstructure evolution 245–246, 252, 257–259
protecting atmosphere 58
pseudoregular solutions 32
pump effect, electromagnetic 249
pure systems, solidification 354–361
pyrometer 58
—two-color 114, 256

q

quartz tube 246
quasi-binary phase diagram 228, 237
quasi-crystals 414–418
quasi-elastic neutron scattering 113–115, 150
quaternary alloys
—Al–Bi–Cu–Sn 11–15
—invariant reactions 13
quenching 75
—as-quenched composite 410

r

radial heat flow 74
radiation, synchrotron, *see* synchrotron radiation
radiogram
—LC method 134–135
—X-ray, *see* X-ray radiography
random packing, dense 175–176
rapid quenching 75, 89, 277–295
—Nd–Fe 280–288
—Nd–Ga–Fe 288–295
rapid solidification 299
—Ni–Al–Zr 376–378
—solute trapping 345
—ternary alloys 367–368
rare-earth magnets 245
Rayleigh instabilities 39, 45, 364
Rayleigh speed 321
reactions, chemical, *see* chemical reactions
Redlich–Kister form 58
Redlich–Kister/Muggianu-type extrapolation 12
reduced gravity 360
reflection, Bragg 91
—Bragg 199
relaxation time, inverse 116
remanence 293
renormalization theory 36
Reynolds number 247–248, 357
Reynolds transport theorem 330–331
rhombohedral Ni–Fe 277
rod-like structure, hexagonal 349
room-temperature demagnetization curves 287–288
rotation
—crucible 247–248
—forced, *see* forced rotation technique

s

S-Voronoi tessellation 174
SAED, *see* selected area electron diffraction
samples, equilibration and preparation 75–76
SANS, *see* small-angle neutron scattering
saturation polarization 245
scanning calorimetry (DSC), differential, *see* differential scanning calorimetry
scattering
—isotropic 266
—neutrons, *see* neutron scattering
—static light 190–193
—ultra small-angle X-ray, *see* ultra small-angle X-ray scattering
scattering law 114
scattering length density (SLD), magnetic 272–273
Scherrer constant 199
screened Coulomb potential 185
SDAS, *see* secondary dendritic arm spacing

second law of thermodynamics 332–336
secondary dendritic arm spacing (SDAS) 252–255, 259
sedimentation, differential 198
"seed" phase 221
segregation, interfacial 98
segregation coefficient 303–304
Seidel smoothing operator, Gauss– 306–307
Sekerka instability, Mullins– 301
selected area electron diffraction (SAED) 285–287
selection, spacing 379–381
self-absorption 114
self-diffusion, activation energy 120
self-diffusion coefficient 116–118
 —Arrhenius plot 137
 —mode coupling theory 124–125
self-diffusivity 111
self-dissociation, solvents 186
self-lubricating bearings 19
semi-grandcanonical ensemble (SGMC) 142, 145
separation, phase 408
SGMC, see semi-grandcanonical ensemble
shadowgraphs 59
sharp-interface model 325
 —binary alloys 362–364
 —extended theory 240
 —one-component systems 354–356
 —ternary alloys 367–369
shear cell experiment 139
shear modulus 187–189, 194–196
 —time evolution 198
shear viscosity 150–151
short-range order
 —fcc 175–176
 —icosahedral 123, 175–176
 —liquid Zr–Ni 120–124
 —topological 121
Si, interfacial tension 24
signal-to-noise ratio, neutron scattering 113
silanol groups 195
silica aerogel 41–42
silica suspensions, colloidal 191–192
silver, see Ag
simplex, Nelder–Mead 173
 —Nelder–Mead 180
simulation
 —binary alloys 141–156
 —buoyancy-driven convection 379
 —compositional stress 306–308
 —fibrous binary eutectic growth 390
 —floating zone facility 250
 —forced rotation technique 247–248
 —MD, see molecular dynamics simulation
 —microscopic temperature field 77
 —Monte Carlo, see Monte Carlo simulations
 —nonisothermal PF 343
 —PF, see phase-field simulations
 —transient growth 402–403
 —univariant growth 390
 —Yukawa systems 187
single-component colloidal melts, solidification 185–211
single convection roll 251
single-phase material, external interfaces 88–97
single step crystallization 235
singularity, critical 148
SLD, see scattering length density
small-angle neutron scattering (SANS) 265–268
 —Zr–Ti–Nb–Cu–Ni–Be 409, 417–418
smoothing operator, Gauss–Seidel 306–307
Sn
 —Al–Bi–Cu–Sn 11–15
 —In–Bi–Sn 395–400
 —interfacial tension 24
 —ternary alloys 6–8, 11
soft magnetic phase 263–265
 —Nd–Ga–Fe 291–292
 —Ni–Fe 277
soft metals 19
software, see algorithms and software
solid-liquid interface energy
 —concentration dependence 83–84
 —measurement 78–79
 —ternary alloys 73–86
solid–liquid interfaces
 —atomic attachment kinetics 111
 —molecular layers 215
 —morphological stability 299
solid phase, diffusion 316
solid solution 187
 —ideal 197
solidification
 —colloidal melts 185–211
 —combined motion of fronts 316–318
 —compositional stresses 299–309
 —constitutional effects 362–366
 —dendritic, see dendrite growth, dendritic
solidification
 —diffusion-limited 369

solidification (contd.)
— directional, see directional solidification
— multicomponent multiphase systems 339–351
— planar front 44
— pure (one-component) systems 354–361
— rapid, see rapid solidification
— ternary alloys 366–369
— transient 401–402
— Zr–Ti–Nb–Cu–Ni–Be 409
solidification cells 299, 307
solidification velocity 45–49
solids, elemental 97
solidus curve 52
solutal boundary layer 343
solute current 306
solute trapping 300
— binary alloys 363
— rapid solidification 345
solutions
— ideal 61–62
— pseudoregular 32
— solid 187, 197
solvents, self-dissociation 186
spacing, stability diagram 381
spacing selection, binary alloy 379–381
spectra, X-ray 232
— X-ray 236
spectroscopy
— atomic absorption 133
— electron energy loss 89
spheres
— charged mixtures 196–199
— hard 175–176
spherical cap model 216
spinning, melt 279
stability
— morphological 299
— nanoscale systems 88–105
— thermal 407
stability diagram 46–47, 100
— spacings 381
stacking sequence 399
static interface 163
static light scattering, time-resolved 190–193
stationary coupled growth 398–400
stationary growth morphology 216
stationary morphology formation 387–406
stationary thermocouples 74
stationary three-phase coupled growth 400

stationary univariant growth 403–404
statistical thermodynamics 30
steady-state convection rolls 383
steady-state regime 312–315
stirring
— electromagnetic 246
— two-phase 250
Stokes–Einstein relation 126
Stokes equation, Navier– 218
— Navier– 355
Stokes motions 19
strain, coherency 316
straining, high-pressure torsion 92
stream lines 249
stress
— compositional 299–309
— surface 331
— thermal 308
stress–strain plots 258
stress–strain relationship 302
stress tensor 302, 331
stretching, interface 300
string of pearls 39–40, 44–47
— transition from fibrous structures 48–50
structural entropy 167–168
structure factors
— neutron scattering 207
— total 120
sum rule of Cummings and Blackburn 59
supercooling, capillary 48
superlattice reflex 233
superlattice structure, intermetallic phases 240
surface
— diffusion 300
— energy density 329
— free-surface phenomena 247
— liquidus 6–7
surface charge density 195
surface entropy density 340–341
surface stress 331
surface tension 57, 59
— Gibbs energy 57
— isothermal curves 66
— multicomponent alloys 55–71
susceptibility 143
suspensions
— charged colloidal 186–190
— colloidal silica 191–192
symmetric LJ mixture 144–148
symmetry, fivefold 172
synchrotron radiation 227–244
synchrotron radiation imaging 380

t

tagged particle motion   143
Teflon tubing system   188
temperature
—critical   151
—Curie   251, 263
—"eutectic"   104
—liquidus   56
—melting, see melting temperature
—phase separation   26
temperature dependence
—enthalpy   159
—liquid-liquid interfacial tension   33
—magnetization   264
temperature field, microscopic   77
temperature gradient, see "grain boundary groove in an applied temperature gradient" method
temperature–time profiles   232–235
tensioactivity   33
tensiometry   21
tension
—interfacial, see interfacial tension
—surface, see surface tension
ternary alloys
—Al-based   6–10
—Al–Ni–Zr   165–168
—buoyancy-driven fluid flow   381–383
—liquidus surface   6–7
—sharp-interface model   367–369
—solid-liquid interface energy   73–86
—solidification   366–369
ternary eutectic alloys   349
—directional solidification   387–406
—general aspects   388–389
ternary eutectic point, invariant   389
ternary interaction parameter   56, 62–64
ternary miscibility gap   5–7
ternary monotectic phase diagrams, systematic classification   4–5
tessellation, Voronoi   174–175
theory
—Brener   341, 359
—classical nucleation, see classical nucleation theory
—extended sharp interface   240
—fast crack propagation   319–323
—mode coupling   119, 124–125
—renormalization   36
—see also laws and equations, model
thermal activation, homogeneous nucleation   203
thermal analysis, differential, see differential thermal analysis

thermal conductivity   74, 81–82
thermal expansion   56, 59
—multicomponent alloys   55–71
thermal insulator   41
thermal stability, BMGs   407
thermal stress   308
thermocouples, stationary   74
thermodynamic assessment   391, 395–396
thermodynamic extrapolation, binary data sets   11
thermodynamic forces, influence on diffusion   135–139
thermodynamic modeling, Calphad method   3
thermodynamics
—crystallization   158–162
—irreversible   332
—multiphase melts   3–108
—statistical   30
thick-film solution   133
Thompson–Freundlich-type equations, Gibbs–   98
Thomson coefficient, Gibbs–   79–83
Thomson equation, Gibbs–   78–79, 299–300, 340
Thornton formalism, Bhatia–   120–122
three-phase coupled growth, stationary   400
Ti   119–120
—Zr–Ti–Nb–Cu–Ni–Be   407–420
Ti–Al   245–261
—forced rotation experiments   256–258
Tiller–Grinfeld instability, Asaro–   319
—Asaro–   321
time of flight (TOF) spectrometer   113–116
time-resolved static light scattering, instrumentation   190–193
tin, see Sn
tip curvature   321
TOF, see time of flight
topological contrast   290
topological short-range order   121
torsion straining, high-pressure   92
total structure factors   120
transient growth   400–404
transient solidification   401–402
transition metals, substitution   68
transport coefficients   143–144
transport processes, microscopic   141–156
transport theorem, Reynolds   330–331
trapping, solute   300
—solute   345
triple junction, phase diagrams   312–315

tubing system, Teflon 188
tunable interactions, in charged colloidal suspensions 186–190
two-color pyrometer 114, 256
two-phase layer structure 161
two-phase magnets 292
"two-phase" nucleus 222
two-phase stirrer 250
two-phases model 169

u

ultra small-angle X-ray scattering (USAXS) 186–190, 193–194, 206–208
undercooled melts
— binary 112
— dendrite growth and grain refinement 353–372
— isothermally 348
— phase selection 227–244
— structure 206–207
undercooling
— critical 361
— kinetic 367
— local 77–78
— nucleation 404
— velocity 343
unfaceted nucleus 222–223
univariant eutectic grooves 403
univariant growth 390
— stationary 403–404
unsaturated magnetic systems 265–268
USAXS, see ultra small-angle X-ray scattering

v

vacuum, "elastic" 319
varying convection 245–261
Vegard's law 56
velocity
— growth, see growth velocity
— solidification, see solidification velocity
velocity undercooling 343
vibrating sample magnetometer (VSM) 251, 255, 280
vibration entropy 159, 167
viscosity, shear 150–151
viscous dissipation 356
Voronoi-like cell 43
Voronoi tessellation 174–175

w

Waller factor, Debye– 122
water, milli-Q 190
water cooled jacket 74
wetting phenomena 33–36
Wilhelmy plate method 21
Wilson–Frenkel law, binary colloidal crystals 200–201
Wilson–Frenkel model 142, 153
Würth's approximation 201

x

X-ray diffraction (XRD) 279–282, 287–293
— Zr–Ti–Nb–Cu–Ni–Be 408–409
X-ray radiography
— *in situ* 112
— LC method 134–135
X-ray scattering, ultra small-angle, see ultra small-angle X-ray scattering
X-ray spectra 232
— energy-dispersive 133
— energy dispersive 236

y

Yevick approximation, Percus– 121–123
Young–Laplace equation of capillarity 21–23
Young's modulus 302
Yukawa systems, simulation 187

z

Zernicke form, Ornstein– 147
Ziman formalism, Faber– 120–122
Zn 6
— Ag–Cu–Zn 391–395
zone, floating, see floating zone facility
Zr
— Al–Ni alloys 367–369, 376–378
— Al–Ni–Zr 165–168
— Cu–Zr amorphous alloy 177–178
— Ni–Zr 158–165, 343, 362–364
Zr-based alloys, Zr–Ti–Nb–Cu–Ni–Be, see Zr–Ti–Nb–Cu–Ni–Be
Zr–Nb–Cu–Ni–Al 408
Zr–Ni 119–124
Zr–Ti–Nb–Cu–Ni–Be 407–420
— solidification 409
Zr–Ti–Ni–Cu–Be 119–120
— neutron scattering 113